SO-BXD-935

REVIEWS IN MINERALOGY AND GEOCHEMISTRY

Volume 64   2006

MEDICAL MINERALOGY AND GEOCHEMISTRY

EDITORS:

Nita Sahai   *University of Wisconsin*
                              Madison, Wisconsin

Martin A.A. Schoonen   *Stony Brook University*
                              Stony Brook, New York

COVER FIGURE CAPTIONS. *Top left:* A series of standards used in a reactive oxygen species (ROS) assay. *Top right:* Electron micrograph showing an example of frustrated phagocytosis (internalization of foreign substance by macrophage cells). [Used with permission by Dr. Vincent Castranova, National Institute for Occupational Health and Safety, Morgantown, WV. URL: *http://www.epa.gov/oswer/asbestos_ws/docs/castranova.pdf.*] *Bottom left:* Image of human femur taken using circularly polarized light microscopy, providing information about the orientation of collagen in the bone (and likely mineral as well, since the crystals are believed to orient themselves relative to the collagen). Cooler colors indicate fibers that are more longitudinally oriented (e.g., blues) and warmer colors indicate fibers that are more transversely oriented (e.g., yellows, reds). Courtesy Dr. Haviva Goldman, Drexel University College of Medicine; Bone sample from the Melbourne Femur Collection, University of Melbourne, School of Dental Science. *Bottom right:* An *ab initio* clsuter model of a $CaHPO_4$ nucleus forming at the silicate "three-ring" site on the (001) face of pseudowollastonite (low temperature $CaSiO_3$), a bioactive ceramic. [Reprinted from Sahai and Anseau (2005) *Biomaterials*, Vol. 26, p. 5763-5770, with permission of Elsevier.]

*Series Editor: **Jodi J. Rosso***

MINERALOGICAL SOCIETY OF AMERICA
GEOCHEMICAL SOCIETY

Reviews in Mineralogy and Geochemistry, Volume 64

Medical Mineralogy and Geochemistry

ISSN 1529-6466
ISBN 978-0-939950-76-6

COPYRIGHT 2006

THE MINERALOGICAL SOCIETY OF AMERICA
3635 CONCORDE PARKWAY, SUITE 500
CHANTILLY, VIRGINIA, 20151-1125, U.S.A.
WWW.MINSOCAM.ORG

The appearance of the code at the bottom of the first page of each chapter in this volume indicates the copyright owner's consent that copies of the article can be made for personal use or internal use or for the personal use or internal use of specific clients, provided the original publication is cited. The consent is given on the condition, however, that the copier pay the stated per-copy fee through the Copyright Clearance Center, Inc. for copying beyond that permitted by Sections 107 or 108 of the U.S. Copyright Law. This consent does not extend to other types of copying for general distribution, for advertising or promotional purposes, for creating new collective works, or for resale. For permission to reprint entire articles in these cases and the like, consult the Administrator of the Mineralogical Society of America as to the royalty due to the Society.

MEDICAL MINERALOGY
and GEOCHEMISTRY

64   *Reviews in Mineralogy and Geochemistry*   **64**

FROM THE SERIES EDITOR

The review chapters in this volume were the basis for a two day short course on *Medical Mineralogy and Geochemistry* held at the United States Geological Survey in Menlo Park, California, U.S.A. (December 9-10, 2006) prior to the American Geophysical Union Meeting in San Francisco, California. This meeting and volume were sponsored by the Mineralogical Society of America, Geochemical Society, and the United States Department of Energy.

Any supplemental material and errata (if any) can be found at the MSA website *www.minsocam.org*.

Jodi J. Rosso, Series Editor
West Richland, Washington
November 2006

SHORT COURSE SERIES DEDICATION

Dr. William C. Luth has had a long and distinguished career in research, education and in the government. He was a leader in experimental petrology and in training graduate students at Stanford University. His efforts at Sandia National Laboratory and at the Department of Energy's headquarters resulted in the initiation and long-term support of many of the cutting edge research projects whose results form the foundations of these short courses. Bill's broad interest in understanding fundamental geochemical processes and their applications to national problems is a continuous thread through both his university and government career. He retired in 1996, but his efforts to foster excellent basic research, and to promote the development of advanced analytical capabilities gave a unique focus to the basic research portfolio in Geosciences at the Department of Energy. He has been, and continues to be, a friend and mentor to many of us. It is appropriate to celebrate his career in education and government service with this series of courses.

1529-6466/06/0064-0000$05.00   DOI: 10.2138/rmg.2006.64.0

MEDICAL MINERALOGY
and GEOCHEMISTRY

64   *Reviews in Mineralogy and Geochemistry*   **64**

TABLE OF CONTENTS

1   The Emergent Field of Medical Mineralogy and Geochemistry

Nita Sahai, Martin A. A. Schoonen, H. Catherine W. Skinner

2   The Toxicological Geochemistry of Earth Materials:
An Overview of Processes and the Interdisciplinary Methods
Used to Understand Them

Geoffrey S. Plumlee,
Suzette A. Morman,
Thomas L. Ziegler

3
Metal Speciation and Its Role in Bioaccessibility and Bioavailability

Richard J. Reeder,
Martin A. A. Schoonen,
Antonio Lanzirotti

4   Aluminum, Alzheimer's Disease and the Geospatial Occurrence of Similar Disorders

Daniel P. Perl, Sharon Moalem

5   Potential Role of Soil in the Transmission of Prion Disease

P. T. Schramm, C. J. Johnson,
N. E. Mathews, D. McKenzie,
J. M. Aiken, J. A. Pedersen

6   Interaction of Iron and Calcium Minerals in Coals and their Roles in Coal Dust-induced Health and Environmental Problems

Xi Huang, Terry Gordon,
William N. Rom, Robert B. Finkelman

7   Mineral-Induced Formation of Reactive Oxygen Species

Martin A. A. Schoonen, Corey A. Cohn,
Elizabeth Roemer, Richard Laffers,
Sanford R. Simon, Thomas O'Riordan

8

Bone: Nature of the Calcium Phosphate Crystals and Cellular, Structural, and Physical Chemical Mechanisms in Their Formation

Melvin J. Glimcher

9   Silicate Biomaterials for Orthopaedic and Dental Implants

Marta Cerruti, Nita Sahai

10   Living Cells in Oxide Glasses

Jacques Livage, Thibaud Coradin

Reviews in Mineralogy & Geochemistry
Vol. 64, pp. 1-4, 2006
Copyright © Mineralogical Society of America

1

The Emergent Field of Medical Mineralogy and Geochemistry

Nita Sahai

Department of Geology and Geophysics
University of Wisconsin
Madison, Wisconsin, 53706, U.S.A.
e-mail: sahai@geology.wisc.edu

Martin A. A. Schoonen

Department of Geosciences
Stony Brook University
Stony Brook, New York, 11794-2100, U.S.A.
e-mail: martin.schoonen@stonybrook.edu

H. Catherine W. Skinner

Department of Geology and Geophysics
Yale University
New Haven, Connecticut, 06520-8109, U.S.A.
e-mail: catherine.skinner@yale.edu

Medical Mineralogy and Geochemistry is an emergent, highly interdisciplinary field of study. The disciplines of mineralogy and geochemistry are integral components of cross-disciplinary investigations that aim to understand the interactions between geomaterials and humans as well as the normal and pathological formation of inorganic solid precipitates *in vivo*. Research strategies and methods include but are not limited to: stability and solubility studies of earth materials and biomaterials in biofluids or their proxies (i.e., equilibrium thermodynamic studies), kinetic studies of pertinent reactions under conditions relevant to the human body, molecular modeling studies, and geospatial and statistical studies aimed at evaluating environmental factors as causes for activating certain chronic diseases in genetically predisposed individuals or populations. Examples presented in this volume include the effects of inhaled dust particles in the lung (Huang et al. 2006; Schoonen et al. 2006), biomineralization of bones and teeth (Glimcher et al. 2006), the formation of kidney-stones, the calcification of arteries, the speciation exposure pathways and pathological effects of heavy metal contaminants (Reeder et al. 2006; Plumlee et al. 2006), the transport and fate of prions and pathological viruses in the environment (Schramm et al. 2006), the possible environmental-genetic link in the occurrence of neurodegenerative diseases (Perl and Moalem 2006), the design of biocompatible, bioactive ceramics for use as orthopaedic and dental implants and related tissue engineering applications (Cerruti and Sahai 2006) and the use of oxide-encapsulated living cells for the development of biosensors (Livage and Coradin 2006).

Despite its importance, the area of Medical Mineralogy and Geochemistry has received limited attention by scientists, administrators, and the public. The objectives of the present short-course and volume are to highlight some of the existing research opportunities and challenges, and to invigorate exchange of ideas between mineralogists and geochemists

1529-6466/06/0064-0001$05.00　　　　　　　　　　　　　　DOI: 10.2138/rmg.2006.64.1

working on medical problems and medical scientists working on problems involving geomaterials and biominerals.

It is important to stress that universal physico-chemical principles fundamentally control the complex interactions of solutions, biomolecules, and minerals in the realms of the human body as well as in the natural geological environment (Frausto da Silva and Williams 1991; Williams and Frausto da Silva 1996; Sahai 2003, 2005). In both contexts, the molecular conformations, stabilities and reactivities of (bio)chemical components are critical to understanding specific reaction pathways. For example, silicate glasses and minerals such as Bioglass®, silica gel, wollastonite, pseudowollastonite, diopside, etc. have been shown to bond by the formation of a layer of carbonated hydroxylapatite at the implant/bone interface, making them suitable for use as orthopaedic and dental implant materials (Hench et al. 1971; Kokubo et al. 1982, 1994; Miake et al. 1995; de Aza et al. 1996; Liu and Ding 2001). An early step within this complex series of processes is the dissolution of the silicates by human blood plasma, involving reactions that are very similar to silicate weathering in the environment (Sahai and Anseau 2005; Cerruti and Sahai 2006). At the same time, it is critical to appreciate that biogeochemical reactions *in vivo* are ultimately under cellular and genetic controls, that acellular inorganic experiments or even *in vitro* experiments that include cells cannot entirely account for some of the unexpected, complex biological interactions, and that biominerals, because they are commonly nanocomposite materials, differ significantly in their physical-chemical and biological stability and reactivity compared to their bulk, macroscopic counterparts formed geologically or synthetically (Glimcher et al. 2006).

Studies of the interaction between the human body and geomaterials have been conducted for some time but the advance in analytical techniques holds the promise of major breakthroughs over the next decade. More than six decades ago, V.M. Goldschmidt, the father of Geochemistry, contributed to a study of the effect of olivine on the lungs of rats (King et al. 1945). This study, led by Early Judson King, a prominent pathologist, was one of a series of studies directed at understanding silicosis in foundry workers. Since Goldschmidt's early work, a number of geochemists and mineralogist have made contributions to the medical sciences, but progress has been hampered by the lack of insight at the molecular scale. The potential for uncovering the fundamental reaction mechanisms lies at this molecular level, leading to a deeper understanding of the overall process, and should increase our ability to design new treatment strategies. Recent developments in spectroscopy, molecular modeling, and high-resolution imaging provide powerful new avenues to explore the interaction between minerals and the human body at scales that are well beyond those accessible to Goldschmidt, his colleagues, as well as those that have been working in this area for the last few decades. While new techniques may enable advance, real progress will be made when geochemists, biomineralogists and biomedical researchers collaborate and harness the expertise in these clearly related disciplines. Geochemists and mineralogists are uniquely trained to contribute of this new field because of their knowledge of mineral stability, mineral reactivity, mineral precipitation/dissolution kinetics, and mineral-sorbate interactions, as well as their ability to study complex systems.

While this short course focuses on biogeochemical interactions *in vivo,* other areas may be included within the broader umbrella of "Geology and Human Health" (Skinner and Berger 2003) such as observing and predicting natural disasters. Early warning systems for volcanic eruptions, earthquakes, hurricanes, land slides, tsunamis and other natural disasters are becoming globally available and could save millions of lives. Legends and historic records of such natural disasters as the massive volcanic eruptions that destroyed Pompeii and Herculaneum in 79 AD, the Laki eruptions in 1883-94 that killed almost a quarter of Iceland's population with atmospheric disturbances over months that spread from Europe to Syria, and the eruption of Krakatoa, Indonesia, in 1883, underscore the importance of hazard prediction. While most natural hazards are episodic, catastrophic, and deadly, some disasters, such as volcanic eruptions generating ash, may also present a long-term health burden that could take years to manifest itself as disorders and disease.

Clearly, biogeochemical interactions *in vivo* are complex processes and understanding such processes at the molecular level will require a truly multi-disciplinary, integrated approach leveraging concepts and techniques from inorganic interfacial, aqueous and solid-state chemistry, computational chemistry, genetics, cellular and molecular biology. In addition to these scientific challenges, mineralogists, geochemists and biomedical scientists working in the field of Medical Mineralogy and Geochemistry field must meet the challenges of finding appropriate venues for research funding, peer-review, oral presentation of results and publication of research papers. In this regard, we are deeply appreciative of the forward-looking approach by the Mineralogical Society of America and the Geochemical Society; the Environmental Geochemistry and Biogeochemistry Program, Division of Earth Sciences, National Science Foundation; and Chemical Sciences, Geosciences and Biosciences Division, Office of Basic Energy Sciences, Office of Science, U.S. Department of Energy, and the Crustal Imaging and Characterization Team, United States Geological Survey, in supporting and sponsoring this short-course and volume. In addition we are very grateful for the contributions of the authors as well as the efforts of the reviewers that helped improve the quality of this volume.

REFERENCES

Cerruti M, Sahai N (2006) Silicate biomaterials for orthopaedic and dental implants. Rev Mineral Geochem 64:283-314

de Aza PN, Luklinska ZB, Anseau MR, Guitan F, de Aza S (1996) Morphological studies of pseudowollastonite for biomedical application. J Microscopy 182(1):24-31

Frausto da Silva JJR, Williams RJP (1991) The Biological Chemistry of the Elements: The Inorganic Chemistry of Life. Oxford University Press

Glimcher MJ (2006) Bone: nature of the calcium phosphate crystals and cellular, structural, and physical chemical mechanisms in their formation. Rev Mineral Geochem 64:223-282

Hench LL, Wilson J (1993) An Introduction to Bioceramics. World Scientific Publishing CO

Huang X, Gordon T, Rom WN, Finkelman RB (2006) Interaction of iron and calcium minerals in coals and their roles in coal dust-induced health and environmental problems. Rev Mineral Geochem 64:153-178

King EJ, Rogers N, Gilchrist M, Goldschmidt VM, Nagelschmidt G (1945) The effect of olivine on the lungs of rats. J Pathol Bacteriol 57:488-491

Kokubo T, Cho SB, Nakanishi K, Soga N, Ohtsuki C, Kitsugi T, Yamamuro T, Nakamura T (1994) Dependence of bone-like apatite formation on structure of silica gel. Bioceramics 7:49-54

Kokubo T, Shigematsu M, Nagashima Y, Tashiro M, Yamamuro T, Higashi S (1982) Apatite- and wollastonite-containing glass-ceramics for prosthetic application. Bull Inst Chem Res Kyoto Univ 60:260-268

Liu X, Ding C (2001) Apatite formed on the surface of plasma sprayed wollastonite coating immersed in simulated body fluid. Biomaterials 22:2007-2012

Livage J, Coradin T (2006) Living cells in oxide glasses. Rev Mineral Geochem 64:315-332

Miake Y, Yanagisawa T, Yajima Y, Noma H, Yasui N, Nonami T (1995) High resolution and analytical electron microscopic studies of new crystals induced by a bioactive ceramic (diopside). J Den Res 74:1756-1763

Perl DP, Moalem S (2006) Aluminum, Alzheimer's disease and the geospatial occurrence of similar disorders. Rev Mineral Geochem 64:115-134

Plumlee GS, Morman SA, Ziegler TL (2006) The toxicological geochemistry of earth materials: an overview of processes and the interdisciplinary methods used to understand them. Rev Mineral Geochem 64:5-57

Reeder RJ, Schoonen MAA, Lanzirotti A (2006) Metal speciation and its role in bioaccessibility and bioavailability. Rev Mineral Geochem 64:59-113

Sahai N (2003) The effects of Mg^{2+} and H^+ on apatite nucleation at silica surfaces. Geochim Cosmochim Acta 67:1017-1030

Sahai N (2005) Modeling apatite nucleation in the human body and in the geochemical environment. Am J Sci 305:661-672

Schoonen MAA, Cohn CA, Roemer E, Laffers R, Simon SR, O'Riordan T (2006) Mineral-induced formation of reactive oxygen species. Rev Mineral Geochem 64:179-221

Schramm PT, Johnson CJ, Mathews NE, McKenzie D, Aiken JM, Pedersen JA(2006) Potential role of soil in the transmission of prion disease. Rev Mineral Geochem 64:135-152

Skinner HCW, Berger AR (2003) Geology and Health: Closing the Gap. Oxford University Press

Williams RJP, Frausto da Silva JJR (1996) The Natural Selection of the Chemical Elements. Oxford University Press

Reviews in Mineralogy & Geochemistry
Vol. 64, pp. 5-57, 2006
Copyright © Mineralogical Society of America

2

The Toxicological Geochemistry of Earth Materials: An Overview of Processes and the Interdisciplinary Methods Used to Understand Them

Geoffrey S. Plumlee

U. S. Geological Survey
Box 25046, Denver Federal Center MS 973
Lakewood, Colorado, 80225-0046, U.S.A.
e-mail: gplumlee@usgs.gov

Suzette A. Morman

U. S. Geological Survey
Box 25046, Denver Federal Center MS 964
Lakewood, Colorado, 80225-0046, U.S.A.

Thomas L. Ziegler

Lakewood, Colorado, U.S.A.

INTRODUCTION

A broad spectrum of earth materials have been linked to, blamed for, and/or debated as sources for disease. In some cases, the links are clear. For example, excessive exposures to mineral dusts have long been recognized for their role in diseases such as: asbestosis, mesothelioma, and lung cancers (asbestos); silicosis and lung cancer (silica dusts); and coal-workers pneumoconiosis (coal dust). Lead poisoning, particularly in toddlers and young children, has been conclusively linked to involuntary ingestion of soils or other materials contaminated with lead-rich paint particles, leaded gasoline combustion byproducts, and some types of lead-rich mine wastes or smelter particulates. Waters with naturally elevated arsenic contents are common in many regions of the globe, and consumption of these waters has been documented as the source of arsenic-related diseases affecting thousands of people in south Asia and other regions. Exposure to dusts or soils containing pathogens has been documented as the cause of regionally common diseases such as valley fever (coccidioidomycosis) and much rarer diseases such as anthrax. Links between many other earth materials and specific diseases, although suspected, are less clear or are debated. For example, it has been suggested that geographic clusters of diseases such as leukemia are related to exposures to waters or atmospheric particulates containing organic or metal contaminants; however, for many clusters the exact causal relationships between disease and environmental exposure are difficult to prove conclusively. Even for many diseases in which the causal relationship is clear, such as in asbestosis and mesothelioma triggered by asbestos exposure, the minimum exposures needed to trigger disease, the influence of genetic factors, and the exact mechanisms of toxicity are still incompletely understood and are the focus of considerable debate within the public health community. Hence, understanding the health effects resulting from occupational and environmental exposures to a wide variety of earth materials remains a very active and fruitful area of research.

As noted in previous papers (Fubini and Areán 1999; Plumlee and Ziegler 2003, 2006), the body interacts both physically and chemically with earth materials to which it is exposed

1529-6466/06/0064-0002$10.00   DOI: 10.2138/rmg.2006.64.2

(and which it may take up) via inhalation, ingestion, and dermal absorption. Although these interactions are ultimately physiological, they are also strongly influenced by the physical and chemical properties of the earth materials, coupled with their geochemical reactivities in the variety of water-based, chemically diverse body fluids (respiratory, gastrointestinal, perspiration, blood serum, interstitial, and intracellular fluids) encountered along the different exposure and uptake pathways. As a result, there are significant opportunities for earth scientists working in collaboration with health scientists to contribute expertise in mineralogy, materials characterization, fluid-mineral reactions, and other areas in the quest for a better understanding of the specific links between earth materials and human health. In order to emphasize the importance of such interdisciplinary approaches, Plumlee and Ziegler (2006) defined toxicological geochemistry (TG) as the study of the geochemical interactions between body fluids and earth materials, and how these interactions may influence toxicity.

This chapter first provides an overview of the processes by which earth materials interact with the body. We then describe how earth science expertise and methods can be applied collaboratively throughout the range of heath science characterization and assessment methods commonly used to understand health issues associated with exposure to and uptake of earth materials. This paper is intended to complement and update the discussions presented in Plumlee and Ziegler (2003, 2006). As with these two earlier papers, we hope that readers from both the earth science and health science communities will find this chapter useful in understanding how geochemical processes and principles can be applied to help understand links between human health and exposures to certain earth materials. Our intent is to not imply that all earth materials are hazardous to health, as this is clearly not the case. Rather, we instead hope to show how a healthy dose of earth science expertise can be used to help understand where potential problems exist and help define the nature of the problems. We also hope that this paper and the many excellent other papers presented in this interdisciplinary volume will help identify and inspire future opportunities for fruitful research collaborations between geochemists, other earth scientists, and their counterparts in the health science community.

EARTH MATERIALS LINKED OR POTENTIALLY LINKED TO HUMAN HEALTH

Earth materials can include a broad range of solid, gaseous, or liquid substances that are produced and released by natural earth processes, that are contaminated by and/or released from the earth as a result of human activities, or that are produced from the earth by humans and transformed for use in society (Fig. 1). Examples of earth materials with known or postulated health concerns are listed in Table 1. The health concerns have a variety of origins, and can result from the earth materials themselves (such as mineral dusts), from contaminants or pathogens contained within or carried by the earth materials, and from consumption of waters, plants, or organisms that have picked up contaminants by interacting with earth materials. The potential health concerns are quite diverse, and can include cancers, respiratory diseases, neurological diseases, diseases of the excretory system, secondary diseases such as congestive heart failure, increased susceptibility to pathogen infections, and many others.

Historically, studies examining the toxicity and health effects of earth materials primarily focused on materials to which humans were exposed in workplace settings (such as asbestos or crystalline silica), or on soils or other earth materials containing toxicants of known health concern (such as lead, mercury, arsenic, or hexavalent chromium). Toxicological studies focused, for example, on commercially-produced asbestos or silica, or on the processes controlling toxicity of soluble metal salts or other materials common in workplace exposures. In the last several decades, there have been increasing concerns regarding environmental exposures to earth materials that a) are released into the environment by natural or anthropogenic processes, b) contain potential toxicants in more complex mineralogical forms than simple salts, or c)

Figure 1. There is a continuum of interactions between the earth's geosphere, hydrosphere, biosphere, atmosphere, and anthropogenic activities that all can produce earth materials or environmental materials to which humans can be exposed and that may be of potential health concern.

contain other elements such as antimony, tungsten, selenium, vanadium, or others whose toxicity effects are less well-known.

FACTORS INFLUENCING THE HEALTH EFFECTS OF EARTH MATERIALS

For detailed discussions on the toxicity effects of a wide variety of potential toxicants, the interested reader is referred to several excellent toxicology textbooks (such as Klassen 2001; Sullivan and Krieger 2001). Discussions focused on earth materials are presented by Plumlee and Ziegler (2003, 2006).

There are many different factors that can influence the health effects of earth materials, including:

- Intensity and duration of the exposure (the dose).
- Exposure route.
- Chemical conditions encountered along the exposure route.
- Physical and chemical characteristics of the material.
- The potential microbial or other pathogens present in the material.

Table 1. Examples of earth materials and their sources with known or postulated health effects, modified from Plumlee and Ziegler (2003, 2006).

Material	Examples of potential exposure sources	Primary exposure routes Health effects	References
Asbestos	Dusts from industrial, commercial asbestos products (insulation, brake linings, building products, others) and industrial activities. Dusts from asbestos accessory minerals in other industrial/commercial products (some, but not all, vermiculite, talc, and other products). Dusts from natural asbestos-bearing rocks and soils (via natural weathering and erosion or human disturbance) may provide low-level exposures that may be of concern to chronically exposed populations.	*Inhalation.* Asbestosis; lung cancer; pleural effusion, thickening, plaques; mesothelioma cancer. Associated secondary illnesses include heart failure, other cardiovascular problems, lung infection, possible autoimmune responses. *Ingestion.* Has been proposed as a trigger of GI cancers, other health effects; however, links have not been demonstrated with certainty.	Skinner et al. (1988); Holland and Smith (2001); Pfau et al. (2005); Pan et al. (2005); Dodson and Hammar (2006); Roggli et al. (2006).
Crystalline silica	Dusts generated by mining, sandblasting, and other industrial or workplace activities. Dusts produced by erosion of friable, silica-rich rocks (i.e., some ash flow tuffs, diatomaceous earth deposits) may also play a role in disease.	*Inhalation.* Silicosis, industrial bronchitis with airflow limitation, progressive massive fibrosis. Associated illnesses include opportunistic infections, silica nephropathy, lung cancer.	SSDC (1988); Castranova and Vallyathan (2000); Castranova (2000); Daroowalla (2001)
Coal dust, coal fly ash, coal combustion gases	Dusts, other aerosols, and gases generated by coal mining, processing, and combustion activities.	*Inhalation.* "Black lung disease" includes Coal Worker's Pneumoconiosis (CWP), progressive massive fibrosis, chronic airway obstruction or bronchitis, emphysema; possible silicosis due to intermixed crystalline silica. Sulfur dioxide and volatilized arsenic released as a result of coal combustion have also been linked to health effects.	Castranova and Vallyathan (2000); Castranova (2000); Daroowalla (2001);
Other commercial and industrial mineral dusts	These include talc, kaolinite, other clays, micas, and aluminosilicates. Sources include dusts generated from industrial and commercial activities and products	*Inhalation.* Mineral-specific fibrosis, such as talcosis; also silicosis and asbestosis due to intermixed crystalline silica and asbestos.	Daroowalla (2001)

(table continued on next page)

Table 1. *(continued)*

Material	Examples of potential exposure sources	Primary exposure routes Health effects	References
Natural mineral dusts	Dusts generated from desert areas, dry lake beds, agricultural areas during dry periods.	*Inhalation.* Possible silicosis, asbestosis. Possible uptake of metals or metalloids such as arsenic. Possible exposure to organic contaminants or pathogens in the dusts	Derbyshire (2005)
Urban particulates	Atmospheric particulates generated by petroleum or coal combustion, construction/demolition, industrial emissions, disturbance of urban soils, many others.	*Inhalation.* Possible uptake of heavy-metal and organic toxicants. Inhalation of acid aerosols of sulfur dioxide. Inhalation of irritant and/or oxidant gases such as nitric oxides and ozone. Increased asthma and cardiac stress, decreased lung function.	Costa (2001)
Man-made mineral fibers.	Glass wool (fiberglass); mineral wool (slag, rock wool)	*Inhalation.* Irritation of upper respiratory tract, skin. No ties to lung cancers; lung fibrosis.	Hesterberg et al. (2001)
Cement / concrete dust	Dusts from cement, concrete manufacturing, Concrete dusts generated by demolition, construction activities	*Inhalation, contact with exposed mucous membranes or moist skin.* Irritation of eyes, throat, respiratory tract; ulceration of mucous surfaces. Effects largely tied to alkalinity of the dusts, although some heavy metals such as thallium or hexavalent chromium may be present. Silicosis maybe a concern with long-term exposure to concrete dust.	Sahai (2001)
Edible soils	Soils purposefully consumed by humans as a result of cultural practices or perceived nutritional needs.	*Ingestion.* Potential uptake of heavy metals or other toxicants. Secondary physiological problems associated with consumption of non-digestible mineral material.	Abrahams et al. (2005)

(table continued on next page)

Table 1. *(continued)*

Material	Examples of potential exposure sources	Primary exposure routes Health effects	References
Volcanic ash	Atmospheric particulates generated by eruptions. Natural and anthropogenic disturbance of volcanic ash deposits, such as earthquakes, landslides, construction activities.	*Inhalation.* Irritation of respiratory tract; asthma; potential effects of crystalline silica within ash.	Weinstein and Cook (2005), and their references.
Volcanic gases, vog, laze	Sulfur dioxide, hydrogen fluoride, hydrogen chloride, other acid gases emanating from active volcanoes. Acidic aerosol droplets formed when hot lava contacts seawater and causes it to boil.	*Inhalation, contact with exposed mucous membranes or moist skin.* Irritation of eyes, throat, respiratory tract; ulceration of mucous surfaces. Effects largely tied to acidity of the gases and droplets. If gases are in sufficiently high concentration, toxic effects can also result.	Weinstein and Cook (2005), and their references.
Wildfire smoke, and ash	Particulate matter (including smoke, ash, and fine soil particulates) and gases generated by forest fires.	*Inhalation.* Asthma, irritation of respiratory tract; possible increased susceptibility to infection of respiratory system; possible long term increased cancer risk. Most studies have focused on organic constituents of smoke; less work has been done on mineral particles and heavy metals in smoke.	Reinhardt and Otmar (2000); Reinhardt et al. (2000); Ward (1999); Wolfe et al. (2004)
Pathogens	Soils, and dusts generated from soils, that host pathogens such as bacteria and fungi .	*Inhalation, ingestion.* Asthma and pathogen-specific diseases such as Valley Fever (from the soil fungus *C. Immitis*) and anthrax (*B. anthracis*). *Percutaneous absorption.* Pathogen-related infections can develop through breaks in the skin.	Bultman et al. (2005)

- Biosolubility, biodurability, bioaccessibility, and bioreactivity of the material in the body fluids encountered along the various exposure routes.

- The body's immune response.

- The body's physiological processes that control absorption, distribution, metabolism, and excretion of toxicants.

- Other confounding factors, such as age, gender, genetics, personal habits (i.e., smoking), and personal socioeconomic, health, nutritional status, and dietary cofactors that may promote or counteract the toxic effects of the exposure.

Intensity and duration of exposure (the dose)

To paraphrase an observation first made by Paracelsus in the 1500's, *"everything is a poison, nothing is a poison, it is the dose that makes the poison."* In its simplest definition, the dose is the amount of a potential toxicant that is taken up by an organism over a given period of time, with dose-response indicating that the greater the amount of a toxicant taken up by an organism, the greater the toxicological effect (Rozman and Klaasen 2001). However, toxicity is ultimately more complicated in that it is a function of the amounts of a potential toxicant that are taken up by the body, that survive the body's many clearance and mitigation mechanisms, and that reach the particular site(s) of toxic action within the body. Each substance will have a threshold level above which it will become toxic that depends on the substance, the exposure route, and the individual (Sullivan et al. 2001). Substances can be acutely toxic, if the dose is sufficiently high over a short period of time, and/or chronically toxic, where the dose is lower but over a longer period of time. A wide variety of chemical elements, including major chemical species (sodium, calcium, potassium, etc.), and trace elements (zinc, copper, selenium, chromium, etc.) are essential to the effective functioning of the body, and so disease can result if these elements are deficient in the body; however, they can also become toxic if present in excess in the body (Lindh 2005a,b).

Exposure route

There are several major exposure routes via which individuals can come into contact with potential toxicants or pathogens. These include inhalation, ingestion, and direct contact through unbroken skin (percutaneous), eyes (ocular), or wounds. A given toxicant or pathogen can have very different effects depending upon the exposure route due to the differences in physical processes, physiological processes, and chemical conditions to which it is subjected along each of the pathways (Fig. 2).

Inhalation. A variety of physical, chemical, and physiological processes within the respiratory tract affect the ultimate disposition and health effects of inhaled earth materials (see overviews by Schlesinger 1988, 1995; McClellan and Henderson 1995; Fubini and Areán 1999; Gehr and Heyder 2000). Important disposition mechanisms include deposition, clearance, retention, and, for gases or soluble materials, absorption. The characteristics of an inhaled earth material play a complex but important role in its disposition, including: the type of material inhaled (solid, liquid, gas, or pathogen); the concentrations of the material in the inhaled air; the chemical composition of the material; the solubility and reactivity of the material in the fluids lining the respiratory tract; and, for solids, their shape, size distribution, density, and surface charge.

Solid particles that deposit in approximately the first twelve generations of airways (which are lined by ciliated epithelium) are cleared by mucocilary clearance. Rhythmically-beating cilia waft the particle on a mucociliary escalator to the pharynx, where it is then swallowed. Particles depositing in more distal generations are subject to alveolar clearance, which is believed to occur primarily through phagocytosis by alveolar macrophages.

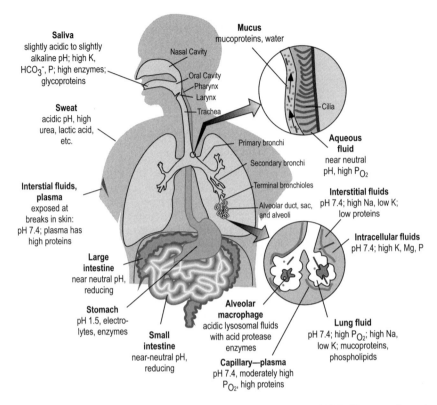

Figure 2. This schematic diagram, modified from Plumlee and Ziegler (2006), illustrates the various exposure routes and the substantial variability in body fluid types and compositions that can be encountered by earth materials during exposure.

The respiratory tract airways are lined with two liquid phases that facilitate deposition and clearance of solid particles: these include a mucous phase containing glycoproteins and phospholipid surfactants, and an underlying aqueous electrolyte phase. Mechanisms of particle deposition include (Fig. 3; Schlesinger 1988, 1995):

- impaction on the airway walls at places where there are rapid changes in airflow direction (such as at branches or constrictions in the airway);

- sedimentation onto the airway walls due to gravity settling;

- electrostatic deposition of freshly generated particles with high surface charges;

- interception of elongated particles by the airway walls;

- deposition of fine particles (<0.2 μm) as a result of brownian diffusion–impaction of air molecules on the particles enhances random motion that increases the likelihood of deposition on airway walls.

Figure 4 illustrates the influence of size in determining particle deposition efficiency in the various levels of the respiratory tract; other factors include, for example individual particle density and breathing patterns in the exposed individual. During normal nasal breathing, the largest inhaled particles (from ~5 to greater than several tens of microns large) are deposited in the nasopharyngeal airways (Fig. 4; ICRP 1995). Progressively smaller particle sizes are more

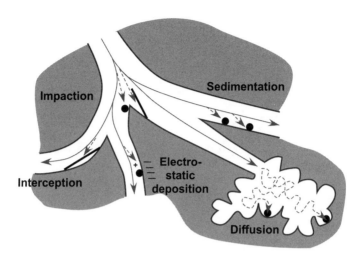

Figure 3. This schematic diagram illustrates various mechanisms by which particles (black) are deposited onto the linings of the respiratory tract. Airflow trajectories are depicted by the solid arrows, and the trajectories followed by deposited particles are shown by dashed arrows. Modified from McClellan (2000).

efficiently deposited in successively deeper portions of the respiratory tract. Particles less than approximately 2 μm (aerodynamic equivalent diameter) in diameter reach the alveoli, the deepest portions of the lungs where the most active exchange of oxygen and carbon dioxide occurs. These very small particles are either trapped in the alveoli or are exhaled.

The airways are close to or at saturation with water (Fubini and Areán 1999). Inhaled particles are therefore fully exposed to water vapor. One effect of the water vapor is to enhance particle clumping, which can enhance deposition higher in the respiratory tract (Schlesinger 1988).

During exercise or conditions where the nasal passages are clogged, or in certain individuals, increased oral breathing can shift the deposition of coarser particles to the upper portions of the trachea, and also can result in increased deposition efficiency of substantially larger particles in the tracheobronchial area (10's of μm) and in the alveoli (3-5 μm) than during normal nasal breathing (Schlesinger 1988). Airway irritation and obstructive airway diseases tend to constrict airways and lead to greater deposition higher in the respiratory tract and reduced deposition in the alveoli (Schlesinger 1988; Schulz et al. 2000). Fibers can behave somewhat differently during inhalation than more equant particles in that, although they tend to flow aerodynamically, wobble along their long axis gives them a larger effective diameter. As a result, only fibers less than approximately 0.5-1.5 μm wide can penetrate into the deepest portions of the alveoli. Also, straighter and shorter fibers tend to escape deposition by interception more readily and so are able to penetrate more deeply into the respiratory tract than curly or long fibers.

Clearance of deposited particles and pathogens from the respiratory tract is accomplished by a variety of mechanisms that are an integral part of the body's immune system, and that can vary as a function of location within the respiratory tract and the physical and chemical nature of the particle (Fig. 3):

- *Increased mucus production:* As hay fever suffers well know, one of the body's immune response mechanisms to inhaled particle overload is to increase production of mucus and fluids in the nasal passageways and upper respiratory tract. This helps

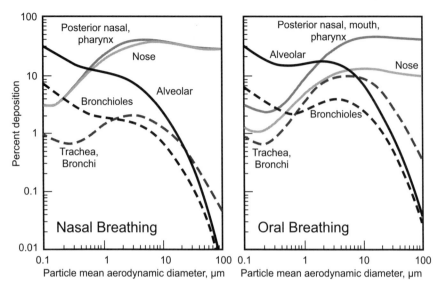

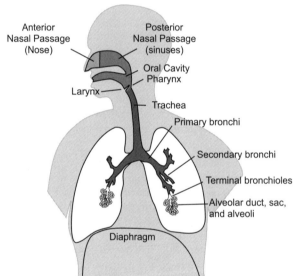

Figure 4. The plots at top show variations in particle deposition efficiency for nasal breathing and oral breathing as a function of region within the human respiratory system, keyed to corresponding regions shown on a schematic diagram (shown below) of the respiratory system. The plots are modified from ICRP (1995) and the schematic is modified from Newman (2001), Plumlee and Ziegler (2003, 2006), and ICRP (1995).

trap particles in the upper parts of the respiratory system and also helps to dilute chemical effects of inhaled particles on adjacent tissues.

- *Coughing:* Particle-laden mucus is cleared from the trachea and bronchi in part by coughing, especially in individuals with respiratory disease or irritation.

- *Mucociliary transport:* The cells lining the trachea, bronchi, and bronchioles are ciliated. Movement of the cilia helps transport the particle-laden mucus up and out of the respiratory tract along the mucociliary escalator, and the cleared mucus is then either expectorated or swallowed.

- *Dissolution in the fluid lining the respiratory tract:* Soluble particles or gases are cleared primarily by dissolution in the near-neutral pH fluid lining the airways at all levels of the respiratory tract. The dissolved materials are relatively efficiently absorbed into the bloodstream.

- *Phagocytosis by airway and alveolar macrophages:* Macrophage cells roam the fluids lining the respiratory tract and engulf (phagocytize) foreign particles (Kreyling and Scheuch 2000). Airway macrophages in the tracheal and bronchial airways play a subsidiary role in particle clearance compared to the mechanisms listed previously, whereas alveolar macrophages play a primary role in particle clearance from the alveoli. The macrophages contain lysosomes with acidic pH and digestive enzymes such as acid hydrolases (Brain 1992; Newman 2001), and so can help dissolve less soluble particles that had not previously dissolved in the fluids lining the respiratory tract. The dissolved material can either be transported out of the macrophage, or may react with cellular components and remain in the macrophage. The macrophages can also transport engulfed particles to the mucociliary elevator, to distal alveolar spaces (sequestration), or through the lung epithelium to the lymph system or the pleural spaces. Once the macrophages phagocytize particles, they also release chemical messengers called cytokines into the surrounding epithelium that recruit other macrophages to the site to aid in the clearance of additional foreign particles (Lehnert 1992). The macrophages, which are 5-10 μm in diameter, can only engulf particles smaller than 2-5 μm (Fubini and Areán 1999). Therefore, long fibers that reach the alveoli are less readily cleared than more equant particles of the same aerodynamic diameter, because they cannot be easily engulfed by the alveolar macrophages. A process called frustrated phagocytosis results when the macrophages that fail to engulf a fiber die and release their cytotoxic chemicals into the surrounding cellular environment, potentially leading to injury to adjacent cells. Fibers or particles with sharp points can also penetrate the macrophages' cell membranes, which also can cause leakage of cytotoxic chemicals into the surrounding alveolar spaces (Fubini and Areán 1999).

- *Particle uptake by airway and alveolar epithelial cells:* Phagocytic uptake of free particles by cells lining the airways and alveoli is typically small compared to other clearance mechanisms, but is enhanced if free particles persist, such as during lung overload conditions where the other mechanisms are overwhelmed (Churg 2000).

- *Particle penetration into the interstitial spaces of the lungs:* Particles can penetrate through the epithelial surface lining the alveoli into the pulmonary interstitial spaces, where they can persist, react with the lung interstital fluids, or be engulfed by interstitial macrophages or interstitial cells.

In spite of these many clearance mechanisms, relatively insoluble particles can be retained in the lungs for extended periods of time following a brief acute exposure. However, there are significant differences in the rates which equivalent particles can be cleared by different species. For example, McClellan (2000) indicates that rats have substantially greater particle clearance rates for respired insoluble particles than do humans; such differences are an important consideration in the extrapolation of inhalation toxicology studies based on rats to make inferences about toxicity of the same particles to humans.

The disposition of inhaled gases in the respiratory tract depends upon the gas species present, their concentrations in the inhaled air, and their solubilities in the water-based fluids lining the respiratory tract. Water-soluble gases (i.e., sulfur dioxide) can be absorbed more readily in the upper portions of the respiratory tract, whereas less soluble gases (i.e., nitrogen oxides) are more likely to persist into the alveolar spaces where gas exchange with blood is most effective (Newman 2001). However, even soluble gases may penetrate into the alveolar spaces if the gases are present in sufficiently high concentrations in inhaled air. Gas absorption in the alveoli occurs primarily by diffusion through the epithelial lining and then dissolution in the blood. Carbon monoxide, hydrogen cyanide, and some other toxic gases preferentially bind with hemoglobin in the blood, thereby precluding uptake of oxygen.

Ingestion. In general, most earth materials are ingested involuntarily, such as through hand-to-mouth contact in toddlers and young children, ingestion of soil particles transported on poorly cleaned foodstuffs, or swallowing of dust particles cleared from the respiratory tract by the mucociliary escalator. In some societies, however, geophagia, the voluntary ingestion of soil, is practiced for nutritional or detoxification purposes (Abrahams 2005).

The gastrointestinal (GI) system includes several major regions that are sequentially traversed by ingested earth materials, each with its own characteristic function, physical processes, and chemical environment. These regions include the mouth, pharynx, esophagus, stomach, small intestine, and large intestine. In addition to its primary function (to intake and digest food, extract energy and nutrients, and expel the remaining waste), the GI tract also is a key component of the body's immune system: for example enzymes and antibodies in the saliva, acid in the stomach, and bacteria in the intestines all help to neutralize ingested pathogens (Coico et al. 2003).

The disposition of ingested earth materials depends upon their particle makeup (the phases present), size distribution, chemical solubility in the fluids encountered along the GI tract, the presence of other material (such as food), and biologically mediated reactions (some of which occur with the participation of resident microbes) that may transform the earth materials and their degradation products (Plumlee and Ziegler 2003, 2006).

Due to physical breakdown during chewing, ingested particles are generally less than 500 μm to 1 mm in size. Chewing helps reduce particle size, and can therefore increase the surface area and reaction rates of the particle in the GI system. Studies of hand-mouth transmission indicate that the largest particles that adhere to hands of toddlers or young children are in the 250-μm size range (EPQ TRW 2000).

Ingested earth materials are progressively subjected to a variety of fluid compositions and chemical conditions along the GI system, starting with near-neutral, enzyme-rich saliva in the mouth, then acidic gastric fluids in the stomach, then near-neutral fluids with pancreatic and bile juices in the small and large intestines (see Table 2, this chapter; also see Plumlee and Ziegler 2003, 2006). The digestive process is initiated by the saliva, which contains enzymes to initiate food breakdown (Table 2). Substantial breakdown and dissolution of ingested substances occurs in the acidic and enzyme-rich fluids of the stomach, whereas most absorption occurs in the intestinal tract (Sipes and Badger 2001). It is important to note that materials that are dissolved in the stomach may not all be absorbed across the intestinal tract lining. For example, those substances that are most readily absorbed by diffusion across the intestinal wall into the bloodstream are in non-ionized, lipid-soluble forms. Further, materials dissolved in the acidic conditions of the stomach may subsequently re-precipitate or adsorb onto solids in the more alkaline environment of the intestines, and then be eliminated as waste. Gastrointestinal absorption of essential trace elements such as zinc, copper, and selenium, as well as toxicants such as lead, can also be enhanced or diminished by the presence or absence of other trace elements and chemicals in the diet (WHO 1996). The venous blood supply of the upper portion of the small intestines (which is responsible for the absorption of nutrients as well ingested toxicants and medications) drains into the portal vein of the liver. The liver can metabolize many of these ingested foreign substances, thereby protecting the rest of the body from exposure, but sometimes at the expense of injury to liver cells (acetaminophen ingestion being the most notorious example).

Exposure via the skin or wounds in the skin. Although the skin is the first line of defense for the immune system against external toxicant insults, percutaneous exposures can be a source of adverse health effects for some earth materials. Some gaseous and liquid chemicals

Table 2. Summary of fluid compositions encountered by earth material particles along the various exposure routes. Most of the fluid compositions are summarized in greater detail by Plumlee and Ziegler (2003, 2006) and references therein. Other citations are listed in the table.

Exposure Route Region	Fluid type(s) composition(s) encountered
Inhalation	
Head airways	**Nasal fluid:** Contains both mucus with glycomucoproteins and phospholipid surfactant, and an electrolyte phase similar to interstitial fluids with pH near 7.4. The fluids are highly oxygenated (P_{O2} as high as 0.2 atm), but contain various reducing organic couples that are out of equilibrium with dissolved O_2.
Trachea, primary bronchi	**Pulmonary fluid:** Contains both mucus with glycomucoproteins and phospholipid surfactant, and an electrolyte phase similar to interstitial fluids with pH near 7.4. The fluids are relatively oxygenated (P_{O2} decreases from 0.2 atm), but contain various reducing organic couples that are out of equilibrium with dissolved O_2.
Bronchioles	**Pulmonary fluid:** Contains both mucus with glycomucoproteins and phospholipid surfactant, and an electrolyte phase similar to interstitial fluids with pH near 7.4. The fluids are relatively oxygenated (P_{O2} decreases to 0.132 atm), but various reducing organic couples are also present that are out of equilibrium with dissolved O_2.
Alveoli	**Pulmonary fluid:** Contains both mucus with glycomucoproteins and phospholipid surfactant, and an electrolyte phase similar to interstitial fluids. The fluids are relatively oxygenated (P_{O2} around 0.132 atm), but contain various reducing organic couples that are out of equilibrium with dissolved O_2.
Macrophages and epithelial cells	**Intracellular fluid:** A pH 7.1, potassium-phosphate-bicarbonate-magnesium-sulfate electrolyte fluid with lesser chloride, calcium, and sodium. Contains a wide variety of organic acids, proteins, and other organic components. Intracellular fluid is more reduced than interstitial fluid, with redox conditions strongly influenced by reduced/oxidized glutathione.
Macrophage lysosomes (phagocytized particles)	**Lysosomal fluid:** Maintains an acid pH around 4.5, and contains acid proteases, acid phosphatases, and other enzymes. We are not aware of literature sources that report the electrolyte composition.
Interstitial tissues adjacent to airways (penetrated by particles from the alveoli and other airways)	**Interstitial fluid:** A pH 7.4, sodium-chloride-bicarbonate electrolyte fluid with: lesser potassium, magnesium, calcium, and sulfate; relatively oxygenated (P_{O2} 0.02 to 0.13 atm); abundant organic acids amino acids, and other organic ligands; substantially lower concentrations of proteins than the plasma.
Blood (particles can penetrate the lung epithelium and capillary walls).	**Plasma:** A pH 7.4, sodium-chloride-bicarbonate electrolyte fluid with lesser potassium, magnesium, calcium, and sulfate; relatively oxygenated (venous P_{O2} around 0.02 atm; arterial P_{O2} around 0.13 atm); abundant organic acids amino acids, and other organic ligands; high concentrations of proteins.

(table continued on next page)

Table 2. (continued)

Exposure Route Region	Fluid type(s) composition(s) encountered
Ingestion	
Mouth	**Saliva:** Dilute electrolyte fluid (Ritschel and Thompson 1983) whose composition depends upon secretion rate. Saliva produced at low secretion rates is slightly acidic (pH 6.6) with higher potassium and high phosphate; saliva produced at higher secretion rates is slightly alkaline (pH 8) with high sodium, bicarbonate, chloride, and phosphate (Thaysen et al. 1954). Saliva contains mucin and a variety of enzymes such as alpha-amylase, lysozyme, and lingual lipase. It also contains antibacterial compounds such as hydrogen peroxide, thiocyanate, and secretory immunoglobulin A.
Stomach	**Gastric fluid:** Sodium-hydronium-chloride electrolyte fluid, with lesser potassium, calcium, and magnesium. The pH is generally maintained near 1.5, but pH values rise due to neutralization by food. Contains pepsin, lipase, lysozyme, and other enzymes, as well as: largely protonated forms of various amino acids and carboxylic acids; mucoproteins; other proteins; and carbohydrates. Although swallowed air may provide some available oxygen, several references suggest gastric juice is rather reducing, perhaps dominated by abundant ascorbic acid - dehydroascorbic acid redox couple (Sobala et al. 1991); several references indicate H_2S is also common (i.e., Keith et al. 2006), but disagree as to the presence or absence of low-molecular-weight sulfur-bearing organics such as cysteine and glutathione.
Intestinal tract	**Intestinal fluid:** Alkaline secretions are added in upper (duodenal) portions of intestines that help raise pH of food-gastric juice mixture (chyme); the pH ranges from around 6 to greater than 7. Bile salts (to aid in digestion and absorption of lipids) and pancreatic juices (which contain enzymes and other organics to help break down proteins and carbohydrates) are also added in the duodenum. Redox conditions are quite reduced, with H_2S stable.
Percutaneous	
Unbroken skin	**Perspiration:** A relatively dilute sodium-chloride electrolyte fluid with lesser potassium and bicarbonate see references in Plumlee and Ziegler (2003). Na and Cl concentrations increase and Na/K ratio decreases with increasing sweat production rate (Freudenrich 1998). Sweat from apocrine glands in armpits and groin contains proteins and fatty acids (which are decomposed by bacteria to produce body odor), whereas sweat from eccrine glands over the rest of the body does not. Sweat also contains metabolic wastes such as urea, uric acid, ammonia, lactic acid, and ascorbic acid.
Wounds in skin	**Plasma:** See description above. **Interstitial fluid:** See description above.

can be readily absorbed directly through the skin, particularly those that are non-ionic or lipophilic, such as methyl mercury. There has recently been increasing discussion in the biomedical literature regarding the potential for solid metallic or metal-bearing nanoparticles (particularly those used in personal care products) to be absorbed directly through the skin and to therefore present a potential source of toxicity (Tinkle et al. 2003; Gulson and Wong 2006). Some gaseous, aqueous, or solid earth materials can react with the skin or with perspiration on the skin, resulting in allergic reactions, skin irritation, or chemical burns. Examples include allergic reactions (contact dermatitis) triggered by jewelry containing nickel, and caustic burns caused by wet alkaline solids such as cement.

Breaks or wounds in the skin can provide direct access for toxicants and pathogens contained in earth materials to tissues and the bloodstream. For example, it has been postulated that Kaposi's sarcoma endemic in parts of Africa results from dermal exposure to iron-rich soils developed on mafic volcanic rocks; farmers that till the soil barefoot take up micron-sized clay- and iron-rich particles through pores or abrasions in the skin of their feet, leading to dermal damage and impaired immunity to pathogens (Ziegler 1993). Further details of these and other health issues linked to dermal exposures to earth materials are provided in Plumlee and Ziegler (2003, 2006).

Physical and chemical characteristics of the earth material

The previous discussions of exposure route illustrated that the physical and chemical characteristics of an earth material can strongly influence how it is taken up by the body, its disposition within the body, and its health effects.

Physical characteristics. The physical characteristics of an earth material primarily include its form—whether it is solid, aqueous, gaseous, or non-aqueous liquid. For solid materials important physical characteristics also include whether the solid is a glass or crystalline, and its mineralogy, density, size, shape (morphology), and surface charge. These physical characteristics ultimately influence how the materials interact both physically and chemically with the body, and how readily the material can be taken up by the body. For example, a variety of solid-based toxicants can have substantial health impacts if they are inhaled or ingested, but a smaller number of solid toxicants have impacts through dermal contact or through breaks in the skin. In contrast, fluids can trigger effects through dermal contact, inhalation of fine droplets, or ingestion. As discussed in the previous sections, size and density influence the depth to which an inhaled particle can penetrate the respiratory tract. Particle shape can influence the aerodynamic properties of a particle as it is being inhaled, and can influence its clearance from the respiratory tract.

Chemical characteristics. Important chemical characteristics include the types (organic, inorganic), amounts, and speciation of potential toxicants in the earth material as it is delivered to the body. The speciation can include how the toxicant occurs within the material (i.e., tied up within the crystal structure or sorbed onto the surface of a solid), and the oxidation state or chemical speciation of potentially toxic elements within the material (for example ferrous and/or ferric iron; arsenate, arsenite, and/or organo-arsenic; mercurous, elemental, mercuric, and/or methylmercury; hexavalent and/or trivalent chromium). Many elements such as mercury or arsenic have substantially different behavior in the body and, as a result, different toxicity effects, depending upon their oxidation state and chemical speciation. Examples will be presented in a later section; see also the discussion in Reeder et al. (2006, this volume).

Potential microbial, parasitic pathogens present in the earth material

It is well-known that water can be the host to a wide variety of bacterial, viral, protozoan, and multicellular pathogens that are the source of illness in humans, and that these pathogens in water can be present both naturally and incidentally as a result of inputs from humans or other organisms. Soils, dusts, and some other solid earth materials are also a substantial reservoir for a variety of microbes, microscopic multicellular organisms, and prions (agents of transmissible spongiform encephalopathies, or TSEs); see the review by Bultman et al. (2005). A small percentage of these are pathogenic, including soil bacteria such as *Bacillus anthracis* (which causes anthrax), soil fungi such as *Coccidioidies immitis* (*C. Immitis*, the etiological agent of valley fever), viruses, protozoa such as *Cryptosporidium parvum*, and parasitic helminthes (microscopic worms, such as hookworms and round worms). These pathogens can be permanent (complete their entire life cycle in the soil), periodic, transient, or incidental (resulting from human or animal activities) inhabitants of the soil. Depending upon the particular pathogen, exposure can result from inhalation of dusts generated from soil (Griffin et al. 2002), direct or incidental (i.e., on foodstuffs) ingestion of soil, dermal contact with soil, and/or ingestion of water containing pathogens washed from soil. Many pathogens (such as soil fungus spores, viruses, bacteria, and bacterial spores) are in the appropriate size range to be inhaled and to penetrate to the alveoli, where they encounter warm, moist, nutrient-rich conditions that can promote pathogen development and absorption into the blood stream. Many multi-celled parasites are in the ingestible size range.

The occurrence of microbial or other pathogenic organisms in soils (and therefore in dusts produced from the soils) is strongly influenced by the physical and chemical characteristics of the soil, along with climate, vegetation, and the extent of activities of humans, wildlife, or other organisms. Pathogens that form spores (such as *B. Anthracis*, and *C. Immitis*) have the ability to survive extended dry periods in the soils. Pathogens that lay eggs or that form cysts or oocysts (including worms and protozoa such as *Giardia lamblia*) can also survive for extended periods of time in soil, with the duration enhanced by, but not a requirement of, elevated soil moisture.

Important physical and chemical characteristics of soil that influence the species of pathogens present include pH, water content, and the presence or absence of abundant organic matter, clays, or soluble salts. *C. Immitis* fungus is endemic in much of the arid U.S. southwest and in parts of South America where climates are seasonally dry and winters are short. It competes successfully against other microbes in soils that are alkaline, and that have abundant pore spaces, low clay and organic matter contents, high salinity, abundant evaporative salts, and, possibly high boron contents (Bultman et al. 2005). The boron, tied up in soluble borate salts, may act as a microbicide that inhibits bacterial growth.

Fluid compositions and chemical environments present along the exposure route

The various body fluids are largely aqueous electrolyte solutions with a complex but variable array of organic components such as amino acids, sugars, other organic acids, low- to high-molecular weight proteins, lipids, and enzymes. As indicated in the exposure route section and summarized in Table 2 and Figures 5 and 6, an earth material to which the body is exposed can encounter substantially different chemical environments and fluid compositions between and along the various exposure routes. Several fluid types encountered along ingestion and dermal exposure routes (saliva, gastric fluids, and sweat) can also vary substantially in pH and electrolyte amounts and proportions as a function of their secretion rates (Thaysen et al. 1954; Freudenrich 1998), thereby adding further complexity to the chemical conditions potentially encountered along these exposure routes.

Geochemists familiar with fluid-mineral reactions will immediately recognize that these differences in chemical composition create a strong likelihood that a particular earth material and any contained or adsorbed toxicants may have quite different chemical responses in the body depending upon the exposure route (Plumlee and Ziegler 2006). For example, most minerals or other phases are likely to be more soluble and more readily dissolved in the acid conditions of the stomach than in the near-neutral pH fluids lining the lungs. Differences in electrolyte compositions also influence the solubilities of some solid phases between and along the different exposure routes. An obvious example is the substantially greater stability

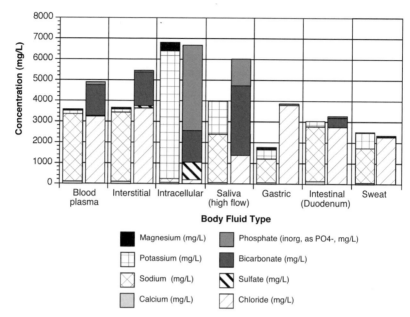

Figure 5. Concentrations of the major inorganic electrolyte species can vary substantially between different body fluid types. These variations play an important role in the relative solubility of a variety of minerals and earth material components along the different exposure routes. Modified from Plumlee and Ziegler (2006).

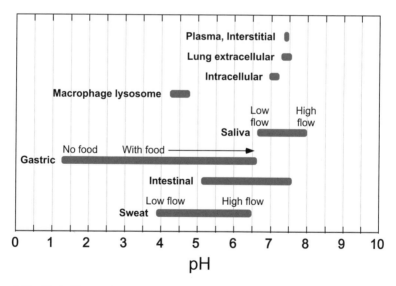

Figure 6. The ideal pH range for proper physiological function of most of the body is a narrow range around 7.4, and so the body employs a variety of physiological mechanisms to maintain the pH of the plasma and interstitial fluids within this range. However, other body fluids are maintained to different pH values depending upon physiological function, such as the gastric fluids, intracellular fluids, and macrophage lysosomal fluids. The pH of gastric fluids, sweat, and saliva are known to vary as a function of secretion rate and, in the case of the gastric foods, with or without the presence of food.

of carbonate minerals in the near-neutral, bicarbonate-rich lung and interstitial fluids than in the acidic, bicarbonate-poor stomach acids. Differences in phosphate levels between the different fluids likely also play a role in the relative solubility of various phosphate minerals; for example, lead phosphates would be expected to be substantially less soluble in the lung, interstitial, and macrophage lysosomal fluids than in the phosphate-poor gastric fluids.

The chemical species dissolved or desorbed from earth materials can also undergo a variety of chemical transformations in the body, such as chelation with inorganic or organic ligands, changes in oxidation/reduction (redox) state, sorption onto other less soluble phases, and re-precipitation as new phases that are more stable under the ambient chemical conditions (Plumlee and Ziegler 2006). Geochemical concepts developed to explain element mobility in the environment (Smith and Huyck 1999; Smith 2006) are also useful to help understand these chemical transformations and element mobility along the different exposure routes in the body. The nature and extent of these chemical transformations can differ extensively between the exposure routes, and also vary sequentially along an exposure route. One obvious example is the increase in pH encountered during the transition from the stomach to the intestines, which likely leads to the precipitation of secondary phases such as carbonates or oxyhydroxides in the intestine. Sorption onto solid particles in the more alkaline conditions of the intestinal tract may further reduce the mobility in and absorption by the body of metals such as lead whose sorption characteristics are strongly enhanced at near-neutral pH. In contrast, the pH increase from the stomach to the intestines may enhance the mobility of some metals that are effectively chelated by organic ligands, which become deprotonated and therefore more available for chelation; it may also enhance the mobility of potentially toxic oxyanion species such as arsenate, selenate, chromate, and tungstate that desorb as the pH increases. Another example is the increase in acidity and decrease in overall redox state as particles insoluble in the near-neutral alveolar fluids are phagocytized by the alveolar macrophages, which can enhance the solubilization of mineral particles such as oxides or oxyhydroxides. However, this enhanced solubilization may be counterbalanced by other chemical conditions within the macrophages. For example, even though uranium oxide particles are noted to dissolve in rat macrophage phagolysosomal fluids, the released uranium most likely reprecipitates within the macrophages as a secondary phosphate phase due to the high levels of intracellular phosphate (Kreyling et al. 1992).

Redox environment. By expending energy, the body can drive or exploit myriad organic and bioinorganic chemical reactions to maintain proper physiologic function in an oxygenated environment with which it is well out of chemical equilibrium. Nowhere is this better illustrated than by the multiple oxidation-reduction reactions that the body manipulates. There are a number of different redox couples or systems that are thought to variably influence the redox environments that are encountered by earth materials along the various exposure routes. Important examples include oxidized/reduced glutathione (GSSG/GSH), ascorbic acid (which at plasma pH includes ascorbate, ascorbyl radical, and dehydroascorbic acid), cysteine-cystine, thioredoxin, nicotinamide adenine dinucleotide phosphate ($NADP^+$/NADPH), and pyruvate-lactate (Buettner 1993; Shafer and Buettner 2001, 2003). Multiple couples (many of which are interlinked chemically) may be active in a given region of the body, with the overall redox environment and Eh in that region due to the combined action of the different couples present (Shafer and Buettner 2001, 2003). The relative importance of each redox couple in each environment is a function of its electrode potential (Fig. 7) and the reservoir size for the pair in the particular body fluid (Shafer and Buettner 2001, 2003). Most literature sources indicate that the overall Eh of the plasma is generally poised in the area of −100 millivolts, due to the combined action of redox couples or systems such as cysteine-cystine, dehydroascorbate-ascorbate and others. In contrast, the intracellular Eh is likely dominated by the glutathione GSSG/GSH couple, due to its substantially greater abundance in the intracellular fluids than the other couples. However, the NADP+/NADPH couple is also present and serves to recycle the GSSG/GSH couple due to its overall lower Eh (Fig. 7). As a result, the intracellular redox

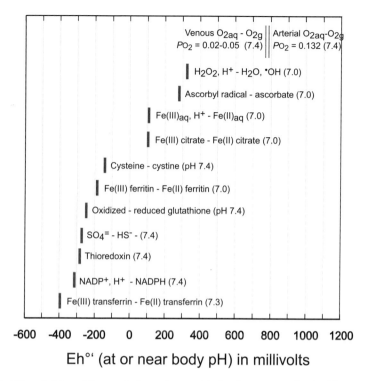

Eh°' (at or near body pH) in millivolts

Figure 7. Although it is well-known in the environmental chemistry community that Eh measurements are fraught with uncertainties (poisoning of the electrode, uncertainty in which of many possible redox couples are actually controlling the measurement, etc.), Eh is commonly used in the biochemical literature to compare the relative oxidation-reduction potential of various biologically active redox couples. This figure plots the Eh for a number of different organic and metal-organic redox couples that are known to occur within the body, compared to inorganic ferrous-ferric iron, sulfate-sulfide, dissolved oxygen, and reactive oxygen species couples (data from: Buettner 1993; Shafer and Buettner 2001, 2003, and the thermodynamic database contained in the Geochemists Workbench chemical speciation and reaction-path modeling program; Bethke 1996). Redox couples with higher Eh values would tend to oxidize couples with a lower redox potential. However, as noted by Schoonen et al. (2006, this volume), the relative Eh for each of the redox couples shown on the plot is for equal activities of the two species in solution; hence, particularly for the reactive oxygen species couples, if one of the pair occurs in the body at substantially lower concentrations than the other, this would shift the relative Eh defined by the couple *in vivo* to a substantially different value.

environment is substantially more reduced than that of the plasma, with values closer to −200 to the −240 millivolts of the glutathione couple. The GSSG/GSH ratio in the cellular environment is increasingly viewed as a key indicator of cellular redox condition and overall health, with elevated concentrations of the oxidized relative to the reduced form indicative of oxidative stress (Shafer and Buettner 2003).

As shown by Figure 7, the Eh values defined by partial pressures of oxygen in arterial (P_{O_2} = 0.132 atm) and venous (P_{O_2} = 0.02–0.053 atm) plasma indicate that the plasma has, even in its most oxygen-depleted state in the veins, high dissolved oxygen concentrations that are well out of redox equilibrium with these various organic redox couples (Plumlee and Ziegler 2003, 2006). Sulfur, iron, and other elements with multiple oxidation states can be similarly out of internal equilibrium and equilibrium with dissolved oxygen in the plasma. Metallothionein proteins and other organic species containing sulfhydril (HS) groups are common in sulfate-rich plasma or interstitial fluids. Hemoglobin, a protein, similarly contains tightly bound ferrous iron even though the iron atoms in hemoglobin are those with which transported oxygen associates or dissociates (Rhoades and Pflanzer 1992; Taylor and Williams 1995).

Literature sources indicate that the redox environment of the gastric fluids is quite reduced, with H_2S rather than HSO_4^- stable (Keith et al. 2006), and an Eh of around -200 millivolts at pH 1.5-2 (Davis et al. 1992). The predominant redox couples in the stomach are not specified in literature sources encountered to date; however, Sobala et al. (1991) indicate that both ascorbic acid and dehydroascorbic acid are abundant in the gastric fluids, and Keith et al. (2006) indicate that a number of thiol-containing species such as cysteine and glutathione are also present. Oxygen may be present transiently in the stomach from swallowed air; however, as with the plasma, the oxygen is substantially out of redox equilibrium with the active organic redox couples. The overall redox environment of the intestines is likely to also be quite reduced with reduced sulfur forms stable. A number of literature sources on the intestines focus on the glutathione redox couple; for example Assimakopoulos et al. (2006) suggest that oxidative stress, manifested by elevated levels of the oxidized glutathione species, is linked to intestinal obstructive jaundice.

Biosolubility and bioreactivity of earth materials in body fluids

From the geochemist's perspective, the toxicological literature can be interpreted to indicate that geochemical reactions of earth materials with the body's fluids and tissues plays a substantial role in their toxicity and health effects. It is thus useful to classify earth materials based on their relative biosolubility and bioreactivity in the various body fluids and chemical environments encountered *in vivo* (Table 3; see also Plumlee and Ziegler 2003, 2006, and extensive references cited therein).

Biosolubility is the extent to which an earth material dissolves in the body's fluids. *Biodurable* materials are those that are generally bioinsoluble or sparingly biosoluble in body fluids, and so cannot be cleared rapidly by chemical dissolution. *Biopersistence* is the extent to which a substance can resist all chemical, physical, and other physiological clearance mechanisms in the body.

It is important to differentiate thermodynamic versus kinetic drivers of biosolubility. Some minerals are biodurable because there is no thermodynamic driver for them to dissolve, meaning that the composition of the particular body fluid is thermodynamically close to saturation or supersaturated with respect to the mineral. Examples of biodurable minerals with which the plasma electrolyte composition is calculated using chemical speciation modeling to be near or above saturation include calcium phosphate (the primary component of bone), calcium carbonate, quartz, and various sodium-rich zeolites such as mordenite (Plumlee and Ziegler 2003, 2006). In contrast, the plasma, interstitial fluid and intracellular fluid compositions are all substantially undersaturated with a wide variety of biodurable aluminosilicate minerals such as chrysotile, various amphiboles, and other minerals, which suggests that these minerals are biodurable because of kinetic controls on dissolution. As proposed by Jurinski and Rimstidt (2001), chrysotile asbestos fibers are thought to dissolve more rapidly *in vivo* than rod-shaped amphibole asbestos fibers because chrysotile's scroll-like crystal structure causes it to unroll as it dissolves, thereby exposing more surface area and allowing greater dissolution rates.

The *bioaccessibility* of a potential toxicant in a substance is the extent to which it can be extracted from the substance by the body (Hamel et al. 1999). Bioaccessibility is related to biosolubility in that toxicants in biosoluble materials are also bioaccessible. However, earth materials can also contain bioaccessible toxicants that are sorbed or otherwise loosely bound to the surfaces of relatively bioinsoluble particles.

Bioaccessibility is commonly confused in the literature with *bioavailability*, which is defined by toxicologists to be the fraction of an administered dose of a toxicant that is absorbed

Table 3. Cations and metals are typically distributed among a number of different forms in the body, listed here in generally decreasing order of exchangeability. The body's major electrolyte cations occur largely as labile complexes with electrolyte anionic species or carboxylic acids, or as the simple ion. The simple ionic forms of most metals typically are present in very low concentrations due to the abundance of complexing ligands available in the body's fluids. Information from Taylor and Williams (1995), modified from Plumlee and Ziegler (2003, 2006).

Form	Description
Labile inorganic complexes or simple ions	Complexes with electrolytes such as chloride, bicarbonate. May also include rarer metal-hydroxy complexes.
Labile amino acids, amino acid chains (peptides), other organic acids	Complexes with: amino acids (i.e., cysteine, histidine); peptides (chain of two or more amino acids, such as the tri-peptide glutathione); other organic acids (i.e., citrate, lactate, ascorbate). Also mixed ligand complexes involving more than one amino acid and/or other organic acid (i.e., cysteinate-histidinate)
Labile proteins	Metals complexed with lower molecular-weight proteins are typically viewed as relatively exchangeable. Examples include: transferrin (iron); albumin (copper, zinc, other metals).
High molecular weight proteins	Metal complexes with HMW proteins are typically are viewed as relatively inert and non-exchangeable in the body once formed. However, their contained metals can be released if the proteins are degraded through disease or the body's normal protein breakdown and recycling mechanisms. Include, for example: metallothioneins such as ceruloplasmin, a copper protein, and α_2 macroglobulin, a zinc protein; ferritin (an Fe[III]-storage protein); hemoglobin (Fe[II]).
Solids	Cations or metals incorporated in to teeth, bones, cuticles, etc.

via an exposure route, reaches the bloodstream, and is transported in the body to a site of toxicological action. For most potential toxicants contained in earth materials, the following relationship is generally correct:

$$bioavailability < bioaccessibility < total\ concentration$$

This relationship shows that equating the total concentration of a toxicant in a substance to its bioavailability is the worst-case scenario, and is scientifically valid only in the rare circumstances when the toxicant is completely released from the substance and is in the appropriate chemical form to permit complete absorption by the body (Reeder et al. 2006, this volume; Ruby et al. 1999; Hamel et al. 1999). However, site-specific risk assessments for metals such as lead commonly assume that the metal is 100% bioavailable to allow for a substantial margin of safety (S. Rodenbeck, oral comm. 2006).

A given earth material will likely differ in its biosolubility or toxicant bioaccessibility depending on the exposure route and the particular body fluids with which it is in contact. For example, a broader variety of minerals are biosoluble in the gastrointestinal system than in the respiratory system because the acid conditions of the stomach promote mineral dissolution. Dissolution rates of phagocytized particles in the relatively acidic (pH 4.5) macrophage phagolysosomal fluids are greater than in the near-neutral lung and interstitial fluids (Kreyling and Scheuch 2000).

Casteel et al. (2006), Ruby et al. (1999) and Plumlee and Ziegler (2003, 2006) present a number of examples of how particle mineralogy and morphology influence the biosolubility of earth materials and the bioaccessibility of their potential toxicants. For example, lead tied up in cerrussite (a very acid-soluble lead carbonate), lead oxides, and lead sorbed onto atmospheric aerosols generated by lead-zinc smelting is substantially more bioaccessible than lead tied up in galena and various lead-phosphate minerals. Very small particles and botryoidal or acicular crystals may be more readily solubilized than coarser, more equant particles of the same mineral due to the greater surface area per unit mass available for reaction with the body's fluids. Particle surface chemistry and other surface phenomena may be quite important in influencing dissolution rates of biodurable minerals (Guthrie 1997).

Bioreactivity is the extent to which an earth material can react with body fluids or tissues to trigger tissue damage or to alter key fluid parameters such as pH, Eh, and/or concentrations of major electrolytes, trace metals, organic species, and redox-active species. Although not specified as such, the toxicological literature seems to describe a continuum between what can be described as *acute bioreactivity* and *chronic bioreactivity* (Plumlee and Ziegler 2006). Acutely bioreactive materials are those that can trigger rapid, substantial changes in fluid chemistry, and possibly cause damage to tissues along the exposure route. Examples include earth materials that produce acids (such as acidic volcanic gases, or acid-generating soluble salts in mine wastes) or those that produce caustic alkalinity (such as cement dust or concrete). In contrast, chronically bioreactive materials are biodurable, and so react primarily at their surfaces with, and slowly release chemicals into, the surrounding fluids and tissues. Examples include crystalline silica or asbestos, which are thought to cause toxicity by generating reactive oxygen species.

Finally, it is important to note that a wide variety of solid earth materials are typically complex mixtures of many different mineral phases, each of which has its own particular biosolubility and bioreactivity characteristics (Fig. 8). For example, the dusts generated by the collapse of the World Trade Center (WTC) in September 2001 were a very complex mixture containing inhalable to respirable particles of acutely bioreactive concrete, various metals or metal alloys (steel, zinc, lead, bismuth), biosoluble gypsum from wallboard, biodurable chrysotile asbestos fibers and crystalline silica, moderately biodurable glass fibers, window glass, and many others (Meeker et al. 2005; Plumlee et al. 2005). As noted by Plumlee and Ziegler (2003, 2006) and Plumlee et al. (2005), the reaction of readily solubilized and reactive WTC components such as concrete particles with lung fluids could theoretically enhance the chemical stability of less readily solubilized components (such as glass fibers and asbestos fibers) from the dusts. The crystalline silica literature notes that the toxicity of crystalline silica dusts is diminished if the dusts contain other aluminosilicate minerals (SSDC 1988). However, we have not encountered in the literature any other appreciable discussions of whether the chemical behavior *in vivo* and resulting toxicity effects of a substance can be modified when that substance is part of a complex multi-substance mixture. Pathogens may also be a part of the complex earth material mixture. Plumlee and Ziegler (2003, 2006) speculated whether borate salts inhaled with *C. Immitis* spores might enhance their viability in the lungs by suppressing macrophage activity. Aside from this speculation, we similarly have not encountered to date in the toxicological or immunological literature any discussion of how chemical reactions of the earth materials hosting the pathogens with the body's fluids might influence the viability of the pathogens *in vivo*.

Immune system mechanisms

The particle clearance mechanisms discussed in an earlier section of this chapter are one small part of the body's complex defense system that helps protect against foreign substances, pathogens, parasites, and mutated cells. A detailed discussion of the immune system is well beyond the expertise of the authors, but can be found in many excellent textbooks on the topic (i.e., Coico et al. 2003). Mechanisms of the immune system that have the greatest influence on earth materials and their contained toxicants include:

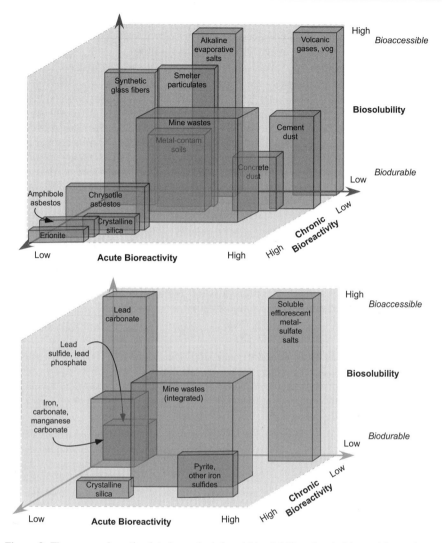

Figure 8. The upper schematic plot shows the inferred biosolubility, chronic bioreactivity, and acute bioreactivity of general classes of earth materials. Many types of earth materials (i.e., mine wastes, volcanic ash, soils) can contain a complex variety of minerals having quite different biosolubilities and bioreactivities, so the particular location of a given earth material on the plot should be considered as an averaged approximation. As one example, the lower plot shows the relative biosolubility and chronic and acute bioreactivity of the various mineral components of mine wastes.

- *Protective barriers.* The skin is one example, but the mucous membranes can also be considered as a physical barrier that helps prevent access of foreign substances and pathogens into the body.

- *Chemical barriers.* These include, for example, the acidic gastric fluids and the digestive enzymes contained in saliva, tears, and gastric fluids. The body's increased production of fluids in the nasal passages and upper respiratory tract in response to the influx of foreign substances is also a chemical defense mechanism that helps dilute chemical impacts of the substance.

- *Microbiological barriers.* The intestines contain a number of different bacteria that compete with pathogens for sustenance, thereby helping to diminish multiplication of pathogens into sufficiently large numbers to trigger illness.

- *Phagocytic activity.* The respiratory macrophages discussed previously one of several types of autonomous phagocytic cells that patrol the body to identify and engulf foreign particles or microbes. The macrophages are in part physical barriers, as they can help isolate foreign substances or pathogens and transport them away from physiologically sensitive areas. They also are chemical barriers in that they can help digest substances with their acidic lysosomal fluids rich in cytotoxic acid proteases and reactive oxygen species. Cytokines (chemical messengers) released from the macrophages can also recruit additional macrophages to the site of particle influx. High burdens of particles and recruited macrophages can lead to increased inflammation of the aveoli, decreased particle clearance, and decreased lung function. Macrophages that are penetrated by sharp particles (Churg 2000; Fattman et al. 2004) or that undergo necrosis (which occurs if they are unsuccessful in engulfing particles such as fibers, or if the engulfed particles are toxic) can release their cytotoxic chemicals into the surrounding environment, triggering irritation of or toxicity to surrounding tissues that can in turn lead to diseases such pulmonary fibroses.

Physiological processes by which potential toxicants from earth materials are absorbed, distributed, metabolized, and/or eliminated by the body (ADME)

There are many processes by which toxicants are absorbed, modified, stored, and eliminated; for detailed descriptions, interested readers are referred to toxicology overview volumes and texts such as Sullivan and Krieger (2001), and Klassen (2001). The brief and greatly simplified discussion here is intended to illustrate how the chemical form of a potential toxicant as released from an earth material plays a role in its ADME. Examples for specific metals and other toxicants contained in earth materials are given in Reeder et al. (2006, this volume) and Plumlee and Ziegler (2003).

Absorption refers to the passage of a potential toxicant through one of the body's various membranes into the circulatory system. It is generally considered to be the chemical passage of a dissolved substance, but as shown previously, solid particles can also penetrate the linings of the respiratory tract and be taken into the circulatory system. The chemical speciation (Reeder et al. 2006, this volume) of a potentially toxic substance as it is derived from an earth material, as well as any chemical transformations that the substance undergoes through reactions with the body's fluids (including dissolution, chelation, precipitation, changes in redox state), strongly influence its absorption.

There are a wide variety of organic ligands in the body that are available to chelate (complex) metals and cations released from earth materials, including (listed in decreasing order of lability; Table 3): simple ions or labile inorganic complexes; labile complexes with amino acids, amino acid chains (peptides), other organic acids (i.e., carboxylic acids such as ascorbate, citrate, lactate), or mixed ligands; labile low-molecular-weight proteins (such as transferrin); generally non-labile, high-molecular-weight proteins (such as ferritin, hemoglobin, metallothioneins); and solids (bones, teeth). Complexation can diminish a metal's tendency to sorb onto solids or precipitate as a solid, both of which can inhibit absorption. Most literature sources suggest that the chemical species that are most readily absorbed through the body's membranes into the circulatory system and that pass most readily pass through the cell membranes, are those with net-neutral charge (Taylor and Williams 1998) or those that mimic the chemical behavior of other physiologically important species. An example of the latter is hexavalent chromium ($Cr[VI]$), which, similar to sulfate and phosphate species, can be transported via facilitated diffusion through nonspecific anion channels.

Changes in redox state as a metal is delivered from the earth material along the exposure route can also influence its dissolution from the earth material and its absorption by the body. Redox processes can enhance the solubility of solids, such as the enhanced dissolution of metal oxide particles by reduction of their contained metals at the particle surface via ascorbate or other organic reductants. In the case of absorption, changes in redox state are particularly important if the resulting redox form is more or less readily complexed, is present in a more absorbable form, or has diminished solubility. The relative $Eh^{\circ\prime}$ (at or near the appropriate ambient pH) of various redox couples, such as shown in Figure 7, may provide some idea of the relative redox stability of a particular metal or metal particle relative to the body's various redox systems. However, the extent and rates of particular of redox reactions may depend on a variety of factors such as the ambient chemical environment, the availability of the species constituting the dominant redox system(s) for the metal in the particular body fluid, and kinetics of redox reactions. For example, Cr(VI) is more readily reduced under the acid conditions of the gastric fluids than in the near-neutral pH lung fluids, interstitial fluids, and plasma (ATSDR 2000). Cr(III) produced in the stomach by Cr(VI) reduction then most likely precipitates out in the intestines and is not absorbed into the bloodstream. In the lungs, Cr(VI) is relatively soluble in the lung fluids, and is readily absorbed into the bloodstream. It can also persist in the plasma in spite of the many abundant organic reducing agents, and can then cross cell barriers, where it is more readily reduced in the more reducing, glutathione-rich intracellular environment.

Distribution refers to the transport of the potentially toxic substance by the circulatory system to various organs or tissues. As with absorption, transport within the body to specific target organs or tissues may be restricted by various membrane barriers such as the blood-brain barrier. Toxicant forms that are most readily exchangeable (Table 3) are also most readily available for distribution to the tissues and organs (Sipes and Badger 2001).

Metabolism refers to the biochemical transformations that potential toxicants undergo within the tissues or organs, which may include binding, storage, generation of toxic metabolites (also called toxic bioactivation), and detoxification of the toxicant and/or its metabolites (Sipes and Badger 2001). Important storage areas include adipose tissue for lipophilic substances, and the bones for fluoride, lead, and strontium. The liver is one of the primary organs in which generation of toxic metabolites and detoxification occurs, primarily due to its high content of enzymes used to catalyze biotransformation. However, the kidneys, lungs, stomach, intestine, gonads, and skin can also host a variety of metabolic transformations.

Elimination is the mechanism by which the body rids itself of toxicants, toxic metabolites, and metabolic wastes. Although the primary excretion pathway is renal, some materials may be excreted in the bile, feces, milk of lactating women, respired air, sweat, hair, nails, and saliva.

Extent to which earth materials or their contained toxicants can generate reactive oxygen species

Reactive oxygen species (ROS) generated by earth materials or by a variety metals released from earth materials *in vivo* can be an important source of toxicity. Discussed in detail by Schoonen et al. (2006, this volume), ROS are intermediate oxidation state species that are formed by the incomplete oxidation of molecular oxygen and a variety of other redox-sensitive organic species (such as ascorbate) present in the body. Examples of ROS include superoxide ($O_2^{\bullet-}{}_{aq}$), hydrogen peroxide (H_2O_2), and hydroxyl radical ($^\bullet OH_{aq}$). Intermediate oxidation state forms of organic species such as ascorbyl radical can also form. The body exploits some ROS to its benefit; for example, the macrophages can generate ROS to help attack engulfed particles and microbes. However, ROS are implicated in a broad variety of toxicity effects such as oxidative damage of lipids, proteins, and DNA, which can in turn lead to a wide variety of diseases such as cancer (Kawanishi 1995).

Earth materials may generate ROS by a variety of mechanisms (Fubini and Areán 1999; Aust et al. 2002; Schoonen et al. 2006, this volume). A variety of metals are noted in the literature to participate in redox reactions that produce ROS, including iron, manganese, copper, chromium, nickel, titanium, and others. The reactions through which metals can generate ROS are noted to occur either in solution (i.e., if the metals are dissolved from earth materials) or if the metals are structurally bound to particle surfaces. In fact, reaction rates involving structurally bound metals can be considerably faster than those involving dissolved metals, because coordination of the metals with anionic species on the particle surface shift the metal redox couples to effectively lower Eh values, making them more effective electron donors (Schoonen et al. 2006, this volume).

A variety of studies have demonstrated that pyrite (iron disulfide), nickel sulfide, and other metal sulfides can be highly effective at production of hydrogen peroxide and ROS, and breakdown of RNA (Borda et al. 2004; Cohn et al. 2006; Schoonen et al. 2006, this volume). Iron sulfides in coal are increasingly suspect as a contributor to coal-workers' pneumoconiosis, due to the generation of ROS by sulfide particles or by iron released from the sulfide particles or sulfide oxidation products in the coal (Huang et al. 1998 2005). Oxidation of iron sulfides *in vivo* may also generate acid that could be a source of irritation (Plumlee and Ziegler 2006).

Grinding-induced surface structural defects, particularly on freshly ground mineral particles, can generate a variety of ROS (Fubini and Areán 1999). Although crystalline silica is best known for this effect, a variety of other minerals have also been investigated, such as metal oxides, sulfides, asbestos, and zeolites (see references in Schoonen et al. 2006, this volume).

ROS can also be generated when the body's clearance mechanisms fail to clear inert particles from the lungs. Alveolar macrophages activated by foreign particles produce and release into the surrounding alveolar environment a variety of ROS and chemicals that recruit additional macrophages to the site. Macrophages that fail to clear particles also release a variety of cytotoxic chemicals into their surrounding environment. All of these activities contribute to inflammation and can, in the case of biodurable or biopersistent particles, lead to long-term opportunities for DNA damage and resulting toxicity

Susceptibility of the exposed individual

Many confounding factors influence an individual's susceptibility to toxicants contained in or released from earth materials. These can include genetics, age, gender, socioeconomic status, nutritional status, general health status, and personal habits such as smoking.

Genetics, age, and gender. It is increasingly recognized, but in some specific cases debated, that genetics and age can play an important role in an individual's response to environmental and workplace toxicants. For example, genetic factors may influence the susceptibility of individuals to heavy metals by affecting production of enzymes involved in the synthesis and cycling of glutathione, metallothioneins, and other sulfhydril compounds used by the body to help detoxify heavy metals (Gochfeld 1997). Immature and elderly individuals generally appear to be more susceptible to metal toxicity than mature adults. For example, absorption of lead is substantially greater in the intestines of immature organisms than in adult organisms, and the vulnerability of their immature organs to metal toxicity may be substantially greater as well, particularly with respect to neurotoxins that can affect the developing neurosystem (Gochfeld 1997). Production of metallothioneins and other metal detoxifiers may also diminish with age. The role of gender has been shown to be a potential link to disease, particularly as it relates to differences in practices between males and females of a population (that would presumably lead to differences in exposure to a particular workplace or environmental toxicant); possible physiological differences in metal metabolism between males and females are discussed briefly by Gochfeld (1997).

In a study of a population exposed to elevated arsenic in drinking water, Meza et al. (2006)

found that small differences in a particular gene could be correlated with differences in the arsenic metabolites found in the urine of children, but not adults. This indicates that different individuals might metabolize and detoxify arsenic in different manners, and further suggests that the differences in arsenic metabolism may also be influenced by age. .

A scan of the medical literature and internet discussion boards quickly encounters lengthy discussions of the possible links between mercury exposure and genetic susceptibility to autism. One published hypothesis is that children with autism may have a genetically based, diminished ability to eliminate mercury from their bodies that results from exposure to mercury *in utero* or as infants. However, this is the focus of continuing debate (Stehr-Green et al. 2003; Mutter et al. 2005). As these discussions are focused on the exposures to mercury contained in a particular vaccine, their implications for potential exposures to mercury contained in earth materials remain to be determined.

Dogan et al. (2006) studied exposures to erionite (a fibrous zeolite associated with asbestos-related disease) in several Turkish villages with extremely high rates of mesothelioma cancer. Because the mesothelioma occurrences are noted to occur only in specific houses, the original hypothesis was that individuals in these households were exposed to a more carcinogenic variety of the erionite. However, mineralogical studies indicated no significant differences in the erionite between affected and unaffected households, and pedigree studies indicated that malignant mesothelioma was prevalent in some families and not others. Dogan et al. (2006) interpreted these results to indicate that mesothelioma development occurs in genetically predisposed individuals, and that genetics therefore plays a direct role in mineral fiber carcinogenesis.

Socioeconomic, health, and nutritional status. The overall health and nutritional status of an individual is thought to influence susceptibility to toxicants (WHO 1996; Gochfeld 1997). For example, nutritional deficiencies of calcium, iron, proteins, and phosphate can enhance absorption of cadmium and lead from the intestine. Some diseases of organs such as the kidneys and liver that are involved with metal metabolism can enhance the body's susceptibility to metal toxicity. Conditions such as chronic asthma or others that diminish airflow to the alveoli may actually diminish particle deposition and resulting toxicity in the lungs. Socioeconomic status can influence overall health and nutritional status, and may influence the extent of environmental exposures to toxicants.

Smoking. Smoking can have several important influences on an individual's susceptibility to toxicants. Cigarette smoke contains a variety of potentially toxic metals (such as chromium), organics, and particulates, and is therefore a well-known trigger of disease in its own right. Nicotine can suppress the concentration of reduced glutathione in the liver, leading to diminished metal detoxification. Smokers exposed to excessive levels of particulates such as silica or asbestos have a combined lung cancer risk that is greater than for smoking or particle exposure alone (Holland and Smith 2001). Inhaled mineral particles such asbestos may also interact with the chemicals in cigarette smoke to enhance toxicity (Fattman et al. 2004); for example, asbestos fibers that have sorbed toxic organic chemicals from cigarette smoke may carry these toxicants to the tissues or lymph system, thereby allowing prolonged contact of the absorbed toxicants with the surrounding cells.

Summary—an overview of how earth materials can cause toxicity

Some earth materials (for example, asbestos, crystalline silica, and oxides of some metals such as chrome) are toxic because they are not readily cleared by the body, and thereby can persist and/or accumulate at the exposure site or elsewhere in the body. Toxic effects for these types of substance in part result from the body's failed attempts to detoxify and/or excrete them. Macrophages that unsuccessfully engulf insoluble particles can die, releasing their contained cellular toxins to the surrounding cells, leading to tissue death and scar tissue buildup (fibrosis). Inflammation and other immune responses to solid earth materials may also play a role in disease; for example, chronic beryllium disease is a cell-mediated immune response that triggers lung disease (Goyer and Clarkson 2001). In addition, biodurable substances may exhibit chronic bioreactivity that leads to adverse effects. For example, reactions of chemicals (such as iron) contained in biodurable substances may produce free radicals and other ROS, which can in turn compromise cell integrity or lead to long-term accumulative damage to cellular DNA, and potential diseases such as cancer.

The toxicities of substances that encounter the body in bioaccessible form (those that are readily released from the earth materials into the body fluids) depend upon the exposure route, the dose, the chemical form of the substance at exposure, and the processes that chemically transform the substance during absorption, transport, and metabolism. There are myriad ways in which toxicants released from earth materials can cause chronic and/or acute toxicity. These can include, for example: generation of reactive oxygen species; interference with the body's antioxidant processes and mechanisms; impairment of cellular function; replacing essential elements in key physiological chemicals (such as cadmium replacing calcium in bone); and many others.

Earth materials can also generate toxicity if they are acutely bioreactive, whereby they react chemically with the body's fluids and tissues to the extent that they trigger extensive changes in body fluid composition and tissue damage. The tissue damage may of itself be problematic if it is sufficiently extensive to interfere with physiological function. Damage to protective tissues such as the skin may also provide an entrance point for pathogens. Examples of acutely bioreactive earth materials include: liquids such as acids (i.e., acid gas condensates from volcanoes or automobile exhaust) or alkalis (i.e., liquid sodium hydroxide drain cleaner); gases such as acid gases produced by volcanoes and fossil fuel combustion; alkaline solids such as cement or concrete dust; and acid-generating solids such as soluble metal-acid sulfate salts formed by the evaporation of acid-mine drainage.

Soils, dusts, and other earth materials can host pathogens, and can serve as carriers for the pathogens into the body (e.g., Bultman et al. 2005). Relatively little work appears to have done that investigates whether the form of the earth material with which a particular pathogen is associated may influence the pathogen's ability to persist and thrive *in vivo*.

INTEGRATING EARTH AND HEALTH SCIENCE METHODS TO ASSESS THE HEALTH EFFECTS OF EARTH MATERIALS

A spectrum of earth and health science analytical methods (Table 4) have been or can be applied collaboratively to help understand the toxicological geochemistry and health effects of earth materials. The earth science methods include a wide range of laboratory, computer, and field techniques commonly used to:

- Characterize the physical and chemical characteristics, materials makeup, geochemical speciation, and geochemical reactivity of earth materials;

- Map spatial variations in the occurrence, chemical composition, and other characteristics of earth materials;

- Model the chemical speciation and chemical evolution of water-based fluids in response to mineral-fluid interactions or other chemical drivers;

- Fingerprint sources of potential toxicants or other materials that are derived from earth materials and found in the environment or human body.

The health science methods range from the global, national, or regional scale examined with epidemiological databases and biomonitoring studies, down to the microscopic scale

Table 4. Earth and health science methods that have been or may be used to help understand the health effects of a variety of earth materials. In general, we have tried to provide references for each that illustrate use of the method in a health-related case study. General references are cited in the case of methods for which we are unaware of health-related case studies. "Source" samples are those of soils, dusts, or other earth materials that are potential sources of exposure and toxicity to humans or other organisms.

Method	Use(s) and notes	Studies
Geologic maps and mineral occurrence databases	Show the occurrences (or distribution of rock types that may have potential to host occurrences) of asbestos, fibrous zeolites, or other potentially toxic earth materials. Can be linked to epidemiological studies of disease occurrence tied to geology-sourced toxicants such as asbestos.	**Naturally occurring asbestos:** Churchill and Hill (2000); Van Gosen (2005, 2006a,b); Pan et al. (2005); Schenker et al. (2006); Brodkin et al. (2006).
Geochemistry surveys and geospatial databases for soil, stream sediment, water quality, plants, etc.	Map the distribution of chemical elements in soils stream sediments, waters, plants, and other media. Useful for mapping regional variations and local anomalies away from baseline trends.	**Natural Soils:** Garrett (2005) **Urban soils:** Mielke (1999), Mielke et al. (2004, 2006) **Arsenic in ground water:** Ryker (2001)
Remote sensing techniques	Use plane- or satellite-based sensors to remotely map the distribution and characterize the makeup of earth material sources in the environment.	**Naturally occurring asbestos:** Swayze et al. (2004) **World Trade Center dusts:** Clark et al. (2005)
Phase contrast microscopy (PCM), polarized light microscopy (PLM)	Use transmitted light microscope to identify phases present (based on optical properties), and particle morphology. Sample types include bulk source samples, samples taken from air filters, tissue samples, others	**Asbestos:** Millette (2006); Roggli and Coin (2004); Dodson (2006).
X-Ray Diffraction (XRD)	Identifies crystalline phases present in bulk source samples, using qualitative to semi-quantitative data reduction routines. Can readily identify nearly all major mineral components of multi-phase mixtures, but some minerals may have interferences from other minerals that preclude their certain identification. Can detect minerals present in mixtures in amounts generally greater than 1-2 weight %.	**Asbestos:** Meeker et al. (2003) **World Trade Center dusts:** Meeker et al. (2005)
Scanning electron microscopy (SEM)	Determines identity, morphology, and chemical composition of solid phases down to less than 1 micron in size based on semi-quantitative energy-dispersive x-ray analysis. Also used to map microscopic distribution of chemical elements in a sample. Sample types include bulk source samples, dust samples from air filters, tissue samples, plants, others.	**Asbestos:** Meeker et al. (2003); Roggli and Coin (2004); Millette (2006); Dodson (2006) **World Trade Center dusts:** Meeker et al. (2005) **Depleted uranium fragments in tissue samples:** Todorov et al. (in press)
Electron probe microanalysis (EPMA)	Provides quantitative and qualitative chemical composition data on spots within mineral grains or other solid phases. Requires polished section or grain mount for optimum results. Mostly for source samples.	**World Trade Center dusts:** Meeker et al. (2005) **Libby asbestos:** Meeker et al. (2003)

(table continued on next page)

Table 4. *(continued)*

Method	Use(s) and notes	Studies
Transmission electron microscopy (TEM)	Integrates electron diffraction and SEM capabilities at a microscopic, single-crystal scale. Sample types include small amounts of bulk source samples, dust samples from air filters.	**Asbestos:** Millette (2006); Dodson (2006); Roggli and Coin (2004)
Atomic force microscopy (AFM) and Vertical Scanning Interferometry (VSI)	Mapping of and measurement of changes in microscopic surface topography of solid samples. Can be used in mineral dissolution studies (such as physiologically-based extraction tests) to measure changes in surface topography that occur in response to mineral-fluid interactions.	**General:** Cama et al. (2005); Arvidson et al. (2003)
X-ray absorption spectroscopy (XAS) and Extended X-ray absorption fine structure (EXAFS) spectroscopy	Measurement of microscopic variations in the chemical form (speciation) and distribution of potentially toxic elements in solid samples. Sample types include bulk source materials, dusts, soils, and others. We are not aware of this type of analysis being used yet for foreign earth materials in tissue samples.	**Mine wastes and tailings:** Foster et al. (1998); Kim et al. (2004). **General:** Reeder et al. (this volume)
ICP-MS, ICP-AES, and other analytical chemistry methods	Used to measure the bulk chemical composition of waters, solids, plants, or other earth materials. Solids require acid digestion prior to analysis.	**World Trade Center dusts:** Plumlee et al. (2005) **General:** Taggart (2002); Vutchkov et al. (2005)
Particle sampling methods	Used to obtain appropriately sized (from a physiological perspective) source materials, dust samples, soil samples, etc., for regulatory particle counting, toxicity tests, physiologically-based extraction tests, and other types of health tests.	**General:** Hinds (1999); Baron and Willeki (2001)
In vivo toxicity tests	Measure toxicity effects or exposure indicators as a result of the direct exposure (via appropriate exposure routes) of living animals under controlled laboratory conditions to variable doses of toxicants over time. Uncertainties in how well tests on other species of animals reproduce actual physiological conditions and processes in the human body. Respiratory tests can include instillation (direct placement of materials in trachea of subject animals) or inhalation (subject animals are exposed to an air stream with specific material doses).	**Asbestos, respiratory:** Fattman et al. (2004); Johnson and Mossoman (2001).
In vivo bioaccessibility assessments (uptake monitoring)	Conducted on individuals exposed to potential toxicants to assess the extent to which the toxicants have been absorbed, transported, and metabolized by the body. Can be used to assess whole-body bioaccumulation of the material of concern as well as the particular tissues in which material accumulation takes place	**General:** Plumlee and Ziegler (2006, and references therein) **Lead in soils:** Casteel et al. (2006) **Arsenic in various earth materials:** Buchet et al. (1995)
Pathology analysis of tissues and contained earth materials	Provides a way of directly measuring or assessing toxicant uptake by, interactions with, and toxic effects upon the body, through study of tissue and other samples collected as part of autopsy, biopsy, or other methods such as bronchiolar lavage.	**General:** Centeno et al. (2005a,b) **Asbestos:** See chapters in Roggli et al. (2004); Dodson and Hammar (2006). **Depleted uranium:** Todorov et al. (in press)

(table continued on next page)

Table 4. (*continued*)

Method	Use(s) and notes	Studies
In vitro physiologically-based bioaccessibility or biodurability tests	Used to analyze and measure rates of chemical reactions between earth materials and various simulated body fluids, including simulated saliva, gastric, intestinal, lung, and alveolar macrophage fluids. The goal is to understand the types and rates of chemical dissolution or alteration reactions that earth materials and other substances might undergo in the body via various exposure routes. The tests can vary substantially in: physical design; particle size, shape, and other characteristics of materials tested; fluid compositions; fluid-solid proportions; test duration; and other parameters. Uncertainties include (1) how well the tests reproduce actual (complex and dynamic) conditions in the body; and (2) how well the predicted results (such as particle dissolution rates, or types and relative abundances of trace elements solubilized from the particles) can be readily extrapolated to infer toxicity responses *in vivo*.	**General:** Plumlee and Ziegler (2003, 2006) **Lead in soils and mine wastes:** Drexler amd Brattin (2006); Ruby et al. (1999); Hamel et al. (1999); Oomen et al. (2002, 2006). **Asbestos and talc:** Johnson and Mossman (2001); Jurinski and Rimstidt (2001); Werner et al. (1995); Mattson (1994a,b); Eastes et al. (1996, 2000a,b). **World Trade Center dusts:** Plumlee et al. (2005)
In vitro toxicity, uptake, or chemical assay tests	Used to model the effects of toxicants on cultures of living cells or tissues. A carrier medium (such as fetal bovine serum) containing given concentrations, or doses, of a particular toxicant are added to cell cultures (cell lines) approximating the types of cells affected during actual exposure (i.e., lung epithelial cells, alveolar macrophages, etc.) Indicators of toxicity, cellular uptake, or other chemical indicators (such as iron release, cytokine generation, etc.) are then measured after specified periods of time. Uncertainties include how comparable test results are to those of *in vivo* toxicity tests, and how well *in vitro* tests reproduce actual physiological conditions and processes in the human body.	**General:** Plumlee and Zeigler (2006) **Asbestos, respiratory:** Fattman et al. (2004); Johnson and Mossman (2001); Yang et al. (2006). **Lead in soils, ingestion:** Oomen et al. (2003) **Coal fly ash and other materials, respiratory:** Aust et al. (2002)
Biomonitoring studies	Measure potential toxicants and/or their metabolites in the blood, hair, urine, saliva, tissues and/or other body components of a target population. One goal of such biomonitoring is to assess the proportion of a target population that has been exposed to particular contaminants. Another goal is to establish baseline levels of exposures in a target population.	**General:** NHANES (2006); CDC (2006) **Wildfire:** Wolfe et al. (2004)
Toxicant uptake modeling	EPA's IEUBK (Integrated Exposure Uptake Biokinetic) and similar lead uptake models estimate blood-lead concentrations in children that might result from their exposure to lead from various sources in the environment.	**Lead in soils:** EPA TRW (1999) **Lead in edible soils:** Abrahams et al. (2005)
Epidemiology studies	Correlate disease occurrences and patterns in populations with geospatial data and external parameters such as possible environmental exposures	**Asbestos:** Sporn and Roggli (2004); Lemen (2006); NIOSH WoRLD (2006); Pang et al. (2005); Brodkin et al. (2006); Schenker et al. (2006)

examined by pathology studies of tissue samples. They also include a wide variety of *in vitro* and *in vivo* methods to assess toxicity and other measures of physiological processes.

Rather than discuss each one of these methods in detail, we will instead provide examples of past or future opportunities for earth-health science collaboration to help understand potential health affects of earth materials that contain asbestos and heavy metals such as lead. For more detailed discussions of a number of the methods listed in Table 4, the interested reader is referred to Plumlee and Ziegler (2003, 2006).

ASBESTOS

Asbestos is likely the most recognized and best studied of the earth materials known to cause adverse health effects. Asbestos exposure has been clearly linked to asbestosis, lung cancer, pleural effusions, pleural thickening, pleural plaques, and mesothelioma cancer, and secondary diseases such as cardiovascular problems (Roggli et al. 2004); causal linkages between asbestos exposure and diseases such as gastrointestinal and laryngeal cancers have been proposed in the past but are currently debated (see discussion in Rolston and Oury 2004).

Over the many decades since its health effects were first recognized, asbestos has been the focus of a very large number of studies utilizing many of the earth and health science methodologies listed in Table 4; review papers on the topic, which are themselves too numerous to list here, commonly list many hundreds of references. However, in spite of the decades of study, there is still extensive debate in the mineralogical, toxicological, pathological, and epidemiological literature regarding many aspects of asbestos terminology and toxicity. Two recent textbooks, Roggli et al. (2004) and Dodson and Hammer (2006), provide excellent summaries of the current state of knowledge regarding asbestos and associated health issues. However, these and myriad other literature sources on asbestos also illustrate many areas of continuing debate, remaining data and knowledge gaps, and, as a result, opportunities for future interdisciplinary research.

Lowers and Meeker (2002) presented a revealing summary of the many different ways in which asbestos and related terms have been described and defined in the scientific literature and regulations. In its historical usage, asbestos is a commercially and industrially derived term describing several silicate minerals that form long, very thin mineral fibers, which combine to form fiber bundles commonly showing splayed ends; when crushed, the fiber bundles split into individual fibers that typically show evidence of flexibility (Fig. 9) (Skinner et al. 1988; Van Gosen 2005, 2006a,b). Most current regulatory definitions of asbestos include

---

Figure 9 (*on facing page*). A. Hand sample photograph of chrysotile asbestos, locality unkown, showing classic commercial asbestos characteristics. B. Field sample photograph (courtesy of T. Hoefen, S. Vance) of relatively friable masses of fibrous and asbestiform amphibole from Libby, Montana. Many microscopic fibers were easily abraded from the samples during transport to the lab. C. SEM photomicrographs of standards commonly used in asbestos-related toxicity testing (UICC A asbestos chrysotile, NIEHS/RTI K48 asbestos crocidolite, and UICC asbestos amosite). Note that the UICC amosite asbestos sample shows a wide range in aspect ratio and morphology (asbestiform versus blocky or acicular) (photos by H. Lowers, courtesy of T. Ziegler). White scale bars are all 50 microns long. D. SEM photomicrograph of fibrous erionite with fragments of volcanic tuff. Scale bar is 50 μm long (photo by J. Dyken). E. SEM photomicrograph of amphiboles intergrown with vermiculite from Libby, Montana, displaying a substantial range in morphologies (photo by H. Lowers, from Meeker et al. 2003). F. Transmitted light photomicrograph of ferruginous bodies cored by asbestos, found in lung digestate taken from an individual who perished from asbestosis; the individual is thought to have been exposed to asbestos from Libby, Montana at a vermiculite expansion plant. The bodies are interpreted to form as a direct result of the interactions of alveolar macrophages with fibers. Photo reproduced from a color plate in Wright et al. (2001), with permission the American Journal of Respiratory and Critical Care Medicine, American Thoracic Society.

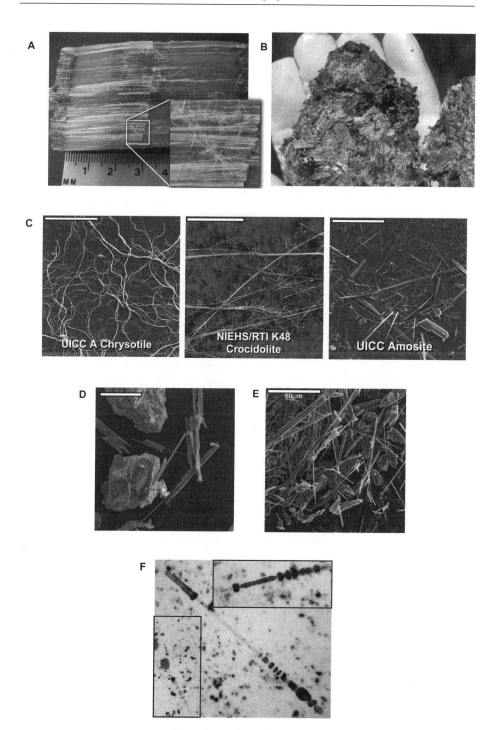

Figure 9. *caption on facing page*

the most commonly used commercial varieties of several different minerals: chrysotile (the asbestiform variety of serpentine), and; the asbestiform varieties of the amphiboles riebeckite (commercially called crocidolite asbestos), cummingtonite-grunerite (commercially called amosite asbestos), anthophyllite (anthophyllite asbestos), actinolite (actinolite asbestos), and tremolite (tremolite asbestos).

Historically, the adverse health impacts of exposure to asbestos-bearing dusts were primarily documented in workers in a variety of trades involved with the mining, processing, or handling of commercial asbestos. Recent attention to high levels of asbestos-related disease via both workplace and environmental exposures to amphibole fibers intergrown with vermiculite mined at Libby, Montana has generated renewed discussions about the specific characteristics of amphibole fibers that influence toxicity (Dearwent et al. 2000; Wylie and Verkouteren 2000; Lybarger et al. 2001; Meeker et al. 2003). It has also led to increased attention to potential health effects resulting from occupational and environmental exposures to dusts generated by mining of mineral deposits that contain asbestos as a natural contaminant (such as Libby), or by natural weathering or human disturbance of certain rock types that contain naturally occurring asbestos (termed NOA) (Renner 2000; Pan et al. 2005; Raloff 2006; Van Gosen 2005, 2006a,b).

Epidemiological studies of worker cohorts exposed to asbestos in the workplace are common in the medical literature, and some epidemiological studies have also been carried out to investigate environmental exposures to commercial asbestos (i.e., see reviews such as Lemen 2006). These epidemiological studies have been used for a variety of applications, such as understanding the prevalence of different types of asbestos disease as a function of exposure and asbestos type. National-scale epidemiological data and derivative maps showing rates of asbestos-related diseases such as asbestosis (Fig. 10) clearly identify areas of elevated disease occurrence in urban areas where workplace exposures were likely greatest, but also indicate other areas of elevated disease incidence where workplace exposures are not as likely.

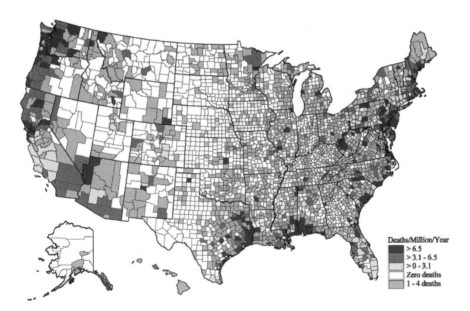

Figure 10. Epidemiological map showing national age-adjusted rates of asbestosis-related mortality by county for U.S. residents age 15 and over, 1970-1999. Reproduced from NIOSH WoRLD (2006).

Pathology studies of tissue samples obtained by autopsy, biopsy, or bronchoalveolar lavage have been used to provide a wide variety of insights into health aspects of asbestos (papers in Roggli et al. 2004; Centeno et al. 2005a,b; Dodson 2006). These include: assessing if an individual has been exposed to asbestos (Fig. 9F), interpreting physiological processes that affect asbestos in the body, diagnosing asbestos-related diseases such as asbestosis or mesothelioma, interpreting if a particular lung cancer case is due to asbestos-related exposure, understanding the type(s) of asbestos to which an individual has been exposed, comparing the asbestos burden in the tissues of exposed and non-exposed populations, and understanding the progression of asbestos-related diseases.

Many different animal-based (*in vivo*) and lab-based (*in vitro*) toxicity tests have contributed substantially to the study and current understanding of asbestos toxicity (see reviews by Johnson and Mossman 2001, and Fattman et al. 2004). *In vivo* inhalation toxicity tests expose subject animals to known concentrations of fibers in an airstream, and *in vivo* implantation toxicity tests involve the direct implantation of fibers in the trachea, pleura, or other tissues of subject animals. The toxicity effects (such as fibrosis, tumor growth, or cell necrosis) after various exposure duration periods are then assessed using pathology studies of the subject animal tissues. In part due to the expense of *in vivo* toxicity testing, a substantial amount of asbestos toxicity testing has shifted in recent years to *in vitro* methods. Cultures of lung epithelial cells, alveolar macrophage cells, or other appropriate target cells are dosed with a suspension of asbestos fibers. Various indications of toxicity are then measured, such as the percent of viable cells remaining at the end of the test (compared to a control line with no added fibers), and the concentrations of various cytokines or other cytoplasmic enzymes produced by the cells. Important uncertainties of these toxicity tests are how well the physiologic processes and endpoints of exposure, dose/response, and toxicity measured in the animal and lab tests can be extrapolated to quantify similar processes and endpoints in humans. One advantage of *in vivo* and *in vitro* toxicity tests is that the test material can be chosen to address specific questions, such as the potential toxicity of short versus long fibers, of amphibole versus chrysotile asbestos, and of asbestos fibers versus cleavage fragments or prismatic or blocky varieties of the same mineral.

In vitro cellular assays also provide useful insights into various physiological processes and cellular function, such as the production of cytokines in response to asbestos exposure, and resulting impacts on surrounding cells *in vitro* and *in vivo* (i.e., Yang et al. 2006).

A variety of *in vitro* acellular tests have been applied to asbestos. A number of studies have examined the generation of reactive oxygen species by iron contained in or released by asbestos (Aust and Lund 1990; Lund and Aust 1992; Fubini and Areán 1999). *In vitro* biodurability tests have been used to examine the chemical reactions of asbestos, cleavage fragments, other mineral particles, and glass fibers with various acids, simulated lung fluids, simulated macrophage lysosomal fluids, and other fluids such as serum-based cell line carrier fluids (Walker 1981; Churg et al. 1989, 1993; Sébastien et al. 1989; Hume and Rimstidt 1992; Werner et al. 1995; van Oss et al. 1999; Jurinski and Rimstidt 2001; Johnson and Mossman 2001; Ziegler et al. 2002). Although such *in vitro* tests cannot reproduce in any detail the complex chemical conditions present in the various regions of the body, they can provide useful comparisons of the relative dissolution rates of different asbestos minerals in proxies for body fluids, and useful insights as to how the fibers might react chemically in the lung or macrophage environment.

Chemical speciation and reaction path modeling (Bethke 1996) has been used to help understand potential solubility constraints on the biodurability of asbestos and other minerals *in vivo*, and to help interpret the results of in vitro biodurability tests (see Wood et al. 2006, and summary and references in Plumlee and Ziegler 2006).

Current views on how asbestos causes toxicity

Although there are still many areas of debate (see next section), studies such as those cited in the previous section have led to a general understanding of how asbestos is toxic. Thin (<0.5 to 1 µm) fibers can flow aerodynamically in to the bronchioles and alveoli, where they lodge by interception. Longer fibers (generally >5-10 µm) cannot be phagocytized by the alveolar macrophages, and so are not as readily cleared as shorter fibers.The biopersistent fibers are not readily cleared by physical breakage or by dissolution in the lung fluids or the more acidic fluids of the macrophage lysosomes, and so can persist *in vivo* for decades. The fibers can also penetrate into the lung tissues and adjacent pleura as a result of the expansion and contraction of the lungs during breathing, and possibly as a result of transportation by the lymph system.

The asbestos fibers are thought to cause toxicity over the long term by a variety of mechanisms. The macrophages release cytokines, which recruit additional macrophages to the site. This influx of macrophages also produces an inflammation response of the local tissues. Macrophages that fail to clear the fibers die and release their cytotoxic chemicals (acid proteases, reactive oxygen species) into the surrounding environment, which can trigger toxicity effects on the surrounding cells. Fibrosis, the buildup of scar tissue, results from the overload of the lung environment by fibers and the lung's failed attempts to clear the fibers.

Several mechanisms are thought to result in carcinogenesis. One of the cytokines released by the alveolar macrophages, TNF-α, is thought to help protect the surrounding cells from short-term toxicity from the fibers, thereby enhancing the fibers' longer-term ability to trigger cellular damage (Yang et al. 2006). The fibers can penetrate the nuclei of the macrophage and lung cells (Churg 2000), where they can physically disrupt the cellular DNA, potentially leading to DNA breaks and mutations. Reactive oxygen species produced by reactions between body fluids and the fibers (and possibly constituents released from the fibers) can also lead to DNA disruption and potential carcinogenesis; iron associated with the asbestos fibers is commonly thought to be an important contributor to the formation of ROS (Aust and Lund 1990; Fubini and Areán 1999).

Aspects of asbestos toxicity that are still under debate

In spite of decades of research into the factors that influence the toxicity of asbestos, there are several areas of continuing debate in the health and mineralogical literature. A few examples (with appropriate references illustrating differences of opinion) are discussed next.

The relative pathogenicity of chrysotile versus amphibole asbestos. A number of epidemiological, toxicological, and pathological studies have generally indicated that amphibole asbestos and erionite are more potent from a carcinogenic standpoint than chrysotile asbestos (i.e., Holland and Smith 2001; Fattman et al. 2004; Sporn and Roggli 2004). Frequently cited pathology studies (e.g., Churg et al. 1984, 1993) found elevated levels of amphibole asbestos fibers (particularly tremolite) in the lung burden of chrysotile miners and workers diagnosed with mesothelioma. These findings were interpreted to indicate that the elevated risk of developing mesothelioma in chrysotile workers was due to natural contamination of the chrysotile ores by tremolite fibers. The amphibole fibers were postulated to have been enriched in the lung burden (and hence associated with increased risk of mesothelioma development) relative to chrysotile fibers as a result of several mechanisms. It is interpreted that amphibole fibers (which tend to be straighter and flow more aerodynamically than the more curly chrysotile fibers) can penetrate deeper into the lungs than the chrysotile. It is also interpreted that amphibole fibers can also penetrate into the surrounding lung tissues and membranes more readily than longer chrysotile fibers, where they can trigger cancers such as mesothelioma.

Results of *in vitro* biodurability tests and other lines of evidence indicate that the amphibole asbestos minerals and erionite are less readily dissolved in lung, interstitial, and phagolysosomal fluids than chrysotile asbestos. This has been interpreted to suggest that amphibole asbestos and erionite fibers therefore can persist for longer periods of time in the lungs and adjacent tissues than chrysotile, thereby imparting a greater potential to trigger fibrosis and cancer (Sébastien et al. 1989; Churg et al. 1989, 1993; Johnson and Mossman 2001).

However, a variety of recent case studies and reviews (i.e., Suzuki and Yuen 2001, 2002; Lemen 2006; Dodson 2006) cite evidence such as the presence of chrysotile asbestos fibers in mesothelioma tumors, and the occurrence of chrysotile asbestos without amphibole asbestos in the lung burden of some individuals with mesothelioma, to indicate that both chrysotile and amphibole asbestos forms are pathogenic. Asbestos researchers who conclude that chrysotile is as pathogenic as amphibole asbestos (i.e., Lemen 2006) interpret that the more rapid leaching of chrysotile relative to amphiboles in the body is an indication of both its greater degree of bioreactivity *in vivo* (and hence its ability to induce cellular damage), and the presumed increased tendency of the partially leached fibers to break into many shorter fibers that would be available to migrate into adjacent tissues or cells, and potentially trigger toxicity.

A middle ground in the discussion may be the view summarized by Sporn and Roggli (2004): "*...it is clear that sufficient exposure to chrysotile may result in the development of mesothelioma, but in contrast to the commercial amphiboles, low level exposures are not likely to increase risk.*"

The potential pathogenicity of short asbestos fibers. A rather traditionally held view in the asbestos toxicological community has been that longer, thinner fibers (>8 μm long, <0.25 μm wide) are more pathogenic than shorter, wider fibers (e.g., Stanton and Wrench 1972; Stanton et al. 1981). The presumption has been that shorter fibers are more rapidly cleared from the lungs than an equivalent number of longer fibers (Dodson 2006).

However, several recent studies of fiber burden in lung and mesothelioma tissues of individuals diagnosed with mesothelioma cancer have found that the majority of the fibers in the tissues were less than 5 microns long, and that chrysotile is a common fiber type (Suzuki and Yuen 2001, 2002; Dodson et al. 2003; Dodson 2006). One interpretation of these results is that shorter chrysotile fibers can also penetrate and migrate to the pleural and peritoneal spaces, and can therefore trigger cancer in these regions.

As noted by Dodson et al. (2003), a workshop on fiber toxicology research needs (summarized by Dement 1990) indicated that much additional information beyond fiber dimension was needed to more fully assess potential fiber toxicity, including surface area, solubility in lung fluid, trace metal and trace organic content, surface charge at physiological pH, and surface reactivity. Dodson et al. (2003) then went on to point out that such information is typically not fully developed in asbestos toxicology studies.

The potential health effects of microscopic, non-commercial amphibole fibers that do not fit all of the compositional or morphological characteristics of regulated asbestos. Several amphibole minerals (as classified mineralogically on the basis of composition) have been noted to occur in a fibrous or asbestiform habit and have been associated with mesothelioma and other asbestos related diseases, but are not specifically listed in asbestos regulations. These include winchite and richterite (e.g., found at Libby, Montana; Meeker et al. 2003) and fluoredenite (e.g., found in Biancavilla, Italy; Giagnfagna et al. 2003). Meeker et al. (2003) also found a range of microscopic morphologies in Libby amphiboles from asbestiform to acicular, prismatic, and blocky. In many Libby samples, acicular prismatic fibers (a number of which are quite straight and show no evidence of flexibility) are commonly observed in SEM images to be parting or cleaving from larger but still microscopic blocky or prismatic amphibole particles. Abundant straight, acicular fibers (including many in Fig. 9E) are also commonly observed by SEM in debris sloughed from friable field samples (Fig. 9B) that had not been subjected to grinding, suggesting that the fibers are single crystals that were somewhat loosely bound within larger masses (G. Meeker, oral comm., 2006). Because the most clearly asbestiform fibers (as defined for commercial purposes) constitute a relatively small proportion of the total amphibole fibers at Libby (Meeker et al. 2003), this merits the question as to whether the truly asbestiform fibers are the only morphology contributing to the toxicity of the Libby amphiboles. Similar microscopic, non-asbestiform but fibrous or acicular morphologies are rather common in amphiboles from other localities, including in a number of standards used in toxicity testing (Ziegler et al. 2002), suggesting that the same question deserves further consideration beyond Libby.

Cleavage fragments constitute a specific type of broken mineral particle most likely produced primarily by grinding, milling or physical impacts of other human or natural activities on larger crystals. Amphibole cleavage fragments have not been regulated by OSHA since 1992 (see regulatory-focused discussions in NIOSH 1990; OSHA 1992). A recent review paper (Ilgren 2004) argues emphatically for the substantially lower toxicity of cleavage fragments compared to asbestos, citing previous studies and many oral communications. Various toxicity studies cited by Ilgren (2004) have concluded that amphibole cleavage fragments do not pose substantial health risks compared to those of their asbestiform counterparts (e.g., Davis et al. 1991). In contrast, other studies suggest that grinding of silicates (e.g., Fubini and Areán 1999) and "acicular" amphiboles (e.g., Palekar et al. 1979) (which should enhance formation of cleavage fragments) increases their *in vitro* bioreactivity, hemolytic activity, and(or) cytotoxicity. When considered individually and in total, details of these toxicology studies and other studies not cited here indicate perhaps greater uncertainties exist than those conveyed by the Ilgren (2004) review, both in terms of study methodologies and some interpretations regarding the toxicity of cleavage fragments. These uncertainties include, for example: the relative reactivities and toxicities of freshly ground versus aged cleavage fragments; the relative toxicities of predominantly acicular versus predominantly blocky cleavage fragment populations of the same amphibole type; whether differences in sample preparation (e.g. grinding in water versus air) influence surface reactivity and toxicity results; relative dissolution rates of cleavage fragments versus fibers as a function of surface area and composition (Walker 1981, and discussion in Plumlee and Ziegler 2006); and, how major- and trace-element compositional variability may complicate interpretations of toxicity differences between cleavage fragments and asbestiform varieties of the same amphibole.

The potential health risks associated with exposures to dusts containing naturally occurring asbestos. As noted earlier, there is currently substantial interest in understanding the health risks associated with environmental or work-related exposures to naturally occurring asbestos dusts that are released by the natural or human disturbance of asbestos-containing rocks. A recent study (Pan et al. 2005) interpreted epidemiological data for mesothelioma occurrences in the context of geologic maps keyed to the occurrence of potential NOA-bearing ultramafic rocks in California (Fig. 11), and concluded that individuals living on or near ultramafic rock outcrops had a higher risk of developing meothelioma. This study has been the focus of subsequent discussion in the journal (i.e., Brodkin et al. 2006 comment; Schenker et al. 2006 reply). Also, ultramafic rocks are not the only rock type in California known to contain asbestos (see discussions in Van Gosen et al. 2004; Van Gosen 2005, 2006b), so restricting the analysis to only ultramafic rocks may not completely estimate the links between mesothelioma and NOA.

Opportunities for further integrated health and earth science research on asbestos

A review of the asbestos literature from an earth science perspective readily leads to the conclusion that earth science methods and techniques can continue to contribute in many ways to health studies that address the many remaining questions regarding potential health effects of asbestos and other fibrous minerals.

An ongoing effort to map asbestos occurrences and their geologic environments of formation across the nation (Van Gosen et al. 2004; Van Gosen 2005, 2006b) will provide geologic

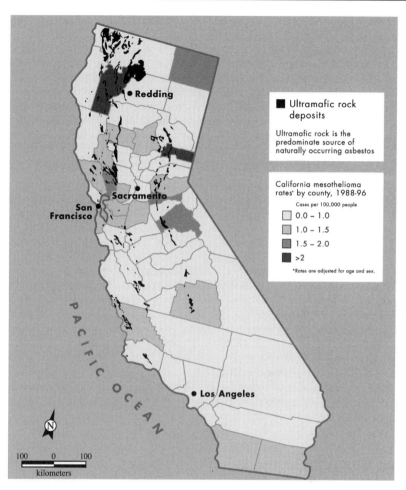

Figure 11. Combined epidemiological and geologic map showing spatial correlations between ultramafic rock and mesothelioma rates by county in California, developed by Pan et al. (2005), based on geologic information compiled by Churchill and Hill (2000). Reproduced from a graphic prepared by UC Davis (2006).

information that can be used to help interpret the potential contributions of NOA to disease using national-scale epidemiological data (Fig. 10), such as the state-scale study by Pan et al. (2005) (Fig. 11). The geologic occurrences of erionite—such as those noted by Sheppard (1996), in sedimentary rocks of the western United States—will also need to be factored in to epidemiological interpretations of asbestos-related disease.

The Dogan et al. (2006) study mentioned earlier provides another excellent example of how mineralogical characterization can facilitate the interpretation of epidemiological data. Similar studies that investigate the mineralogy, morphology, chemical composition, and accessory mineral content of asbestos as a function of the different geologic environments are also needed to help in the interpretation of regional- to national-scale asbestos epidemiological data sets. Epidemiological studies would also benefit greatly from detailed modern mineralogical and geochemical characterization of the materials in air samples collected from the breathing zones of potentially affected individuals.

Remote sensing techniques that enable mineralogical characterization can also provide useful information on the distribution of potential asbestos-forming minerals within regions geologically favorable for NOA. For example, Swayze et al. (2004) used AVIRIS, a hyper-spectral remote sensing platform flown at 12,000 feet altitude, to successfully map rock outcroppings containing serpentinite and tremolite within mapped ultramafic rock units, and to map roads paved with quarried serpentinite well outside the mapped ultramafic rock units. Although this remote sensing technique can provide information of value to more local-scale epidemiological studies, it must be kept in mind that the technique so far cannot discriminate asbestiform versus non-asbestiform varieties of the same mineral, Further, it only maps minerals present at the immediate ground surface and requires extensive ground verification and field checking in collaboration with field geologists and mineralogists (Swayze et al. 2004).

Traditional mineralogical characterization methods have long been a crucial component of asbestos pathology studies, and have primarily been used to determine the mineralogical type and morphology of asbestos fibers found in tissue or BAL samples. There are interesting opportunities for the application of emerging mineralogical characterization technologies to asbestos pathology. For example, field emission scanning electron microscopy (FESEM) should permit higher resolution characterization of the surfaces of fibers obtained from tissue burden, and may provide further indications of the extent of chemical interactions between the fibers and physiological fluids (such as leaching or other chemical breakdown of the fibers). Detailed textural and compositional analysis of the interface between fibers and their coatings in asbestos bodies using FESEM may provide insights into the question whether fibers are active or passive participants in the formation of the asbestos bodies, by looking for evidence of leaching, replacement or other fiber-fluid reactions. High-resolution characterization of fibers in tissue sections using techniques such as X-ray absorption spectroscopy (Table 4) or Raman microspectroscopy (Centeno et al. 2005b) should be extremely helpful in understanding the distribution and (in the case of XAS) oxidation state of potentially ROS-generating species such as iron, chromium, and others in the fibers and adjacent tissues, thereby increasing the understanding of how fibers and their contained metals may contribute to chronic bioreactivity *in vivo*.

Earth science methods and techniques can contribute in several ways to enhanced interpretation of *in vivo* and *in vitro* toxicity tests and assays. Hochella (1993) cites studies such as Nolan et al. (1991), in which the same mineral from different localities shows a variable range of carcinogenicity in laboratory animals. Johnson and Mossman (2001) summarize results of some studies that have found that the fiber length and biological activity can vary substantially between different chrysotile samples collected from different geological localities. Nonetheless, it is rather apparent that many modern toxicological studies still do not recognize the potential for, account for, or rule out possible variability in test results introduced by geologic variability or differences in sample processing methodologies. As noted by Plumlee and Ziegler (2003, 2006) and Ziegler et al. (2002), potential geological variability within and between asbestos test materials includes differences in morphology, trace element content, oxidation state of redox-active species (i.e., Fe, Mn, Cr, As, W, others), accessory minerals, contaminants introduced by processing, and other characteristics. These differences can occur between samples of the same mineral from different samples in the same locality, between samples of the material from different localities of the same geologic environment, and between samples of the same mineral collected from different geologic environments of formation (i.e., as described by Van Gosen 2006b). Hence detailed mineralogical and chemical characterization of the source materials used for asbestos-related toxicity studies is crucial, as is factoring of important geological and mineralogical characteristics into the interpretation of the toxicological results. There is also a role for mineralogical, chemical, and redox characterization of fibers and asbestos bodies (as well as their elemental constituents) in tissue samples and cell cultures at the end of the toxicity testing. When compared to the original mineralogical and chemical characteristics of the source materials used in the testing, coupled with information determined in various physiological or cellular assays, this information should provide valuable insights into fiber-fluid or fiber-tissue reactions and their potential links to toxicity.

UPTAKE OF HEAVY-METAL TOXICANTS FROM EARTH MATERIALS

A considerable amount of attention has been given in the health literature to the toxicity effects of lead, arsenic, mercury, chromium, cadmium, nickel, aluminum, and other metals (such as manganese, vanadium, tungsten, and others) in forms to which workers are commonly exposed in the workplace, or in highly soluble salt forms that are easily absorbed and that therefore can be readily assessed for toxicity effects. However, with the exception of lead, arsenic, and mercury, the uptake of heavy metals from earth materials and their resulting health effects have typically received less attention. We will use lead as an example of the ways that the earth and health science methods summarized in Table 4 have been or can applied to understand the uptake of heavy metals from earth and environmental materials.

Uptake of lead from earth and environmental materials

It has been readily demonstrated by a diverse array of studies over the years that lead, depending upon its form, can be readily taken up, absorbed, and cause toxicity via ingestion of lead-bearing soils, mine wastes, smelting byproducts, leaded gasoline combustion byproducts, and other earth or environmental materials. Likely mechanisms of exposure include involuntary ingestion of particles via hand to mouth contact, involuntary ingestion of deposited particles cleared from the respiratory tract by mucociliary action, ingestion of soil particles on poorly cleaned vegetables, and, more rarely, geophagia.

Biomonitoring. Biomonitoring studies have traditionally focused on measurements of blood lead levels (BLL), such as in individuals who demonstrate symptoms of lead toxicity or who have potentially been exposed to a known source of lead in the environment. If elevated BLL (at present, BLL > 10 µg/dL are considered elevated) are found, then investigations are typically undertaken to determine potential sources for the lead exposure; these source characterization studies routinely have involved measurements of bulk lead concentrations in the material(s) of interest, and have increasingly involved more detailed mineralogical characterization or chemical speciation studies to determine the form(s) of the lead. For example, Mielke (1999) and Mielke et al. (2004, 2006, and references therein) noted that 20-30% of the children living in inner-city New Orleans prior to Hurricane Katrina had elevated blood lead levels and that there was a direct correlation between child blood lead and residence in census tracts with elevated soil lead. They also noted a direct correlation between elevated soil lead and elevated lead in interior house dusts. The source of the lead in the soils and blood is likely lead in paint (dispersed into the soils as a result of paint removal practices from building exteriors), and possibly remnant leaded gasoline combustion products remaining in the soils.

There are a variety of shortcomings of BLL biomonitoring, such as the short half life (21-30 days) of lead in blood that precludes determination of non-recent exposures, and the remobilization of lead from the skeleton, which can obscure blood lead from recent exposures. These shortcomings are leading to the application of analytical methods from the earth sciences to examine records of past lead exposure recorded in the bones or teeth. For example, Arora et al. (2006) found that the spatial distribution of lead in teeth dentine reflected blood lead levels measured at birth and one year of age, using laser-ablation ICP-MS and confocal laser scanning microscopy; they concluded that the spatial distribution of lead in human primary teeth may be useful in obtaining temporal information on possible lead exposure during pre- and neonatal periods.

Tracking sources of blood lead using earth science methods. Radiogenic lead isotope analysis methods were developed originally for earth science applications, but have now become a common tool in the biomonitoring toolkit. They are used quite widely and effectively to infer exposure sources for blood lead (Gulson et al. 1994, 1996a). They also can aid in the understanding of physiological processes by which lead is taken up, stored, and remobilized in the body (i.e., Gulson et al. 1996b).

In vivo lead uptake studies. Whereas biomonitoring can provide information on past or ongoing lead exposures, it is also highly desirable to be able to anticipate potential lead uptake that may result from future or present but unrecognized exposures to lead in a variety of specific source materials. Due to the complexity of chemical conditions in the human body, *in vivo* tests are viewed by toxicologists as perhaps the most useful means to assess and predict potential absorption of lead by children from various source materials. Casteel et al. (2006) provided a very recent summary of *in vivo* lead uptake tests in which juvenile swine were fed various lead-contaminated soils and soluble reference material (lead acetate) twice a day for 15 days. The resulting lead uptake, measured as relative bioavailability (RBA, compared to the amount of lead uptake from soluble lead acetate reference material) was assessed though measurements of blood lead during the study course and analysis of lead in liver, kidney, and bone samples collected at sacrifice following the administration period. The soil materials from various mining- and smelting-related superfund sites were well characterized to understand the mineral phase, particle size, and matrix association of the lead in each of the samples. The study found that the RBA of lead from the soils is a strong function of the lead mineralogy (Fig. 12): lead that is tied up in cerussite (lead carbonate) and with manganese oxides is highly bioaccessible, whereas lead in galena (lead sulfide), anglesite (lead sulfate) and mixed lead-metal oxides and silicates, and iron sulfates, is relatively non-bioaccessible. Casteel et al. (2006) concluded that their data were not sufficiently well defined that RBA could be predicted based solely on the lead mineralogy and speciation. As discussed by Plumlee and Ziegler (2003, 2006), there are other factors in addition to the mineralogic host, particle size, and associated matrix that may influence biosolubility and bioaccessibility of lead and other metals, such as the texture of the mineral host, and the presence of absence of other trace elements in the mineral.

In vitro bioaccessibility tests. Because *in vivo* studies are quite expensive, time-consuming, and logistically complicated, *in vitro* extraction tests using simulated body fluids (sometimes termed physiologically-based extraction tests, or PBET) have been used to model bioaccessibil-

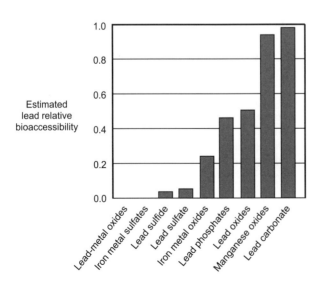

Figure 12. Plot from Casteel et al. (2006) showing the relatively bioavailability of lead from soils (determined by swine uptake studies) as a function of the lead-bearing minerals present in the soil.

ity of lead, other metals, and some organic toxicants in a variety of earth materials. Numerous studies have examined lead bioaccessibility from soils, mine wastes, and smelter byproducts by using a variety of different extraction methods, simulated gastrointestinal leachate fluid recipes, and solid:leachate ratios (i.e., Hamel et al. 1999; Ruby et al. 1999; Oomen et al. 2002, 2003, 2006; Shroder et al. 2004; Drexler and Brattin 2006; references in Plumlee and Ziegler 2003, 2006). Some of these tests model the entire passage through the gastrointestinal tract, by progressively subjecting a sample of the earth material of interest to simulated saliva for several minutes, followed by simulated gastric fluid for 1-2 hours, and then simulated intestinal (duodenal) fluids for 3-4 hours (i.e., Hamel et al 1999; Oomen et al. 2006). These tests can also be modified to simulate lead bioaccessibility and absorption in fasted versus fed conditions (with the addition of infant formula supplemented with sunflower oil), because metal bioaccessibility and absorption is generally greater in fasted than in fed conditions (Oomen et al. 2006).

Shroder et al. (2004) and Drexler and Brattin (2006) summarize a simpler test that was developed in conjunction with, and best models results of, the *in vivo* juvenile swine study discussed in the previous section (Casteel et al. 2006). This test uses a simple simulated gastric leach with hydrochloric acid and glycine; key features are that the pH must be maintained near 1.5, and that the test is most specific for lead.

Oomen et al. (2006) provide an excellent overview of the various tests, including a comparison of the saliva-gastric-intestinal tests versus gastric. For example, tests that only examine metal bioaccessibility in the gastric compartment overestimate total metal bioaccessibility, due to the diminished metal solubility and hence diminished bioaccessibility in the higher pH conditions of the intestinal tract. Also, soils with lead speciation and mineralogy not previously examined, as well as soils containing metals of interest other than lead, must ideally be calibrated against *in vivo* uptake tests to maximize the accuracy of the predicted bioaccessibility values. Although not specifically discussed by Oomen et al. (2006), it is possible that factors such as the amounts of iron-oxide particulates in the earth material may influence lead solubility and absorption in the intestinal tract by providing a substrate for lead sorption.

Lead uptake models have been developed to help estimate potential risk from exposure to lead-rich sources at specific sites. Examples include the US EPA Integrated Exposure Uptake Biokinetic Model for Lead in Children (IEUBK) (EPA TRW 1999, and references therein), and the International Commission on Radiological Protection (ICRP) Pb model (see example of its use and references in Abrahams et al. 2006). These models predict blood lead levels in children that would result from consumption of lead-containing particles from a variety of site-specific sources, including contaminated soils, lead paint chips, dust, foodstuffs, etc. Where data on these sources are not available, default values are assumed. The models also typically assume a default value for lead bioavailability (30% in the IEUBK model), which can be modified with the use of well-constrained bioaccessibility data developed through site-specific animal uptake or *in vitro* bioaccessibility testing (EPA TRW 1999).

Figure 13 compares the total lead content of and amounts of lead leached from a variety of earth materials by a simple gastric leach (our unpublished data, obtained primarily following the protocols of Shroder et al. 2004, and Drexler et al. 2006). Compared are: two samples of lead-zinc mine wastes from the Kabwe, Zambia lead-zinc mine (samples provided by Dr. Gary Krieger); NIST (National Institute of Standards and Technology) Butte, Montana soil standard, affected by copper- and arsenic-rich mine wastes; two urban atmospheric particulate standards collected by the in the mid-1970's in St. Louis, Missouri, and Washington, D.C.; a sample of the fine fraction of ballast lining a subway railbed; a NIST coal fly ash standard; two NIST soil standards with known contamination from lead paint chips; two edible soils purchased in a market in Kabwe (samples from Dr. Gary Krieger); a sample of ash left after a wildfire in ponderosa pine forest, Colorado front range (provided by Dr. Deborah Martin); and a sample of lake-bed dust from Owens Lake, California (provided by Dr. Marith Reheis). The

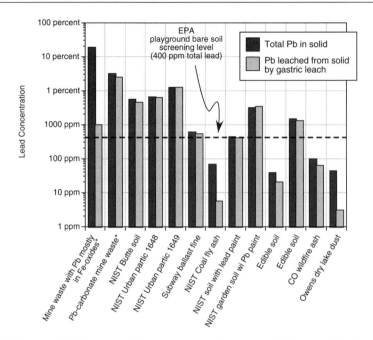

Figure 13. Our unpublished data on bulk metal concentrations and metal bioaccessibility from diverse earth materials indicate (in agreement with studies such as Casteel et al. 2006) that there can be substantial differences between the total lead concentration and lead bioaccessibility between different earth materials. Total metals were determined by ICP-MS following 4-acid digestion. Most of the leach measurements were made using the physiologically based gastric-fluid extraction test found to best reproduce lead uptake by swine (Casteel et al. 2006; Drexler and Brattin 2006); one part solid was added to 100 parts by weight simulated gastric fluid (HCl, glycine), and the mixture rotated at 37 °C for 2 h. Leach tests for samples indicated by the asterisk (*) were made using an NaCl-HCl simulated gastric fluid at 1 part solid to 40 parts gastric fluid. Lead concentrations in the leachates from both tests were recalculated to milligrams leached per kilogram solid for direct comparison to the total lead concentrations. In the NaCl leach tests, the pH shifted to substantially higher values (pH 4-5) due to the higher solid:liquid ratio and abundant carbonates in the samples; as a result, the amounts of lead leached should be an underestimate.

results show that Pb-Zn mine wastes (particularly those with high levels of lead carbonate), urban atmospheric particulates from the 1970's, and soils containing lead paint chips have both high to very high lead contents, as well as high lead bioaccessibility; many samples have lead contents far above the 400 ppm EPA soil screening level for bare playground soil. The urban particulate standards have elevated lead levels and very high lead bioaccessibility due to the presence of combustion byproducts of leaded gas, which was banned in the 1970's; hence most modern urban atmospheric particulates should have substantially lower lead levels than the NIST standards. The two mine waste samples illustrate the difference in lead bioaccessibility between lead largely tied up in iron oxides (less soluble in the gastric fluids) and lead tied up largely in lead carbonates (highly soluble in the gastric fluids). The African edible soil sample with elevated lead was purchased in a market near the now-closed lead-zinc mine from where the two Kabwe mine waste samples were obtained, and so it appears that the edible soil may have been produced from a soil contaminated by mine wastes or from a pre-mining soil developed on lead-mineralized rocks. The elevated lead levels and high bioaccessibility of the lead are obviously a source of health concern, as the soil is consumed mostly by pregnant women at a rate of tens to hundreds of grams per day; this could translate into the consumption of many tens to hundreds of milligrams of bioaccessible lead per day.

Lead uptake from earth materials via inhalation exposure? While many tests have been carried out to investigate the potential uptake of lead from earth materials by ingestion, far fewer have investigated potential lead uptake from earth materials via inhalation exposure (see summary in Plumlee and Ziegler 2003, 2006). We have investigated biosolubility of and metal bioaccessibility from a number of different earth materials using simulated lung and serum-based fluids, and have rather consistently found that lead does not show appreciable bioaccessibility in these simulated fluids (see the example of Owens Lake dusts shown in Fig. 14). In part, this may result from the lesser extent to which lead-bearing phases are solubilized in these pH 7.4 fluids (in spite of the abundance of myriad organic ligands available to complex the lead). However, it also likely results from the reaction of any solubilized lead with phosphate in the fluids to precipitate an insoluble lead phosphate phase. In contrast to lead, however, these tests suggest that oxyanion-forming (and potentially toxic) metalloids

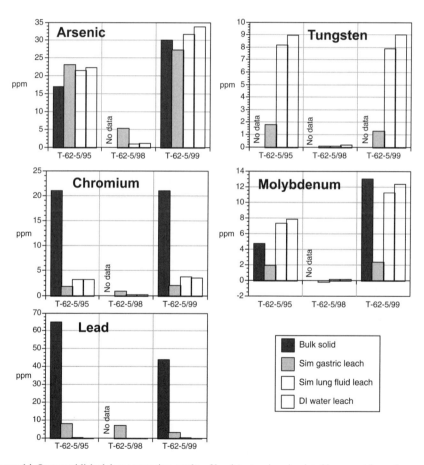

Figure 14. Our unpublished data comparing results of leach tests using simulated lung, gastric, and serum-based leach fluids of dust samples collected near Owens dry lake, CA, show that oxyanion species such as arsenic may be effectively leached by near-neutral pH fluids lining the lungs, due to their enhanced chemical mobility at near-neutral pH values. In contrast, lead is typically not leached by the simulated lung and serum-based fluids, due primarily to lead's limited solubility at pH 7.4 and its reaction with phosphate in the fluids to form insoluble lead-phospate precipitates. Lung and serum-based fluid compositions are listed in Plumlee and Ziegler (2003, 2006). All simulated lung fluid tests were run using a 1:40 solid:liquid ratio for 24 h at 37 °C, with continuous rotation of the sample container.

such as arsenic, chromium, molybdenum, tungsten and antimony (not shown) are relatively bioaccessible in the simulated lung and serum-based fluids (Fig. 14), due to the enhanced mobility of the oxyanion species at pH 7.4. We are currently evaluating metal bioaccessibility from earth materials in acidic simulated lysosomal fluids, as a simplified proxy for potential metal uptake via particle phaogcytosis.

SUMMARY

In this chapter, we have provided a comparatively brief overview of the myriad potential geochemical and biochemical processes that can occur when earth materials come into contact with body fluids via inhalation, ingestion, or percutaneous exposure routes. We have also shown how a wide variety of earth science methods have been and can be integrated with health science methods to better understand the potential health effects that might be associated with exposure to diverse earth materials.

As noted by Plumlee and Ziegler (2003, 2006), it is possible to group individual earth material components according to how they behave *in vivo*, as well as by similarities in how they trigger toxicity:

- Bioreactive earth materials can substantially modify the chemical composition of body fluids and tissues, to produce tissue irritation or more serious alkali or acid burns.

- Solubilization of bioaccessible toxicants from earth materials and their subsequent absorption can produce toxic effects in the body.

- Exposure to earth materials that are insoluble (biodurable) in body fluids can trigger toxic responses as the body attempts to clear the materials.

- Pathogens associated with earth materials can trigger disease.

- The body's immune response to the earth materials or toxicants contained within the earth materials can trigger a toxic response.

All of these different these types of health effects are strongly influenced by the forms in which earth materials are delivered to the body (mineralogy; particle size and morphology; particle solubility, alkalinity, acidity; oxidation state of contained constituents; and others). These health effects all also ultimately require some level of chemical interactions between the earth materials and body fluids.

However, most earth materials are complex mixtures of many different components, each having a particular mix of bioreactivity, biosolubility, and biodurability characteristics. Potentially complex chemical interactions between the various earth material components and the body's fluids may result. Hence, as the potential health effects of complex earth materials are assessed, it should always be kept in mind that the integrated physical and chemical characteristics, chemical behavior *in vivo*, and resulting toxicity effects of the whole material may be substantially different from those of its individual components..

These conclusions implicitly require important roles for the earth scientist in helping to both characterize earth materials and understand geochemical processes in the context of the physiological processes of the human body. Many studies have been carried out to address this complex but fascinating topic; however, as indicated by the sections on asbestos and lead, there are many unresolved questions remaining. There should therefore be ample opportunities for fruitful future collaborations to occur between geochemists, toxicologists, physiologists, epidemiologists, and other biomedical disciplines.

REFERENCES

Abrahams PW (2005) Geophagy and the involuntary ingestion of soil. *In:* Essentials of Medical Geology. Selinus O, Alloway B, Centeno J, Finkelman R, Fuge R, Lindh U, Smedley P (eds) Elsevier, p 435-458

Abrahams PW, Follansbee MH, Hunt A, Smith B, Wragg J (2006) Iron nutrition and possible lead toxicity: an appraisal of geophagy undertaken by pregnant women of UK Asian communities. Appl Geochem 21: 98-108

Arora M, Kennedy BJ, Elhlou S, Pearson NJ, Walker DM, Bayl P, Chan SW (2006) Spatial distribution of lead in human primary teeth as a biomarker of pre- and neonatal lead exposure. Sci Total Environ (*in press*)

Arvidson RS, Ertan IE, Amonette JE, Lüttge A (2003) Rates of calcite dissolution obtained by vertical scanning interferometry. Geochim Cosmchim Acta 67:1623-1634

Assimakopoulos SF, Thomopoulos KC, Patsoukis N, Georgiou CD, Scopa CD, Nikolopoulou VN, Vagianos CE (2006) Evidence for intestinal oxidative stress in patients with obstructive jaundice. Eur J Clin Invest 36:181-187

ATSDR (2000) Chromium toxicological profile. US Department of Health and Human Services, Public Health Service, Agency for Toxic Substances and Disease Registry. Available from: *http://www.atsdr.cdc.gov/toxpro2.html*

Aust AE, Ball JC, Hu AA, Lighty JS, Smith KR, Straccia AM, Veranth JM, Young WC (2002) Particle characteristics responsible for effects on human lung epithelial cells. Health Effects Institute Research Report Number 110, December, 77 pp

Aust AE, Lund LG (1990) The role of iron in asbestos-catalyzed damage to lipids and DNA. *In:* Biological Oxidation Systems, Volume 2. Reddy CC, Hamilton GA, Madyastha KM (eds) Academic Press, p 597-605

Baron PA, Willeki K (eds) (2001) Aerosol Measurement – Principles, Techniques, and Applications. Wiley Interscience

Bethke CM (1996) Geochemical Reaction Modeling-Concept and Applications. Oxford University Press

Borda MJ, Elsetinow AR, Strongin DR, Schoonen MA (2004) A mechanism for the production of hydroxyl radical at surface defect sites on pyrite. Geochim Cosmchim Acta 67:935-939

Brain JD (1992) Mechanisms, measurement, and significance of lung macrophage function. Environ Health Perspect 97:5-10

Brodkin CA, Balmes JR, Redlich CA, Cullen MR (2006) Residential proximity to naturally occurring asbestos: health risk or ecologic fallacy? Am J Respir Crit Care Med 74:573

Buchet JP, Lauwerys RR, Yager JW (1995) Lung retention and bioavailability of arsenic after single intratracheal administration of sodium arsenite, sodium arsenate, fly ash and copper smelter dust in the hamster. Environ Geochem Health 17:182-188

Buettner GR (1993) The pecking order of free radicals and antioxidants: lipid peroxidation, α-tocopherol, and ascorbate. Arch Biochem Biophys 300:535-543

Bultman MW, Fisher FS, Pappagianis D (2005) Chapter 19, The ecology of soil-borne human pathogens. *In:* Essentials of Medical Geology. Selinus O, Alloway B, Centeno J, Finkelman R, Fuge R, Lindh U, Smedley P (eds) Elsevier, p 481-512

CDC (2006) National biomonitoring program. *http://www.cdc.gov/biomonitoring/*

Cama J, Acero P, Ayora C, Lobo A (2005) Galena surface reactivity at acidic pH and 25 °C based on flow-through and *in situ* AFM experiments. Chem Geol 214:309-330

Casteel SW, Weis CP, Henningsen GM, Brattin WJ (2006) Estimation of relative bioavailability of lead in soil and soil-like materials using young swine: Environ Health Perspect 114:1162-1171

Castranova V (2000) From coal mine dust to quartz: mechanisms of pulmonary pathogenicity. Inhalat Toxicol 12:7-14

Castranova V, Vallyathan V (2000) Silicosis and coal workers pneumoconiosis. Environ Health Perspect 108(suppl. 4):675–684

Centeno JA, Mullick FG, Ishak KG, Franks TJ, Burke AP, Koss MN, Perl DP, Tchounwou PB, Pestaner, JP (2005a) Environmental pathology. *In:* Essentials of Medical Geology. Selinus O, Alloway B, Centeno J, Finkelman R, Fuge R, Lindh U, Smedley P (eds) Elsevier, p 563-594

Centeno JA, Todorov T, Pestaner JP, Mullick FG, Jonas WB (2005b) Histochemical and microprobe analysis in medical geology. *In:* Essentials of Medical Geology. Selinus O, Alloway B, Centeno J, Finkelman R, Fuge R, Lindh U, Smedley P (eds) Elsevier, p 725-736

Churchill RK, Hill RL (2000) A general location guide for ultramafic rocks in California: areas more likely to contain naturally occurring asbestos. CA Dept. Conserv., Div. Mines and Geol., DMG Open-File Report 2000–19. *http://www.consrv.ca.gov/*

Churg A (2000) Particle uptake by epithelial cells. *In:* Particle-Lung Interactions. Gehr P, Heyder J (eds) Marcel Dekker, Inc., p 401-436

Churg A, Wiggs B, Depaoli L, Kampe B, Stevens B (1984) Lung asbestos content in chrysotile workers with mesothelioma. Am Rev Respir Dis 130:1042-1045

Churg A, Wright JL, Gilks B, DePaoli L (1989) Rapid short term clearance of chrysotile compared to amosite asbestos in the guinea pig. Am Rev Respir Dis 139:885-890

Churg A, Wright JL, Vedal S (1993) Fiber burden and patterns of asbestos-related disease in chrysotile miners and millers. Am Rev Respir Dis 148:25-31

Clark RN, Swayze GA, Hoefen TM, Green RO, Livo KE, Meeker G, Sutley S, Plumlee G, Pavri B, Sarture C, Boardman J, Brownfield I, Morath LC (2006) Environmental mapping of the World Trade Center area with imaging spectroscopy after the September 11, 2001 attack. *In:* Urban Aerosols and Their Impacts: Lessons Learned from the World Trade Center Tragedy. ACS Book Symposium Series 919. Gaffney JS, Marley NA (eds) Oxford University Press, p 66-83

Cohn CA, Mueller S, Wimmer E, Leifer N, Greenbaum S, Strongin DR, Schoonen M (2006) Pyrite-induced hydroxyl radical formation and its effect nucleic acids. Geochem Trans 7:3, *http://www.geochemicaltrans actions.com/content/7/1/3*

Coico R, Sunshine G, Benjamini E (2003) Immunology: A Short Course, 5th Edition. City University of New York

Costa DL (2001) Air pollution. *In:* Casarett and Doull's Toxicology: The Basic Science of Poisons. Klassen CD (ed) McGraw-Hill, p 979-1012

Daroowalla FM (2001) Pneumoconioses. *In:* Clinical Environmental Health and Exposures 2nd Edition. Sullivan JB Jr, Krieger G (eds) Lippincott Williams and Wilkins, p 538-545

Davis A, Ruby MV, Bergstrom PD (1992) Bioavailability of arsenic and lead in soils from the Butte, Montana, mining district. Environ Sci Technol 26:461-468

Davis JMG, Addison J, McIntosh C, Miller BG, Niven K (1991) Variations in the carcinogenicity of tremolite dust samples of differing morphology. Ann NY Acad Sci 1991:473-490

Dearwent S, Imtiaz R, Metcalf S, Lewin M (2000) Health consultation—mortality from asbestosis in Libby, Montana (report dated December 12, 2000). Agency for Toxic Substances and Disease Registry, *http://www.atsdr.cdc.gov/HAC/pha/libby/lib_toc.html*

Dement JM (1990) Overview: workshop of fiber toxicology research needs. Environ Health Perspect 88:261-268

Derbyshire E (2005) Aerosolic mineral dust and human health. *In:* Essentials of Medical Geology. Selinus O, Alloway B, Centeno J, Finkelman R, Fuge R, Lindh U, Smedley P (eds) Elsevier, p 459-480

Dodson RF (2006) Analysis and relevance of asbestos burden in tissue. *In:* Asbestos: Risk Assessment, Epidemiology, and Health Effects. Dodson RF, Hammar SP (eds) Taylor and Francis, p 39-90

Dodson RF, Atkinson MAL, Levin JL (2003) Asbestos fiber length as related to potential pathogenicity: a critical review. Am J Ind Med 44:291-297

Dodson RF, Hammar SP (eds) (2006) Asbestos: Risk Asssessment, Epidemiology, and Health Effects. Taylor and Francis

Dogan AU, Baris YI, Dogan M, Emri S, Steele I, Elmishad AG, Carbone M (2006) Genetic predisposition to fiber carcinogenesis causes a mesothelioma epidemic in Turkey. Cancer Res 66:5063-5068

Drexler JW, Brattin WJ (2006) An *in vitro* procedure for estimation of lead relative bioavailability: with validation. Hum Ecol Risk Assess. Also see *http://www.colorado.edu/geolsci/legs/invitro1.html (in press)*

EPA TRW (1999) Short sheet: IEUBK model bioavailability variable. U.S. Environmental Protection Agency Technical Review Workgroup for Lead, EPA #540-F-00-010, OSWER #9285.7-38

EPA TRW (2000) Short sheet: TRW recommendations for sampling and analysis of soil at lead (Pb) sites. U.S. Environmental Protection Agency Technical Review Workgroup for Lead, EPA #540-F-00-006, OSWER #9285.7-32, 8 pp. *http://www.epa.gov/superfund/lead/products/sssiev.pdf*

Eastes W, Hadley JG, Bender J (1996) Assessing the role of biological activity of fibres: insights into the role of fibre biodurability, Australia and New Zealand. J Occupat Health Safety 12:381-385

Eastes W, Potter RM, Hadley JG (2000a) Estimating in-vitro glass fiber dissolution rate from composition. Inhalat Toxicol 12:269-280

Eastes W, Potter RM, Hadley JG (2000b) Estimating rock and slag wool fiber dissolution rate from composition. Inhalat Toxicol 12:1127-1139

Fattman CL, Chu CT, Oury TD (2004) Chapter 10, Experimental models of asbestos-related diseases. *In:* Pathology of Asbestos-Associated Diseases, 2nd Edition. Roggli VL, Oury TD, Sporn TA (eds) Springer, p 256-308

Foster AL, Brown GE Jr, Tingle T, Parks GA (1998) Quantitative arsenic speciation in mine tailings using X-ray absorption spectroscopy. Am Mineral 83:553-568

Freudenrich CC (1998) How sweat works. *http://health.howstuffworks.com/sweat.htm*, accessed July, 2006

Fubini B, Areán CO (1999) Chemical aspects of the toxicity of inhaled mineral dusts. Chem Soc Rev 28:373-381

Garrett RG (2005) Natural distribution and abundance of elements. *In:* Essentials of Medical Geology. Selinus O, Alloway B, Centeno J, Finkelman R, Fuge R, Lindh U, Smedley P (eds) Elsevier, p 17-42

Gehr P, Heyder J (eds) (2000) Particle-Lung Interactions, Marcel Dekker, Inc

Gianfagna A, Ballirano P, Bellatreccia F, Bruni B, Paoletti L, Oberti R (2003), Characterization of amphibole fibres linked to mesothelioma in the area of Biancavilla, eastern Sicily, Italy. Mineral Mag 67:1221-1229

Gochfeld M (1997) Factors influencing susceptibility to metals. Environ Health Perspect 105(Suppl 4):12

Goyer RA, Clarkson TW (2001) Toxic effects of metals. *In:* Casarett and Doull's Toxicology: the Basic Science of Poisons, 6th Edition. Klassen CD (ed) McGraw-Hill, p 811-868

Griffin DW, Kellogg CA, Garrison VA, Shinn EA (2002) The global transport of dust. Am Sci 90:228-235

Gulson BL, Mizon KJ, Korsch MJ, Howarth D (1996a) Non-orebody sources are significant contributors to blood lead of some children with low to moderate lead exposure in a major lead mining community. Sci Total Environ 181:223-230

Gulson BL, Pisaniello D, McMichael AJ, Mizon KJ, Korsch MJ, Luke C, Ashbolt R, Pederson DG, Vimpani G, Mahaffey KR (1996b) Stable lead isotope profiles in smelter and general urban communities—A comparison of environmental and blood measures. Environ Geochem Health 18:147-163

Gulson BL, Mizon KJ, Law AJ, Korsch MJ, Davis JJ, Howarth D (1994), Source and pathways of lead in humans from the Broken Hill mining community—an alternative use of exploration methods. Econ Geol 89:889-908

Gulson B, Wong H (2006) Stable isotope tracing—a way forward for nanotechnology. Environ Health Perspect 114:1486-1488

Guthrie GD (1997) Mineral properties and their contributions to particle toxicity. Environ Health Perspect 105(Suppl 5):1003–1011

Hamel SC, Ellickson KM, Lioy PJ (1999) The estimation of the bioaccessibility of heavy metals in soils using artificial biofluids by two novel methods: mass-balance and soil recapture. Sci Total Environ 243/244: 273-283

Hesterberg TW, Anderson R, Bunn WB III, Chase GR, Hart GA (2001) Man-made mineral fibers. *In:* Clinical Environmental Health and Exposures, 2nd Edition. Sullivan JB Jr, Krieger G (eds) Lippincott Williams and Wilkins, p 1227-1240

Hinds WC (1999) Aerosol Technology: Properties, Behavior, and Measurement of Airborne Particles. John Wiley and Sons

Hochella MF Jr (1993) Surface chemistry, structure, and reactivity of hazardous mineral dust. Rev Mineral 28: 275-308

Holland JP, Smith DD (2001) Asbestos. *In:* Clinical Environmental Health and Exposures, 2nd Edition. Sullivan JB Jr, Krieger G (eds) Lippincott Williams and Wilkins, p 1214–1227

Huang X, Fournier J, Koenig K, Chen LC (1998) Buffering capacity of coal and its acid-soluble Fe^{2+} content: possible role in coal workers' pneumoconiosis. Chem Res Toxicol 11:722–729

Huang X, Li W, Attfield MD, Nádas A, Frenkel K, Finkelman RB (2005) Mapping and prediction of coal workers' pneumoconiosis with bioavailable iron content in the bituminous coals. Environ. Health Perspect 113:964-968

Hume LA, Rimstidt JD (1992) The biodurability of chrysotile asbestos. Am Mineral 77:1125-1128

ICRP (1995) Human respiratory tract model for radiological protection. ICRP Publication 66, Elsevier

Ilgren EB (2004) The biology of cleavage fragments: A brief synthesis and analysis of current knowledge. Indoor Built Environ 13:343-356

Johnson NF, Mossman BT (2001) Dose, dimension, durability, and biopersistence of chrysotile asbestos. *In:* The Health Effects of Chrysotile Asbestos: Contribution of Science to Risk-management Decisions. Nolan RP, Langer AM, Ross M, Wicks FJ, Martin RF (eds) The Canadian Mineralogist Special Pubublication 5: 145-154

Jurinski JB, Rimstidt JD (2001) Biodurability of talc. Am Mineral 86:392–399

Kawanishi S (1995) Role of active oxygen species in metal-induced DNA damage. *In:* Toxicology of Metals—Biochemical Aspects, Handbook of Experimental Pharmacology. Goyer RA, Cherian MG (eds) 115: 349–372

Kieth JD, Pacey GE, Cotruvo JA, Gordon G (2006) Experimental results from the reaction of bromate ion with synthetic and real gastric juices. Toxicol 221:225-228

Kim CS, Rytuba JJ, Brown GE Jr (2004) Geological and anthropogenic factors influencing mercury speciation in mine wastes: an EXAFS spectroscopic study. Appl Geochem 19:379-393

Klassen CD (ed) (2001) Casarett and Doull's Toxicology: The Basic Science of Poisons. McGraw-Hill

Kreyling WG, Beisker W, Miaskowski U, Neuner M, Heilmann P (1992) Intraphagolysosomal pH and intracellular particle dissolution in canine and rat alveolar macrophages. *In:* Environmental Hygiene III. Seemayer NH, Hadnagy N (eds) Springer Verlag, p 19-23

Kreyling WG, Scheuch G (2000) Clearance of Particles deposited in the lungs. *In:* Particle-Lung Interactions. Gehr P, Heyder J (eds) Marcel Dekker, Inc., p 323-378

Leake BE, Woolley AR, Arps CES, Birch WD, Gilbert MC, Grice JD, Hawthorne FC, Kato A, Kisch HJ, Krivovichev VG, Linthout K, Laird J, Mandarino JA, Maresch WV, Nickel EH, Rock NMS, Schumacher JC, Smith DC, Stephenson NCN, Ungaretti L, Whittaker EJW, Youzhi G (1997) Nomenclature of the amphiboles: Report of the subcommittee on amphiboles of the International Mineralogical Association, Commission on New Minerals and Mineral Names. Am Mineral 82:1019-1037

Lehnert BE (1992) Pulmonary and thoracic macrophage subpopulations and clearance of particles from the lung. Environ Health Perspect 97:17–46

Lemen RA (2006) Epidemiology of asbestos-related diseases and the knowledge that led to what is known today. *In:* Asbestos: Risk Assessment, Epidemiology, and Health Effects. Dodson RF, Hammar SP (eds) Taylor and Francis, p 201-308

Lindh U (2005a) Uptake of elements from a biological point of view. *In:* Essentials of Medical Geology. Selinus O, Alloway B, Centeno J, Finkelman R, Fuge R, Lindh U, Smedley P (eds) Elsevier Inc., p 87-114

Lindh U (2005b) Biological functions of the elements. *In:* Essentials of Medical Geology. Selinus O, Alloway B, Centeno J, Finkelman R, Fuge R, Lindh U, Smedley P (eds) Elsevier Inc., p 115-160

Lowers H, Meeker GP (2002) Tabulation of asbestos-related terminology: U.S. Geol. Survey Open-File Report 02-458, *http://pubs.usgs.gov/of/2002/ofr-02-458/*

Lybarger JA, Lewin M, Peipins LA, Campolucci SS, Kess SE, Miller A, Spence M, Black B, Weis C (2001) Year 2000 Medical Testing of Individuals Potentially Exposed to Asbestoform [sic] Minerals Associated with Vermiculite in Libby, Montana: a Report to the Community. *http://www.atsdr.cdc.gov/asbestos/doc_phl_testreport.html*

Lund LG, Aust AE (1992) Iron mobilization from crocidolite asbestos greatly enhances crocidolite-dependent formation of DNA single-strand breaks in ϕ X174 RFI DNA. Carcinogenesis 13:637–642

Mattson SM (1994a) Glass fiber dissolution in simulated lung fluid and measures needed to improve consistency and correspondence to *in vivo* dissolution. Environ Health Perspect 102(suppl 5):87–90

Mattson SM (1994b) Glass fibers in simulated lung fluid: dissolution behavior and analytical requirements. Ann Occup Hyg 38:857-877

McClellan RO (2000) Particle interaction with the respiratory tract. *In:* Particle-Lung Interactions. Gehr P, Heyder J (eds) Marcel Dekker, Inc., p 3-66

McClellan RO, Henderson RF (eds) (1995) Concepts in Inhalation Toxicology, 2nd Edition. Taylor & Francis

Meeker GP, Bern AM, Brownfield IK, Sutley SJ, Hoefen TM, Vance JS, Lowers HA (2003) The composition and morphology of prismatic, fibrous, and asbestiform amphibole from the Rainy Creek district, Libby, Montana. Am Mineral, 88:1955-1969

Meeker GP, Sutley SJ, Brownfield IK, Lowers HA, Bern AM, Swayze GA, Hoefen TM, Plumlee GS, Clark RN, Gent CA (2005) Materials characterization of dusts generated by the collapse of the World Trade Center. *In:* Urban Aerosols and Their Impacts: Lessons Learned from the World Trade Center Tragedy. Marley NA, Gaffney JS (eds) ACS Book Series 919, Oxford University Press, p 84-102

Meza MM, Yu L, Rodriguez YY, Guild M, Thompson D, Gandolfi AJ, Klimecki WT (2006) Developmentally restricted genetic determinants of human arsenic metabolism: association between urinary methylated arsenic and CYT19 polymorphisms in children. Environ Health Perspect 113:775-781

Mielke HW (1999) Lead in the inner cities. Am Sci 87:62-73

Mielke HW, Powell ET, Gonzales CR, Mielke PW Jr, Ottesen RT, Langedal M (2006) New Orleans soil lead (Pb) cleanup using Mississippi River alluvium: need, feasibility, and cost. Environ Sci Technol 40:2784-2789

Mielke HW, Wang G, Gonzales CR, Powell ET, Le B, Quach VN (2004) PAHs and metals in the soils of inner-city and suburban New Orleans, Louisiana, USA. Environ Toxicol Pharmacol 18:243-247

Millette JR (2006) Asbestos analysis methods. *In:* Asbestos: Risk Assessment, Epidemiology, and Health Effects. Dodson RF, Hammar SP (eds) Taylor and Francis, p 9-38

Mutter J, Naumann J, Schneider R, Walach H, Haley B (2005) Mercury and autism: accelerating evidence? Neuro Endocrinol Lett 26:439-446

NIOSH (1990) Testimony of the National Institute for Occupational Safety and Health on the Occupational Safety and Health Administration's notice of proposed rulemaking on occupational exposure to asbestos, tremolite, anthophyllite, and actinolite. 29 CFR Parts 1910 and 1926, Docket No. H-033d, presented at the OSHA Informal Public Hearing, May 9, 1990, Washington, D.C.

NIOSH WoRLD (2006) Work-related lung disease (WoRLD) surveillance system. *http://www2a.cdc.gov/drds/WorldReportData/FigureTableDetails.asp?FigureTableID=12*

NHANES (2006) National Health and Nutrition Examination Survey. *http://www.cdc.gov/nchs/nhanes.htm*

Newman LS (2001) Clinical pulmonary toxicology. *In:* Clinical Environmental Health and Exposures, 2nd Edition. Sullivan JB Jr, Krieger G (eds) Lippincott Williams and Wilkins, p 206–223

Nolan RP, Langer AM, Herson GB (1991) Characterization of palygorskite specimens from different geological locales for health hazard evaluation. J Br Indust Med 48:463–475

OSHA (1992) Regulations (Preambles to Final Rules), Occupational Exposure to Asbestos, Tremolite, Anthophyllite and Actinolite, Section 4 - IV. Mineralogical Considerations, *http://www.osha.gov/pls/oshaweb/owadisp.show_document?p_table=PREAMBLES&p_id=785*

Oomen AG, Brandon EFA, Swartjes FA, Sips AJAM (2006) How can information on oral bioavailability improve humanhealth risk assessment for lead-contaminated soils? Implementation and scientific basis. RIVM report 711701042/2006, *http://www.rivm.nl/bibliotheek/rapporten/711701042.pdf*

Oomen AG, Hack A, Minekus M, Zeijdner E, Cornelis C, Schoeters G, Verstraete W, Van De Wiele T, Wragg J, Rompelberg CJM, Sips AJAM, Van Wijnen JH (2002) Comparison of five *in vitro* digestion models to study the bioaccessibility of soil contaminants. Environ Sci Technol 36:3326–3334

Oomen AG, Tolls J, Sips AJAM, Groten JP (2003) *In vitro* intestinal lead uptake and transport in relation to speciation. Arch Environ Contam Toxicol 44:116-124

Palekar LD, Spooner CM, Coffin DL (1979) Influence of crystallization habit of minerals on *in vitro* cytoxicity. Ann NY Acad Sci 1979:673-688

Pan X, Day HW, Wang W, Beckett LA, Schenker MB (2005) Residential proximity to naturally occurring asbestos and mesothelioma risk in California. Am J Respir Crit Care Med 172:1019-1025

Pfau JC, Sentissi JJ, Weller G, Putnam EA (2005) Assessment of autoimmune responses associated with asbestos exposure in Libby, Montana, USA. Environ Health Perspect 113:25-30

Plumlee GS, Hageman PL, Lamothe PJ, Ziegler TL, Meeker GP, Theodorakos P, Brownfield I, Adams M, Swayze GA, Hoefen T, Taggart JE, Clark RN, Wilson S, Sutley S (2005) Inorganic chemical composition and chemical reactivity of settled dust generated by the World Trade Center building collapse. *In:* Urban Aerosols and Their Impacts: Lessons Learned from the World Trade Center Tragedy. Marley NA, Gaffney JS (eds) ACS Book Series 919, Oxford University Press, p 238-276

Plumlee GS, Ziegler TL (2003) The medical geochemistry of dusts, soils, and other earth materials. *In:* Treatise on Geochemistry. Volume 9. Lollar BS (ed) Elsevier, p 263-310

Plumlee GS, Ziegler TL (2006) The medical geochemistry of dusts, soils, and other earth materials. *In:* Treatise on Geochemistry. Volume 9. Lollar BS (ed) Elsevier, online version, *http://www.sciencedirect.com/science/referenceworks/0080437516*

Raloff J (2006) Dirty little secret – asbestos laces many residential soils. Sci News 170:26, *http://sciencenews.org/articles/20060708/bob9.asp*

Reeder RJ, Schoonen MAA, Lanzirotti A (2006) Metal speciation and its role in bioaccessibility and bioavailability. Rev Mineral Geochem 64:59-113

Reinhardt TE, Ottmar RD (2000) Smoke exposure at western wildfires. Research Paper PNW-RP-525, 84,USDA Forest Service

Reinhardt TE, Ottmar RD, Hanneman AJS (2000) Smoke exposure among firefighters at prescribed burns in the Pacific Northwest. Research Paper PNW-RP-526, 54, USDA Forest Service

Renner R (2000) Asbestos in the air. Sci Am February:34

Rhoades R, Pflanzer R (1992) Human Physiology, 2nd Edition. Sanders College Publishing, Harcourt Brace College Publishers

Ritschel WA, Thompson GA (1983) Monitoring of drug concentrations in saliva: a non-invasive pharmacokinetic procedure. Methods Find Exp Clin Pharmacol 5: 511-525

Roggli VL, Coin P (2004) Mineralogy of asbestos. *In:* Pathology of Asbestos-Associated Diseases, 2nd Edition. Roggli VL, Oury TD, Sporn TA (eds) Springer, p 1-16

Roggli VL, Oury TD, Sporn TA (eds) (2004) Pathology of Asbestos-Associated Diseases, 2nd Edition. Springer

Rolston R, Oury TD (2004) Other neoplasia. *In:* Pathology of Asbestos-Associated Diseases, 2nd Edition. Roggli VL, Oury TD, Sporn TA (eds) Springer, p 217-230

Rozman KK, Klaasen CD (2001) Absorption, distribution, and excretion of toxicants. *In:* Toxicology: The Basic Science of Poisons, 6th Edition. Klassen CD (ed) McGraw-Hill, p 107–132

Ruby MV, Schoof R, Brattin W, Goldade M, Post G, Harnois M, Mosby DE, Casteel SW, Berti W, Carpenter M, Edwards D, Cragin D, Chappell W (1999) Advances in evaluating the oral bioavailability of inorganics in soil for use in human health risk assessment. Environ Sci Technol 33:3697-3705

Ryker S (2001) Mapping arsenic in groundwater. Geotimes 46:34, *http://www.agiweb.org/geotimes/nov01/feature_Asmap.html*

Sahai D (2001) Cement hazards and controls: health risks and precautions in using Portland cement. Constr Safety Mag 12, accessed 2006 at *http://www.cdc.gov/eLCOSH/docs/d0500/d000513/d000513.html*

Schenker M, Day H, Beckett L, Pan X (2006) Reply to Brodkin et al., Residential proximity to naturally occurring asbestos: health risk or ecologic fallacy? Am J Respir Crit Care Med 74:573-574

Schlesinger RB (1988) Biological disposition of airborne particles: basic principles and application to vehicular emissions. *In:* Air Pollution, the Automobile, and Public Health. National Academy of Sciences, Health Effects Institute, National Academy Press, p 239-298

Schlesinger RB (1995) Deposition and clearance of inhaled particles. *In:* Concepts in Inhalation Toxicology, 2nd Edition, McClellan RO, Henderson RF (eds) Taylor & Francis, p 1-224

Schoonen MAA, Cohn CA, Roemer E, Laffers R, Simon SR, O'Riordan T (2006) Mineral-induced formation of reactive oxygen species. Rev Mineral Geochem 64:179-222

Schroder JL, Basta NT, Casteel SW, Evans TJ, Payton ME, Si J (2004) Validation of the in vitro gastrointestinal model (IVG) to estimate relative bioavailable lead in contaminated soils. J Environ Qual 33:513-521

Schulz H, Brand P, Heyder J (2000) Particle deposition in the respiratory tract. *In:* Particle-Lung Interactions. Gehr P, Heyder J (eds) Marcel Dekker, Inc., p 229-290

Sébastien P, McDonald JC, McDonald AD, Case B, Harley R (1989) Respiratory cancer in chrysotile textile and mining industries: exposure inferences from lung analysis. Br J Indust Med 46:180–187

Shafer FQ, Buettner GR (2001) Redox environment of the cell as viewed through the redox state of the glutathione disulfide/glutathione couple. Free Rad Biol Med 30:1191-1212

Shafer FQ, Buettner GR (2003) Redox state and redox environment in biology. *In:* Signal Transduction by Reactive Oxygen and Nitrogen Species: Pathways and Chemical Principles. Forman HJ, Fukuto J, Torres M (eds) Kluwer Academic Publishers, p 1-14

Sheppard RA (1996), Occurrences of erionite in sedimentary rocks of the western United States. USGS Open-File Report 96-018

Sipes IG, Badger D (2001) Principles of toxicology. *In:* Clinical Environmental Health and Exposures, 2nd Edition. Sullivan JB Jr, Krieger G (eds) Lippincott Williams and Wilkins, p 49–67

Skinner HCW, Ross M, Frondel C (1988) Asbestos and Other Fibrous Minerals. Oxford University Press

Smith KS (2006) Strategies to predict metal mobility in surficial mining environments. Rev Eng Geol (*in press*)

Smith KS, Huyck HLO (1999) An overview of the abundance, relative mobility, bioavailability, and human toxicity of metals. *In:*, The Environmental Geochemistry of Mineral Deposits: Part A. Processes, Techniques and Health Issues. Plumlee GS, Logsdon MJ (eds) Soc Econ Geol Rev Econ Geol 6A:29-70

Sobala GM, Pignatelli B, Schorah CJ, Bartsch H, Sanderson M, Dixon MF, Shires S, King RF, Axon AT (1991) Levels of nitrite, nitrate, N-nitroso compounds, ascorbic acid and total bile acids in gastric juice of patients with and without precancerous conditions of the stomach. Carcinogenesis 12:193-198

Sporn TA, Roggli VL (2004) Mesothelioma. *In:* Pathology of Asbestos-Associated Diseases, 2nd Edition. Roggli VL, Oury TD, Sporn TA (eds) Springer, p 104-168

Stanton MF, Wrench C (1972) Mechanisms of mesothelioma induction with asbestos and fibrous glass, J Natl Canc Inst 48:797-821

Stanton MF, Layard M, Tegeris E, Miller E, May M, Morgan E, Smith A (1981) Relation of particle dimension to carcinogenicity in amphibole asbestoses and other fibrous minerals. J Natl Canc Inst 67:965-975

SSDC (1988) Diseases associated with exposure to silica and nonfibrous silicate minerals (Silicosis and Silicate Disease Committee). Arch Pathol Lab Med 112:673–720

Sullivan JB Jr, Krieger G (eds) (2001) Clinical Environmental Health and Exposures, 2nd Edition. Lippincott Williams and Wilkins

Sullivan JB Jr, Levine RJ, Bangert JL, Maibach H, Hewitt P (2001) Clinical dermatoxicology. *In:* Clinical Environmental Health and Exposures, 2nd Edition. Sullivan JB Jr, Krieger G (eds) Lippincott Williams and Wilkins, p 182–206

Suzuki Y, Yuen SR (2001) Asbestos tissue burden study on human malignant mesothelioma. Indust Health 39: 150–160

Suzuki Y, Yuen SR (2002) Asbestos fibers contributing to the induction of human malignant mesothelioma. Ann NY Acad Sci 982:160-176

Stehr-Green P, Tull P, Stellfeld M, Mortenson PB, Simpson D (2003) Autism and Thimerosal-containing vaccines: lack of consistent evidence for an association. Am J Prev Med 25:101-106

Swayze GA, Higgins CT, Clinkenbeard JP, Kokaly RF, Clark RN, Meeker GP, Sutley SJ (2004) Preliminary report on using imaging spectroscopy to map ultramafic rocks, serpentinites, and tremolite-actinolite-bearing rocks in California: U.S. Geol. Survey Open-File Report 2004-1304 and California Geological Survey Geologic Hazards Investigation 2004-01, *http://pubs.usgs.gov/of/2004/1304/*

Taggart JE Jr (ed) (2002) Analytical methods for chemical analysis of geologic and other materials. U.S. Geological Survey Open-file Report 02-0223, version 5.0, *http://pubs.usgs.gov/of/2002/ofr-02-0223/*

Taylor DM, Williams DR (1995) Trace Element Medicine and Chelation Therapy. The Royal Society of Chemistry

Taylor DM, Williams DR (1998) Bio-inorganic chemistry and its pharmaceutical applications. *In:* Smith and Williams' Introduction to the Principles of Drug Design and Action. Smith HJ (ed) Harwood Academic, p 509–538

Thaysen JH, Thorn NA, Schwartz IL (1954) Excretion of sodium, potassium, chloride and carbon dioxide in human parotid saliva. Am J Physiol 178:155-159

Tinkle SS, Antonini JM, Rich BA, Roberts JR, Salmen R, DePree K, Adkins EJ (2003) Skin as a route of exposure and sensitization in chronic beryllium disease. Environ Health Perspect 111:1202-1208

Todorov TI, Ejnik JW, Mullick FG, Centeno JA (2006) Chemical and histological assessment of depleted uranium in tissues and biological samples. *In:* Depleted uranium: properties, uses, and health consequences, AC Miller, ed, Lewis publishers (*in press*)

UC Davis (2006) On the absestos trail. Synthesis, Fall-Winter 2006, UC Davis Cancer Center, accessed 8/2006 at *http://www.ucdmc.ucdavis.edu/synthesis/Archives/fall_winter_06/features/outreach.html*

Van Gosen BS (2005) Reported historic asbestos mines, historic asbestos prospects, and natural asbestos occurrences in the Eastern United States. U.S. Geol. Survey Open-File Report 2005-1189, *http:// pubs.usgs.gov/of/2005/1189/*

Van Gosen BS (2006a) Using the geology of asbestos deposits to predict the presence or absence of asbestos in mining and natural environment. *In:* Proceedings of the 42nd Forum on the Geology of Industrial Minerals, Asheville, North Carolina, May 7-13, 2006. JC Reid (ed) North Carolina Geological Survey Information Circular 34 [CD-ROM], 412-432

Van Gosen BS (2006b) Reported historic asbestos prospects and natural asbestos occurrences in the Central United States. U.S. Geol. Survey Open-File Report 2006-1211, *http://pubs.usgs.gov/of/2006/1211/*

Van Gosen BS, Lowers HA, Sutley SJ, Gent CA (2004) Asbestos-bearing talc deposits, southern Death Valley region, California. *In:* Betting on Industrial Minerals—Proceedings of the 39th Forum on the Geology of Industrial Minerals, Reno/Sparks, Nevada, May 18-24, 2003. Castor SB, Papke KG, Meeuwig RO (eds) Nevada Bureau of Mines and Geology Special Publication 33:215-223

Vutchkov M, Lalor G, Macko S (2005) Inorganic and organic geochemistry techniques. *In:* Essentials of Medical Geology. Selinus O, Alloway B, Centeno J, Finkelman R, Fuge R, Lindh U, Smedley P (eds) Elsevier Inc., p 695-723

van Oss CJ, Naim JO, Costanzo PM, Giese RF Jr, Wu W, Sorling AF (1999) Impact of different asbestos species and other mineral particles on pulmonary pathogenesis. Clays Clay Min 47:697–707

WHO (1996) Trace Elements in Human Nutrition and Health. World Health Organization in collaboration with the Food and Agricultural Organization of the United nations and the International Atomic Energy Agency, 343 pp

Walker JS (1981) Asbestos and the Asbestiform Habit of Minerals. Unpub. M.S. thesis, University of Minnesota, 175 pp

Weinstein P, Cook A (2005) Volcanic emissions. *In:* Essentials of Medical Geology. Selinus O, Alloway B, Centeno J, Finkelman R, Fuge R, Lindh U, Smedley P (eds) Elsevier Inc., p 563-594

Werner AJ, Hochella MF Jr, Guthrie GD Jr, Hardy JA, Aust AE, Rimstidt JD (1995) Asbestiform riebeckite (crocidolite) dissolution in the presence of Fe chelators: implications for mineral-induced disease. Am Mineral 80:1093–1103

Wolfe MI, Mott JA, Voorhees RE, Sewell CM, Paschal D, Wood CM, McKinney PE, Redd S (2004) Assessment of urinary metals following exposure to a large vegetative fire, New Mexico, 2000. J Expo Anal Environ Epidemiol 14:120-128

Ward DE (1999) Smoke from wildland fires. *In:* Health Guidelines for Vegetation Fire Events, Lima, Peru, 6-9 October 1998, Background papers. World Health Organization, p 70-85

Wood S, Taunton A, Normand C, Gunter M (2006) Mineral-fluid interaction in the lungs: Insights from reaction-path modeling. Inhal Toxicol 18:975-984

Wright RS, Abraham JL, Harber P, Burnett BR, Morris P, West P (2002) Fatal asbestosis 50 years after brief high intensity exposure in a vermiculite expansion plant. Am J Resp Crit Care Med 165:1–5

Wylie AG, Verkouteren JR (2000) Amphibole asbestos from Libby, Montana, aspects of nomenclature. Am Mineral 85:1540-1542

Yang H, Bocchetta M, Kroczynska B, Elmishard AG, Chen Y, Liu Z, Bubici C, Mossman BT, Pass HL, Testa JR, Franzoso G, Carbone M (2006) TNF-α inhibits asbestos-induced cytotoxicity via a NF-κB-dependent pathway, a possible mechanism for asbestos-induced oncogenesis. Proc Nat Acad Sci 103:10397-10402

Ziegler JL (1993) Endemic Kaposi's sarcoma in Africa and local volcanic soils. Lancet 342:1348-1351

Ziegler TL, Plumlee GS, Lamothe PJ, Meeker GP, Witten ML, Sutley SJ, Hinkley TK, Wilson SA, Hoefen TF, Brownfield IK, Lowers H (2002) Mineralogical, geochemical and toxicological variations of asbestos toxicological standards and amphibole samples from Libby, MT. *In:* Geol Soc Am Abst Progr, 2002 Annual Meeting, *http://gsa.confex.com/gsa/htsearch.cgi*

Reviews in Mineralogy & Geochemistry
Vol. 64, pp. 59-113, 2006
Copyright © Mineralogical Society of America

3

Metal Speciation and Its Role in Bioaccessibility and Bioavailability

Richard J. Reeder, Martin A. A. Schoonen

Department of Geosciences and Center for Environmental Molecular Science
Stony Brook University
Stony Brook, New York, 11794-2100, U.S.A.
e-mail: rjreeder@stonybrook.edu, martin.schoonen@stonybrook.edu

Antonio Lanzirotti

Consortium for Advanced Radiation Sources
University of Chicago
Chicago, Illinois, 60637, U.S.A.
e-mail: lanzirotti@bnl.gov

INTRODUCTION

Metals play important but varied roles in human health. Some metals are required for normal metabolic function, with optimal amounts for maximum benefit. Others are only known to cause toxic effects. Most of our knowledge of the function of metals in human health has been acquired in the last 100 years. However, evidence of adverse health effects attributed to metal exposures dates back to early civilizations. For example, it has been deduced that extensive mining and smelting of lead and its widespread use in the Roman Empire caused significant incidence of lead poisoning (Nriagu 1983; Hong et al. 1994).

The source of metals in the environment ultimately can be traced back to their occurrence primarily in rocks, with their release to soil, water, and air facilitated by weathering processes. Consequently, the natural occurrences of metals in soils and waters are strongly correlated to the varied distribution of rock types and the compositions of the constituent minerals. More than 2000 years ago the Greek physician Hippocrates recognized relationships between disease and location, illustrating that environmental factors influenced human health. Today there are many known geographic patterns of disease that have been correlated with properties of soils or waters, or even aerosol particles. It has been difficult to demonstrate cause-effect relationships for many correlations, and efforts to relate total concentration of a metal contaminant to toxic impact have proven difficult (Davies et al. 2005). This point illustrates the essential concept that the total amount of an element in an environmental setting is not necessarily a good measure of its potential health threat.

Within the last two centuries human activities have been highly effective in redistributing metals on local, regional, and even global scales. This has contributed to a greater exposure to humans. We have also changed the chemical forms of metals, sometimes with unfortunate consequences that include enhanced mobility in the environment as well as creation or enhancement of more toxic forms.

Nearly three-quarters of the elements in the periodic table are classified as metals. Inasmuch as all but a few of them occur in nature, it is probably correct to say that each one has (or will be found to have) an important role with regard to environmental health. In this paper, we focus primarily on a few of the so-called *heavy metals* that are known to be

 DOI: 10.2138/rmg.2006.64.3

associated with adverse health effects. In a broad sense the heavy metals can be considered to include metals with atomic number greater than 20 (Ca), but many readers will know that a consistent definition has not found universal acceptance (Duffus 2002). Heavy metals should not necessarily be equated with toxicity, however, as a number of them are essential for human health in small amounts. Heavy metals have widespread occurrence as trace or minor constituents of soils, rocks, and waters, as well as living organisms (including humans). Commonly included among the heavy metals are arsenic, which is a metalloid, and selenium, a non-metal. Depending on factors such as oxidation state, electron configuration, ionic radius, and the presence of various ligands, metals exhibit a rich variety of coordination compounds in aqueous, solid, and even gaseous forms. The fact that the particular chemical form of a metal strongly influences its chemical behavior, mobility in the environment, uptake by organisms, and toxicity is certainly justification for characterizing and understanding metal speciation, not only in the environment but also in the body.

METAL SPECIATION CONCEPTS

Speciation refers to aspects of the chemical and physical form of an element. Oxidation state, stoichiometry, coordination (including the number and type of ligands), and physical state or association with other phases all contribute to define speciation. These properties govern the chemical behavior of elements, whether in environmental settings or in human organs, and play a crucial role in determining toxicity. The focus on metal speciation in this chapter reflects the varied roles it plays in human health. Metals such as iron and zinc are essential for metabolic function, but can be toxic in excess. Others, like cadmium and lead, have no known beneficial function and pose health risks even at low levels of exposure and uptake. The amount of exposure or uptake is obviously a key factor in assessing adverse health impacts, and defines the field of toxicology. However, the metal speciation is also a critical factor in determining toxicity. For example, inorganic dissolved mercury ($Hg^{2+}_{(aq)}$) and methyl mercury chloride ($CH_3HgCl_{(aq)}$) are both considered to be toxic, but the properties and behavior of the latter make it a significantly greater health threat (NRC 2000). Another example is illustrated by the two common oxidation states of chromium in soils and water. Hexavalent chromium in the form CrO_4^{2-} is soluble in water, making it mobile, and readily taken up by organisms. This form is also a known carcinogen (ATSDR 2000). Trivalent chromium tends to be insoluble, often forming hydroxide solids, and is considered an essential element in small amounts. We will see later that the pathways of uptake involve processes that may alter speciation, thereby changing a toxic form into a benign one, or the reverse.

Although the concept of speciation is now widely appreciated in many fields, there have been few efforts to provide a formal definition. Bernhard et al. (1986) pointed out that usage varies among different fields, ranging from evolutionary changes to distinctions based on chemical state. A Molecular Environmental Science Workshop convened by the U.S. Department of Energy in 1995 cited at least five aspects important for defining speciation: element identity, physical state, oxidation state, chemical formula, and detailed molecular structure (DOE 1995). Both reports emphasized the importance of techniques available for determining these properties and the limitations they place in our ability to identify and distinguish species.

Several Divisions within the International Union of Pure and Applied Chemistry (IUPAC) addressed speciation concepts in detail (Templeton et al. 2000), recommending that its usage in chemistry be restricted to distribution of chemically distinct species: "Chemical compounds that differ in isotopic composition, conformation, oxidation or electronic state, or in the nature of their complexed or covalently bonded substituents, can be regarded as distinct chemical species." Although very similar to current usage in the Earth and environmental sciences, the IUPAC notion of speciation includes the isotopic identity of the element. Isotopic identify (i.e., mass differences within the nucleus) may have only a marginal effect on chemical behavior, and then only when mass differences are large, such as with light isotopes. However, isotopic identity could also be significant in instances where particular isotopes have special characteristics, such as radioactivity.

The IUPAC concept of speciation does not draw attention to the significance of the physical state or association in distinguishing species. Earth and environmental scientists are particularly aware of the importance of physical associations. For example, occurrences of Cs^+ as an aquo ion in a soil water solution, sorbed at the surface of a mineral, or exchanged into an interlayer site in a smectite all represent distinct species, even though they may share the same oxidation state, coordination, ligands, and other local properties.

This last example also highlights the fact that multiple species of a metal commonly co-exist in real systems. Where different species of a given metal share some property, such as oxidation state, it can be difficult to distinguish among them. In fact our ability to characterize metal speciation is dependent on the techniques used and the information they provide. For example, the use of a technique that gives direct information regarding oxidation state may fail to distinguish metal species having the same oxidation state but are complexed by different ligands. Identification of species using a mass spectrometry method may fail to distinguish oxidation state, but also may require sample processing that could alter speciation. Such problems become amplified in complex solutions such as lung or gastric fluids, which contain numerous organic components, including peptides, amino acids, and phospholipids. Hence, species assessment is operationally defined, and no single technique provides information concerning all relevant aspects that define a species. Advantages can often be gained by use of several complementary techniques for species characterization. In view of the fact that many of the metals most relevant for human health occur at very low concentrations, adequate characterization may pose serious challenges. Later is this chapter, we consider some of the more commonly used techniques for assessment of metal speciation.

SIGNIFICANCE OF SPECIATION

The chemical and physical aspects that define speciation of a metal control its reactivity, including its solubility and uptake behavior, and, in many circumstances, toxicity. Solubility and uptake behavior, in turn, influence mobility of the metal in the environment, and therefore constrain pathways of exposure to organisms, including humans. During exposure the metal speciation directly influences absorption across a physiological membrane, which allows entry into systemic circulation. A transformation in speciation may occur in biological fluids (e.g., lung or gut fluids) prior to any absorption, however, which may affect absorption and subsequent toxicity. Within organ systems detoxification processes may further alter speciation and toxicity, and also influence transport, excretion, and storage. This oversimplified description illustrates the importance of metal speciation over the entire spectrum of process impacting the metal's fate from weathering to human impact. Readers are referred to Plumlee and Ziegler (2003) and Plumlee et al. (2006) for more a comprehensive discussion of these aspects.

The dependence of toxicity on speciation is now well known. The behavior of a metal may be completely changed by its oxidation state or its association with specific ligands, as exemplified by the contrasting toxicities of methylmercury and inorganic mercury species. The metalloid tin also shows markedly different health threats depending on its association with specific ligands. Neither metallic nor inorganic forms of tin present a health problem in small amounts; in fact, SnF_2 is a common additive of toothpaste. However, many organotin compounds, which are predominantly created by human industrial processes, are highly toxic (ATSDR 2005c). Tributylin tin, widely used as a biocide and antifouling agent for seagoing vessels since the 1970s (de Mora 1996), is a potent ecotoxicant (Alzieu 1996; Maguire 1996), persists in marine environments (Diez et al. 2002; Sudaryanto et al. 2004, 2005), accumulates in tissue of fish and shellfish (Alzieu 1996; Sudaryanto et al. 2004), and may cause adverse health effects in humans (Kafer et al. 1992; Dopp et al. 2004).

One of the complicating aspects of speciation is that each species exhibits a distinct behavior, making generalizations about stability and reactivity difficult. In this chapter, we do not attempt to provide a comprehensive review of metal speciation. Instead, our approach is to illustrate important aspects of metal speciation on human health using selected examples.

As noted already, the total concentration of a particular element in any system, environmental or human, is not necessarily a good indicator of its potential health impact. Although this concept has been widely embraced by the research community and acknowledged by regulatory agencies, its impact on development of regulatory standards in the US has been limited. Even in the toxicology field, many bioassays do not consider speciation of metals. It is noteworthy, for example, that current methods for analysis of metals in soils (e.g., EPA 3050b) are designed to recover the more soluble metal fraction by use of acid digestion. As speciation techniques become more widely used and as understanding of the differences in behavior among species, including transformations, improves, it is likely that agencies will take greater account of these factors in formulating regulations. This is a critical area of research to which geochemists and mineralogist will be able to make important contributions.

ROLE OF METALS IN HUMAN HEALTH AND METAL TOXICITY CONCEPTS

The human body requires the uptake of several essential metals for its proper function. As briefly summarized in Appendix 1, a number of heavy metals are known to be essential. The roles of some of the essential heavy metals listed in Appendix 1, such as vanadium and tungsten, are not fully known. For others, including arsenic and tin, essential roles have been suggested, but not demonstrated. Some metals play a role in active centers of metalloenzymes. In fact, for metals such as cobalt this may represent the dominant "species" in the human body. Other metals, such as chromium(III) and vanadium, are metabolized within the body to form low molecular weight compounds that play a role in glucose metabolism.

The dose-response curve for an *essential* metal, schematically shown in Figure 1a, has a characteristic optimal range, flanked by suboptimal ranges. For some metals, a deficiency may be expressed as a specific disease. For example, chromium is important in the human metabolic system. A lack of chromium(III) disrupts glucose metabolism and can lead to obesity, diabetes, and cardiovascular disease, as well impairment of the reproductive system (Appendix 1).

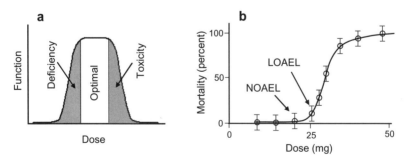

Figure 1. Schematic dose-response curves for (a) an essential metal and (b) a non-essential toxic metal. NOAEL and LOAEL are *no observed adverse effect level* and *lowest observed adverse effect level* (see Table 1).

It is important to note that deficiency in a metal may be caused by several factors. A diet lacking an essential metal is a common cause. For example, in China outbreaks of Keshan disease, a type of heart disease, are restricted to well defined geographic regions (Fordyce 2005). The occurrence of this disease is associated with a lack of dietary selenium. It is within this context that speciation is important. Crops grown on soil must be able to extract the selenium from the soils. Within Zhangjiakou District, Hebei Province, China, Keshan Disease has a high prevalence despite the fact that there is significant total selenium in the soil. In fact, a study in this region showed that the prevalence of the disease is not correlated with a lack of selenium in the soil as might be expected (Johnson et al. 2000). Rather, the cause for the selenium deficiency in the diet is a result of the fact that the soil-bound selenium is not in a form that is available to the plants. Soils in affected areas are rich in organic matter, and it is hypothesized that selenium is strongly bound with the organic fraction (Johnson et al. 2000). Alternatively, the organic matter may promote reduction of selenate to selenite, the latter commonly being more strongly sorbed to iron hydroxides in soils (Hartikainen 2005). This example illustrates the importance of speciation—in this instance in soils—rather than total concentration in understanding the incidence of a disease.

While geological factors have been shown to contribute to diseases, such as Keshan Disease, additional confounding factors can lead to complex patterns in the prevalence of a disease. The interaction between metals (or metalloids) is one of the confounding factors. Nutritional status is known to affect toxicity. For example, anemia—low iron status—promotes the uptake of nickel and manganese. Hence, individuals affected by anemia are at a higher risk for adverse effects of nickel and manganese uptake compared to healthy individuals exposed to the same nickel and manganese levels. Conversely, exposure to methylmercury has been shown to inhibit uptake of selenium, an antioxidant (Norling et al. 2004). A second major confounding factor is genetic predisposition. Genetic disorders that disrupt the uptake or transport of essential metals or formation of antioxidants are more difficult to diagnose and remedy. For example, Hallervorden-Spatz syndrome (HSS) is a neurodegenerative disease caused by a genetic defect that disrupts the function of ferritin, an iron storage protein. HSS leads to accumulation of iron in the brain and is thought to cause oxidative stress (Zhou et al. 2001).

Exposure and uptake of toxic metals, or even essential metals at levels in excess of the optimal range, can lead to adverse health effects. Exposure history is an important factor in evaluating the toxicity of a metal. Toxicologists distinguish between acute and chronic toxicity. Acute toxicity is that associated with short-term exposure to a toxicant, sometimes in lethal doses. Chronic toxicity is that associated with long-term exposure, and is usually the type most relevant to environmental toxicants. Studies based on rats or other laboratory animal models are commonly used to assess adverse health effects and to establish the *no observed adverse effect level* (NOAEL) and the *lowest observed adverse effect level* (LOAEL). A schematic dose-response curve for a toxic metal or substance is shown in Figure 1b. The outcomes of these studies form the basis for regulations. An overview of the methodology behind such studies is given by the U.S. EPA (*http://www.epa.gov*). Table 1 defines common terms used in such studies. Toxicity data for metals and other susbstances are available (on-line) from the Agency for Toxic Substance and Disease Registry (ATSDR) within the U.S. Department of Health and Human Services (*http://www.atsdr.cdc.gov*). For a web-based introduction to the basic concepts in toxicology and its terminology the reader is referred to the National Library of Medicine (National Institutes of Health) web page: *http://sis.nlm.nih.gov/enviro/toxtutor/Tox1/amenu.htm*.

BIOAVAILABILITY AND BIOACCESSIBILITY CONCEPTS

The term *bioavailability* is much used throughout the literature. There is a general understanding that the meaning addresses the potential for a substance to interact with an organism. With the use of this term now widespread among many disciplines, confusion

Table 1. Some abbreviations and terms relating to toxic effects of metals.

Abbreviation	Definition
NOAEL	No-observed-adverse-effect level. The actual doses (levels of exposure) used in studies that showed no observable adverse effects to the organism.
LOAEL	Lowest-observed-adverse-effect level; LOAELS have been classified into "less serious" or "serious" effects. "Serious" effects are those that evoke failure in a biological system and can lead to morbidity or mortality (e.g., acute respiratory distress or death). "Less serious" effects are those that are not expected to cause significant dysfunction or death, or those whose significance to the organism is not entirely clear.
MRL	Minimal Risk Level; an estimate of daily <u>human</u> exposure to a substance that is likely to be without an appreciable risk of adverse effects (noncarcinogenic) over a specified duration of exposure
CEL	Cancer effect level
LD$_{50}$	Lethal dose that leads to 50% mortality

Note: Other web resources that provide definitions of terms used in toxicology are *http://extoxnet.orst.edu/ tibs/standard.htm* and *http://www.atsdr.cdc.gov/glossary.html*.

sometimes develops as specific meanings emerge from a particular context, discipline, or method of study. An example is the meaning of bioavailability shared by the toxicology and pharmacology fields, where it refers specifically to the fraction of an administered dose that is absorbed into the organism's circulatory system or into the organ where an effect occurs (Ruby et al. 1999; NRC 2003). The reason for such a restricted definition is clear upon consideration of the methodologies typically employed to evaluate efficiency of uptake of a drug or a toxicant. For example, a study might involve assays of blood levels of a given toxicant to identify peak plasma concentration and half-life resulting from specific oral dosages. Even this definition of bioavailability could find disfavor, since substances absorbed through the gastrointestinal tract of humans first circulate through the liver, where metabolism may limit the amount released to general systemic circulation.

This view of bioavailability is of little value to the soil geochemist examining the fraction of a metal in a soil that becomes soluble and mobile during a sequential extraction procedure. In fact, there may be no universally acceptable definition of bioavailability, a prospect that may be most evident where operational definitions are desired. The lack of a common view of bioavailability was addressed by a recent report from the National Research Council (2003) in the context of soil and sediment contaminants, and summarized by Ehlers and Luthy (2003). We refer the reader to this publication for a more detailed description of various definitions of bioavailability and the rationale for their use. Rather than propose a working definition, the NRC study recommended the adoption of a process-based view, with *bioavailability processes* being defined as "the individual physical, chemical, and biological interactions that determine the exposure of plants and animals to chemicals associated with soils and sediments," which is shown schematically in Figure 2.

While a process-oriented view will be intuitive for many geochemists, the NRC perspective is purposely limited on the organism side to exclude processes following transport of a substance across the biological membrane. Because of the context relating to soil and sediment contaminants, it places emphasis on those factors and processes that make the substance "available" to the organism, that is, in a form that can be transported across the organism's biological membrane. This is most commonly interpreted to be a soluble form. However, it is also possible that very small solids or colloidal particles could be transported across some

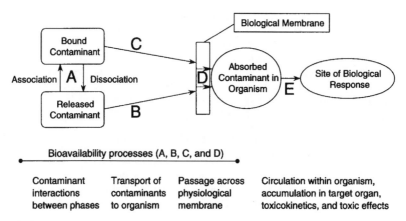

Figure 2. Schematic diagram illustrating the range of bioavailability processes as defined in the NRC study (2003). [Reprinted with permission from *Bioavailability of Contaminants in Soils and Sediments,* Copyright (2003) by The National Academies of Sciences, Courtesy of the National Academies Press, Washington, D.C.]

membranes, such as the linings of lungs. Ruby et al. (1996, 1999) used the term *bioaccessibility* to represent the fraction of a toxicant (or substance) that becomes soluble within the gut or lungs and therefore available for absorption through a membrane. The amount actually absorbed, which according to the view described above reflects bioavailability, may be less than the soluble fraction. This concept of bioaccessibility carries over to the release or solubilization of soil- and sediment-associated metals in environmental systems external to the organism. Selective sequential extraction procedures are familiar examples of protocols for solubilizing certain constituents in geomaterials (e.g., Tessier et al. 1979; Scheckel et al. 2003, 2005). This concept of bioaccessibility also applies to *in vitro* studies that assess the release (solubility) of solid-bound metals in simulated biological fluids, as models of human lung, gastric, or intestinal fluid. Such physiologically based extraction tests (PBET) are being evaluated as *in vitro* alternatives to more costly *in vivo* studies using animal models (Ruby et al. 1996, 1999; EPA 2005). Plumlee and Ziegler (2003) describe bioaccessibility tests and comparisons with bioavailability for a variety of earth materials. Similar *in vitro* approaches have been widely used in pharmacological studies to assess drug release (e.g., Lin and Lu 1997).

Agreement on the usage of bioavailability, bioaccessibility, and related terms seems unlikely in the near future, including within the environmental geochemistry and mineralogy fields. In response to the NRC report, Semple et al. (2004) suggested temporal and/or spatial distinctions between bioavailable and bioaccessible compounds. They proposed that the term bioavailable compound be used where the substance "is freely available to cross an organism's cellular membrane from the medium the organism inhabits at a given time." They proposed using bioaccessible to indicate a compound that "is available to cross an organism's cellular membrane from the environment, if the organism has access to the chemical," indicating that such a compound could subsequently become bioavailable once the proximity of the compound to the organism allows. Other authors have proposed the term *geoavailability* to represent the fraction of a toxicant or contaminant in a geologic material that becomes soluble or mobile as a result of various biogeochemical processes (Smith and Huyck 1999). This focus on terminology may appear to be only marginally relevant to the main topic. However, as Ehlers and Luthy (2003) note, bioavailability concepts may very well enter into risk assessment in the future and play an important regulatory role. Hence, we can expect further discussion of these concepts and related terminology, and it is incumbent on geoscientists to participate in this dialog.

In this chapter we refrain from recommending a preferred usage, and merely alert the reader to the potential for confusion. However, having noted that *bioaccessibility* can be generalized to include solubilization of metals in geomaterials, we tend to follow that usage here and retain the distinction with *bioavailability,* which we will use to refer to absorption across a physiological membrane. As emphasized in the following section, we believe there is merit in taking a broad view of the role of metal speciation, encompassing processes from weathering and transport in geomedia, through exposure and uptake by humans, to fate within organs.

PATHWAYS OF METAL UPTAKE— FROM SOIL, WATER, AND AIR TO HUMAN ORGANS

Historically, the gulf between the geosciences and the health sciences has tended to constrain the research activities in each community to address different parts of the broader subject of environmental health. Geoscientists have typically limited their attention to environmental processes that govern the bioaccessibility of metals up to and sometimes including exposure. Health scientists have naturally focused on the health impact, generally beginning with exposure. Clearly both communities recognize the continuum of processes that link the geologic and health aspects. In this chapter, the primary goal is to illustrate the role of metal speciation spanning both environmental and physiological processes.

General exposure pathways are described by Plumlee et al. (2006) elsewhere in this volume (also see Plumlee and Ziegler 2003). Exposure pathways for metals are the same as for other toxicants (*aka* xenobiotics): ingestion (gastrointestinal tract), inhalation (respiratory tract), and dermal contact. A particular metal may be present in a solid, a liquid, or a gas phase. The physical form commonly dictates the nature of the exposure. For example, arsenic that desorbs from iron oxide-coated sands in an aquifer will enter groundwater that may be a source of drinking water. Here, ingestion of dissolved arsenic is the main exposure pathway. Ingestion is also possible for metals in solid forms, including food and soil particles. Certain associations of a metal may also dictate exposure. For example, inhalation would be the primary exposure pathway for arsenic associated with airborne particles (e.g., in mineral dust). If a fraction of inhaled particles is transported by mucociliary action to the pharynx, then ingestion may become a secondary exposure pathway.

Environmental processes

Geochemists generally consider that metals are released to the human environment initially by weathering of rocks and are subsequently modified by various processes operating at or near Earth's surface, both natural and anthropogenic; these processes can enhance the environmental mobility of metals or lead to their sequestration. Human activities have been especially effective in redistributing metals and modifying their form, including speciation.

There are numerous physical, geochemical, and biologic processes that influence the behavior of metals in surficial settings. Much of the attention has focused on processes that mobilize or immobilize metals, since mobility generally facilitates exposure. A dissolved metal can be transported by fluid flow, eventually entering a water supply or being taken up within a food chain. In contrast, a metal that precipitates as a coating on mineral grains in a soil or aquifer is immobilized, and may be effectively eliminated from exposure unless remobilized by a subsequent process. In many circumstances, bioaccessibility in geomaterials is equated to the occurrence of a metal in a dissolved form. However, there are examples in which solid forms may have significant potential for exposure. Metals that are associated with small particles that are mobile may have greater potential for exposure. For example, colloids and airborne particles are mobile and both have been shown to have associated metals. Metals contained in the bottom sediments in streams, lakes, and marine settings may also enter the food chain if they are scavenged by bottom feeders or may even become remobilized upon interacting with gut juices.

The most important processes that control the mobility of metals in the environment include dissolution/precipitation, complexation with ligands, sorption/desorption by solids, biotransformation, uptake by soil and aquatic biota, and reduction/oxidation (redox) (Fig. 3). We provide only a brief overview of these processes here, as there are a number of existing sources that offer detailed reviews relating to metals (e.g., O'Day 1999; Traina and Laperche 1999; Brown and Parks 2001; Warren and Haack 2001; Sparks 2003).

Dissolution and precipitation involving metal species are subject to both thermodynamic and kinetic controls. Using the chemical analysis of a water as input, aqueous speciation programs, such as MINTEQA2 (U.S. EPA), PHREEQC (Parkhurst and Appelo 1999), and The Geochemists Workbench (*http://www.rockware.com*) (Bethke 2002), facilitate calculation of saturation states of the water with respect to a wide range of minerals and solids. However, there are many important environmental phases for which thermodynamic stability data are lacking, including many amorphous phases. Moreover, some of the phases controlling metal solubility in nature are solid solutions, and satisfactory solution models are available for relatively few environmentally relevant phases. Owing to a variety of possible kinetic factors, it is important to recognize that supersaturation does not guarantee that precipitation will occur. For example, inhibitors, often sorbed metals, may introduce kinetic constraints that limit precipitation (and dissolution). Plumlee et al. (2006) and Plumlee and Ziegler (2003) describe kinetic factors that influence the dissolution of asbestos and other mineral phases in human fluids.

Metal solubility is also strongly coupled to complexation in the aqueous phase. Metals that have a high affinity for either organic or inorganic ligands may exhibit significantly increased solubility through formation of complexes. The high affinity of UO_2^{2+} for dissolved CO_3^{2-} is an environmentally important example. In the absence of dissolved CO_3^{2-}, the total solubility of UO_2^{2+} controlled by equilibrium with the mineral schoepite ($UO_3 \cdot 2H_2O$) at neutral pH is 3.4 µM. With 1 mM total dissolved carbon dioxide in the solution the total UO_2^{2+} solubility is tenfold higher (56 µM) because of carbonate complexation.

Aqueous complexation may also influence the uptake of metals by aquatic and soil organisms. For example, dissolved Cu^{2+} and Ni^{2+} are readily taken up by some aquatic phytoplankton, whereas uptake is extremely limited when these same metals are strongly complexed with organic ligands or natural organic matter. Although many studies of aquatic organisms have shown correlations between metal uptake and the fraction of the metal occurring as a free ion

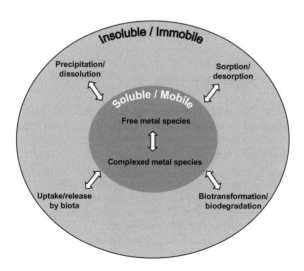

Figure 3. Dominant processes that control the mobility of metals in the environment. Oxidation/reduction may be associated with any of these processes, depending on the metal.

(i.e., the free-ion activity model)(Morel 1983; Campbell 1995), relatively little is known of the actual chemical form of the metal taken up (cf. Hudson 1998; Sunda and Huntsman 1998).

In systems that contain solid phases, sorption processes, involving transfer of dissolved metals to solid surfaces, represent some of the most important controls on dissolved metal concentrations (e.g., Brown et al. 1999; Sparks 2003). Sorption can be considered to include adsorption (i.e., accumulation of ions at the surface), surface precipitation (formation of a distinct phase at the surface), and co-precipitation (incorporation of ions into a phase, commonly as it precipitates). Sorption is generally most effective at removing a dissolved metal from solution when the metal is present at relatively low concentration and the available solid surface area is high. However, sorption also depends on several solution properties, including pH, ionic strength, and the presence of complexing ligands or competing species. Properties of the solid-liquid interface, including surface charge (Hochella and White 1990) and surface site coordination (Reeder 1996; Elzinga and Reeder 2002) are also important factors.

Generally, metal cations exhibit increasing adsorption as pH increases, as a response to decreasing proton charge on the surface. The change in adsorption efficiency typically occurs over a narrow pH range and is commonly referred to as the *adsorption edge*. The position of this edge with respect to pH may vary for different metals on the same sorbent, which generally reflects differing affinities of the metals for surface sites or different sorption mechanisms. For a given metal cation sorbing on different solid phases the pH range of the adsorption edge typically differs because of different surface charge properties and different surface sites. Adsorption of anions is usually greatest at low pH and decreases with increasing pH, also reflecting electrostatic properties of the surface. Anion species of acids may exhibit sorption maxima in their pH dependence, usually coinciding with their pK_a values and demonstrating differences in sorption behavior among the more and less protonated species. Metal cation sorption may also be influenced by complexation, which may vary with pH and ligand concentration. Uranium again provides a good example. U(VI), occurring as the UO_2^{2+} aqueous species, shows a rapid increase in adsorption on ferrihydrite with increasing pH (in the pH range 3.5-5.5), which is typical behavior for a cation (Fig. 4). However, at higher pH (7.5-9.5), U(VI) adsorption decreases abruptly as a result of the formation of uranyl carbonate anion complexes having low affinity for the surface (Waite et al. 1994). Increasing the dissolved carbonate concentration causes the edge at higher pH to shift to lower pH values due to increased carbonate complexation, thereby resulting in a narrower pH range of maximum adsorption.

Spectroscopic investigations have demonstrated that different types of surface complexes may form, including inner-sphere and outer-sphere types (Fig. 5). Although multiple factors may be important, inner-sphere surface complexes are generally more strongly bound to surfaces than outer-sphere complexes, and therefore may be less susceptible to desorption, or release to solution. For more information, readers are referred to the excellent reviews of sorption by Anderson and Rubin (1981), Davis and Kent (1990), Stumm (1992), Sparks (2003), and Sposito (1984, 2004).

Biotransformations commonly involve reduction/oxidation (redox) processes. Because large solubility differences sometimes exist between different oxidation states of a metal, bacterially mediated reduction or oxidation may be highly effective in controlling metal concentrations in environmental solutions. Uranium again provides a useful illustration. Over the environmentally relevant pH range, U(IV) is relatively insoluble, often forming oxide phases such as uraninite, UO_2. Uranium(VI) is relatively soluble and is generally considered the mobile form of uranium. Bacteria present in soil systems have been shown to reduce dissolved U(VI) to U(IV), resulting in the formation of uraninite (e.g., Lovley et al. 1991; Fredrickson et al. 2000). This process may effectively immobilize uranium in the subsurface. However, in oxidizing conditions, uraninite may become re-oxidized, resulting in its remobilization as U(VI) (e.g., Senko et al. 2002).

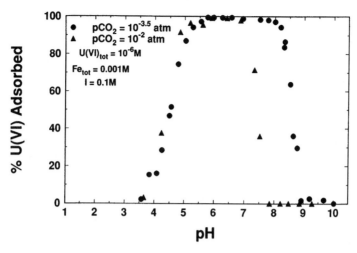

Figure 4. Uranium(VI) adsorption behavior on ferrihydrite as a function of pH and at different total CO_2 concentrations. The decrease in adsorption above pH 7.5-8 is attributed to formation of aqueous uranyl carbonate complexes that have low affinity for sorption. [Reprinted with permission from Waite et al., *Geochimica et Cosmochimica Acta*, Vol. 58, Fig. 6, p. 5470. Copyright (1994) Elsevier.]

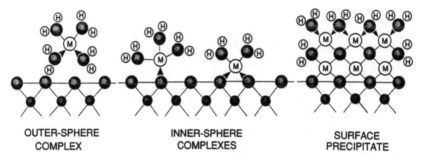

OUTER-SPHERE COMPLEX INNER-SPHERE COMPLEXES SURFACE PRECIPITATE

Figure 5. Diagrams showing outer- and inner-sphere adsorption complexes and a surface precipitate. [Reproduced from Brown (1990).]

Redox processes may also occur without biological mediation, however, kinetics are commonly sluggish. In surface environments important electron donors/acceptors include organic matter and compounds containing iron, manganese, and sulfur. Many redox processes occur at surfaces of solids and are associated with sorption processes. Adsorption of metals onto surfaces can change their electronic structure and promote redox reactions that are inhibited when the metal is dissolved. For example, it has been demonstrated that reduction of Cr(VI) co-adsorbed with Fe(II) onto an Fe(III)-oxide mineral substrate is significantly faster compared to the homogenous reaction between these two metal species (Buerge and Hug 1999). The same phenomenon has been observed for the reaction between Fe(II) and uranyl (Liger et al. 1999). While the effect of sorption on the kinetics of electron transfer reactions of transition metals in aquatic systems has received some attention, this phenomenon is likely to also play a role within the human body where transition metals may be ingested or inhaled as adsorbed species on mineral dust (Schoonen et al. 2006). Further information on electron transfer reactions in the environment is provided in reviews by Bartlett and James (1993), Sparks (2003), and Schoonen and Strongin (2005).

Because the processes described above are the dominant controls on metal mobility in the near-surface environment, they are also among the most important factors governing the environmental exposure of humans to metals. Next we consider the processes associated with metals following exposure.

Internal processes

After ingestion, inhalation, or dermal contact, the fate and impact of a metal are affected by absorption, distribution, metabolism (or biotransformation), and elimination. Collectively, these processes, commonly referred to as ADME, define the field of *toxicokinetics* (Guidotti 2005). The behavior of metals throughout these internal processes is strongly dependent on speciation. Moreover, metal speciation typically changes during these processes. The health effect of a metal toxicant, including the mechanism of toxicity, is also dependent on speciation and constitutes the field of *toxicodynamics*. We do not discuss toxicodynamics, except to note a few well known examples. Moreover, the level of material that we present here is very basic, and interested readers are encouraged to consult more comprehensive reviews. A useful starting point for geoscientists is the Environmental Health and Toxicology web page of the National Library of Medicine (National Institutes of Health) (*http://sis.nlm.nih.gov/enviro.html*).

Absorption involves transport of the metal across a physiological membrane (e.g., commonly a phospholipid bi-layer). This may occur via different mechanisms (e.g., passive, facilitated, or active transport) depending on the substance and its chemical form as well as the cell type (Dawson and Ballatori 1995; Foulkes 2000). Absorption of Cr(VI), for example, occurs via facilitated transport, following the sulfate and phosphate pathway, whereas Cr(III) transport is primarily by passive diffusion and much less efficient. Methylmercury has a significantly greater absorption efficiency than inorganic ionic mercury, which is mainly attributable to the higher lipid solubility of the methylated form. Some of the properties of metals and other substances that influence absorption are listed in Table 2. Metal speciation may be altered before absorption occurs, for example in the presence of gastric or lung fluid, or in mucus. Foulkes (2000) has emphasized that, because of their high affinity for complexing with proteins and other biological molecules in internal fluids, most non-essential heavy metals are

Table 2. Properties of a metal or substance that influence absorption at physiological membranes.

Property	Effect
Concentration/dose	Fractional absorption may vary with concentration, including inversely.
Molecule size and charge	Dependent on transport mechanism (passive, facilitated, or active). For passive diffusion, neutral charge and small size favor absorption.
Competing antagonists	Competition depending on transport mechanism.
pK_a of acids	Nonionized form of some acids more readily absorbed.
Lipid solubility	Lipid-soluble species more readily absorbed.
Particle size/surface area (solid)	Smaller particles solubilized more rapidly.
Phase identity (solid)	Relates to solubility and association of metal.
Solubility (solid)	More soluble solids generally allow greater absorption.
Sorption state (solid)	Metal sorbed on solid surface more readily released.
Matrix components (solid)	Other components in solid may enhance/reduce absorption.

transported across membranes as complexes, rather than as free metals. Transformations may also include changes in metal oxidation state. An example is the partial reduction of Cr(VI) to Cr(III), which begins in the saliva and gastric fluid and influences absorption. Absorption may be influenced by a number of factors, including the fed state. For example, in rats orally ingested Cr(VI) is absorbed more readily in the fasted than in the fed state (O'Flaherty 1996).

After absorption, a metal undergoes distribution and metabolism (also known as biotransformation). Some metals are distributed to all or most tissues and organs (e.g., arsenic), whereas others concentrate in specific organs, which may not be the target organs. Lead, for example concentrates in bone, yet its primary toxicity is in brain function. Both transport and storage depend on chemical form. Methylmercury, for example, readily crosses the blood-brain barrier, whereas inorganic mercury (Hg^{2+}) does not. Metal transport in the blood commonly involves an equilibrium partitioning between plasma and proteins, with some metal species preferentially entering red blood cells. More than 90% of the methylmercury in the blood enters red blood cells and binds with hemoglobin (NRC 2000). Metals absorbed from the gastrointestinal tract do not pass directly into general systemic circulation, but first enter portal circulation where some toxicants are metabolized by the liver.

Metabolism encompasses all of the biotransformations that modify the form (i.e., speciation) of a toxicant. These are typically enzymatic processes occurring with multiple steps and intermediate metabolites. Metabolites may be more or less biochemically active than the original substance. Generally metabolism serves to detoxify a toxicant, often by creating a chemical form that is more readily eliminated. For example, the methylation of arsenic in the liver facilitates its elimination and is considered a detoxification mechanism.

Elimination of metals occurs primarily through urine, feces, and by exhalation. Water soluble forms are excreted readily in urine. The liver secretes some metals (e.g., lead and mercury) into bile, which are then eliminated with feces or remain in enterohepatic circulation (Guidotti 2005). Some metals are stored in various tissues, and may provide evidence of exposure over time. For example, analysis of hair serves as a method for assessing chronic exposure to mercury. The primary storage site for lead is in bone and its analysis serves as a cumulative biomarker for exposure.

These ADME processes are usually unique to the particular species of metal, and clearly involve complex processes. Quantitative modeling of these processes is a major focus in the toxicokinetics field. Such *physiologically based pharmacokinetic* (PBPK) models have been developed for a number of metal toxicants, and are being improved as new data permit (O'Flaherty 1998). Inasmuch as quantitative modeling of environmental processes for metals is also well advanced, it may not be long before comprehensive models are developed that address both environmental and internal fate of metals.

The range of environmental and internal processes that we have described is shown conceptually in Figure 6. In the following section we illustrate the ways in which speciation influences bioaccessibility and bioavailability through four examples of metals/metalloids: arsenic, chromium, lead, and mercury.

ROLE OF METAL SPECIATION: ARSENIC

Arsenic has become the focus of worldwide concern following the recognition of its widespread and devastating health impact in many developing regions, most notably Bangladesh and West Bengal. The occurrence of arsenic in soils and aquatic systems, including the sediments and groundwater in the Bengal Basin region, and the geochemical processes that influence arsenic mobility have been reviewed recently by Plant et al. (2005) and Smedley and Kinniburgh (2002, 2005). Recent reviews of arsenic toxicity and human health effects

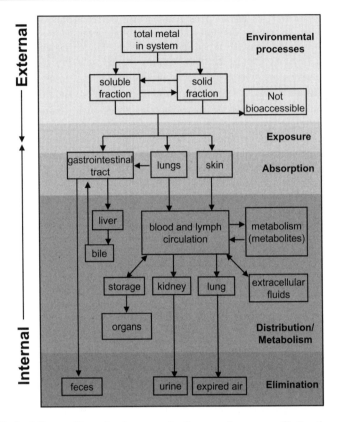

Figure 6. Idealized diagram illustrating the range of pathways and processes affecting the environmental and physiological behavior of metals. [Adapted from the National Library of Medicine (*http://www.sis.nlm.nih.gov/enviro/toxtutor/Tox2/a11.htm*).]

are given by NRC (1999, 2001), IPCS (2001), and ATSDR (2005a). Here, we provide a brief overview of the role of arsenic speciation as it relates to mobility in the environment, human exposure, bioavailability, and health effects.

Arsenic in the environment

Arsenic is not an abundant element in Earth's crust, with an average concentration in crustal rocks of ~2 ppm (Wedepohl 1995). However, it has a strong association with sulfide-bearing mineral deposits and especially with pyrite (FeS_2), which is widespread. Arsenic also exhibits an association with iron oxide and hydroxide minerals, which are abundant as weathering products. Common sources of arsenic in the environment are listed in Table 3. Arsenic occurs in the 3−, 1−, 0, 3+ and 5+ oxidation states, but in nature 1−, 3+, and 5+ oxidation states dominate. Arsenide minerals and sulfides in which As(I−) is a common substituent typically exhibit low solubilities in reducing natural waters. However, oxidative weathering results in the formation of arsenite [As(III)] and arsenate [As(V)] species; these occur as oxyanions or neutral species and may be quite soluble depending on pH and other solution properties. In solution, redox potential (Eh) and pH are the dominant controls on As speciation. An Eh-pH stability diagram is shown in Figure 7. Coexistence of As(III) and As(V) species in natural waters has been interpreted to indicate that electron transfer occurs, if not redox equilibrium. However, rates of electron transfer may be slow and strongly influenced

Table 3. Common sources of arsenic in the environment.

- Volcanic emissions and hot springs
- Dissolved in groundwater from interaction with rock (mobilization from igneous and sedimentary sources, oxidative dissolution of arsenic-bearing sulfide minerals)
- Mining waste, pH-mediated mine effluents, and tailings ponds
- Arsenic-containing pesticides (sodium arsenite or lead arsenate)
- Organic arsenic compounds as herbicides: monosodium methanoarsonate (MSMA), disodium methanoarsonate (DSMA), arsenic acid, and dimethylarsenic acid
- Waste from industrial metal smelting processes
- Leaching of wood preservatives: chromated copper arsenate (CCA) and ammoniacal copper arsenate
- Combustion of fossil fuels in electrical power plants
- Arsenic dusts and gases released during cement manufacture
- Animal waste management from feed additives in poultry (roxarsone to control coccidiosis and promote growth)
- Arsenic trioxide waste from glass manufacturing process (pre-1970)

by bacterial activity (e.g., Cullen and Reimer 1989; Smedley and Kinniburgh 2005). The acid dissociation constants (K_a), expressed here as pK_a values, for H_3AsO_4 are 2.22, 6.98, and 11.53. Below pH ~7 the dominant As(V) species is $H_2AsO_4^-$ and above pH ~7 $HAsO_4^{2-}$ is dominant, except at very low or high pH (Fig. 8). The pK_a values for H_3AsO_3 are 9.23, 12.13, and 13.4, so that $H_3AsO_3^0$ is the dominant dissolved inorganic As(III) species in most natural waters.

In view of its affinity for sulfur, As(III) may form thio-complexes (cf. Wilkin et al. 2003). Methylation may also occur in the environment as a result of microbial processes, resulting in formation of several As(III) and As(V) methyl species, including monomethylarsonous acid [MMA(III)], dimethylarsinous acid [DMA(III)], monomethylarsinous acid [MMA(V)], and dimethylarsinic acid [DMA(V)] (Fig. 9). Anderson and Bruland (1991) reported the formation of dimethylarsenate [DMA(V)], $(CH_3)_2AsO_2^-$, on a seasonal basis to become the dominant dissolved As species in a freshwater reservoir, followed by breakdown to arsenate. Bright et al. (1996) also reported methylated arsenic species in lake sediment pore waters, with formation presumed to be associated with bacterial activity. Other organic species include arsenobetaine and arsenocholine, which are forms commonly present in food. In the majority of natural waters, however, inorganic arsenic forms are most abundant (Smedley and Kinniburgh 2002).

Human exposure to arsenic is primarily through ingestion of water and food, although airborne particulates containing arsenic may be locally important. For this example, we focus on exposure through drinking water and to a lesser extent on ingestion of arsenic-containing solids. Consequently, solubility of arsenic is the main geochemical factor influencing exposure in this circumstance. Additionally, arsenic associated with particles that are mobile could also contribute to exposure. As noted earlier, numerous geochemical processes control metal solubility and mobility, including mineral precipitation/dissolution, sorption, and biotransformation. Metal speciation is a critical aspect of all these processes.

Arsenic sorption. Because dissolved arsenic concentrations are typically very low, even in contaminated systems, sorption processes can be highly effective in limiting As concentration and thereby controlling mobility and bioaccessibility. A number of studies have demonstrated that As(III) and As(V) species may sorb strongly with oxide and hydroxide minerals, especially iron, aluminum, and manganese oxyhydroxides. Sorption is strongly dependent on pH and

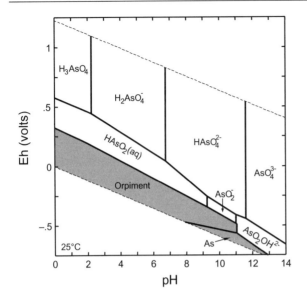

Figure 7. Eh-pH predominance diagram for arsenic in the presence of sulfur at 25 °C ($As_{tot} = 10^{-6}$ M; $S_{tot} = 10^{-4}$ M).

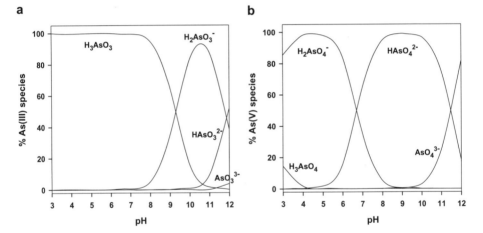

Figure 8. Aqueous speciation diagrams for (a) As(III) and (b) As(V) systems at 25 °C ($As_{tot} = 10$ ppb, and 1 mM NaCl).

other solution properties. Arsenite and arsenate exhibit very different sorption behaviors as a function of pH, and in some systems sorption is sensitive to ionic strength. As an example, Arai et al. (2001) compared As(III) and As(V) sorption on γ-alumina over the pH range 3-10 (Fig. 10). Arsenate sorption increases with decreasing pH, over the range from pH 9.5 (near the point of zero charge, PZC) to 4.5. This behavior is typical for sorption of anion species (e.g., Hingston 1981). Arsenate sorption is not observed to be sensitive to ionic strength; this has been interpreted to be suggestive of formation of inner-sphere surface complexes, which was confirmed using EXAFS spectroscopy (Arai et al. 2001).

Arsenite exhibits less overall sorption than arsenate on γ-alumina, with a broad maximum at pH ~8.5 (see also Goldberg 2002). A decrease in arsenite sorption is observed above pH 9 and likely reflects the change in dominant aqueous speciation to an anion, $H_2AsO_3^{1-}$. Over

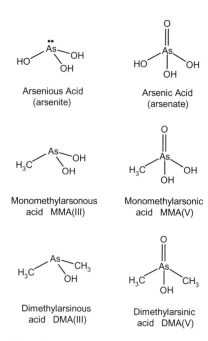

the pH range 5-9, arsenite sorption is decreased at higher ionic strength, which is suggestive of outer-sphere surface complexation. *In situ* EXAFS spectroscopy showed that inner-sphere As(III) complexes dominate at pH < 5.5 and a mixture of inner- and outer-sphere complexes exists at pH > 5.5 (Arai et al. 2001). Several studies have shown that arsenate and phosphate compete for sorption sites, owing to similar chemical behaviors (e.g., Jain and Loeppert 2000). Consequently arsenate sorption may be significantly depressed where phosphate is elevated.

There are numerous studies of As(III) and As(V) sorption on other mineral and oxide surfaces that readers should consult for further insight (e.g., Fuller et al. 1993; Waychunas et al. 1993; Goldberg 2002; Smedley and Kinniburgh 2002; Stollenwerk 2003).

Arsenic redox behavior. Much of the arsenic present in surface environments has been derived from the oxidative weathering of reduced species in sulfide mineral deposits and primary rocks (Smedley and Kinniburgh 2002). Because of the relatively low solubility of reduced arsenic, much of the interest in redox behavior, from the perspective of health impact,

Figure 9. Common inorganic and methylated As(III) and As(V) species. [Adapted from O'Day (2006).]

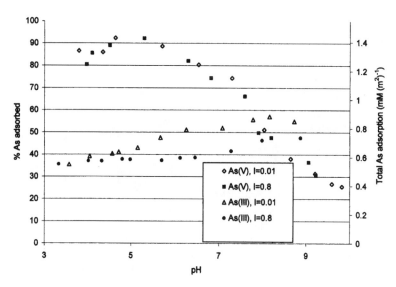

Figure 10. The pH and ionic strength dependence of As(III) and As(V) sorption on γ-alumina. [Reprinted with permission from Arai et al., *Journal of Colloid and Interface Science*, Vol. 253, Fig. 1, p. 83. Copyright (2001) Elsevier.]

has focused on transformations between As(III) and As(V) species. Both oxidation states are common in soils and near-surface waters, and it is not uncommon to have both As(III) and As(V) species co-existing in solutions and in solids (Hering and Kneebone 2002). The Eh-pH diagram shows that the predominance fields for species of both oxidation states coincide with Eh-pH conditions typical of many environmentally important settings (Fig. 7), so that redox transformations can be expected. However, in most cases As(III)/As(V) redox behavior is controlled by interaction with minerals, microbes, or organic matter that serve as electron donors or acceptors. As is often the case in systems containing multiple redox couples, the As(III)/As(V) couple is commonly not in equilibrium with other redox couples (Spliethoff et al. 1995; Hering and Kneebone 2002). This reflects widely differing kinetics of electron transfer and strong dependence on mechanism. Hence As(III) species may persist in oxidizing systems and As(V) may persist in reducing conditions (cf. Inskeep et al. 2002).

Important oxidants for As(III) include Mn(IV) oxides (e.g., birnessite), titanium oxides, H_2O_2, and possibly dissolved ferric species (Foster et al. 1998; Manning et al. 2002; Voegelin and Hug 2003). Photochemical oxidation may also be important (Inskeep et al. 2002). Dissolved oxygen has been shown not to be an effective oxidant, except at high pH (e.g., Manning and Goldberg 1997). Dissolved sulfide has been shown to reduce As(V), with formation of intermediate sulfide complexes (Rochette et al. 2000). Many of the important redox processes occur at mineral surfaces and are associated with sorption. As noted above, arsenate is often associated with iron and aluminum hydroxides, and is typically more strongly sorbed by these phases than arsenite at circum-neutral and acidic pH conditions. An important mechanism of release of sorbed As(V) involves its reduction to As(III). This may occur either by arsenate reduction at the surface with subsequent As(III) release to solution, or by reductive dissolution of a ferric hydroxide sorbent followed by reduction of arsenate to arsenite as shown in Figure 11 (Inskeep et al. 2002). Reductive dissolution of iron (hydr)oxides (with sorbed arsenate) is thought to be a factor causing the elevated dissolved arsenic concentrations in groundwater in Bangladesh (e.g., Smedley and Kinniburgh 2002).

Microbial activity has been shown to cause both reduction and oxidation of arsenic species. We do not describe these reactions here, but refer readers to the overview of Inskeep et al. (2002).

Arsenic precipitation/dissolution. Precipitation and dissolution may also provide important constraints on As solubility. In many aquatic systems, however, arsenic concentrations

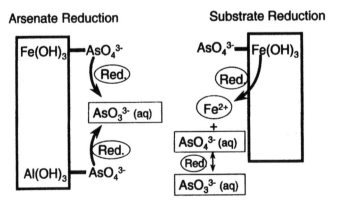

Figure 11. Schematic diagram showing reductive release mechanisms for As(III). [From Inskeep et al. (2002), *Environmental Chemistry of Arsenic.* p. 191, Fig. 4. Reproduced by permission of Routledge/ Taylor & Francis Group LLC. Copyright (2002).]

remain very low so that supersaturation with respect to arsenic oxides/hydroxides or to arsenite or arsenate salts is uncommon. However, high concentrations of cations may allow supersaturation and precipitation. Some remediation strategies rely on precipitation to remove arsenic from solution, such as wastewater. Additives such as lime, fly ash, Portland cement, and ferrihydrite have been used to induced formation of arsenic phases (e.g., Moon et al. 2004). Calcium arsenite and arsenate phases have received some attention as possible precipitates for immobilizing arsenic (e.g., Bothe and Brown 1999). However, the effectiveness of precipitation as a remediation strategy depends on the solubility of the solid. Overviews of important arsenic-containing phases and minerals are given by Cullen and Reimer (1989), Nordstrom and Archer (2003), and O'Day (2006). Foster (2003) describes arsenic speciation in a number of solid phases, including that for arsenic bound with oxides and hydroxides. Arsenic mineral solubilities vary widely according to arsenic oxidation state, composition, crystal structure. The total dissolved arsenic concentration and the concentrations of individual arsenic species in equilibrium with different arsenic-containing phases depend on the overall solution composition, including pH, ionic strength, T, and P. Aqueous speciation programs facilitate such calculations and provide an equilibrium distribution of aqueous species. Nordstrom and Archer (2003) have critically evaluated the thermochemical data available for selected arsenic phases and aqueous species. In general, As(III) and As(V) oxides are moderately soluble in most aqueous solutions. As_2O_3 occurs as arsenolite and claudetite, whereas As_2O_5 is not known as a mineral. Arsenic sulfides, such as realgar and orpiment, tend to be relatively insoluble in most reduced solutions, as are metal arsenides and arsenic sulfosalts. Metal arsenates and arsenites exhibit a range of solubilities depending on the metal involved and other factors (cf. Sadiq 1997). As described below, solubility is important in controlling the bioavailability of arsenic-containing solids following ingestion or inhalation.

Arsenic coprecipitation. Arsenic may also coprecipitate with mineral phases that are forming, thereby providing an effective mechanism for removing arsenic from solution. For example, coprecipitation with ferrihydrite has been shown to significantly reduce dissolved arsenate (e.g., Fuller et al. 1993; Richmond et al. 2004). The similarity of AsO_4^{3-} to other tetrahedral oxoanions, such as PO_4^{3-} and SO_4^{2-}, should allow As(V) substitution (e.g., Myneni et al. 1997; Foster 2003). In addition As commonly substitutes for S in pyrite and other sulfides.

Arsenic in the body

Arsenic absorption. Once ingested, dissolved As(III) and As(V) are both readily absorbed in the human gastrointestinal tract. Studies that systematically compare the relative absorption efficiencies of dissolved As(III) and As(V) in humans from oral exposure are lacking. However, an NRC review cited studies of humans and animals indicating 80-90% absorption for dissolved As(III) and As(V) doses (NRC 1999). MMA(V) and DMA(V) are also readily absorbed in the gastrointestinal tract of humans (ATSDR 2005a), as is arsenobetaine from ingestion of fish (IPCS 2001).

Arsenic present in or associated with solid forms generally shows lower absorption efficiency than dissolved arsenite or arsenate, especially for phases exhibiting low solubility. There are few systematic studies of gastrointestinal absorption (oral bioavailability) in humans for different As compounds. Studies in animal models have been summarized by ATSDR (2005a) and IPCS (2001). Although significant differences exist between different animals, solubility of the arsenic source is the most significant factor influencing absorption. For example, gastrointestinal absorption in rodents is low for relatively insoluble GaAs (<22%) compared to more soluble sodium arsenite and sodium arsenate (80-90%) (Yamauchi et al. 1986; IPCS 2001).

Gastrointestinal absorption from arsenic-containing soils and sediments has also been studied using animal models. The summary given by IPCS (2001) shows great variation depending on the animal model and the characteristics of the soil (not surprisingly). A series of recent studies sponsored by the U.S. EPA (Region 8) have examined the relative bioavailability of arsenic in various soils, sediments, and mine waste material using a juvenile swine model (*http://www.epa.gov/region8/r8risk/hh_rba.html*). These *in vivo* studies compared absorption of arsenic in different solid forms relative to that of sodium arsenate, which is expected to be completely soluble in gastric fluid. A summary of this work (EPA 2005) reports that relative bioavailabilities range from 10 to 60% for the substances examined. Lowest bioavailabilities were observed when arsenic was present as As_2O_3 or in reduced forms, such as arsenides or As-containing sulfides. Higher bioavailability was observed for As-sorbed onto metal oxyhydroxides, such as FeOOH. Owing to the complex and heterogeneous nature of the test materials used in these studies the speciation of the arsenic was not fully characterized. Nevertheless, the wide range of relative bioavailabilities observed underscores the importance of speciation. Furthermore, the task of characterizing the speciation in such materials represents a challenge for geochemists and mineralogists.

Arsenic metabolism. The most important metabolic pathway of inorganic arsenic in humans involves methylation. The liver is the most important site of methylation. The likely *in vivo* mechanisms of methylation involve reduction of As(V) to As(III) and oxidative methyl transfer to produce monomethylarsonic acid [MMA(V)] and dimethylarsinic acid [DMA(V)] as illustrated in Figure 12. Methylation is enzymatically controlled, and S-adenosylmethionine is the primary methyl donor (Zakharyan et al. 1995). Glutathione is a likely electron donor for the reduction steps (Scott et al. 1993). The process does not go to completion as indicated by the presence of As(III), MMA(V), and DMA(V) in human urine (e.g., Le 2002), which serve as important biomarkers for arsenic exposure. Methylation has been found to be variable among individuals, by gender, and according to diet (Vahter 2000). Methylation is highly variable among other mammals (Vahter 2002).

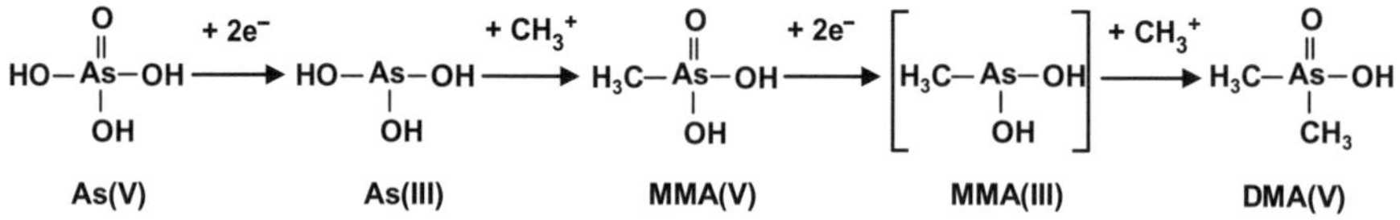

Figure 12. Summary of the human arsenic methylation process involving reduction and oxidative addition of methyl groups. [Reproduced with permission from *Environmental Health Perspectives*, Le XC et al. (2000).]

Methylation has long been considered a detoxification mechanism for arsenic, as MMA(V) and DMA(V) are less reactive and less toxic than inorganic arsenic (Styblo et al. 2000; Vahter 2002), and methylation facilitates elimination through urine and decreases retention (NRC 1999). However, several recent studies have noted the release and persistence of the intermediate metabolite MMA(III), which is highly reactive and possibly more toxic than inorganic As(III) (Styblo et al. 2000; NRC 2001). Therefore, methylation should not be considered solely as a detoxification process (NRC 2001).

Arsenic elimination. The principal pathway by which arsenic is eliminated is through the urine. Studies have shown that approximately 33-38% of an ingested arsenic dose is eliminated in the urine within 48 h and 45-58% within 4-5 days (Tam et al. 1979; Buchet et al. 1981). Elimination of ingested As(V) is slightly more rapid than for As(III) (Pomroy et al. 1980). MMA(V) and DMA(V) are eliminated more rapidly: 75-78% within 4 days (Buchet et al. 1981).

Arsenic toxicity. Mechanisms of arsenic toxicity are still not well understood, and a description of proposed models is beyond the scope of this chapter. Excellent summaries of the vast spectrum of adverse health effects associated with arsenic exposure are provided by the NRC reviews (1999; 2001), the Agency for Toxic Substances and Disease Registry (2005a) within the U.S. Center for Disease Control (*http://www.atsdr.cdc.gov*), and the IPCS (2001), sponsored by the World Health Organization (*http://www.inchem.org/documents/ehc/ehc/ehc224.htm*). At least one contributing reason for the variety of health effects associated with arsenic is the fact that arsenic is transported to and retained to some degree in all major organs, as indicated by postmortem findings.

It is commonly stated that the toxicity from arsenic exposure varies in the order As(III) > As(V) >> MMA(V) ≅ DMA(V) (and other organic forms). Ingested MMA(V) and DMA(V) are metabolized less and eliminated more rapidly and to a greater degree than inorganic arsenic (IPCS 2001; ATSDR 2005a). Arsenic present in fish, primarily arsenobetaine, apparently undergoes little metabolism and is readily eliminated (Le 2002). At high doses, studies have shown that more arsenic is retained after ingestion of As(III) than As(V) (NRC 1999). A significant factor contributing to the greater toxicity of As(III) is its greater solubility in lipids and its ability to cross cell membranes more readily than As(V) species (Schoolmeester and White 1980), in part due to its neutral charge at physiological pH. Studies using laboratory animals have generally shown lower LD_{50} concentrations for As(III) ingestion compared to As(V) ingestion (IPCS 2001), which supports the greater toxicity of the As(III) form. As noted above, there is recent evidence suggesting that the toxicity of the metabolite MMA(III) may be more toxic that inorganic As(III). However, the ATSDR (2005a) has emphasized that studies based on laboratory animals do not provide good quantitative models for human toxicity.

Finally, many related factors are known or thought to modify arsenic toxicity. For example, diet has been shown to influence arsenic toxicity (Peraza et al. 1998) and more specifically the degree of methylation (e.g., Steinmaus et al. 2005), which may be one factor that explains variability of health effects among individuals or groups. Another aspect of interest is the observation that co-exposure to selenium may reduce the toxicity of arsenic (e.g., Levander 1977). The mechanism for this effect is under investigation, but may involve the formation of an arsenic-selenium complex either with reduced toxicity or more rapid elimination. Recent studies by Gailer and co-workers (2000, 2002) have identified the formation of a seleno-bis (S-glutathionyl) arsinium complex, $[(GS)_2AsSe]^-$, in rabbits injected with arsenite and selenate, with rapid excretion to bile.

ROLE OF METAL SPECIATION: CHROMIUM

Chromium is one of the more abundant heavy metals, with an average crustal concentration of 126 ppm (Wedepohl 1995). There is a long history of mining and processing of chromium, fueled by numerous industrial applications. Chromium is an important component in steel and other alloys, paints, magnetic recording tape, electroplating, wood preservative, and leather tanning, and serves as an anticorrosive agent in water-cooling systems. Other sources of chromium in the environment are listed in Table 4. Its widespread use has been accompanied by releases to the environment that pose persistent health hazards. Chromium is among the metals included in the ATSDR/EPA National Priority List of hazardous substances, and is present at a majority of sites on the CERCLA National Priority List (*http://www.atsdr.cdc.gov/cercla*). The environmental behavior and toxicology of chromium have been studied extensively. Readers are directed to the reviews of the environmental behavior of chromium by Rai et al. (1989), Fendorf (1995), and Kimbrough et al. (1999). Summaries of the environmental toxicology of chromium are given in ATSDR (2000), IPCS (1988), EPA (1998a,b), and O'Flaherty et al. (2001).

Chromium in the environment

Chromium may exist in oxidation states ranging from −II to VI. In the near-surface environment, the III and VI oxidation states are most important. It is well known that Cr(III)

Table 4. Some common sources of chromium in the environment.

- Occupational exposure from production of chromate, stainless-steel, chrome plating
- Air emissions and water effluents from ferrochrome production, ore refining, tanning industries, chemical manufacturing industries (e.g., dyes for paints, rubber and plastic products), metal-finishing industries (e.g., chrome plating), manufacturing of pharmaceuticals, wood, stone, clay and glass products, electrical and aircraft manufacturing, steam and air conditioning supply services, cement production
- Air emissions from incineration of refuse and sewage sludge
- Combustion of oil and coal
- Oxidation and leaching from stainless steel into a water-soluble form
- Motor vehicle exhaust and emissions from automobile brake linings and catalytic converters
- Tobacco smoke

is an essential micronutrient, playing a role in glucose metabolism. In contrast, Cr(VI) is considered a known carcinogen (by inhalation route) and is associated with both acute and chronic health effects (by inhalation and ingestion), as well as being a contact allergen (EPA 1998a; ATSDR 2000). Cr(IV) and Cr(V) are thought to play a role in the mechanism of toxicity and have been associated with production of reactive oxygen species and carbon-based radicals that may interact with DNA (cf. EPA 1998a; Gaggelli et al. 2002). In minerals, water, and most soils, chromium is dominantly coordinated with oxygen. Cr(III) shows a strong preference for octahedral coordination. The Cr^{3+} aqua ion undergoes hydrolysis, yielding $Cr(OH)^{2+}$, $Cr(OH)_2^+$, $Cr(OH)_3^0$, and $Cr(OH)_4^-$ species depending on pH (Fig. 13). Cr(III) may also form aqueous complexes with organic and inorganic ligands. Cr(VI) occurs almost exclusively in tetrahedral coordination with oxygen, as the chromate anion (CrO_4^{2-}). The pK_{a2} value for chromic acid, H_2CrO_4, is ~6.5, so that $HCrO_4^-$ dominates in solutions below pH 6.5, and CrO_4^{2-} dominates above (Fig. 13). Cr(VI) also occurs as dichromate, $Cr_2O_7^{2-}$, but this species only become important at millimolar Cr concentrations and greater.

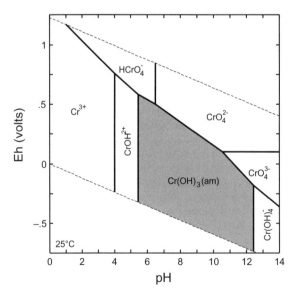

Figure 13. Eh-pH predominance diagram for aqueous chromium at 25 °C (Cr_{tot} = 1 µM).

Interconversion between Cr(III) and Cr(VI) depends on oxidation potential and pH (Fig. 13). Inasmuch as chromium is usually present at very low concentration in soils, surface sediments, and aquatic systems, other redox couples, commonly involving iron or organic matter, are usually important in controlling chromium oxidation state. Chromium occurring in natural minerals is dominantly as Cr(III), as in the important ore mineral chromite ($FeCr_2O_4$). The occurrence of Cr(VI) species and compounds is usually a result of anthropogenic activities or oxidative weathering.

Solubility and dissolution/precipitation. There are significant differences in the behavior of Cr(III) and Cr(VI) species under most surface conditions. Cr(III) compounds and minerals are largely insoluble except at low and very high pH. Over the pH range 6-10.5 precipitation of crystalline or amorphous $Cr(OH)_3$ or $(Fe,Cr)(OH)_3$ effectively limits dissolved Cr(III) concentrations to values below the current EPA drinking water MCL of 0.1 mg/L (Rai et al. 1987; Sass and Rai 1987), thereby limiting the oral exposure of this Cr species through water.

In contrast, Cr(VI) is highly soluble over the pH range of most natural waters. Chromate salts of sodium, potassium, magnesium, and calcium are highly soluble and rarely limit environmental Cr(VI) concentrations. Only $PbCrO_4$ (crocoite) is relatively insoluble. This striking difference in solubility between Cr(III) and Cr(VI) species has the unfortunate consequence that the essential micronutrient, Cr(III), is largely absent as a dissolved species in water while the toxic form, Cr(VI), is soluble and therefore potentially mobile. This solubility difference, however, also presents a potential remediation strategy based on conversion of soluble Cr(VI) to insoluble Cr(III).

Chromium reduction/oxidation. Owing to the significant difference in Cr(III) and Cr(VI) mobilities in the near-surface environment, redox processes are important for influencing chromium bioaccessibility and exposure to organisms. Although pH dependent, the Cr(III/VI) redox potential is relatively high, and CrO_4^{2-} is a strong oxidant. Ferrous iron, organic matter, and sulfides have been shown to reduce Cr(VI) to Cr(III) readily even in the presence of dissolved oxygen (James and Bartlett 1983a,b; Rai et al. 1989; Fendorf 1995; Patterson et al. 1997). Reduction by Fe(II) has been shown to result in the formation of $(Fe^{III},Cr^{III})(OH)_3$, which is insoluble and stable (Eary and Rai 1988). Cr(VI) reduction may be facilitated at surfaces of minerals containing Fe(II). For example, magnetite and ferrous biotite surfaces have been shown to reduce Cr(VI) to Cr(III) (Ilton and Veblen 1994; Peterson et al. 1996; Peterson et al. 1997). Fendorf et al. (2000) also demonstrated that Cr(VI) may be reduced by bacterial processes even in aerobic conditions.

Rai et al. (1989) and Fendorf (1995) report several studies in which Cr(III) was rapidly oxidized to Cr(VI) by manganese oxides, which are common constituents in soils. The effectiveness of this process, however, may be limited by formation of a hydrous $Cr(OH)_3$ or CrOOH precipitate at the MnO_2 surface (Fendorf et al. 1992; Charlet and Manceau 1993).

Chromium sorption. Owing to the high mobility of Cr(VI) in environmental systems much attention has been focused on the sorption behavior of this Cr species and its effectiveness in reducing mobility. As anion species (CrO_4^{2-} and $HCrO_4^-$), Cr(VI) exhibits greatest sorption at low pH (Fig. 14). Above pH 7 sorption may not be effective in removing Cr(VI) from solution (Zachara et al. 1987, 1989, 2004). In view of their nearly ubiquitous occurrence, iron oxides and hydroxides are likely to be some of the most effective sorbents. Chromate sorption may be strongly diminished as a result of competition for available surface sites by other anion species, such as carbonate, phosphate, and sulfate (e.g., Zachara et al. 1987).

Chromium in the body

Chromium exposure. The previous discussion focused primarily on environmental processes that control the behavior of different chromium species in soils, sediments, and aquatic systems. These influence human exposure mainly through ingestion of water, but may

Reeder, Schoonen, Lanzirotti

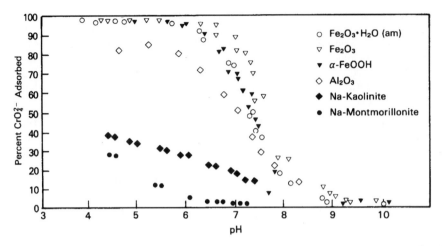

Figure 14. The pH dependence of chromate adsorption on different mineral and solid sorbents. [Reprinted with permission from Rai D et al., *The Science of the Total Environment*, Vol. 86, Fig. 4, p. 21. Copyright (1989) Elsevier.]

also relate to accidental ingestion of soil or inhalation of mineral dusts that contain chromium. The largest source of chromium intake for the general population is from food, where it occurs in the Cr(III) state (WHO 2003). Inhalation of Cr-containing airborne particles from industrial emissions, both Cr(III) and Cr(VI), may be locally significant, but most instances are associated with occupational exposures. Exposure may also occur through dermal contact. Here, we restrict the following discussion to illustrate examples where chromium speciation is relevant to health effects. Much of this information has been taken from the comprehensive reviews provided by IPCS (1988), EPA (1998a,b), ATSDR (2000), and WHO (2003). Readers are also referred to O'Flaherty (1996) and O'Flaherty et al. (2001).

Absorption and distribution. The efficiency of chromium absorption in the gastrointestinal tract is relatively low, and contrasts with the much greater absorption of arsenic noted in the previous section. Cr(VI) exhibits greater absorption than Cr(III), but both depend on the chemical form and other factors, including food and nutritional status. After oral exposure, most studies have shown absorption of soluble Cr(III) of ~0.5%, whereas absorption of soluble Cr(VI) compounds was 2-7% (ATSDR 2000; EPA 1998a,b), with a similar absorption efficiency for dissolved Cr(VI). However, Cr(VI) is partially reduced to Cr(III) in gastric fluid (and elsewhere in the body) (De Flora et al. 1987), which decreases the total absorption owing to the lower absorption efficiency of Cr(III). Essentially no absorption was observed following oral exposure of Cr(III) as insoluble Cr_2O_3 (Finley et al. 1996). Absorption of dietary Cr(III) is approximately 0.5-2% (EPA 1998b), and absorption of Cr(III) in the form Cr(III) picolinate (the form commonly used in vitamins) is as much as 3% (Gargas et al. 1994). Studies using lab animals generally support the greater absorption efficiency of Cr(VI) over Cr(III). Cr(III) absorption, however, may be enhanced when Cr(III) is complexed by organic ligands (e.g., oxalate) (ATSDR 2000) or when present in biologically active complexes (Mertz and Roginski 1971; O'Flaherty 1996).

Studies of inhalation exposure also show greater absorption of Cr(VI) than Cr(III) in the lungs. This has been confirmed by studies using lab animals, which have also indicated that the efficiency of chromium absorption in the lungs may be significantly greater than in the GI tract. Studies using a rat model (summarized in ATSDR 2000) showed 53-85% absorption from Cr(VI) particles (including clearance to the pharynx and into the GI tract) and 5-30%

absorption from Cr(III) particles. Lab animal studies also show that absorption depends strongly on the solubility of the inhaled compound, with greater absorption for more soluble phases (Bragt and van Dura 1983). This supports the view that dissolution of a solid to a soluble form is normally required for absorption across a physiological membrane.

The greater absorption of Cr(VI) compared to Cr(III) is largely due to the facilitated transport of chromate across cell membranes, following the same anion channel pathway as sulfate and phosphate (Wiegand et al. 1985). Absorption of Cr(III) is limited, occurring only by passive diffusion and/or phagocytotsis, which are much less effective. In the blood, Cr(VI) is able to cross the membrane of red blood cells, where it is rapidly reduced to Cr(III) and interacts with proteins. In contrast, Cr(III) is largely restricted to the plasma (WHO 2003). The greater tendency of Cr(VI) to cross cell membranes is also reflected in chromium distribution following exposure. One study using a mouse model showed that Cr was detected only in the liver following one year of exposure to Cr(III) chloride (a soluble form). In contrast, Cr was detected in all organs for mice exposed to soluble Cr(VI) for the same period (ATSDR 2000).

O'Flaherty (1996) has suggested that because of its low absorption efficiency the factors that influence bioaccessibility of any particular environmental chromium source are likely to be the single most important determinant of toxicity.

Metabolism and elimination. The essential role of Cr(III) is associated with a biologically active Cr(III) complex that is involved in glucose metabolism (Anderson 1986). Cr(VI) is not stable in the body, and a variety of electron donors, such as ascorbate and glutathione, cause reduction to Cr(III) species (De Flora et al. 1987). This proceeds throughout the body, including in saliva, gastric and lung fluids, blood, and in major organs. This reduction of Cr(VI) can be considered a defense or detoxification mechanism. However, Cr(V) and Cr(IV) intermediates are formed during reduction; these have been associated with formation of reactive oxygen species, and may be involved in the mechanism of toxicity (Gaggelli et al. 2002; Levina et al. 2003).

Absorbed chromium is mostly excreted as Cr(III) complexes through the urine within a period of several hours to several days (ATSDR 2000). The absence of Cr(VI) in urine, even after exposure to Cr(VI), indicates that reduction is complete within this time frame. Some chromium is retained in tissue and bone for periods on the order of months or longer. The large chromium fraction that is not absorbed following oral exposure is eliminated in the feces.

ROLE OF METAL SPECIATION: LEAD

Lead in the environment

Lead poisoning is one of the most common and serious environmental issues in industrialized nations, particularly with regard to its effect on the cognitive development of young children. Although lead occurs in the environment naturally, the vast majority of the instances of elevated lead levels in the environment that are of concern for human health are the result of human activity. Anthropogenic sources of lead include the mining, smelting, and refining of lead ore, emissions from coal and oil combustion, emissions from combustion of leaded gasoline, lead-based paints and solders, lead arsenate pesticides, and waste incineration. Much of the review of lead in the environment provided here is taken from the U. S. Department of Health and Human Service's Agency for Toxic Substances and Disease Registry on the effects of lead in the environment and its toxicology (ATSDR 2005b). In addition to the ATSDR overview, excellent reviews are also provided by Baxter and Frech (1995) and Ryan et al. (2004). Here, we provide a brief overview of the role of lead speciation as it relates to mobility in the environment and human exposure, bioavailability, and health effects.

Lead as a mineral is extremely rare (Wedepohl 1995). The most common form of lead in the Earth's crust is galena (PbS), but other important ore minerals include cerussite ($PbCO_3$), anglesite ($PbSO_4$), and minium (Pb_3O_4). Lead minerals are commonly found in association with zinc, copper, and iron sulfides and are commonly associated with gold and silver ores. Lead can also occur as a trace element in coal, oil, and wood. The majority of the lead used in industry today comes from recycling. Lead occurs in both the 2+ and 4+ oxidation states, but in nature the 2+ oxidation state dominates.

Lead species in natural waters. In most ground and surface waters, the solubility of lead compounds is generally low. The solubilities are dependent on water pH, the presence of coexisting ionic species and ligands, water salinity, and the organic matter content. Solubility tends to be highest in soft, acidic water. Eh-pH diagrams for the system $Pb-CO_2-H_2O$ are shown in Figures 15a and 15b for 5 ppb and 207 ppb Pb_{tot}, respectively (the EPA action level for Pb in drinking water is 15 ppb; 207 ppb = 1 μM). Lead can exist as the Pb^{2+} ionic species at pH < 7.5 in fresh water, but readily complexes with dissolved carbonate at pH > 5.4 and forms lead carbonates, $PbCO_3$ and $Pb_2(OH)_2CO_3$, limiting its solubility (Long and Angino 1977). In natural waters lead will readily precipitate in the form of various lead hydroxides, carbonates, sulfates and phosphates, so that the amount of lead remaining dissolved in solution is typically low (Mundell et al. 1989). In highly oxidizing waters, such as may be found in municipal water systems, lead may precipitate as plattnerite (PbO_2). In surface waters most of the lead that is in suspension occurs as colloidal and undissolved particles. In seawater lead carbonate complexes are also prevalent, but lead chloride complexes and surface complexes with iron and manganese oxides can also occur (Long and Angino 1977; Elbaz-Poulichet et al. 1984).

Lead species in the atmosphere. The largest influx of lead to the environment has historically come from emissions to the atmosphere due to automotive and industrial sources. Lead particles are emitted from smelters primarily in the form of elemental lead and lead-sulfur compounds, $PbSO_4$,

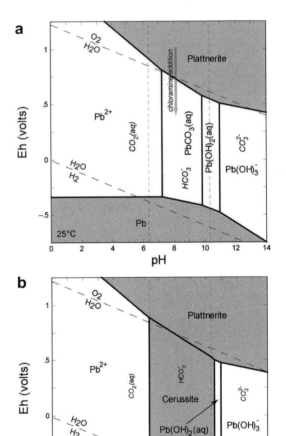

Figure 15. Eh-pH predominance diagrams for lead in the system $Pb-CO_2-H_2O$, 25 °C. (a) Pb_{tot} = 5 ppb (0.024 μM), C_{tot} = 18 ppm (1.5 mM); (b) Pb_{tot} = 207 ppb (1 μM), C_{tot} = 18 ppm.

PbO·PbSO$_4$, and PbS (Corrin and Natusch 1977; EPA 1986; Spear et al. 1998). Oil and coal combustion has also been found to release PbCl$_2$ and PbO (Nerin et al. 1999). Prior to 1990 nearly 90% of all lead emissions were the direct result of combustion of leaded gasoline. In 1984 the U.S. Environmental Protection Agency mandated the gradual phase-out of the use of lead alkyls in gasoline and banned their use entirely in 1996 (EPA 1996). Today, industrial emissions from metal smelting operations and battery manufacturing are the largest sources of lead to the atmosphere.

The vast majority of the lead emissions from leaded gasoline are in the form of inorganic particulates such as lead bromochloride. These halogenated lead compounds are formed from the combustion of gasoline as the tetra-alkyl lead additives react with halogenated lead scavenger compounds (EPA 1985). Less than 10% of the lead emitted from combustion of leaded gasoline is emitted as organolead compounds such as vapor phase lead alkyls. However, tetra-alkyl lead vapors are strongly photoreactive, and quickly react to form trialkyl and dialkyl lead compounds, and eventually inorganic lead oxides (Eisenreich et al. 1981). In direct sunlight the half-life of tetraethyl lead vapors is approximately 2 hours (DeJonghe and Adams 1986). These lead particles are then deposited to either land or water through both wet and dry deposition. Their fine particle size and solubility in acidic solutions make these particles a particularly potent source of bioaccessible lead.

Lead species in soils. Once deposited to soils, the speciation of lead can take several different forms depending on the soil type. Key factors controlling its speciation in soils include soil pH, the organic matter content, and the soil cation exchange capacity (NSF 1977; Reddy et al. 1995). All of these factors, along with the mineral form, particle size, and association with other mineral phases, affect speciation of lead in soils and its potential for solubilization, which would influence its bioaccessibility and bioavailability (Fig. 16; Ruby et al. 1999).

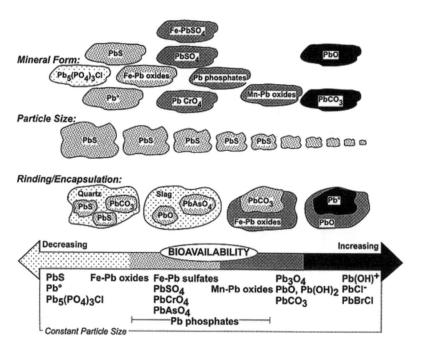

Figure 16. Properties of lead species and compounds and their effects on lead bioaccessibility and bioavailability in soils. [Reprinted with permission from Ruby et al. (1999). Copyright (1999) American Chemical Society.]

Sorption processes play a very important role in lead speciation in soils. In particular, iron(III) and manganese(III/IV) oxides and hydroxides, which are ubiquitous in most soils, are very effective sorbents for lead (O'Reilly and Hochella 2003). A number of studies (e.g., O'Reilly and Hochella 2003; Villalobos et al. 2005) have demonstrated that Mn-oxides in particular are very efficient sorbents of Pb. Figure 17 shows the measured lead sorption capacities for a variety of synthetic Fe- and Mn-oxides. Synthetic birnessite, in particular, can sorb large amounts of lead from solution, almost an order of magnitude more than Fe-oxides. EXAFS studies have indicated that Pb^{2+} forms inner-sphere complexes within the birnessite inter-layer (Morin et al. 1999; Matocha et al. 2001; Manceau et al. 2002).

Soil organic matter can also play a vital role in determining lead speciation in soils. Soil humic and fulvic substances, for example, have been shown to tightly bind lead (Pinheiro et al. 1999; Christl et al. 2005; Newton et al. 2006). Although humic substances contain several major functional groups (primarily carboxylic groups ~80% and phenolic groups ~20%), it is generally thought that Pb exhibits a strong affinity for the carboxylic groups (COOH) of the humic substances (Stevenson 1994; Xia et al. 1997). X-ray absorption studies of Pb-humate systems indicate that Pb can form inner-sphere complexes with the humic component and that the structure of the binding site does not change appreciably as a function of pH (Xia et al. 1997).

For example, Morin et al. (1999) examined smelter-contaminated soils from Evin-Malmaison, France, analyzing both tilled and wooded soil samples. They showed that although the source of Pb contamination was similar, differences in the chemical forms of Pb existed depending on soil type. In the wooded soil ~40% of the Pb was found to exist as outer-sphere complexes, ~10% as Pb^{2+} adsorbed on hydrous Fe oxides or oxyhydroxides, and ~50% as Pb^{2+} inner-sphere complexes bound to organic matter. By contrast, in the tilled soils Pb was present almost entirely as inner-sphere complexes adsorbed on hydrous Fe and Mn oxides. The higher amount of organo-Pb^{2+} and exchangeable outer-sphere Pb^{2+} complexes in the wooded soil was attributed to their higher organic matter content (6.4 wt% vs. 1.5 wt% TOC in the tilled soils).

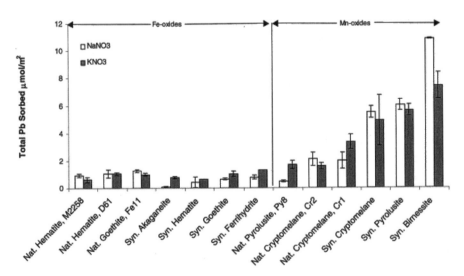

Figure 17. Average total lead sorption by various Mn and Fe-oxide minerals. Studies were conducted using NaNO₃ (white bars) and KNO₃ (gray bars) background electrolyte. [Reprinted with permission from O'Reilly and Hochella, *Geochimica et Cosmochimica Acta*, Vol. 67, Fig. 6, p. 4479. Copyright (2003) Elsevier.]

However, the binding of lead with humic components can be strongly affected by the presence of other major ions such as Ca^{2+}, Mg^{2+}, and Al^{3+}. Mota et al. (1996) demonstrated that by increasing Al^{3+} concentration, it was possible to also increase the concentration of free lead, showing that aluminum competes with lead for the binding sites of humic substances. This can be an important effect in soils containing gibbsite, where Al^{3+} concentrations may be high. Pinheiro et al. (1999) demonstrated a similar competitive binding of Pb^{2+} and Ca^{2+} to fulvic acid. Whereas solid organics can be quite efficient in diminishing lead mobility, aqueous organics, by contrast, can act to enhance lead mobility. Organic acids produced by plants and soil microorganisms can be particularly effective in chelating lead and enhancing mobility in soils.

Soil phosphate content also plays a role in controlling lead speciation. A number of studies of Pb-contaminated soils have shown that precipitation of Pb^{2+} as lead phosphate is effective in sequestering lead in soils (Davis et al. 1993; Cotter-Howells et al. 1994; Ruby et al. 1994; Juillot 1998; Traina and Laperche 1999; Morin et al. 2001). This is particularly important in evaluating lead bioaccessibility, inasmuch as members of the pyromorphite group [$(Pb,Ca)_5$ $(PO_4)_3(OH,Cl,F)$] are among the most stable and least soluble lead minerals under surface environmental conditions (Nriagu 1973, 1974, 1984) and may form rapidly when adequate phosphate is present (Ruby et al. 1994). Their formation effectively limits the mobility of Pb in some soils. Cotter-Howells (1996) reported on the presence of lead phosphate in urban and roadside soils, demonstrating that these phases formed in the soil as a weathering product of Pb-bearing grains. The low solubility of pyromorphite has also advanced the use of phosphate amendments to lead-contaminated soils to promote pyromorphite formation.

There is also some indication that soil organisms may be able to biogenically precipitate pyromorphite in contaminated environments. Jackson et al. (2005) demonstrated that soil nematodes placed in a KCl-NaCl medium along with $Pb(NO_3)_2$ immobilize the Pb through pyromorphite precipitation. Given the high density and turnover of roundworms in soil, biogenic pyromorphite formation may influence Pb mobility and cycling in soils.

Lead species in paints and solders. The primary mode of lead exposure in children is through oral ingestion of Pb-containing products, such as lead-based paint and solder. Lead has been used as a pigment since ancient times and has been added to paint in a variety of forms. Perhaps the most common pigment is so-called "lead white," which contains a lead carbonate phase [$2PbCO_3 \cdot Pb(OH)_2$]. The popularity of lead white is due to its superior opacity. Other commonly used lead-based pigments include 'red lead' (Pb_3O_4 or $2PbO \cdot PbO_2$), lead-tin yellow (Pb_2SnO_4), and Naples yellow [$Pb(SbO_3)_2$ or $Pb(SbO_4)_2$]. The amount of lead in these pigments may be very high, commonly more than 38% of the dried weight of the paint.

In solders, lead is usually present as an alloy with tin. Electronic solders are usually 60% tin and 40% lead by weight in order to produce a near-eutectic mixture. Plumbing solders typically contain a higher proportion of lead. Corrosion of lead-containing plumbing materials in water distribution systems can release lead to drinking water, but the speciation of the lead in the water can vary dependent on municipal water chemistry. Renner (2004) highlights some of the complexities in evaluating the high levels of lead found in drinking water from some areas in Washington, D.C. In most natural waters, lead occurs in the Pb(II) oxidation state. Municipal drinking water, however, tends to be highly oxidizing due to the addition of significant amounts of chlorine. These oxidizing conditions promote the stability of Pb(IV) species such as PbO_2 (Fig. 15), which has been observed as a common scaling product in municipal water pipes in Washington, D.C. These scales remain stable and insoluble as long as the water remains oxidizing. However, many municipalities have switched from the use of chlorine to chloramine to comply with the EPA's 1998 Disinfection Byproducts Rule, which restricts disinfection byproducts in water. Chloramine use lowers the oxidizing potential of drinking water, so that PbO_2 scale may become soluble.

Lead in the body

Lead exposure levels. The toxicological effects of lead exposure have been well publicized and are generally familiar to most readers. Humans are often exposed to lead levels above that naturally occurring in soil or dust, and the most common contaminant sources and species have been discussed above. A 1999 Swedish study (Baecklund et al. 1999) characterized lead blood levels (PbB) in 176 men and 248 women, aged 49–92 years. Blood lead levels ranged from 0.56 to 15 µg Pb/dL (median 2.7 µg Pb/dL). The concentration of lead in the blood (PbB) is the most commonly used metric of absorbed dose for lead. The average PbB in the U.S. is ~2 µg/dL (0.1 µM or 20 ppb). The U.S. Centers for Disease Control (CDC) estimates that 890,000 children in the U.S. under 6 years of age may still have unsafe PbB and has established a 10 µg/dL threshold for elevated blood-lead level (CDC 1985).

Although elevated lead exposure is generally harmful, the greatest health risks from lead exposure are to preschool-age children and pregnant women and their fetuses. The most profound effects of lead exposure are neurological. Lead affects virtually every neurotransmitter system in the brain, and lead exposure has been linked to schizophrenia. Ancient Romans used lead acetate as a sweetener in wine, which has been thought to have caused a high incidence of dementia. Other medical conditions symptomatic of lead exposure include nephropathy (kidney damage), acute abdominal pain (lead colic or painter's colic), anemia, bluish discoloration of gums (gingival deposition of lead sulfide, *a.k.a.* Burtonian lines), gout (Saturnine gout), cardiac toxicity, endocrine effects, reproductive damage, and miscarriage. The EPA has classified elemental lead and inorganic lead compounds as probable human carcinogens.

Lead speciation and toxicokinetics. The toxicokinetics of lead in the human body have been extensively studied, but the role of speciation in toxicity is not completely understood. Studies have generally demonstrated that both organic and inorganic forms of lead are readily absorbed through inhalation, although factors such as particle size, solubility, and speciation may play important roles in determining lead absorption rates (ATSDR 2005b). Particles with diameters <1 µm are particularly well absorbed through inhalation exposure (~95% absorption for inorganic lead) since they can be deposited in alveolar regions of the lungs and can then be absorbed through extracellular dissolution or ingestion by phagocytic cells (ATSDR 2005b). Measurements of clearance rates for submicron particles of both inorganic lead, such as lead oxide and lead nitrate (Chamberlain et al. 1978), and organic species, such as tetraethyl or tetramethyl lead (Heard et al. 1979), are consistent with about 60-80% absorption 48 hours after initial deposition in the respiratory tract.

Ingestion is also a relatively effective pathway for lead absorption, and there are clear indications that the form of the ingested lead and the biochemistry of the gastrointestinal tract at the time of ingestion affect absorption rates. Dermal absorption appears not to be a particularly efficient pathway for the absorption of inorganic lead species. However, animal studies have shown that organic lead is well absorbed through the skin (ATSDR 2005b) and is much more readily absorbed in general. The available data on organic (i.e., alkyl) lead compounds indicate that some of the toxic effects of alkyl lead are mediated through metabolism to inorganic lead (EPA 1985).

Of the lead absorbed by the body, the majority ends up in mineralized tissues such as bone and teeth (Barry 1981, 1975): ~95% in adults and ~70% in children. The elimination half-life for inorganic lead in blood is approximately 30 days, whereas for bone it is approximately 27 years. Thus, PbB values only reflect exposure history for the previous few months, whereas lead in bone is considered a cumulative biomarker. These effects are a reflection of the proteins that bind lead in the body—proteins that naturally bind Ca and Zn (Godwin 2001). The two proteins that have seen the most study are synaptotagmin, a Ca^{2+} binding membrane protein widely expressed in the central and peripheral nervous system, and δ-aminolevulinic acid dehydratase or ALAD, a zinc enzyme that catalyzes the second step of heme synthesis (Figure 18). Heme is the portion of hemoglobin that carries oxygen in the blood from the lungs to the rest of the body. In both these cases, the Pb^{2+} ion has a much higher binding affinity than either Ca^{2+} or Zn^{2+}. For example, consider that the CDC level for elevated PbB is 0.5 μM, whereas the average total concentration of zinc in human plasma is about 17 μM and calcium roughly 10^{-6} to 10^{-3} M (Godwin 2001). In the ALAD enzyme, lead binding to a three-cysteine site in the enzyme is ~500 times greater than zinc binding. When Pb^{2+} displaces the active Zn^{2+} ion present in the metalloenzyme, the enzyme is rendered useless, inhibiting hemoglobin production and resulting in anemia. In the synaptotagmin protein, Pb^{2+} appears to bind ~1000 times more strongly than Ca^{2+} (Bouton et al. 2001). Here, substitution of Pb^{2+} interferes with calcium-mediated signal transduction in neurotransmitters.

There are indications that at least for lead introduced via ingestion, its speciation in the presence of other metal components affects its absorption and toxicity. Simply relying on the total concentration level as a measure of potential threat is inadequate. For example, Ruby et al. (1999) examined lead gastrointestinal absorption using a juvenile swine model. Nineteen lead-bearing substrates containing similar amounts of total lead from eight hazardous waste sites were fed to test animals (Fig. 19), after which PbB was measured serially. At the completion of the study, samples of blood, bone (femur), liver, and kidney were collected and analyzed for lead concentration. This was then used to produce an estimate of relative lead bioavailability. Samples from carbonate-rich soils (Jasper, MO) yielded high lead bioavailability values. Soils derived from tailings, smelter slags, and soil from mining sites had the lowest relative bioavailability, while soils from the vicinity of smelters generally yielded higher values. It is important to bear in mind, however, that in large part this is a function of lead mineralogy; mine sites with high levels of cerrussite ($PbCO_3$) have tailings and soils with greater lead mobility than mine sites rich in galena (PbS).

Rat models also demonstrate differences in absorption depending on the form of lead ingested. Bone and tissue lead levels increase in a dose-dependent manner for animals receiving lead in their diet in the form of lead acetate, lead sulfide, or lead-contaminated soil. However, the bioavailability of lead sulfide was approximately 10% that of lead acetate (Freeman et al. 1996).

The presence of food in the gastrointestinal tract tends to decrease absorption of lead, but likely this is compositionally dependent as well as being a reflection of the biochemistry of the

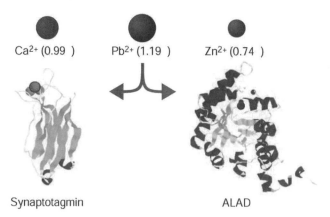

Ca^{2+} (0.99) Pb^{2+} (1.19) Zn^{2+} (0.74)

Synaptotagmin   ALAD

Figure 18. Lead binds with protiens that naturally bind calcium (synaptotagmin) and zinc (ALAD) in spite of their larger size. [Reprinted from Godwin, *Current Opinion in Chemical Biology,* Vol. 5, Fig. 1, p. 224. Copyright (2001) Elsevier.]

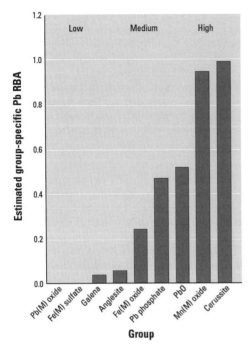

gastrointestinal tract during exposure (Rabinowitz et al. 1980; Heard and Chamberlain 1982; Blake and Mann 1983; Blake et al. 1983; James et al. 1985; Maddaloni et al. 1998). James et al. (1985) demonstrated that fasting adults given a tracer dose of lead acetate absorbed approximately 63% of the lead while subjects that were given a meal absorbed only ~3%. In particular, the presence of calcium and phosphate in the gastrointestinal tract tends to depress the absorption of ingested lead, as may the presence of oxalates and phytates (Heard and Chamberlain 1982; Blake and Mann 1983; Blake et al. 1983). For example, *in vitro* studies (Scheckel and Ryan 2003) have shown that when various Pb compounds ($PbCl_2$ and Pb paint) are introduced into a simulated stomach and gastrointestinal experimental system in the presence of cola soft drinks that contain phosphoric acid, rapid *in vitro* formation of pyromorphite occurs. The formation of pyromorphite within the stomach cavity may minimize Pb absorption through the higher pH gastrointestinal tract. Likewise, food in the stomach cavity may increase stomach pH so that less Pb is solubilized prior to reaching the GI tract. There are also animal studies indicating that zinc

Figure 19. Relative lead bioavailabilities (Pb RBA) of various groups of solid phases, as determined from *in vivo* studies using a juvenile swine model. Pb(M) oxide = Pb oxide, Pb/As oxide, Pb silicate, Pb vanadate; Fe(M) sulfate = Fe/Pb sulfate, Pb sulfosalts; Fe(M) oxide = Fe/Pb oxide, Zn/Pb silicate; Mn(M) oxide = Mn/Pb oxide. [Reproduced with permission from *Environmental Health Perspectives*, Casteel et al. (2006).]

and iron deficiency may enhance lead absorption. Gastrointestinal absorption of lead occurs by acid solubilization (Ellenhorn and Barceloux 1988) and it seems that at least some lead transport across the digestive mucosa is similar to that of calcium (Gilman et al. 1990).

In considering gastrointestinal absorption of lead, researchers have developed a number of *in vitro* digestion models that compartmentalize the digestive process (Fig. 20; Oomen et al. 2003a,b). In order for lead to be absorbed within the gastrointestinal tract it must be mobilized from its matrix and present in a form capable of being transported across the intestinal epithelium. Gastric juices in the stomach can be quite acidic, with pH values as low as 1, which favors release of lead to a soluble form that makes it available for absorption (Fig. 15). Many studies assume that lead absorption primarily occurs in the small intestine, where pH values are 5-7.5 and lead is likely to be complexed (Oomen et al. 2003a). *In vitro* studies (Zhang et al. 1998; Oomen et al. 2003a) have found that the change from simulated stomach to intestinal conditions causes a rapid and complete transformation of lead to lead phosphates, such as chloropyromorphite [$Pb_5(PO_4)_3Cl_{[s]}$], and lead-bile complexes. As discussed earlier, lead phosphates would be expected to have very low solubilities and their formation could limit the concentration of soluble Pb; in fact these *in vitro* studies have demonstrated that the soluble Pb fraction in the simulated intestinal fluid is negligible. However, Oomen et al. (2003a) also demonstrated that the lead phosphates that are formed are labile, as also was shown by Feroci et al. (1995) for the lead-bile complexes.

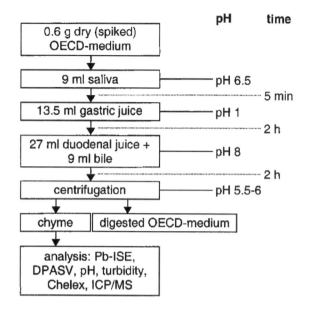

Figure 20. Schematic representation of an *in vitro* digestion procedure and subsequent analysis. [Reproduced with kind permission of Springer Science and Business Media from Oomen et al. (2003b).]

ROLE OF METAL SPECIATION: MERCURY

Mercury is an example of a metal that is far more toxic as an organometal species than in either metallic form or as an inorganic ionic species. Methylmercury (MeHg), in particular, poses a serious health threat. Monomethylmercury is the primary species of concern, although dimethylmercury has been detected in the marine environment. It has been proposed that one of the primary factors contributing to the high toxicity of MeHg is its high lipid solubility, which facilitates its transport across physiological membranes. Significantly, MeHg readily crosses the blood-brain barrier, where it has greatest impact. Unlike arsenic discussed above, the methylation of mercury occurs in the environment rather than within the body.

In this section we briefly summarize the sources of mercury in the environment, processes that control mercury speciation in the environment, their absorption and metabolism by humans, and the differences in toxicity between the methylated and inorganic forms. Detailed reviews of the environmental chemistry of mercury (Ullrich et al. 2001; Tchounwou et al. 2003; Fitzgerald and Lamborg 2004), the distribution and toxicity of MeHg (NRC 2000; Yokoo et al. 2003; Dopp et al. 2004; Norling et al. 2004), bioavailability of MeHg (NRC 2000; Tchounwou et al. 2003), and mode of action of mercury and mercury compounds have been published recently (Langford and Ferner 1999; NRC 2000; Schettler 2001; Tchounwou et al. 2003; Zheng et al. 2003; Fitsanakis and Aschner 2005; Garcia et al. 2005).

Mercury in the environment

Mercury, a Group IIb element in the periodic table, is rare in the Earth's crust, with average abundance estimated to be approximately 40 ppb (Wedepohl 1995). Organic-rich sediments and soils are enriched in mercury due to the uptake of mercury by biota, while sandstones and sedimentary carbonates contain on average 30 and 40 ppb Hg, respectively (Fitzgerald and Lamborg 2004). Plants take up atmospheric mercury through their leaves and soil-bound mercury through their roots (Tomiyasu et al. 2005). Once taken up, the mercury is essentially sequestered (Greger et al. 2005) and mercury remains associated with buried organic material. US coal contains on average 170 ppb mercury, much of it associated with pyrite and to a lesser extent the organic matrix (Tewalt et al. 2005). Most of the mercury in coal pyrite is derived from degradation of organic compounds as the initial organic-rich sediment undergoes diagenesis. Mercury ores are formed from hydrothermal fluids. Cinnabar, HgS, is commonly the most abundant ore mineral in mercury deposits. Mercury has been used in ancient Chinese cultures and was mined by the Greeks and Romans for use as pigments and ointments. The modern-day, widespread use of mercury in gold mining, medical equipment, chemical processing, fluorescent lamps, and relays has led to a high demand for mercury.

Weathering of Hg-containing rocks and volcanic emissions are the most important natural processes that release inorganic mercury into the biosphere. Unlike weathering, volcanic emissions are episodic. For example, a recent study of an ice core retrieved from the Fremont Glacier in Wyoming, USA, shows a fivefold increase in atmospheric mercury deposition associated with the Krakatau eruption in 1883 (Schuster et al. 2002).

Mercury is a redox element, with Hg(0), Hg(I), and Hg(II) as stable oxidation states in the presence of water. Mercury is unusual in the fact that it is a liquid in its elemental state, a form in which it is readily volatilized. Furthermore, liquid mercury is stable over a wide range of redox conditions. The solubility of liquid mercury in water in the form of $Hg^0_{(aq)}$ is ~10^{-7} M at 25 °C (Glew and Hames 1971). However, $Hg^0_{(aq)}$ is volatile and readily lost to the atmosphere. In sulfur-rich, reducing environments the formation of insoluble mercury sulfide phases (cinnabar and metacinnabar) limits the solubility of mercury. In oxidizing environments the solubility of mercury is limited by the formation of mercury oxides. In natural environments, where the total concentrations of mercury are often vanishingly low, the solution concentration is controlled by complexation with organic ligands and sulfide on one hand and sorption onto organic matter and particulate matter on the other hand (Reddy and Aiken 2001).

Once mercury enters aquatic systems—either as dissolved species, associated with particulate matter, or as elemental mercury—it can become methylated (Fitzgerald and Lamborg 2004). An idealized cycle illustrating the major processes of mercury methylation in freshwater lakes is shown in Figure 21. Methylation of mercury is closely tied to the sulfur redox cycle. In anaerobic environments, sulfate-reducing bacteria promote methylation. It is thought that the formation of sulfide by sulfate-reducing bacteria leads to the formation of a suite of mercury-sulfide complexes (Ullrich et al. 2001). As shown in Figure 22, species such as $Hg(HS)_2^0{}_{(aq)}$ and $HgS_2^{2-}{}_{(aq)}$ account for as much as 10% of the aqueous mercury

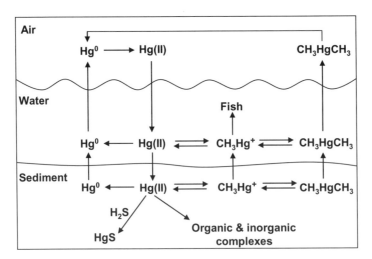

Figure 21. Idealized cycle of mercury methylation in freshwater lake environments. [Adapted from Winfrey and Rudd (1990).]

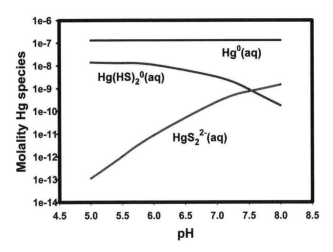

Figure 22. Aqueous inorganic Hg speciation in equilibrium with liquid metallic mercury in the presence of 0.1 mM sulfidic sulfur at 25 °C. Calculation based on MinteqV4 database provided with Phreeqc 2.12.5. The formation of metacinnabar and cinnabar has been suppressed for these calculations.

species in an anaerobic, sulfide-containing system in equilibrium with liquid mercury. Some of the neutral inorganic complexes, such as $Hg(HS)_2^0{}_{(aq)}$ and $Hg^0{}_{(aq)}$, can readily diffuse into the cells of sulfate-reducing bacteria where they are subsequently methylated as part of a cellular defense mechanism (Fitzgerald and Lamborg 2004). Methylation takes place under reducing conditions where hydrogen sulfide or bisulfide is the dominant sulfur species (see Fig. 23). Methylated mercury species are labile and can be demethylated leading to a complex, dynamic mercury cycle with significant fluxes of mercury between the atmospheric and aquatic reservoirs. A fraction of the methylated mercury is taken up in the aquatic food chain, bioaccumulates, and resists degradation (Norling et al. 2004).

While there is consensus that the global cycles of mercury have been significantly altered by anthropogenic emissions, there is considerable uncertainty about the magnitude of some

Figure 23. Eh-pH predominance diagram for mercury in the presence of sulfur at 25 °C; $Hg_{tot} = 10^{-9}$ M, $S_{tot} = 10^{-4}$ M. Formation of cinnabar and metacinnabar have been suppressed. Calculations performed with Geochemist's Workbench using LNLL database supplemented with thermodynamic data for $Hg^0{}_{(aq)}$, $Hg(OH)_{2(aq)}$, $Hg(HS)_2^0{}_{(aq)}$, $HgS(HS)^-{}_{(aq)}$, and $HgS_2^{2-}{}_{(aq)}$ taken from MinteqV4 database.

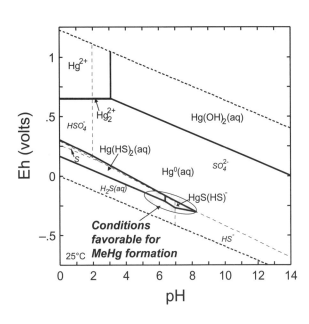

of the mercury fluxes (Fitzgerald and Lamborg 2004). It is estimated that anthropogenic mercury emissions account for about 66% of the total emissions. Table 5 summarizes natural and human inputs to the environment. Among anthropogenic inputs, the single most important process by far is fossil fuel burning, particularly burning of coal (Nriagu and Pacyna 1988; Pacyna and Pacyna 2002). While mercury emissions from coal power plants in Europe and the US have dropped substantially over the last decade, the rapid economic development in India and China has led to a significant increase in emissions in Asia (Pacyna and Pacyna 2002). Mercury released by fossil fuel burning is predominantly in the form of gaseous elemental mercury. A complex set of reactions in the atmosphere lead to the conversion of elemental mercury into ionic mercury, which is deposited into aquatic systems via a combination of wet

Table 5. Summary of relevant toxicokinetic data for three important mercury species[#].

MeHg	Hg(0)	Hg(II)$_{(aq)}$
Exposure		
Fish, marine mammals, crustaceans, animals and poultry fed fish meal	Dental amalgams, fossil fuels, occupational exposure, incinerators	Oxidation of elemental mercury or demethylation of MeHg; deliberate or accidental poisoning with $HgCl_2$
Absorption		
Inhalation: Vapors of MeHg absorbed. Oral: Approximately 95% of MeHg in fish readily absorbed from GI tract. Dermal: In guinea pigs, 3-5% of applied dose absorbed in 5 hr	Inhalation: ~80% of inhaled vapor Hg^0 dose readily absorbed in lungs. Oral: GI absorption of metallic Hg is poor; any released vapor in GI tract converted to mercuric sulfide and excreted. Dermal: Absorption of Hg^0 vapor through human skin is very low relative to inhalation absorption	Inhalation: Aerosols of $HgCl_2$ absorbed. Oral: 7-15% of ingested dose of $HgCl_2$ absorbed from the GI tract; absorption proportional to water solubility of mercuric salt; uptake by neonates greater than adults. Dermal: In guinea pigs, 2-3% of applied dose of $HgCl_2$ absorbed
Distribution		
Distributed throughout body since lipophilic; ~1-10% of absorbed oral dose of MeHg distributed to blood; 90% of blood MeHg in red blood cells; readily crosses blood-brain and placental barriers	Rapidly distributed throughout the body since it is lipophilic. Readily crosses blood-brain and placental barriers	Highest accumulation in kidney; fraction of dose retained in kidney is dose dependent. Does not readily penetrate blood-brain or placental barriers
Metabolism		
Slowly demethylated to Hg^{2+}	Hg(0) is oxidized in tissue and blood to Hg^{2+}	Hg^{2+} not methylated in tissue, but process may proceed in GI tract mediated by gut microorganisms
Excretion		
About 1% of burden released, mostly via bile and feces.	Excreted as Hg^0 in exhaled air, sweat, and saliva, and as mercuric Hg in feces and urine	Excreted in urine and feces; also excreted in saliva, bile, sweat, exhaled air, and breast milk

[#]Abbreviated from Table 2-2 in (NRC 2000)

and dry deposition (Fitzgerald and Lamborg 2004). Disposal of mercury-containing devices and incineration of mercury-containing waste contribute to its release into atmosphere. The use of liquid mercury in the extraction of gold by amalgamation has led to the release of large quantities of mercury directly into aquatic environments (Eisler 2004). A recent study has shown that methylation of mercury in wetlands increases with atmospheric sulfate loading (Jeremiason et al. 2006). This finding suggests that acid rain deposition is a confounding factor in the bioaccessibility of mercury in the environment.

Mercury in the body

Consumption of fish is the main source of exposure to mercury for humans. An estimated 95% of methylmercury contained in fish or shellfish is absorbed, with a significant fraction accumulating in the brain (Table 5). MeHg is slowly demethylated in the body to inorganic Hg(II), which is more readily excreted, primarily in the bile and then into feces. The ratio of MeHg to inorganic Hg(II) in tissue varies with duration of exposure and length of time after absorption. Postmortem studies have shown that as much as 80% of the mercury in the brain is inorganic Hg(II) (IPCS 1990).

Oral ingestion of liquid mercury results in very low absorption in the gastrointestinal tract. Less than 10% absorption is reported for oral ingestion of inorganic Hg(II) contained in food (Elinder et al. 1988). Inhalation of mercury vapor results in high absorption in the alveoli of the lungs, as much as 80%. Once dissolved in the blood, Hg(0) is enzymatically oxidized to Hg(II) and is partly associated with red blood cells. Hence, metallic mercury is considered a Fenton metal (Valko et al. 2005).

The ability to cross the blood-brain barrier contributes to the toxicity of methylmercury (Garcia et al. 2005). The mechanism(s) of toxicity are not fully understood (IPCS 1990). Poisoning by MeHg is known as Minamata Disease, named after an infamous incident in which thousands of people were exposed to fish contaminated with MeHg released from an industrial source. Currently, informal gold mining in the Amazon is affecting native Indians who rely on fish as a source of protein (Lodenius and Malm 1998; Hylander et al. 2000; Maurice-Bourgoin et al. 2000). An estimated 2000 tonnes of mercury has been released into the Amazon since 1980 (Malm 1998). Some of this mercury is transformed into MeHg which enters into the aquatic food chain. Population studies among Amazonian Indian communities show that individuals in communities eating contaminated fish are more likely to develop neurological diseases (Dolbec et al. 2000; Counter 2003) and some have mild symptoms of Minamata disease (Lodenius and Malm 1998).

METHODS FOR CHARACTERIZATION OF METAL SPECIATION

The choice of methods for speciation analysis typically depends on which metals need to be analyzed, which species are expected, and the physical forms of the materials to be analyzed (e.g., liquids vs. solids). The approaches that can be used in speciation analysis vary broadly since distinct chemical species encompass chemical compounds that can "differ in isotopic composition, conformation, oxidation or electronic state, or in the nature of their complexed or covalently bound substituents" (Templeton et al. 2000). The following overview of commonly used techniques is not meant to be either definitive or comprehensive. The number of techniques available to the researcher today is truly staggering and the techniques are rapidly evolving. The techniques that we highlight here reflect some of the research examples presented in this chapter. These brief descriptions should serve as a starting point for readers in assessing the techniques that will be most suitable for their given problems. It is also important to recognize that often it is necessary to use several techniques to obtain a more complete view of metal speciation (e.g., Scheinost et al. 2002).

For more comprehensive descriptions readers are referred to the "Handbook of Elemental Speciation: Techniques and Methodology" by Cornelis et al. (2003) and to review articles by Tack and Verloo (1995), Caruso et al. (2003), and D'Amore et al. (2005). But it must be emphasized that in applying any of these techniques particular attention should be given to ensure that sampling, sample storage, and sample preparation do not alter either the metal concentrations or other chemical properties for which the speciation analysis is being conducted.

Equilibrium modeling

Frequently, equilibrium modeling is a critical step in establishing the identity of chemical species likely to be present under given conditions, particularly in aquatic systems or in solutions. Equilibrium modeling generally assumes that all reactions in a system have gone to completion and is thus not typically suited to evaluating reaction kinetics. However, it can be very powerful in modeling equilibria among dissolved, adsorbed, solid, and gas phases as a function of pH and ionic strength. For many metal species, with just a measurement of solution pH and component concentrations a researcher can readily model the equilibrium abundances of metal ion species in the system. Numerous computer software programs exist to facilitate such modeling, more than can be described here. Some of the most commonly used programs include MINTEQA2 (U.S. EPA), PHREEQC (Parkhurst and Appelo 1999), and The Geochemists Workbench (Bethke 2002) (*http://www.rockware.com*). The user can typically input the total component concentrations and any invariant system parameters such as pH, pe, or gas partial pressures. In some cases you can also specify if given mineral phases are presumed to be present at equilibrium. The results of model calculations are not only dependent on user-defined constraints, but also the thermodynamic database used. In programs such as PHREEQC the user can select among several databases. The databases differ in the number of elements and species for which thermodynamic data are included. A substantial caution exists in that such modeling is only as good as the thermodynamic data used and the assumptions made to constrain the modeling.

Sequential extraction of solid species

Sequential extraction protocols in and of themselves are not methods of directly assessing speciation, but rather a means of isolating chemical species for analysis. For some solids, extraction protocols are necessary for effectively measuring their abundance. The number of sequential extraction protocols available for various metal species are numerous and are readily found by searching published articles. Many of these techniques are modifications of the procedure developed by Tessier et al. (1979) for the partitioning of particulate trace metals into five fractions: exchangeable, acid soluble (bound to carbonates), reducible (bound to Fe-Mn oxides), oxidizable (bound to organic matter or sulfides), and residual. In particular, an emphasis is placed on the choice of extraction reagents and their selectivity in each leaching solution. For example ammonium acetate is commonly used to liberate exchangeable metals, and sodium acetate or acetic acid are effective at acidic pH values for selective dissolution of the carbonate fraction.

Caution is needed, however, in applying extraction procedures to systems other than those for which the procedures were initially developed. Several studies have shown that results from sequential extraction procedures can vary dramatically when applied to terrestrial soils and sediments. For example, Scheckel et al. (2003) noted the problem of extracted metal species alteration before, during, and after separation of solids from solution. For soils spiked with Pb and amended with calcium phosphate, they observed significant shifts of extractable Pb to the residual phase during the sequential extraction steps, which were attributed to pyromorphite formation. Similarly, in the evaluation of As species in soils Mihaljevi et al (2003) demonstrated how the use of extraction techniques can be complicated by the occurrence of As extracts in the form of anionic complexes resulting from the dissociation of H_3AsO_4. Moreover, sequential extraction protocols are generally found to be problematic for As and Se (Gruebel et al. 1988).

Electrochemical methods: voltammetry and polarography

Voltammetry allows metal concentrations in solutions to be measured at extremely low levels, often at parts per trillion concentrations. Some organic compounds can also be measured. It is also an excellent tool for measuring metal oxidation states and differentiating between free and bound metal ions in solution. Voltammetry measures the current at a working electrode as a function of the applied potential, where the current is proportional to the analyte concentration in solution. Working electrodes can be made of several conducting materials such as gold, platinum, or carbon. In polarography, a dropping mercury electrode is used to continuously renew the electrode surface. The method is really only directly applicable to the analysis of species in solution and care must be taken to ensure that sample preparation does not induce changes in speciation. Nevertheless, it is a technique well suited for examining metal speciation in aquatic systems and has been widely applied to *in situ* measurements on natural waters (Van Leeuwen et al. 2005). It is also applicable to analysis of metals in bodily fluids, for example the analysis of blood lead levels (Bannon and Chisolm 2001). In cyclic voltammetry the scan rate of the experiment can be "tuned" to the kinetics of the electrochemical reaction, making it also useful for examining redox kinetics.

Liquid membrane techniques

As with voltammetry, liquid membrane techniques are suitable for measuring the concentration and distribution of chemical species in liquids, including natural waters, plant extracts, and bodily fluids. Detection limits are typically on the order of micrograms per liter. Dialysis is a familiar liquid-membrane technique. Different membrane techniques exist based on the type of membrane material. The permeation liquid membrane (PLM) technique, for example, is based on liquid-liquid extraction principles. A polymer membrane is used that supports a hydrophobic ligand as a carrier molecule. The membrane separates the solution to be analyzed (containing a natural ligand) from a receiving, strip solution, setting up a chemical potential gradient. The flux through the membrane can then be measured and related to the concentration of the free ion or labile metal complex. The Donnan membrane technique (DMT) is similar, but uses a cation exchange membrane to measure free ion activity using the Donnan principle (Lampert 1982). Detection limits are typically on the order of milligrams per liter. Here, the analyte solution is separated from an anion containing receiving solution by a negatively charged semipermeable membrane (Zhang and Young 2006). Because of the negative charge of the membrane, cations move across the membrane to the acceptor solution.

Liquid chromatography

There are numerous chromatographic methods used for speciation analysis. In chromatography an analyte is passed through a stationary phase (such as an ion exchange resin) as a gas or liquid, with each component having a distinct retention time. In essence, all chromatographic methods are separation techniques that take advantage of differences in partitioning, adsorption, ion exchange behavior, or molecular size of the analyte species relative to the stationary phase. For the speciation analysis of environmental samples the most commonly employed chromatographic techniques are typically based on some form of liquid chromatography such as high-performance liquid chromatography (HPLC) or ion chromatography. The exact technique to be used is dependent on the species of interest. Liquid and gas chromatographs are also commonly interfaced to element specific detection systems such as mass spectrometers (LCMS and GCMS) or nuclear magnetic resonance systems (LC/NMR). Such coupled devices ('hyphenated' techniques) have detection limits at low-femtomole levels. Such combined systems can elute species sequentially from the chromatographic column directly to the mass spectrometer for compound-specific identification. For example, such techniques have been used in measuring Se speciation in Se-enriched garlic (Dumont et al. 2006).

Capillary electrophoresis

Capillary electrophoresis (CE) can be used for the separation of a wide variety of species, of inorganic and organic ions and of varying ionic radius. The technique relies on ion migration of charged electrical species in an electrolyte in the presence of an electric current. The use of capillaries facilitates automated analysis and peak detection. A number of variants of the technique exist that allow for separation based on size and charge differences between species. As with chromatography, CE systems are typically coupled to detector systems such as UV absorbance detectors, diode detectors, fluorescence detectors, or as hyphenated techniques coupled to mass spectrometers. Advantages of the technique are high separation efficiency, high speed of analysis, and the ability to use very small amounts of sample. As an example, Forte et al. (2005) were able to use CE analysis to attain the simultaneous separation of seven arsenical species: arsenite (As III), arsenate (As V), monomethylarsonic acid (MMA), dimethylarsinic acid (DMA), arsenobetaine (AsBet), arsenocholine (AsCh) and p-arsanilic acid (pAs), as well as quantify the abundance of each species.

Inductively coupled mass spectrometry (ICP-MS)

As described above, ICP-MS systems are now commonly coupled to GC and HPLC separation systems. In fact, HPLC-ICP-MS may arguably be the most widely used technique for speciation analysis in aqueous media. In ICP-MS, plasma is used to atomize and ionize the species in a sample. The species are then identified by their mass-to-charge ratio. Detection limits for ICP-MS are typically better than sub ng/L. ICP-MS can detect a wide range of species containing trace elements, with a very large dynamic range (Rosen and Hieftje 2004). It can thus be used for analysis of a wide range of both trace and major elements from the same sample dilution. However, in this mode serious interference problems can exist, which must be evaluated during analysis. For example, low levels of arsenic cannot be analyzed in high-chloride solutions, such as sea water, using standard ICP-MS due to interferences; as a result, specialized ICP-MS systems, such as high-resolution ICP-MS or dynamic-reaction chamber (DRC) ICP-MS techniques, are needed for analysis of low levels of As, Mo, Se, and some other elements in sea water. There can also be problems with the composition of the solvent injected into the nebulizer leading to plasma instability.

Magnetic spectroscopies (NMR, EPR, and Mössbauer spectroscopy)

Magnetic spectroscopies exploit the magnetic properties of nuclei. Nuclear Magnetic Resonance (NMR) spectroscopy, Electron Paramagnetic Resonance (EPR) spectroscopy, and Mössbauer spectroscopy are the three most familiar magnetic techniques used for speciation analysis. NMR spectroscopy uses the fact that when NMR-active nuclei are placed in a magnetic field of a given strength, they resonate at a specific frequency, dependent on strength of the magnetic field. Since different chemical groups in a compound resonate at slightly different frequencies, the "chemical shift" imparted to the proton spectrum by each component can be used to identify the compound. NMR techniques are restricted to the analysis of non-ferromagnetic elements with spin ½ nuclear isotopes. Detection sensitivities are on the order of 10-10,000 ppm (D'Amore et al. 2005). Electron paramagnetic resonance (EPR) or electron spin resonance (ESR) is analogous to NMR, but detects species that have unpaired electrons. Thus transition-metal and rare-earth species that contain unpaired electrons can readily be analyzed. Organic molecules that contain unpaired electrons can also be analyzed. EPR usually requires microwave-frequency radiation (GHz), while NMR is observed at lower radio frequencies (MHz). NMR and EPR are generally nondestructive techniques as they utilize non-ionizing radiation. Mössbauer spectroscopy utilizes the fact that when some solid samples are exposed to a beam of gamma radiation, the intensity of the beam transmitted through the sample will change as a function of gamma-ray absorption because there is a lack of recoil. Mössbauer analysis is restricted to elements that display this behavior. Most notably for environmental analysis, Fe, Cs, Ba, Ni, Zn, Sn, Sb, and Hg can be analyzed. Some actinides such as Th, Pa, U, Np, and Pu can also be analyzed. Detection limits are on the order of 1-1000 ppm, and since the technique can be used to detect shifts in transition energy around the absorbing atom, it is well suited for detecting changes in valance and site substitutions (D'Amore et al. 2005).

X-ray diffraction and X-ray absorption spectroscopy

A number of X-ray techniques can be used to evaluate metal speciation in environmental and biological materials. The three most commonly used techniques are X-ray diffraction (XRD), X-ray fluorescence (XRF), and X-ray absorption spectroscopy (XAS). Many readers will be familiar with XRD, which provides information on crystal structure and phase identity from the Bragg scattering vectors from X-rays interacting with a periodic array of atoms. It is worth noting that neutron diffraction offers benefits for certain materials, particularly where structural information for protons or water is needed. Although most commonly used for crystalline materials, scattering techniques may also be useful for characterization of non-crystalline materials.

We also assume that many readers are familiar with X-ray fluorescence (XRF), which provides quantitative elemental analysis of materials in solid, liquid, and even gaseous forms. The increased availability of synchrotron X-ray sources, with their extremely high brightness and brilliance, has allowed improvements in spatial resolution and sensitivity of X-ray methods by orders of magnitude relative to laboratory X-ray tube sources (Sutton et al. 2003). Synchrotron-based XRF can detect metal concentrations with a sensitivity of approximately 1 fg. Synchrotron-based XRD with spatial resolution better than 10 μm is now possible, allowing for *in situ* phase identification of complex, heterogeneous materials. X-ray absorption spectroscopy (XAS) is an element-specific technique. Largely restricted to synchrotron sources, XAS is now routinely performed on samples with metal abundances as low as 1 to a few tens of ppm, depending on the metal and the nearest-neighbor atoms. X-ray absorption near-edge structure spectroscopy (XANES) provides direct information on oxidation state and coordination, whereas extended X-ray absorption fine structure spectroscopy (EXAFS) allows characterization of the number and type of neighboring atoms and their distance from the absorbing element. One of the primary advantages of these synchrotron-based X-ray techniques for assessing speciation (compared to those discussed above) is that samples need not be disturbed or destroyed for study; XAS allows direct, *in situ* characterization (Manceau et al. 2003). For example, synchrotron-based XAS methods can identify the chemical speciation and elemental associations of arsenic in dilute (<100 mg kg^{-1}) heterogeneous materials, such as poultry litter, with high spatial resolution (Arai et al. 2003), or measure the speciation of lead within organisms such as soil nematodes *in situ* with no need for species separation, pre-concentration, or pre-treatment (Jackson et al. 2005). Most of the methods previously described require separation or processing (e.g., dissolution for the analysis of non-liquid samples). It is not surprising, therefore, that synchrotron-based techniques have become so widely used in speciation analysis.

FUTURE RESEARCH AND ROLE OF GEOCHEMISTS

Geochemistry and mineralogy have been integral parts of exposure studies for many years. In particular the insight provided into the speciation of metals and the processes that control speciation, mobility, and ultimately bioaccessibility represent important contributions. A new frontier will be the integration of geochemical concepts and methods with molecular toxicology to better understand the mechanisms by which metals cause adverse health effects. One of the challenges is to identify and incorporate all relevant biological ligands into models. For example, ligands such as glutathione (Fig. 24) may be present at millimolar concentrations in different physiological fluids and are known to have a high affinity for certain metal species. However, in many cases the necessary metal complexation equilibria and redox equilibria are not yet known. One possible path forward is to use approximation and extrapolation techniques that have served geochemist well over the last half century to obtain estimates of the most relevant equilibria.

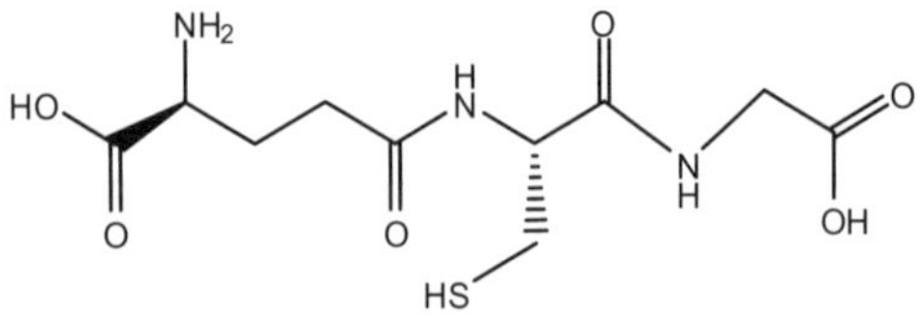

Figure 24. Glutathione is one of many biologically important molecules that complex metal species, with the sulfhydryl group being a primary binding site.

Microspatial speciation analysis of biological tissue samples represents another major challenge. While total metal analysis of tissues is routine, microspatial and in many cases redox-specific microspatial analysis is expected to offer new insights into the role of metals in disease. Instrumental techniques familiar to many geochemists and mineralogists, such as synchrotron-based micro-XRF and micro-XAS, are now being used more commonly for speciation analysis in biological tissues. With continued improvements in spatial resolution and signal quality provided by new generation synchrotron sources, further advances in speciation of metals will offer new opportunities for research.

ACKNOWLEDGMENTS

We thank our many colleagues and students who share an interest in multidisciplinary research. Support for RJR and MAAS was partially provided by the Center for Environmental Molecular Science (NSF grant CHE-0221934). AL was supported by Department of Energy (DOE) - Geosciences (grant DE-FG02-92ER14244). We thank Geoff Plumlee, Blair Jones, and an anonymous reviewer for helpful comments that improved this work. We also thank Nita Sahai for her patience during manuscript preparation.

REFERENCES

Alzieu C (1996) Biological effects of tributyltin on marine organisms. *In:* Tributyltin: Case Study of an Environmental Contaminant. de Mora SJ (ed) Cambridge University Press, p 167-211

Anderson LCD, Bruland KW (1991) Biogeochemistry of arsenic in natural waters: the importance of methylated species. Environ Sci Technol 25:420-427

Anderson MA, Rubin AJ (1981) Adsorption of Inorganics at Solid-Liquid Interfaces. Ann Arbor Sciences

Anderson RA (1986) Chromium metabolism and its role in disease processes in man. Clin Physiol Biochem 4:31-41

Arai Y, Elzinga EJ, Sparks DL (2001) X-ray absorption spectroscopic investigation of arsenite and arsenate adsorption at the aluminum oxide-water interface. J Colloid Interface Sci 235:80-88

Arai Y, Lanzirotti A, Sutton S, Davis J, Sparks D (2003) Arsenic speciation and reactivity in poultry litter. Environ Sci Technol 37:4083-4090

ATSDR (2005a) Toxicological Profile for Arsenic. U.S. Department of Health and Human Services-Agency for Toxic Substances and Disease Registry, *http://www.atsdr.cdc.gov/toxprofiles/tp2.html*

ATSDR (2005b) Toxicological Profile for lead. (Draft for Public Comment). U.S. Department of Health and Human Services-Agency for Toxic Substances and Disease Registry, *http://www.atsdr.cdc.gov/toxprofiles/tp13.html*

ATSDR (2005c) Toxicological profile for tin and tin compounds U.S. Department of Health and Human Services-Agency for Toxic Substance and Disease Registry, *http://www.atsdr.cdc.gov/toxprofiles/tp55.html*

ATSDR (2000) Toxicological Profile for Chromium. U.S. Department of Health and Human Services-Agency for Toxic Substances and Disease Registry, *http://www.atsdr.cdc.gov/toxprofiles/tp7.html*

Badmaev V, Prakash S, Majeed M (1999) Vanadium: a review of its potential role in the fight against diabetes. J Altern Complement Med 5:273-291

Baecklund M, Pedersen NL, Bjorkman L, Vahter M (1999) Variation in blood concentrations of cadmium and lead in the elderly. Environ Res 80:222-230

Bannon DI, Chisolm JJ (2001) Anodic stripping voltammetry compared with graphite furnace atomic absorption spectrophotometry for blood lead analysis. Clin Chem 47:1703-1704

Barceloux DG (1999a) Molybdenum. J Toxicol Clin Toxicol 37:231-237

Barceloux DG (1999b) Nickel. J Toxicol Clin Toxicol 37:239-258

Barry PS (1981) Concentrations of lead in the tissues of children. Br J Ind Med 38:61-71

Barry PS (1975) Letter: Lead levels in blood. Nature 258:775-775

Bartlett RJ, James BR (1993) Redox chemistry of soils. Adv Agronomy 50:151-208

Baxter D, Frech W (1995) Speciation of lead in environmental and biological samples. Pure Appl Chem 67: 615-648

Bernhard M, Brinckman FE, Irgolic KJ (1986) Why Speciation? *In:* The Importance Of Chemical Speciation in Environmental Processes. Bernhard M, Brinckman FE, Sadler PJ (eds) Springer-Verlag, p 7-14

Bethke CM (2002) The Geochemist's Workbench - release 4.0, A user's guide to Rxn, Act2, Tact, React and Gtplot, University of Illinois

Blake KC, Mann M (1983) Effect of calcium and phosphorus on the gastrointestinal absorption of ^{203}Pb in man. Environ Res 30:188-194

Blake KCH, Barbezat GO, Mann M (1983) Effect of dietary constituents on the gastrointestinal absorption of ^{203}Pb in man. Environ Res 30:182-187

Bothe JV, Brown PW (1999) Arsenic immobilization by calcium arsenate formation. Environ Sci Technol 33: 3806-3811

Bouton CM, Frelin LP, Forde CE, Arnold Godwin H, Pevsner J (2001) Synaptotagmin I is a molecular target for lead. J Neurochem 76:1724-1735

Bragt PC, van Dura EA (1983) Toxicokinetics of hexavalent chromium in the rat after intratracheal administration of chromates of different solubilities. Ann Occup Hyg 27:315-322

Bright DA, Dodd M, Reimer KJ (1996) Arsenic in subarctic lakes influenced by gold mine effluent: the occurrence of organoarsenicals and arsenic. Sci Total Environ 180:165-182

Brown GE, Foster AL, Ostergren JD (1999) Mineral surfaces and bioavailability of heavy metals: A molecular-scale perspective. Proc Nat Acad Sci 96:3388-3395

Brown GE, Parks GA (2001) Sorption of trace elements on mineral surfaces: Modern perspectives from spectroscopic studies, and comments on sorption in the marine environment. Int Geol Rev 43:963-1073

Brown GE (1990) Spectroscopic studies of chemisorption reaction mechanisms at oxide-water interfaces. Rev Mineral 23:309-363

Buchet JP, Lauwerys R, Roels H (1981) Comparison of the urinary excretion of arsenic metabolites after a single oral dose of sodium arsenite, monomethylarsonate or dimethylarsinate in man. Int Arch Occup Environ Health 48:71-79

Buerge IJ, Hug SJ (1999) Influence of mineral surfaces on chromium(VI) reduction by iron(II). Environ Sci Technol 33:4285-4291

Campbell PGC (1995) Interactions between trace metals and aquatic organisms: A critique of the free-ion activity model. *In:* Metal Speciation and Bioavailability in Aquatic Systems. Tessier A, Turner DR (eds) John Wiley and Sons, 45-102

Caruso JA, Klaue B, Michalke B, Rocke DM (2003) Group assessment: elemental speciation. Ecotox Environ Safety 56:32-44

Casteel SW, Weis CP, Henningsen GM, Brattin WJ (2006) Estimation of relative bioavailability of lead in soil and soil-like materials using young swine. Environ Health Perspect 114:1162–1171

CDC (1985) Preventing lead poisoning in young children, a statement by the Centers for Disease Control. US Department of Health and Human Services, 44, *http://aepo-xdv-www.epo.cdc.gov/wonder/prevguid/ p0000029/p0000029.asp*

Chamberlain A, Heard C, Little MJ (1978) Investigations into lead from motor vehicles. Harwell, United Kingdom. Phil Trans Royal Soc London A 290:557-589

Charlet L, Manceau A (1993) Structure, formation and reactivity of hydrous oxide particles; insights from X-ray absorption spectroscopy. *In:* Environmental Particles. Buffle J, Leeuwen HPV (eds) Lewis Publishers, p 117-164

Christl I, Metzger A, Heidmann I, Kretzschmar R (2005) Effect of humic and fulvic acid concentrations and ionic strength on copper and lead binding. Environ Sci Technol 39:5319-5326

Cornelis R, Caruso J, Crews H, Heumann K (2003) Handbook of Elemental Speciation, Wiley

Corrin ML, Natusch DFS (1977) Physical and chemical characteristics of environmental lead. *In:* Lead in the Environment. Boggess WR, Wixson BG (eds) National Science Foundation, p 7-31

Cotter-Howells J (1996) Lead phosphate formation in soils. Environ Pollution 93:9-16

Cotter-Howells J, Champness P, Chamock J, Pattrick R (1994) Identification of pyromorphite in mine-waste contaminated soils by ATEM and EXAFS. Eur J Soil Sci 45:393-402

Counter SA (2003) Neurophysiological anomalies in brainstem responses of mercury-exposed children of Andean gold miners. J Occup Environ Med 45:87-95

Cullen WR, Reimer KJ (1989) Arsenic speciation in the environment. Chem Rev 89:713-764

D'Amore JJ, Al-Abed SR, Scheckel KG, Ryan JA (2005) Methods for speciation of metals in soils: A review. J Environ Qual 34:1707–1745

Davies BE, Bowman C, Davies TC, Selinus O (2005) Medical geology: perspectives and prospects. *In*: Essentials of Medical Geology. Selinus O (ed) Elsevier, p 1-14

Davis A, Drexler JW, Ruby MV, Nicholson A (1993) Micromineralogy of mine wastes in relation to lead bioavailability, Butte. Environ Sci Technol 27:1415-1425

Davis JA, Kent DB (1990) Surface complexation modeling in aqueous geochemistry. Rev Mineral Geochem 23:177-260

Dawson DC, Ballatori N (1995) Membrane transporters as sites of action and routes of entry for toxic metals. *In*: Toxicology of metals: biochemical aspects. Goyer RA, Cherian MG (eds) Springer-Verlag, p 53-76

De Flora S, Badolati GS, Serra D (1987) Circadian reduction of chromium in the gastric environment. Mutation Research 192:169-174

de Mora SJ (1996) The tributyltin debate: ocean transportation versus seafood harvesting. *In*: Tributyltin: Case Study of an Environmental Contaminant. de Mora SJ (ed) Cambridge University Press, p 1-20

De Voss JJ, Rutter K, Schroeder BG, Barry CEI (1999) Iron acquisition and metabolism by mycobacteria. J Bacteriol 181:4443-4451

DeJonghe WRA, Adams FC (1986) Biogeochemical cycling of organic lead compounds. Adv Environ Sci Technol 17:561-594

Diez S, Abalos M, Bayona JM (2002) Organotin contamination in sediments from the Western Mediterranean enclosures following 10 years of TBT regulation. Water Res 36:905-918

DOE (1995) Molecular Environmental Science: Speciation, Reactivity, and Mobility of Environmental Contaminants: An Assessment of Research Opportunities and the Need for Synchrotron Radiation Facilities. Stanford Synchrotron Radiation Laboratory, SLAC-R-477

Dolbec J, Mergler D, Sousa Passos CJ, Sousa de Morais S, Lebel J (2000) Methylmercury exposure affects motor performance of a riverine population of the Tapajós river, Brazilian Amazon. Int Arch Occup Environ Health 73:195-203

Dopp E, Hartmann LM, Florea AM, Rettenmeier AW, Hirner AV (2004) Environmental distribution, analysis, and toxicity of organometal(loid) compounds. Crit Rev Toxicol 34:301-333

Duffus JH (2002) "Heavy metals"—A meaningless term? Pure Appl Chem 74:793–807

Dumont E, Ogra Y, Vanhaecke F, Suzuki KT, Cornelis R (2006) Liquid chromatography-mass spectrometry (LC-MS): a powerful combination for selenium speciation in garlic (Allium sativum). Anal Bioanal Chem 384:1196-1206

Eary LE, Rai D (1988) Chromate removal from aqueous wastes by reduction with ferrous ion. Environ Sci Technol 22:972-977

Ehlers LJ, Luthy RG (2003) Contaminant bioavailability in soil and sediment. Environ Sci Technol 37:295A-302A

Eisenreich SJ, Looney BB, Thornton JD (1981) Airborne organic contaminants in the Great Lakes ecosystem. Environ Sci Technol 15:30-38

Eisler R (2004) Mercury hazards from gold mining to humans, plants, and animals. Rev Environ Contam Toxicol 181:139-198

Elbaz-Poulichet F, Holliger P, Huang WW (1984) Lead cycling in estuaries, illustrated by the Gironde Estuary, France. Nature 308:409-414

Elinder CG, Gerhardsson L, Oberdörster G (1988) Biological monitoring of toxic metals — Overview. *In:* Biological Monitoring of Toxic Metals. Clarkson TW, Friberg L, Nordberg GF, Sager PR (eds), Plenum Press, p 1-72

Ellenhorn M, Barceloux D (1988) Medical Toxicology: Diagnosis and Treatment of Human Poisoning. Elsevier

Elzinga EJ, Reeder RJ (2002) X-ray absorption spectroscopy study of Cu^{2+} and Zn^{2+} adsorption complexes at the calcite surface. Geochim Cosmochim Acta 66:3943-3954

EPA (2005) Estimation of Relative Bioavailability of Arsenic in Soil and Soil-Like Materials by In Vivo and In Vitro Methods - USEPA Review Draft. U.S. Environmental Protection Agency, Region 8, Denver, CO

EPA (1998a) Toxicological Review of Hexavalent Chromium. U.S. Environmental Protection Agency

EPA (1998b) Toxicological Review Of Trivalent Chromium. U.S. Environmental Protection Agency

EPA (1996) U.S. Environmental Protection Agency, Federal Registry # 61:3832

EPA (1986) Air Quality Criteria for Lead. U.S. Environmental Protection Agency, Office of Research and Development, Office of Health and Environmental Assessment, Environmental Criteria and Assessment Office

EPA (1985) Lead Exposures in the Human Environment. Environmental Criteria and Assessment Office, U.S. Environmental Protection Agency

Fendorf S, Wielinga BW, Hansel CM (2000) Chromium transformations in natural environments: The role of biological and abiological processes in chromium(VI) reduction. Int Geol Rev 42:691-701

Fendorf SE (1995) Surface reactions of chromium in soils and waters. Geoderma 67:55-71

Fendorf SE, Fendorf M, Sparks DL, Gronsky R (1992) Inhibitory mechanisms of Cr(III) oxidation by δ-MnO$_2$. J Colloid Interface Sci 153:37-54

Feroci G, Fini A, Fazio G (1995) Interaction between dihydroxy bile salts and divalent heavy metal ions studied by polarography. Anal Chem 67:4077-4085

Finley BL, Scott PK, Norton RL (1996) Urinary chromium concentrations in humans following ingestion of safe doses of hexavalent and trivalent chromium: Implications for biomonitoring. J Toxicol Environ Health 48:479-499

Fitsanakis VA, Aschner M (2005) The importance of glutamate, glycine, and gamma-aminobutyric acid transport and regulation in manganese, mercury and lead neurotoxicity. Toxicol Appl Pharmacol 204:343-354

Fitzgerald WF, Lamborg CH (2004) Geochemistry of mercury in the environment. *In*: Treatise on Geochemistry. Lollar SB (ed) Elsevier, p 107-148

Fordyce F (2005) Selenium deficiency and toxicity in the environment. *In*: Essentials of Medical Geology. Selinus O (ed) Elsevier, p 373-415

Forte G, D'Amato M, Caroli S (2005) Capillary electrophoresis speciation analysis of various arsenical compounds. Microchem J 79:15-19

Foster AL (2003) Spectroscopic investigations of arsenic species in solid phases. *In*: Arsenic in Ground Water. Welch AH, Stollenwerk KG (eds) Kluwer Academic Publishers, p 27-65

Foster AL, Brown GE, Parks GA (1998) X-ray absorption fine-structure spectroscopy study of photocatalyzed, heterogeneous As(III) oxidation on kaolin and anatase. Environ Sci Technol 32:1444-1452

Foulkes EC (2000) Transport of toxic heavy metals across cell membranes. Proc Soc Exp Biol Med 223:234-240

Fredrickson JK, Zachara JM, Kennedy DW, Duff MC, Gorby YA, Li S-mW, Krupka KM (2000) Reduction of U(VI) in goethite (α-FeOOH) suspensions by a dissimilatory metal-reducing bacterium. Geochim Cosmochim Acta 64:3085-3098

Freeman GB, Dill JA, Johnson JD, Kurtz PJ, Parham F, Matthews HB (1996) Comparative absorption of lead from contaminated soil and lead salts by weanling Fischer 344 rats. Fundam Appl Toxicol 33:109-119

Fuller CC, Davis JA, Waychunas GA (1993) Surface chemistry of ferrihydrite: Part 2. Kinetics of arsenate adsorption and coprecipitation. Geochim Cosmochim Acta 57:2271-2282

Gaetke LM, Chow CK (2003) Copper toxicity, oxidative stress, and antioxidant nutrients. Toxicology 189:147-163

Gaggelli E, Berti F, D'Amelio N, Gaggelli N, Valensin G, Bovalini L, Paffetti A, Trabalzini L (2002) Metabolic pathways of carcinogenic chromium. Environ Health Perspect Suppl 110:733-738

Gailer J, George GN, Pickering IJ, Prince RC, Ringwald SC, Pemberton JE, Glass RS, Younis HS, DeYoung DW, Aposhian HV (2000) A metabolic link between arsenite and selenite: The seleno-bis(*S*-glutathionyl) arsenium ion. J Am Chem Soc 122:4637-4639

Gailer J, George GN, Pickering IJ, Prince RC, Younis HS, Winzerling JJ (2002) Biliary excretion of [(GS)$_2$AsSe]$^-$ after intravenous injection of rabbits with arsenite and selenate. Chem Res Toxicol 15:1466-1471

Garcia JJ, Martínez-Ballarín E, Millán-Plano S, Allué JL, Albendea C, Fuentes L, Escanero JF (2005) Effects of trace elements on membrane fluidity. J Trace Elem Med Biol 19:19-22

Gargas ML, Norton RL, Paustenbach DJ, Finley BL (1994) Urinary excretion of chromium by humans following ingestion of chromium picolinate. Implications for biomonitoring. Drug Metab Dispos 22:522-529

Gilman A, Rall T, Nies A, Taylor P (1990) Goodman and Gilman's The Pharmacological Basis of Therapeutics. Pergamon Press

Glew DN, Hames DA (1971) Aqueous nonelectrolyte solutions. Part X. Mercury solubility in water. Canadian J Chem 49:3114-3118

Godwin H (2001) The biological chemistry of lead. Curr Opinion Chem Biol 5:223-227

Goldberg S (2002) Competitive adsorption of arsenate and arsenite on oxides and clay minerals. Soil Sci Soc Am J 66:413-421

Goldhaber SB (2003) Trace element risk assessment: essentiality vs. toxicity. Reg Toxicol Pharmacol 38:232-242

Greger M, Wang Y, Neuschutz C (2005) Absence of Hg transpiration by shoot after Hg uptake by roots of six terrestrial plant species. Environ Pollut 134:201-208

Gruebel K, Davis J, Leckie J (1988) The feasibility of using sequential extraction techniques for arsenic and selenium in soils and sediments. Soil Sci Soc Am J 52:390-397

Guidotti TL (2005) Toxicology. *In:* Essentials of Medical Geology. Selinus O (ed) Elsevier, p 595-608

Gurzau ES, Neagu C, Gurzau AE (2003) Essential metals--case study on iron. Ecotoxicol Environ Safety 56: 190-200

Hartikainen H (2005) Biogeochemistry of selenium and its impact on food chain quality and human health. J Trace Elements Med Biol 18:309-318

Heard MJ, Chamberlain AC (1982) Effect of minerals and food on uptake of lead from the gastrointestinal tract in humans. Human Toxicol 1:411-415

Heard MJ, Wells AC, Newton D (1979) Human uptake and metabolism of tetraethyl and tetramethyl lead vapour labeled with [203]Pb. *In:* International Conference on Management and Control of Heavy Metals in the Environment. CEP Consultants, Ltd., p 103-108

Hering JG, Kneebone PE (2002) Biogeochemical controls on arsenic occurrence and mobility in water supplies. *In:* Environmental Chemistry of Arsenic. Frankenberger WT (ed) Marcel Dekker, p 155-181

Hingston FJ (1981) A review of anion adsorption. *In:* Adsorption of Inorganics at Solid-Liquid Interfaces. Anderson MA, Rubin AJ (eds) Ann Arbor Science, p 51-90

Hochella MF, White AF (1990) Mineral-Water Interface Geochemistry. Reviews in Mineralogy, Vol. 23. Mineralogical Society of America

Hong SM, Candelone JP, Patterson CC, Boutron CF (1994) Greenland ice evidence of hemispheric lead pollution 2-millennia ago by Greek and Roman civilizations. Science 265:1841-1843

Hudson RJM (1998) Which aqueous species control the rates of trace metal uptake by aquatic biota? Observations and predictions of non-equilibrium effects. Sci Total Environ 219:95-115

Hylander LD, Pinto FN, Guimarães JR, Meili M, Oliveira LJ, de Castro e Silva E (2000) Fish mercury concentration in the Alto Pantanal, Brazil: influence of season and water parameters. Sci Total Environ 261:9 - 20

Ilton ES, Veblen DR (1994) Chromium sorption by phlogopite and biotite in acidic solutions at 25 °C: Insights from X-ray photoelectron spectroscopy and electron microscopy. Geochim Cosmochim Acta 58:2777-2788

Inskeep WP, McDermott TR, Fendorf S (2002) Arsenic (V)/(III) cycling in soils and natural waters: Chemical and microbial processes. *In:* Environmental Chemistry of Arsenic. Frankenberger WT (ed) Marcel Dekker, p 183-215

IPCS (2001) Arsenic and Arsenic Compounds. *In:* Environmental Health Criteria 224. World Health Organization

IPCS (1990) Methylmercury. *In:* Environmental Health Criteria 101. World Health Organization

IPCS (1988) Chromium. *In:* Environmental Health Criteria 61. World Health Organization

Jackson B, Williams P, Lanzirotti A, Bertsch P (2005) Evidence for biogenic pyromorphite formation by the nematode Caenorhabditis elegans. Environ Sci Technol 39:5620-5625

Jain A, Loeppert RH (2000) Effect of competing anions on the adsorption of arsenate and arsenite by ferrihydrite. J Environ Qual 29:1422-1430

James BR, Bartlett RJ (1983a) Behavior of chromium in soils. VI. Interactions between oxidation-reduction and organic complexation. J Environ Qual 12:173-176

James BR, Bartlett RJ (1983b) Behavior of chromium in soils. VII. Adsorption and reduction of hexavalent forms. J Environ Qual 12:177-181

James HM, Hilburn ME, Blair JA (1985) Effects of meals and meal times on uptake of lead from the gastrointestinal tract in humans. Human Toxicol 4:401-407

Jeremiason JD, Engstrom DR, Swain EB, Nater EA, Johnson BM, Almendinger JE, Monson BA, Kolka RK (2006) Sulfate addition increases methylmercury production in an experimental wetland. Environ Sci Technol 40:3800-3806

Johnson CC, Ge X, Green KA, Liu X (2000) Selenium distribution in the local environment of selected villages of the Keshan Disease belt, Zhangjiakou District, Hebei Province, People's Republic of China. Appl Geochem 15:385-401

Juillot F (1998) Localisation et speciation de l'arsenic, du plomb et du zinc dans des sites et sols contaminés. Comparaison avec un sol développé sur une anomalie géochimique naturelle en plomb. Ph.D. Dissertation, Université Paris, Paris

Kafer A, Zoltzer H, Krug HF (1992) The stimulation of arachidonic acid metabolism by organic lead and tin compounds in human HL-60 leukemia cells. Toxicol Appl Pharmacol 116:125-132

Kimbrough DE, Cohen Y, Winer AM, Creelman L, Mabuni C (1999) A critical assessment of chromium in the environment. Crit Rev Environ Sci Technol 29:1-46

Köhrl J, Brigelius-Flohé R, Böck A, Gärtner R, Meyer O, Flohé L (2000) Selenium in biology: facts and medical perspectives. Biol Chem 381:849-864

Lampert JK (1982) Measurement of trace cation activities by Donnan membrane equilibrium and atomic adsorption analysis. Ph.D. Dissertation, University of Wisconsin, Madison

Langford N, Ferner R (1999) Toxicity of mercury. J Hum Hypertens 13:651-656

Le XC (2002) Arsenic speciation in the environment and humans. *In:* Environmental Chemistry of Arsenic. Frankenberger WT (ed) Marcel Dekker, p 95-116

Le XC, Ma M, Lu X, Cullen WR, Aposhian HV, Zheng B (2000) Determination of monomethylarsonous acid, a key arsenic methylation intermediate, in human urine. Environ Health Perspect 108:1015-1018

Levander OA (1977) Metabolic interrelationships between arsenic and selenium. Environ Health Perspect 19: 159-164

Levina A, Zhang L, Lay PA (2003) Structure and reactivity of a chromium(V) glutathione complex. Inorg Chem 42:767-784

Liger E, Charlet L, Cappellen PV (1999) Surface catalysis of uranium(VI) reduction by iron(II). Geochim Cosmochim Acta 63:2939-2955

Lin JH, Lu AYH (1997) Role of pharmacokinetics and metabolism in drug discovery and development. Pharmacol Rev 49:403-449

Lindh U (2005) Biological function of the elements. *In:* Essentials of Medical Geology. Selenius O, Alloway B, Centeno JA, Finkelman RB, Fuge R, Lindh U, Smedley P (eds) Elsevier Academic Press, p 115-160

Lodenius M, Malm O (1998) Mercury in the Amazon. Rev Environ Contam Toxicol 157:25-52

Long DT, Angino EE (1977) Chemical speciation of Cd, Cu, Pb, and Zn in mixed freshwater, seawater, and brine solutions. Geochim Cosmochim Acta 41:1183-1191

Lovley DR, Phillips EJP, Gorby YA, Landa ER (1991) Microbial reduction of uranium. Nature 350:413-416

Maddaloni M, Lolacono N, Manton W, Blum C, Drexler J, Graziano J (1998) Bioavailability of soilborne lead in adults, by stable isotope dilution. Environ Health Perspect Suppl 106:1589-1594

Maguire RJ (1996) The occurrence, fate, and toxicity of tributyltin and its degradation products in freshwater environments. *In:* Tributyltin: Case Study of an Environmental Contaminant. de Mora SJ (ed) Cambridge University Press, p 94-138

Malm O (1998) Gold mining as a source of mercury exposure in the Brazilian Amazon. Environ Res 77:73-78

Manceau A, Lanson B, Drits VA (2002) Structure of heavy metal sorbed birnessite. Part III: Results from powder and polarized extended X-ray absorption fine structure spectroscopy. Geochim Cosmochim Acta 66:2639-2663

Manceau A, Marcus M, Tamura N (2003) Quantitative speciation of heavy metals in soils and sediments by synchrotron X-ray techniques. Rev Mineral Geochem 49:341-428

Manning BA, Fendorf SE, Bostick B, Suarez DL (2002) Arsenic(III) oxidation and arsenic(V) adsorption reactions on synthetic birnessite. Environ Sci Technol 36:976-981

Manning BA, Goldberg S (1997) Adsorption and stability of arsenic(III) at the clay mineral-water interface. Environ Sci Technol 31:2005-2011

Matocha CJ, Elzinga EJ, Sparks DL (2001) Reactivity of Pb(II) at the Mn(III,IV) (oxyhydr)oxide-water interface. Environ Sci Technol 35:2967-2972

Maurice-Bourgoin L, Quiroga I, Chincheros J, Courau P (2000) Mercury distribution in waters and fishes of the upper Madeira rivers and mercury exposure in riparian Amazonian populations. Sci Total Environ 260:73-86

Mertz W, Roginski EE (1971) Chromium metabolism: The glucose tolerance factor. *In:* Newer Trace Elements in Nutrition. Mertz W, Cornatzer WE (eds) Dekker, p 123–153

Mihaljevi M, Po avi M, Ettler V, Šebek O (2003) A comparison of sequential extraction techniques for determining arsenic fractionation in synthetic mineral mixtures. Anal Bioanal Chem 377:723-729

Moon DH, Dermatas D, Menounou N (2004) Arsenic immobilization by calcium-arsenic precipitates in lime treated soils. Sci Total Environ 330:171-185

Morel FMM (1983) Principles of Aquatic Chemistry. Wiley-Interscience

Morin G, Juillot F, Ildefonse P, Calas G, Samama J, Chevallier P, Brown G (2001) Mineralogy of lead in a soil developed on a Pb-mineralized sandstone (Largentière, France). Am Mineral 86:92–104

Morin G, Ostergren J, Juillot F, Ildefonse P, Calas G, Brown G (1999) XAFS determination of the chemical form of lead in smelter-contaminated soils and mine tailings: Importance of adsorption processes. Am Mineral 84:420-434

Mota AM, Rato A, Brazia C, Goncalves MLS (1996) Competition of Al^{3+} in complexation of humic matter with Pb^{2+}: A comparative study with other ions. Environ Sci Technol 30:1970-1974

Mundell JA, Hill KR, Weaver JW (1989) In situ case history: leachable lead required precipitation immobilization. Hazard Waste Manage 12:23-27

Myneni SCB, Traina SJ, Logan TJ, Waychunas GA (1997) Oxyanion behavior in alkaline environments: Sorption and desorption of arsenate in ettringite. Environ Sci Technol 31:1761-1768

Nerin C, Domeno C, Garcia JI, et al. (1999) Distribution of Pb, V, Cr, Ni, Cd, Cu and Fe in particles formed from the combustion of waste oils. Chemosphere 38:1533-1540

Newton K, Amarasiriwardena D, Xing B (2006) Distribution of soil arsenic species, lead and arsenic bound to humic acid molar mass fractions in a contaminated apple orchard. Environ Pollut 143:197-205

Nordstrom DK, Archer DG (2003) Arsenic thermodynamic data and environmental geochemistry. *In:* Arsenic in Ground Water. Welch AH, Stollenwerk KG (eds) Kluwer Academic Publishers, p 1-25

Norling P, Wood-Black F, Masciangioli TM (2004) Water and Sustainable Development. National Academies Press

NRC (2003) Bioavailability of Contaminants in Soils and Sediments: Processes Tools, and Applications. National Academies Press

NRC (2001) Arsenic in Drinking Water: 2001 Update. National Academies Press

NRC (2000) Toxicological Effects of Methylmercury. National Academies Press

NRC (1999) Arsenic in Drinking Water. National Academies Press

Nriagu J (1984) Formation and stability of base metal phosphates in soils and sediments. *In:* Phosphate Minerals. Nriagu JO, Moore PB (eds) Springer-Verlag, p 318-329

Nriagu JO (1983) Occupational exposure to lead in ancient-times. Sci Total Environ 31:105-116

Nriagu JO (1974) Lead orthophosphates-IV. Formation and stability in the environment. Geochim Cosmochim Acta 38:887-898

Nriagu JO (1973) Lead orthophosphates-II. Stability of chloropyromorphite at 25 °C. Geochim Cosmochim Acta 37:367-377

Nriagu JO, Pacyna JM (1988) Quantitative assessment of worldwide contamination of air, water and soils by trace metals. Nature 333:134-139

NSF (1977) Transport and distribution in a watershed ecosystem. *In:* Lead in the environment: Chapter 6. Boggess W (ed) National Science Foundation, p 105-133

O'Day PA (2006) Chemistry and mineralogy of arsenic. Elements 2:77-83

O'Day PA (1999) Molecular environmental geochemistry. Rev Geophys 37:249–274

O'Flaherty EJ (1998) Physiologically based models of metal kinetics. Crit Rev Toxicol 28:271-317

O'Flaherty EJ (1996) A physiologically based model of chromium kinetics in the rat. Toxicol Appl Pharmacol 138:54-64

O'Flaherty EJ, Kerger BD, Hays SM, Paustenbach DJ (2001) A physiologically based model for the ingestion of chromium(III) and chromium(VI) by humans. Toxicol Sci 60:196-213

O'Reilly SE, Hochella MF (2003) Lead sorption efficiencies of natural and synthetic Mn and Fe-oxides. Geochim Cosmochim Acta 67:4471–4487

Oomen AG, Tolls J, Sips AJAM, Groten JP (2003a) In vitro intestinal lead uptake and transport in relation to speciation. Arch Environ Contam Toxicol 44:116-124

Oomen AG, Tolls J, Sips AJAM, Van den Hoop MAGT (2003b) Lead speciation in artificial human digestive fluid. Arch Environ Contam Toxicol 44:107-115

Oteiza PI, Mackenzie GG (2005) Zinc, oxidant-triggered cell signaling, and human health. Mol Aspects Med 26:245-255

Pacyna EG, Pacyna JM (2002) Global emission of mercury from anthropogenic sources in 1995. Water, Air, Soil Pollution 137:149-165

Parkhurst DL, Appelo CAJ (1999) User's guide to PHREEQC (version 2)--A computer program for speciation, batch-reaction, one-dimensional transport, and inverse geochemical calculations. Report 99-4259, US Geological Survey

Patterson RR, Fendorf S, Fendorf M (1997) Reduction of hexavalent chromium by amorphous iron sulfide. Environ Sci Technol 31:2039-2044

Peraza MA, Ayala-Fierro F, Barber DS, Casarez E, Rael LT (1998) Effects of micronutrients on metal toxicity. Environ Health Perspect Suppl 106:203-216

Peterson ML, Brown GE, Parks GA (1996) Direct XAFS evidence for heterogeneous redox reaction at the aqueous chromium/magnetite interface. Colloids Surf A 107:77-88

Peterson ML, Brown GE, Parks GA, Stein CL (1997) Differential redox and sorption of Cr (III/VI) on natural silicate and oxide minerals: EXAFS and XANES results. Geochim Cosmochim Acta 61:3399-3412

Pinheiro JP, Mota AM, van Leeuwen HP (1999) On lability of chemically heterogeneous systems: Complexes between trace metals and humic matter. Colloids Surf A 151:181-187

Plant JA, Kinniburgh DG, Smedley PL, Fordyce FM, Klinck BA (2005) Arsenic and Selenium. *In:* Environmental Geochemistry. Lollar BS (ed) Elsevier, p 17-66

Plumlee GS, Morman SA, Ziegler TL (2006) The toxicological geochemistry of earth materials: an overview of processes and the interdisciplinary methods used to understand them. Rev Mineral Geochem 64:5-57

Plumlee GS, Ziegler TL (2003) The medical geochemistry of dusts, soils and other earth materials. *In:* Treatise on Geochemistry. Lollar BS (ed) Elsevier, p 263-310

Pomroy C, Charbonneau SM, McCullough RS, Tam GKH (1980) Human retention studies with [74]As. Toxicol Appl Pharmacol 53:550-556

Rabinowitz MB, Kopple JD, Wetherill GW (1980) Effect of food intake and fasting on gastrointestinal lead absorption in humans. Am J Clin Nutrit 33:1784-1788

Rai D, Eary LE, Zachara JM (1989) Environmental chemistry of chromium. Sci Total Environ 86:15-23

Rai D, Sass BM, Moore DA (1987) Chromium(III) hydrolysis constants and solubility of chromium(III) hydroxide. Inorg Chem 26:345-349

Reddy KJ, Wang L, Gloss SP (1995) Solubility and mobility of copper, zinc and lead in acidic environments. Plant Soil 171:53-58

Reddy MM, Aiken GR (2001) Fulvic acid-sulfide ion competition for mercury ion binding in the Florida Everglades. Water, Air, Soil Pollution 132:89-104

Reeder RJ (1996) Interaction of divalent cobalt, zinc, cadmium, and barium with the calcite surface during layer growth. Geochim Cosmochim Acta 60:1543-1552

Renner R (2004) Plumbing the depths of D.C.'s drinking water crisis. Environ Sci Technol 38:224A-227A

Richmond WR, Loan M, Morton J, Parkinson GM (2004) Arsenic removal from aqueous solution via ferrihydrite crystallization control. Environ Sci Technol 38:2368-2372

Rochette EA, Bostick BC, Li G, Fendorf S (2000) Kinetics of arsenate reduction by dissolved sulfide. Environ Sci Technol 34:4714-4720

Rosen AL, Hieftje GM (2004) Inductively coupled plasma mass spectrometry and electrospray mass spectrometry for speciation analysis: applications and instrumentation. Spectrochim Acta B 59:135-146

Ruby MV, Davis A, Nicholson A (1994) In situ formation of lead phosphates in soils as a method to immobilize lead. Environ Sci Technol 28:646-654

Ruby MV, Davis A, Schoof R, Eberle S, Sellstone CM (1996) Estimation of lead and arsenic bioavailability using a physiologically based extraction test. Environ Sci Technol 30:422-430

Ruby MV, Schoof R, Brattin W, Goldade M, Post G, Harnois M, Mosby DE, Casteel SW, Berti W, Carpenter M, Edwards D, Cragin D, Chappell W (1999) Advances in evaluating the oral bioavailability of inorganics in soil for use in human health risk assessment. Environ Sci Technol 33:3697-3705

Rudel H (2003) Case study: bioavailability of tin and tin compounds. Ecotoxicol Environ Safety 56:180-189

Ryan JA, Scheckel KG, Berti WR, Brown SL, Casteel SW, Chaney RL, Hallfrisch J, Doolan M, Grevatt P, Maddaloni M, D. M (2004) Reducing children's risk from lead in soil. Environ Sci Technol 38:18A-24A

Sadiq M (1997) Arsenic chemistry in soils: An overview of thermodynamic predictions and field observations. Water, Air, Soil Pollution 93:117-136

Sass BM, Rai D (1987) Solubility of amorphous chromium(III)-iron(III) hydroxide solid solutions. Inorg Chem 26:2228 - 2232

Scheckel KG, Impellitteri CA, Ryan JA, McEvoy T (2003) Assessment of a sequential extraction procedure for perturbed lead-contaminated samples with and without phosphorus amendments. Environ Sci Technol 37:1892-1898

Scheckel KG, Ryan JA (2003) In vitro formation of pyromorphite via reaction of Pb sources with soft-drink phosphoric acid. Sci Total Environ 302:253-265

Scheckel KG, Ryan JA, Allen D, Lescano NV (2005) Determining speciation of Pb in phosphate-amended soils: Method limitations. Sci Total Environ 350:261-272

Scheinost A, Kretzscmar R, Pfister S, Roberts D (2002) Combining selective sequential extractions, X-ray absorption spectroscopy, and principal component analysis for quantitative zinc speciation in soil. Environ Sci Technol 36:5021-5028

Schettler T (2001) Toxic threats to neurologic development of children. Environ Health Perspect Suppl 109: 813-816

Schoolmeester WL, White DR (1980) Arsenic poisoning. Southern Med J 73:198-208

Schoonen MAA, Cohn CA, Roemer E, Laffers R, Simon SR, O'Riordan T (2006) Mineral-induced formation of reactive oxygen species. Rev Mineral Geochem 64:179-221

Schoonen MAA, Strongin DR (2005) Catalysis of electron transfer reactions at mineral surfaces. *In:* Environmental Catalysis. Grassian V (ed) CRC Press, p 37-60

Schuster PF, Krabbenhoft DP, Naftz DL, Cecil LD, Olson ML, Dewild JF, Susong DD, Green JR, Abbott ML (2002) Atmospheric mercury deposition during the last 270 years: A glacial ice core record of natural and anthropogenic sources. Environ Sci Technol 36:2303-2310

Scott N, Hatelid KM, MacKenzie NE, Carter DE (1993) Reactions of arsenic(III) and arsenic(V) species with glutathione. Chem Res Toxicol 6:102-106

Semple KT, Doick KJ, Jones KC, Burauel P, Craven A, Harms H (2004) Defining bioavailability and bioaccessibility of contaminated soil and sediment is complicated. Environ Sci Technol 38:228A-231A

Senko JM, Istok JD, Suflita JM, Krumholz LR (2002) *In situ* evidence for uranium immobilization and remobilization. Environ Sci Technol 36:1491-1496

Shim H, Harris ZL (2003) Genetic defects in copper metabolism. J Nutr 133:1527S-1531S

Smedley PL, Kinniburgh DG (2005) Arsenic in Groundwater and the Environment. *In:* Essentials of Medical Geology. Selinus O (ed) Elsevier, p 263-299

Smedley PL, Kinniburgh DG (2002) A review of the source, behaviour and distribution of arsenic in natural waters. Appl Geochem 17:517-568

Smith KS, Huyck HLO (1999) An overview of the abundance, relative mobility, bioavailability, and human toxicity of metals. *In:* The Environmental Geochemistry of Mineral Deposits: Part A Processes, Techniques and Health Issues. Plumlee GS, Logsdon MJ (eds) Society of Economic Geologists, p 29-70

Sparks DL (2003) Environmental Soil Chemistry, 2nd ed. Academic Press

Spear TM, Svee W, Vincent JH, Stanisich N (1998) Chemical speciation of lead dust associated with primary lead smelting. Environ Health Perspect 106:565-571

Spliethoff HM, Mason RP, Hemond HF (1995) Interannual variability in the speciation and mobility of arsenic in a dimictic lake. Environ Sci Technol 29:2157-2161

Sposito G (2004) The Surface Chemistry of Natural Particles. Oxford University Press

Sposito G (1984) The Surface Chemistry of Soils. Oxford University Press

Steinmaus C, Carrigan K, Kalman D, Atallah R, Yuan Y, Smith AH (2005) Dietary intake and arsenic methylation in a U.S. population. Environ Health Perspect 113:1153-1159

Stevenson F (1994) Humus Chemistry: Genesis, Composition, Reactions. Wiley

Stollenwerk KG (2003) Geochemical processes controlling transport of arsenic in groundwater: A review of adsorption. *In:* Arsenic in Ground Water. Welch AH, Stollenwerk KG (ed) Kluwer Academic Publishers, p 67-100

Stumm W (1992) Chemistry of the Solid-Water Interface. Wiley

Styblo M, Del Razo LM, Vega L, Germolec DR, LeCluyse EL, Hamilton GA, Reed W, Wang C, Cullen WR, Thomas DJ (2000) Comparative toxicity of trivalent and pentavalent inorganic and methylated arsenicals in rat and human cells. Arch Toxicol 74:289-299

Sudaryanto A, Takahashi S, Iwata H, Tanabe S, Ismail A (2004) Contamination of butyltin compounds in Malaysian marine environments. Environ Pollut 130:347-358

Sudaryanto A, Takahashi S, Iwata H, Tanabe S, Muchtar M, Razak H (2005) Organotin residues and the role of anthropogenic tin sources in the coastal marine environment of Indonesia. Marine Pollut Bull 50: 226-235

Sunda WG, Huntsman SA (1998) Processes regulating cellular metal accumulation and physiological effects: Phytoplankton as model systems. Sci Total Environ 219:165-181

Sutton SR, Bertsch PM, Newville M, Rivers M, Lanzirotti A, Eng P (2003) Microfluorescence and microtomography analyses of heterogeneous earth and environmental materials. Rev Mineral Geochem 49:429-483

Tack FM, Verloo MG (1995) Chemical speciation and fractionation in soil and sediment heavy metal analysis: A review. Int J Environ Anal Chem 59:225-238

Tam GKH, Charbonneau SM, Bryce F, Pomroy C, Sandi E (1979) Metabolism of inorganic arsenic ([74]As) in humans following oral ingestion. Toxicol Appl Pharmacol 50:319-322

Tchounwou PB, Ayensu WK, Ninashvili N, Sutton D (2003) Environmental exposure to mercury and its toxicopathologic implications for public health. Environ Toxicol 18:149-175

Templeton D, Ariese F, Cornelis R, Danielsson L, Muntau H, Leeuwen HV, Lobinski R (2000) Guidelines for terms related to chemical speciation and fractionation of elements. Definitions, structural aspects, and methodological approaches. Pure Appl Chem 72:1453-1470

Tessier A, Campbell PGC, Bisson M (1979) Sequential extraction procedure for the speciation of particulate trace metals. Anal Chem 51:844-851

Tewalt SJ, Bragg LJ, Finkelman RB (2005) Mercury in U.S. Coal—Abundance, Distribution, and Modes of Occurrence. US Geological Survey Fact Sheet 095-01

Tomiyasu T, Matsuo T, Miyamoto J, Imura R, Anazawa K, H. S (2005) Low level mercury uptake by plants from natural environments--mercury distribution in Solidago altissima L. Environ Sci 12:231-238

Traina SJ, Laperche V (1999) Contaminant bioavailability in soils, sediments, and aquatic environments. Proc Nat Acad Sci 96:3365-3371

Tudor R, Zalewski PD, Ratnaike RN (2005) Zinc in health and chronic disease. J Nutr Health Aging 9:45-51

Ullrich SM, Tanton TW, Abdrashitovab SA (2001) Mercury in the aquatic environment: A review of factors affecting methylation. Crit Rev Environ Sci Technol 31:241–293

Uriu-Adams JY, Keen CL (2005) Copper, oxidative stress, and human health. Mol Aspects Med 26:268-298

Vahter M (2002) Mechanisms of arsenic biotransformation. Toxicol 181-182:211-217

Vahter M (2000) Genetic polymorphism in the biotransformation of inorganic arsenic and its role in toxicity. Toxicol Lett 112-113:209-217

Valko M, Morris H, Cronin MTD (2005) Metals, toxicity and oxidative stress. Curr Med Chem 12:1161-1208

Van Leeuwen HP, Town RM, Buffle J, Cleven RFMJ, Davison W, Puy J, van Riemsdijk WH, Sigg L (2005) Dynamic speciation analysis and bioavailability of metals in aquatic systems. Environ Sci Technol 39: 8545-8556

Villalobos M, Bargar J, Sposito G (2005) Mechanisms of Pb(II) sorption on a biogenic manganese oxide. Eviron Sci Technol 39:569-576

Voegelin A, Hug SJ (2003) Catalyzed oxidation of arsenic(III) by hydrogen peroxide on the surface of ferrihydrite: An *in situ* ATR-FTIR study. Environ Sci Technol 37:972-978

Vyskocil A, Viau C (1999) Assessment of molybdenum toxicity in humans. J Appl Toxicol 19:185 - 192

Waite TD, Davis JA, Payne TE, Waychunas GA, Xu N (1994) Uranium(VI) adsorption to ferrihydrite: Application of a surface complexation model. Geochim Cosmochim Acta 58:5465-5478

Walsh CT, Sandstead HH, Prasad AS, Newberne PM, Fraker PJ (1994) Zinc: health effects and research priorities for the 1990s. Environ Health Perspect Suppl 102:5 - 46

Warren LA, Haack EA (2001) Biogeochemical controls on metal behaviour in freshwater environments. Earth-Science Rev 54:261-320

Waychunas GA, Rea BA, Fuller CC, Davis JA (1993) Surface chemistry of ferrihydrite: Part 1. EXAFS studies of the geometry of coprecipitated and adsorbed arsenate. Geochim Cosmochim Acta 57:2251-2269

Wedepohl HK (1995) The composition of the continental crust. Geochim Cosmochim Acta 59:1217-1232

WHO (2003) Chromium in Drinking-water. World Health Organization, 8. *http://www.who.int/water_sanitation_health/dwq/chemicals/chromium.pdf*

Wiegand HJ, Ottenwalder H, Bolt HM (1985) Fast uptake kinetics in vitro of $^{51}Cr(VI)$ by red blood cells of man and rat. Arch Toxicol 57:31-34

Wilkin RT, Wallschlaeger D, Ford RG (2003) Speciation of arsenic in sulfidic waters. Geochem Trans 4:1-7

Winfrey MR, Rudd JWM (1990) Environmental factors affecting the formation of methylmercury in low pH lakes. Environ Toxicol Chem 9:853-869

Xia K, Bleam W, Helmke P (1997) Studies of the nature of Cu^{2+} and Pb^{2+} binding sites in soil humic substances using X-ray absorption spectroscopy. Geochim Cosmochim Acta 61:2211-2221

Yamauchi H, Takahashi K, Yamamura Y (1986) Metabolism and excretion of orally and intraperitoneally administered gallium arsenide in the hamster. Toxicol 40:237-246

Yokoo E, Valente J, Grattan L, Schmidt S, Platt I, Silbergeld E (2003) Low level methylmercury exposure affects neuropsychological function in adults. Environ Health: A Global Access Sci Source 2:8

Zachara JM, Ainsworth CC, Brown GE, Catalano JG, McKinley JP, Qafoku O, Smith SC, Szecsody JE, Traina SJ, Warner JA (2004) Chromium speciation and mobility in a high level nuclear waste vadose zone plume. Geochim Cosmochim Acta 68:13-30

Zachara JM, Ainsworth CC, Cowan CE, Resch CT (1989) Adsorption of chromate by subsurface soil horizons. Soil Sci Soc Am J 53:364-373

Zachara JM, Girvin DC, Schmidt RL, Resch CT (1987) Chromate adsorption on amorphous iron oxyhydroxide in the presence of major groundwater ions. Environ Sci Technol 21:589-594

Zakharyan R, Wu Y, Bogdan GM, Aposhian HV (1995) Enzymatic methylation of arsenic compounds: Assay, partial purification, and properties of arsenite methyltransferase and monomethylarsonic acid methyltransferase of rabbit liver. Chem Res Toxicol 8:1029-1038

Zhang H, Young S (2006) Characterizing the availability of metals in contaminated soils. II. The soil solution. Soil Use Manage 21:459-467

Zhang P, Ryan J, Yang J (1998) In vitro soil Pb solubility in the presence of hydroxyapatite. Environ Sci Technol 32:2763-2768

Zheng W, Aschner M, Ghersi-Egea JF (2003) Brain barrier systems: a new frontier in metal neurotoxicological research. Toxicol Appl Pharmacol 192:1-11

Zhou B, Westaway SK, Levinson B, Johnson MA, Gitschier J, Hayflick SJ (2001) A novel pantothenate kinase gene (PANK2) is defective in Hallervorden-Spatz syndrome. Nat Genet 28:345-349

APPENDIX 1.

Heavy metals and metalloids that are known or thought to be essential for human health.

Metal	Forms in body	Biological Roles	Deficiency-related diseases	Exposure-related diseases
Arsenic	Inorganic forms of arsenic are transformed into methylated arsenic compounds (MMA and DMA). The transformation of As(III) to MMA and DMA takes place primarily in the liver.	Unclear if As is an essential element.	No cases of arsenic deficiency in humans have ever been reported. Animals fed a diet with unusually low concentrations of arsenic did not gain weight normally. They also became pregnant less frequently than animals fed a diet containing a normal amount of arsenic. Furthermore, the offspring from these animals tended to be smaller than normal, and some died at an early age.	Inorganic forms of arsenic are far more toxic than methylated forms. As(III) reacts with sulfhydryl groups in proteins and inactivates enzymes. It is thought that As(III) interferes with DNA repair. As(V) is also genotoxic, but its toxicity may be related to its similarity to phosphate. Arsenic exposure is known to lead to skin lesions, bladder cancer, and neurological diseases. As(III) and As(V) can cross the placenta and cause fetotoxicity, decreased birth weight, and congenital malformations.
Chromium	Cr(III) in the form of a Cr(III)-oligo-peptide compound (chromodulin) (Lindh 2005) Cr(VI) is reduced to Cr(III) after uptake.	Required for normal glucose, fat, and protein metabolism	Deficiency leads to impaired glucose tolerance, fasting hyperglycemia, glucosuria, elevated percent body fat, decreased lean body mass, maturity-onset diabetes, cardiovascular disease, decreased sperm count, and impaired fertility.	Cr(VI) enters cells through sulfate channels and reacts with antioxidants. Reactions with glutathione lead to a Cr(V) compound that promotes formation of reactive oxygen species within cell, which promotes carcinogenesis.
Cobalt	Co(II) in several metalloenzymes	Cofactor in vitamin B_{12}, component of metal center in several metalloenzymes	Little documentation of diseases related to lack of Co, except amnesia, lack of red blood cells. Deficiency in vitamin B_{12} may lead to depression and adversely affect the nervous system. .	Inhalation can lead to respiratory irritation, diminished pulmonary function, wheezing, asthma, pneumonia, fibrosis, and lung cancer. Dermal contact can lead to eczema. May cause memory loss and other neurological problems. Cobalt dissolved in cells is thought to promote the formation of reactive oxygen species. Exposure to alloys containing Co and W appear to increase toxicity. Co may also interfere with DNA repair system.

Metal	Forms in body	Biological Roles	Deficiency-related diseases	Exposure-related diseases
Copper	$Cu(II)/Cu(I)$ in 30+ enzymes	Copper in several important enzymes and proteins. For example, Cu is active component in Superoxide Dismutase (SOD), which protects against reactive oxygen species. The Cu-containing protein Ceruloplasmin accounts for 95 % of Cu in human serum.	Copper deficiency is rare. A case involving children in Peru showed that copper deficiency leads to low counts in neutrophils (type of white blood), which puts individuals at a higher risk for bacterial infections (Goldhaber 2003). In addition, other symptoms included anemia and bone mineralization. Animal studies suggest that copper deficiency leads to poor growth, anemia, and degeneration of the central nervous system. Individuals without ceruloplasmin protein (a genetic disorder) develop diabetes, retinal degeneration and neurodegeneration (Shim and Harris 2003).	Inhalation can lead to respiratory irritation, including coughing, sneezing, thoracic pain, runny nose, and fibrosis. Exposure to copper dust can lead to headaches, vertigo, and drowsiness. Excessive accumulation as a result of a genetic abnormality is known as Wilson's disease. Excessive copper exposure inhibits zinc absorption and leads to zinc deficiency. Copper is suspected to be a Fenton metal on the basis of its ability to cycle between Cu(II) and Cu(I) oxidation states (Gaetke and Chow 2003; Uriu-Adams and Keen 2005). Excess Cu may play a role in the development of Alzheimer disease (Uriu-Adams and Keen 2005).
Iron	Ferritin protein, which contains a nano-size Fe^{III} oxhydroxide core, hemoglobin, and several other iron-containing metalloenzymes	Major redox cycling element in body. Hemoglobin is critical to the transfer of oxygen to tissue (Gurzau et al. 2003).	Iron deficiency is common and leads to anemia, or low red blood cell counts, which impairs efficient oxygenation of tissue. Low-iron status leads also to increased uptake of other metals, such as Mn and Ni. Hence, low iron status increases Ni and Mn toxicity for a given exposure.	Acute excessive iron exposure by ingestion leads to vomiting and diarrhea (Goldhaber 2003). Genetically-induced excessive accumulation of iron Hallervorden-Spatz Syndrome leads to hemochromatosis (Zhou et al. 2001). High iron exposure leads to oxidative stress and promotes TB (De Voss et al. 1999).
Manganese	Component of metalloenzymes. Mn present as Mn(II), Mn(III), and Mn(IV) in enzymes.	Mn is an essential nutrient and plays a role in bone mineralization, protein and energy metabolism, metabolic regulation, cellular protection from damaging free radical species, and the formation of glycosaminoglycans.	Mn deficiency is rare. Studies in which human subjects received a low-Mn diet showed slow hair and nail growth, a decrease in clotting proteins, and onset of dermatitis.	High acute exposures lead to manganism, a Parkinson-like, neurological disease.

Metal	Forms in body	Biological Roles	Deficiency-related diseases	Exposure-related diseases
Molybdenum	Mo(IV), Mo(V), Mo(VI)	Co-factor in several enzymes that regulate metabolism of carbon, nitrogen, and sulfur (Barceloux 1999a).	Molybdenum is a trace essential metal and deficiencies are uncommon. Animal studies suggest that Mo deficiency leads to reduced weight gain, decreased food consumption, and impaired reproduction. (Goldhaber 2003)	Little data are available on the human toxicity of molybdenum. A gout-like syndrome (arthritis) and pneumoconiosis (black lung disease) have been associated with excessive concentrations of molybdenum (Barceloux 1999a). Kidney failure has been observed in rats exposed to a high-Mo diet. (Goldhaber 2003). Individuals with inadequate intake of copper are at higher risk for adverse effects of molybdenum exposures (Vyskocil and Viau 1999)
Nickel	Ni-containing enzymes	Component of several metalloenzymes, including Superoxide Dismutase (SOD), which protects against reactive oxygen species.	Trace essential metal. No reports of human nickel deficiency could be located. Rats and chicks on a Ni-deficient diet developed liver problems. (Barceloux 1999b)	Genotoxic metal. Thought to interact with DNA, leading to inhibition of gene expression. Complexation of nickel with low-molecular weight ligands, as well as proteins and peptides, may convert dissolved Ni(II) into a Fenton metal and induce the formation of ROS. Allergic contact dermatitis is a common allergic response among humans to dermal exposure to nickel metal (Barceloux 1999b)
Selenium	Selenoprotein	Antioxidant, Selenoproteins, including glutathione peroxidase	Disturbance of selenoprotein expression or function is associated with deficiency syndromes (Keshan and Kashin-Beck disease), might contribute to tumorigenesis and atherosclerosis, is altered in several bacterial and viral infections, and leads to infertility in male rodents (Köhrl et al. 2000)	High chronic exposure leads to selenosis: symptoms include hair loss and brittle nails.(Goldhaber 2003)
Tin	unknown	Component of gastrine, a stomach-stimulating peptide hormone (Rudel 2003).	No studies addressing Sn deficiencies could be located.	Organotin compounds are lipid soluble and far more toxic than inorganic tin compounds. Exposure to organotin compounds at high dose levels leads to acute neurological effects, such as memory loss. Organotin compounds are thought to cause disruption of cell-signaling in the brain, programmed cell death, and cell death with increasingly high dose exposures. Little is known about the effect of lower dose levels. Dissolved inorganic tin is thought to be a Fenton metal. Inorganic tin is sequestered in bone.

Metal	Forms in body	Biological Roles	Deficiency-related diseases	Exposure-related diseases
Tungsten	W(IV)	Component of some metalloenzymes; role in humans not clear.	No studies addressing W deficiencies could be located.	Tungsten oxide fibers are capable of generating hydroxyl radicals in human lung cells in vitro, thought to contribute to the development of pulmonary fibrosis in hard metal workers.
Vanadium	V(V): VO_3^- V(IV): VO^{2+}	Unclear if V is an essential metal (Badmaev et al. 1999); possibly an antioxidant (Lindh 2005) and may be involved in development of skeleton and teeth.	Rats and chicks showed reduced growth and impairment of reproductive system. V deficiency-related diseases are not known to occur in humans	In rats, mice, and hamsters, it has been established that V(IV) and V(V) compounds are developmental and reproductive toxicants.
Zinc	Zn(II): over 300 known Zn-containing enzymes	Zinc-containing enzymes are important in respiration (SOD), in gene expression, DNA repair, and in programmed cell death (Tudor et al. 2005)	Chronic dietary zinc deficiency is common; it is estimated that 4 million people in the U.S. are affected. Initial symptoms are loss of taste and smell. Chronic severe deficiency leads to immune disorders (Walsh et al. 1994). Zinc deficiency leads to oxidative stress (Oteiza and Mackenzie 2005). Suboptimal zinc status may lead to bronchial asthma, rheumatoid arthritis, and Alzheimer's disease (Tudor et al. 2005).	Highly excessive inhalation exposures can lead to metal fume fever. Excessive Zn exposure may be neurotoxic (Walsh et al. 1994)

Reviews in Mineralogy & Geochemistry
Vol. 64, pp. 115-134, 2006
Copyright © Mineralogical Society of America

4

Aluminum, Alzheimer's Disease and the Geospatial Occurrence of Similar Disorders

Daniel P. Perl and Sharon Moalem

Department of Pathology, Neuropathology Division
Mount Sinai School of Medicine
One Gustave L. Levy Place
New York, New York, 10029, U.S.A.
e-mail: daniel.perl@mssm.edu

INTRODUCTION

Over the years, aluminum has been associated with a number of neurodegenerative diseases, however, it is the association of the metal with Alzheimer's disease that has attracted the greatest attention and discussion in both the scientific literature and the lay press. In this chapter, we will discuss the background for this association and its implications for an understanding of its role in the etiopathogeneisis of age-related neurodegenerative disorders. Aluminum, although the most abundant metal in the earth's crust, comprising 8% of the geologic mantle, is unique among the elements of abundance in that it is not considered to be essential for life. Most of the earth's aluminum exists in forms that are not soluble at physiologic pH and are therefore relatively unavailable to mammalian species (Martin 1986). Aluminum-containing compounds are widely used in a large number of commercial products, are constituents of many food additives and medications and are extensively employed as a deflocculant in the purification of water delivered to a large number of cities throughout the world. However, it should be understood that most ingested aluminum remains unabsorbed by the body and there are little in the way of aluminum stores in the tissues of mammalian species. This is particularly true of the brain where normal concentrations of the element are around 1 part per million. Despite this, in a number of neurodegenerative disorders, focal accumulations of aluminum have been identified using a number of different analytic techniques. When localized, these aluminum deposits have been identified within the specific pathologic lesions by which these diseases are characterized, especially neurofibrillary tangles. Such lesions consist of intraneuronal protein aggregates which once formed cannot be properly broken down by the nerve cells and eventually cause their destruction. Aluminum, with its high charge and small ionic radius, would be expected to bind tightly to the constituent proteins which make up these inclusions and to cross-link them. The very presence of this element, with its high binding capacity, would suggest that aluminum may play a role in either the etiology or, more likely, the pathogenesis of these conditions through the stabilization of intraneuronal proteins. However, despite considerable investigation the source of this aluminum and its means of entry into the brain remain unclear.

ALUMINUM AND ALZHEIMER'S DISEASE

The association of aluminum and Alzheimer's began in 1965 with the simultaneous publications of Terry and Peña (1965) and of Klatzo, Wisniewski and Streicher (1965) reporting that when the brains of rabbits are directly exposed to aluminum-containing compounds, such as Holt's adjuvant, the animals rapidly developed neurofibrillary lesions that

1529-6466/06/0064-0004$05.00 DOI: 10.2138/rmg.2006.64.4

appeared remarkably similar to the neurofibrillary tangles of cases of Alzheimer's disease. Neurofibrillary tangles represent one of the two hallmark pathologic lesions, along with the senile or neuritic plaques, by which the disease is characterized and diagnosed pathologically (see Fig. 1). In order to make an autopsy diagnosis of Alzheimer's disease, the presence of neurofibrillary tangles must be documented by the pathologist, and the degree of involvement by this lesion in certain regions of the brain is correlated with the extent of cognitive failure in the patient. Holt's adjuvant is a paste composed primarily of aluminum oxide and it was quickly learned by these researchers that similar results could be obtained by direct application to the rabbit brain of a wide range of aluminum-containing salts. Indeed, even aluminum metal filings could produce the experimental neuronal lesions.

The application of Holt's adjuvant to the cerebral cortex of rabbits had been used as an experimental model of focal seizures (epilepsy) since the 1930's but this was the first time someone had published on the morphologic changes produced by such aluminum exposure. The aluminum-related tangles induced in experimentally exposed rabbits were widespread, involving cerebral cortex, brainstem and even spinal cord neurons. The rabbit tangles were initially characterized as being composed of straight neurofilaments with a different electron microscopic appearance than the paired-helical filaments seen in the neurofibrillary tangles of cases of Alzheimer's disease. At the time, much was made of these differences but subsequent studies by Savory and colleagues (Savory et al. 1995; Huang et al. 1997; Rao et al. 1998) have shown that, while different, the aluminum-induced tangles do actually share many constituents with those of man.

This experimental work stimulated Crapper-McLachlan and colleagues in Toronto to measure bulk aluminum concentrations in brain tissues derived from autopsies obtained from Alzheimer's disease patients and compare those measured within specimens similarly obtained from normal elderly control subjects (Crapper et al. 1973). These workers reported the presence of significantly increased aluminum concentrations in the Alzheimer's disease brain tissues. A subsequent study showed a wide variation of the regional aluminum concentrations in additional Alzheimer's disease brain specimens with higher values tending to be found in brain areas with

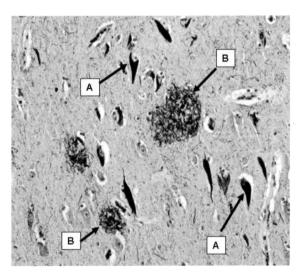

Figure 1. Histologic appearance of brain tissue from a patient with Alzheimer's disease showing the presence of numerous neurofibrillary tangles (A) and senile plaques (B). (Hippocampus, modified Bielschowsky stain).

more extensive neurofibrillary tangle involvement (Crapper et al. 1976). However, all these studies were performed using bulk tissue analyses of small samples of brain tissue and therefore did not allow localization of the excess aluminum on the cellular or even subcellular level of resolution. It was the studies of Perl and Brody (1980), using electron probe microanalysis that addressed this issue. We analyzed frozen sections of formalin fixed brain tissues that had been stained with silver impregnation stains to delineate the tangle-bearing and non-tangled neurons in the specimens. Silver stains have been used by neuropathologists for almost 100 years to identify the presence of neurofibrillary tangles. Using a combination of both energy-dispersive X-ray analysis and wave-length dispersive X-ray analysis, they demonstrated that aluminum-related signals were detected in the probe sites directed to the tangle-bearing neurons but were rarely seen in the adjacent non-tangled neurons. This work provided the first demonstration that the association between aluminum with Alzheimer's disease extended to the cellular level of resolution and indeed showed evidence that aluminum was concentrated in the cells bearing the neurofibrillary tangles, a cardinal lesion by which the disease is defined.

Around the same time, a group in Newcastle-Upon-Tyne, England, began a series of microprobe studies of senile plaques in cases of Alzheimer's disease. The senile plaque, with its central core of beta-amyloid, is the other cardinal lesion (along with the neurofibrillary tangle) by which Alzheimer's disease is defined neuropathologically. The English group reported finding high concentrations of aluminum and silicon in the cores of senile plaques (Candy et al. 1986). They interpreted their microprobe findings to mean that the core of the senile plaque consisted primarily of a dense deposit of aluminosilicates. How aluminosilicate deposits could occur in the brain was left unclear but they considered the central core deposit to form a "seed" for further fibrillization of the beta amyloid which accompanies this pathologic lesion. Over the years, a number of laboratories have attempted to confirm this finding using a number of different analytic techniques but have failed to demonstrate any significant concentrations of either aluminum or silicon in senile plaque cores of cases of Alzheimer's disease (Stern et al. 1986; Landsberg et al. 1992).

Good and co-workers (Good et al. 1992b) introduced the use of laser microprobe mass analysis (LAMMA) to this field and in the process provided more precise localization and elemental identification of the association of aluminum and the neurofibrillary tangle. The LAMMA instrument uses a high energy pulsed Nd/YAG laser which is focused by the objective lens of an optical microscope to perforate a semithin plastic embedded tissue section. This perforation produces ions which are accelerated by a charged ion lens into the tunnel of a time-of-flight mass spectrometer. By analyzing the resultant time-of-flight spectra, with each laser perforation of the tissue section the LAMMA instrument is capable of detecting virtually all of the positively charged elements throughout the atomic table. For most elements the detection limits of the LAMMA instrument are on the order of a few parts per million and the laser perforation can be focused to a probe site of approximately 1 μm in diameter. The intensity of the mass spectral signal is, for the most part, proportional to the concentration of the element being detected. The semithin tissue section is stained with toluidine blue and is transluminated to allow precise visualization of histologic details through a standard light microscope. Finally, the LAMMA has a collinear low energy continuous pilot laser which allows the operator to direct the high energy pulsed laser to specific sites for analysis.

Using this very powerful laser microprobe technology, excess aluminum was detected within the neurofibrillary tangle-bearing neurons and these aluminum accumulations were more specifically localized to the neurofibrillary tangle, itself (see Fig. 2). Furthermore, in addition to aluminum, all other positively charged ions were analyzed in these probe sites. In evaluating these data, the only other element which consistently showed significantly higher signal intensity in the probe sites directed to the neurofibrillary tangles was iron (see Table 1). In this study, each of ten consecutively accessioned cases of Alzheimer's disease

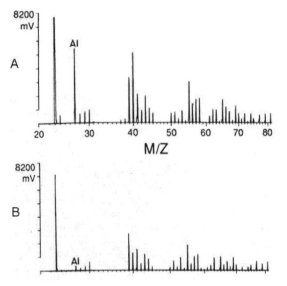

Figure 2. (A) Partial mass spectrum obtained from a LAMMA probe site directed to the neurofibrillary tangle within a tangle bearing hippocampal neuron of an Alzheimer's disease brain specimen. Note the prominent mass 27 peak related to the presence of aluminum in this probe site. (B) Partial mass spectrum obtained from a LAMMA probe site directed to the neuronal cytoplasm adjacent to the probe site displayed in Figure 2A. This portion of cytoplasm is not involved in the neurofibrillary tangle within this cell. Note the relative absence of a mass 27 aluminum-related peak.

Table 1. Analysis of Variance (ANOVA) comparing neurofibrillary tangle-derived data with other cellular compartments in hippocampal neurofibrillary tangle-bearing neurons of ten consecutively obtained Alzheimer's disease autopsied brain specimens. F calculations for aluminum and iron-related peak intensities are shown along with other elemental masses obtained with a p value of <0.05. From Good et al. (1992b).

Alzheimer's Disease Case	Aluminum-related (mass 27) F value, p value	Iron-related (mass 56) F value, p value	Other elements detected with $p < 0.05$
1	$F(1,60) = 113, p <0.005$	$F(1,60) = 40, p <0.005$	Zr
2	$F(1,40) = 213, p <0.005$	$F(1,40) = 55, p <0.005$	Pt
3	$F(1,36) = 79, p <0.005$	$F(1,36) = 29, p <0.005$	
4	$F(1,20) = 29, p <0.005$	$F(1,20) = 5.6, p <0.05$	Li, Ti, Zr
5	$F(1,32) = 72, p <0.005$	$F(1,32) = 28, p <0.005$	
6	$F(1,36) = 33, p <0.005$	$F(1,36) = 79, p <0.005$	K
7	$F(1,44) = 132, p <0.005$	$F(1,44) = 29, p <0.005$	Li, Si, P, S
8	$F(1,32) = 51, p <0.005$	$F(1,32) = 7.4, p <0.01$	
9	$F(1,32) = 394, p <0.005$	$F(1,32) = 20, p <0.005$	F, La, Hf, Ta
10	$F(1,36) = 68, p <0.005$	$F(1,36) = 20, p <0.005$	W

showed evidence of significantly increased concentrations of aluminum and iron within the neurofibrillary tangles of the probed neurons. These prominent aluminum and iron-related signals were not encountered in the adjacent non-tangled neurons of the Alzheimer's disease cases or in the non-tangled neurons derived from four normal age-matched control cases that had been prepared in an identical fashion. In conjunction with this study, unstained, unfixed tissue sections obtained from regions with tangle-bearing neurons also demonstrated prominent aluminum-related signals. This important piece of data demonstrated that the aluminum being detecting in the neurofibrillary tangle-bearing neurons was original to the tissues and could not represent the results of contamination through the processes of fixation, plastic embedding or toluidine blue staining of the specimens.

In 1999, Murayama and colleagues published an important study further demonstrating the association of aluminum and the neurofibrillary tangles of cases of Alzheimer's disease (Murayama et al. 1999). These workers showed that Morin staining of the neurofibrillary tangles of cases of Alzheimer's disease could be abolished by pretreatment of the sections with desferioxime, a known aluminum chelator. Morin is a relatively sensitive histochemical tissue stain for aluminum. The study also reported that desferioxime chelation of Alzheimer's disease tissue sections abolished immunostaining of the neurofibrillary tangles using antibodies raised against hyperphosphorylated forms of *tau*. *Tau*, a cytoskeletal structural protein, is a major constituent of the neurofibrillary tangle and in the setting of this disease shows evidence of excess phosphorylation of many of its constituent amino acids. Murayama and colleagues findings suggest that the aluminum found in the neurofibrillary tangle is likely bound to these abnormal phosphorylation sites. It is very likely that the presence of aluminum on these sites causes cross-linking and stabilization of this constituent protein. This important study provides a possible mode of action for aluminum in the pathogenesis of Alzheimer's disease. Furthermore, since the changes reported in the staining reactions had all been induced by chemical removal of aluminum from frozen tissue sections, these results serve as an important independent confirmation of the original microprobe results of aluminum being present in the tissues and localized to the neurofibrillary tangle-bearing neurons. They further eliminate concerns that others have expressed about possible contamination of the tissues through staining procedures and/or its fixation prior to the above-mentioned microprobe studies.

IRON, ALUMINUM AND PARKINSON'S DISEASE

Parkinson's disease is the second-most common age-related neurodegenerative disorder. The disease was first described in 1817 by James Parkinson (Parkinson 1955) in his remarkably astute monograph entitled *An Essay on Shaking Palsy*. Parkinson's disease is characterized by three cardinal signs and symptoms, namely, bradykinesia, or slowness of movement, muscular rigidity and a characteristic resting tremor. The tremor has a 4-7 Hz frequency and in the hands is typically described as "pill-rolling" in nature. The tremor of Parkinson's disease is particularly notable at rest and tends to diminish dramatically when the limbs are carrying out a purposeful motion. Patients also suffer from postural instability with a classic stooped posture and shuffling gait. Although many areas of the brain are affected in the disease, the major site of neuronal degeneration is the substantia nigra pars compacta of the midbrain. Neuronal degeneration of this portion of the brain produces dysfunction of a series of interconnected brain regions referred to collectively as the basal ganglia, a brain structure which is important for the control of motor movements. The nerve cells in the substantia nigra pars compacta produce dopamine as their neurotransmitter and project to another part of the basal ganglia, the striatum, which consists of the caudate and the putamen. As a byproduct of their dopamine metabolism, these neurons of the substantia nigra pars compacta contain neuromelanin, a brown-black pigment which gives rise to its normal black appearance to the naked eye. With the severe degeneration of the dopamine-producing neurons in this location, the substantia nigra pars compacta of patients with Parkinson's disease has a distinctly pale appearance. Microscopically, there is prominent loss of the normal neuronal population, although even in the most severely affected patient, some neurons will remain intact. Within those remaining neurons one encounters inclusion bodies referred to as Lewy bodies in honor of their discoverer in 1912, Dr. Frederich Lewy. The Lewy bodies are rather large spherical inclusions that are present in the neuronal cytoplasm and consist primarily of a protein called alpha synuclein (Spillantini et al. 1997), although other proteins are also encountered within them (Fig. 3).

There has been a long history of studies documenting evidence of increased iron concentrations in portions of the basal ganglia of cases of Parkinson's disease. These studies have included magnetic resonance imaging (MRI) localization of living Parkinson's disease patients

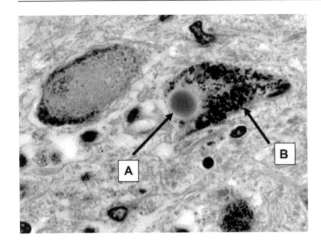

Figure 3. Histologic appearance of a substantia nigra pars compacta neuron derived from a patient with Parkinson's disease showing the presence of a Lewy body (A) and numerous neuromelanin granules (B). (Hematoxylin and eosin stain).

(Olanow and Drayer 1987; Drayer 1989) as well as analytic studies of autopsy-derived brain specimens (Earle 1968; Dexter et al. 1989; Youdim et al. 1989; Sofic et al. 1991). In the 1990's, several studies reported the results of microprobe elemental analysis of the substantia nigra pars compacta neurons of cases of Parkinson's disease. One study found (Good et al. 1992a) that the neuromelanin in remaining nigral neurons of Parkinson's disease patients showed prominent evidence of accumulation of both iron and aluminum (see Fig. 4). Neuromelanin has been shown to have metal binding sites (Enochs et al. 1994; Bridelli et al. 1999; Zecca et al. 2002; Double et al. 2003). Furthermore, microprobe studies of Lewy bodies of cases of Parkinson's disease also showed the presence of iron and aluminum within the inclusion body, itself, again suggesting selective binding of these two metals (Hirsch et al. 1991; Jellinger et al. 1992).

The remaining intact neurons in the substantia nigra pars compacta of cases of Parkinson's disease show distinct evidence that they have undergone oxidative damage and it has been suggested that the presence of such iron deposits may promote oxidative damage as a mechanism for the neuronal destruction seen in association with Parkinson's disease (Olanow

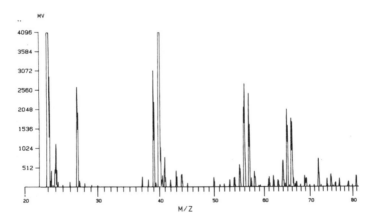

Figure 4. Representative partial mass spectrum obtained from LAMMA probe site directed to neuromelanin granules of sustantia nigra pars compacta neuron of a case of Parkinson's disease showing the presence of prominent peaks at mass 27 (aluminum-related) and mass 56 (iron-related). [Used with permission of Elsevier from Good et al. (1992) *Brain Res.* 593:343-346.]

and Perl 1994; Jenner and Olanow 1996; Good et al. 1998; Castellani et al. 2002; Perry et al. 2002). The brain normally contains a relatively large amount of iron. Much of that iron is confined to portions of the basal ganglia. It is felt that in the case of Parkinson's disease, the iron becomes redistributed and accumulates with the nerve cells targeted for degeneration. Once again, the source of the aluminum that is also encountered is unknown. However, it has been noted that when aluminum is present in association with iron, aluminum enhances iron's capacity to promote oxidative damage by many fold (Gutteridge et al. 1985).

ALUMINUM-IRON AND AGE-RELATED NEURODEGENERATIVE DISORDERS, GENE-ENVIRONMENTAL INTERACTIONS

When one considers the possible risks related to environmental exposure to toxins, including toxic metals, rarely is much thought given to the genetics of the individual being exposed. Recently, increasing awareness and understanding of gene-environment interactions has been growing. Most of this work today falls under the umbrella of pharmacogenomics (Roden 2001). This enhanced awareness began when researchers realized that given the exact dose per weight of a specific drug there was a rather marked range of responses among various hosts. Upon further investigation the different responses observed were found to be the result of differing genotypes which were usually associated with polymorphisms of specific genes coding for enzymes responsible for metabolism of those drugs. Polymorphisms of one such gene, CYP2D6, belonging to the cytochrome P450 family, is thought to affect the metabolism of at least 25% of all pharmaceuticals (Poolsup et al. 2000). The presence of a variety of polymorphisms should not be surprising since different groups of people were historically exposed to differing environments, including various toxins present in the food and water they consumed. From an evolutionary perspective thus there would be differing pressures to be able to cope with environmental challenges since there is always a trade-off between needed nutrients and vitamins and their associated toxicities usually acquired upon their acquisition.

All vertebrate species require significant amounts of iron. The metal iron is a classic example of such a paradigm since in acquiring this essential element an organism must deal with iron's ability to generate harmful reactive oxygen species (ROS), the stress of which can shorten both the viability and life of cells and tissues. Humans, of course, are not an exception when it comes to the acquisition and maintenance of a healthy balance of stored iron and its ability to generate ROS. There are some conditions, genetic hemochromatosis being one, where excessive absorption of iron from the diet leads to deposition of the metal in many organs leading to their damage over time, as a result of the aforementioned actions of ROS. Most cases of hemochromatosis, an autosomal recessive disorder, are due to maternal and paternal inheritance of the HFE mutation (Feder et al. 1996; Eijkelkamp et al. 2000; Bomford 2002). In some ethnic groups the HFE carrier state is quite common and having the C282Y, H63D and other polymorphism in association with the HFE hemochromatosis gene leads such carriers to a situation in which there are higher levels of absorption of environmental iron (Kuhn 1999; Fodinger and Sunder-Plassmann 1999; Worwood 2005). In environmental conditions where there is a lack of bioavailable environmental iron, having the iron loading hemochromatosis phenotype would therefore be beneficial (Datz et al. 1998). However, problems arise when the bioavailability of iron begins to supersede its usefulness. This is because individuals heterozygous or homozygous for the super-absorbing polymorphism will accumulate higher levels of environmental iron than those individuals who might have the wild-type polymorphism for this gene. The higher level of iron loading leads to its deposition of the metal in organs such as liver, pancreas and even heart. Over time, the deposition of iron in these tissues is thought to make such individuals more susceptible to a variety of conditions such as cirrhosis and cancer of the liver, diabetes and heart disease (Mura et al. 2000; Thorburn et al. 2002). If detected early these complications can be avoided through therapeutic bleeding and avoidance of dietary iron.

What has not been fully explored is whether aluminum, which chemically behaves in a similar fashion to iron, is absorbed at higher rates in those hemochromatotic individuals who hyperaccumulate iron. Since both iron and aluminum share many observed chemical behavioral characteristics, aluminum's absorption and storage through the transferrin-ferritin system into the human body is thought to follow that employed by iron. This might initially seem like an esoteric problem, but when one considers the reports of up to 30% of individuals from western European descent have one or more of the hemochromatosis mutations, the absorption of environmental aluminum must be suspected in these susceptible individuals (Moalem et al. 2002).

Unfortunately, at the present time research has not been done to assess the degree to which aluminum loading occurs in individuals who carry the hemochromatosis genotype. The importance of this glaring oversight is that many of the epidemiological/environmental studies have ignored the genotype of their cohort. There have been numerous studies trying to ascertain whether chronic exposure to aluminum, particularly through drinking water sources, is associated with an increased prevalence of Alzheimer disease (Martyn et al. 1997; Rondeau 2002). Since many of these studies were conducted in countries with a high degree of hemochromatosis one has to wonder if the genes associated with increased loading of metals could be the source of confounding and conflicting results.

Using the absorption of iron as a case study it is clear that there are different genetic components that might contribute to how the human body metabolizes and stores metals and the importance of the study of these interactions from both the point of view of organism and environment and of course how they interact.

GUAM AMYOTROPHIC LATERAL SCLEROSIS/ PARKINSONISM-DEMENTIA COMPLEX AND RELATED FOCI

It appears clear that aluminum accumulates in neurofibrillary tangles associated with cases of Alzheimer's disease and Lewy bodies encountered in Parkinson's disease. However, the source of the aluminum that has been detected remains unknown, nor is it clear how the element gains access to the central nervous system. To investigate this further, we turned to studies of a unique focus of neurodegenerative diseases characterized by severe and widespread neurofibrillary tangles, namely amyotrophic lateral sclerosis and parkinsonism-dementia complex of Guam.

Although Alzheimer's disease represents the prototype disorder in which neurofibrillary tangles are seen, there are a number of other nervous system diseases in which this neuronal lesion is also encountered (Wisniewski et al. 1979). Some of these examples are common diseases in which neurofibrillary tangles are uncommonly observed while others are relatively rare disorders in which neurofibrillary tangles represent a relatively reproducible feature. One of the disorders that has been of particular interest is the high incidence focus of amyotrophic lateral sclerosis/parkinsonism-dementia complex encountered on the island of Guam in the western Pacific. For a number of reasons, studies of Guam have been considered integral to any consideration of the causes of neurofibrillary tangles formation and the role of this pathologic lesion in a number of analogous neurodegenerative disorders.

Guam is the largest island in the western Pacific, measuring 549 km^2 in area and located approximately 6100 km west of Honolulu, Hawaii, 2100 km south of Tokyo, Japan and a similar distance east of Manila, the Philippines (see Fig. 5). Guam is the largest and southernmost island of the Marianas archipelago, a chain of 15 islands. Guam is home to an indigenous native population referred to as Chamorros, who currently number approximately 60,000 individuals. The Chamorros make up the major ethnic group living on Guam (total population is currently about 170,000 people), with the remainder of the inhabitants on the island being made up primarily of a large number of United States military personnel and their dependents, migrants from the Philippines and Korea and peoples originating from several of the other neighboring islands.

Due to its strategic location, Guam played a significant role in late phases of the action against Japan in World War II. Towards the end of the Pacific campaign of World War II, Dr. Harry Zimmerman, a Navy physician assigned to the island, noted the presence of an inordinately high prevalence of amyotrophic lateral sclerosis among the Chamorro natives (Zimmerman 1945). Zimmerman noted in a memo to his superiors in Washington, D.C., that he had personally observed the presence of 6 or 7 cases of the disease that were in the process of being treated in the local hospital on Guam. Furthermore, during a single month, Zimmerman, trained as a pathologist, confirmed the diagnosis of two cases at autopsy. Amyotrophic lateral sclerosis is a neurodegenerative disorder involving the motor system and leads to progressive paralysis of the voluntary musculature. Following the completion of the war, Zimmerman's observations were confirmed by other workers (Koerner 1952; Arnold et al. 1953). This led to further studies which led to the recognition that amyotrophic lateral sclerosis, a relatively uncommon form of neurodegenerative disease,

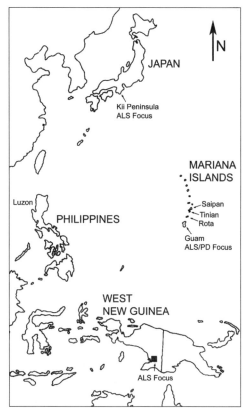

Figure 5. Map of the western Pacific region showing the location of the three high risk foci for amyotrophic lateral sclerosis and parkinsonism-dementia complex, namely, Guam (and Rota), the Kii Peninsula on the island of Kyushu in Japan and southwestern New Guinea. [Used with permission of Elseveir from Garruto and Yase (1986) *Trends Neuroscience* 9:368-374.]

was 100 to 1000 times more prevalent among the Chamorros of Guam than had been seen anywhere else in the world (Kurland and Mulder 1954; Kurland 1958).

The cases of amyotrophic lateral sclerosis among the native Chamorros of Guam, with progressive weakness and paralysis related to degeneration of both the upper and lower motor neurons, were identical clinically to those seen in populations elsewhere in the world (Kurland and Mulder 1954; Kurland et al. 1956). Amyotrophic lateral sclerosis victims on Guam show muscular weakness leading to progressive motor paralysis and typically survive for 2-4 years after the diagnosis is made, again, paralleling that seen in cases throughout the world. However, when autopsies were performed on affected patients these studies revealed that the Guam cases showed both the neuropathologic features seen in cases of amyotrophic lateral sclerosis elsewhere (degeneration of both upper and lower motor neurons) plus the additional and unique feature of large numbers of neurofibrillary tangles in many brain structures (Malamud et al. 1961). Neurofibrillary tangles, as mentioned above, is a classic feature of Alzheimer's disease, but these changes are not encountered in cases of amyotrophic lateral sclerosis, as it is seen elsewhere in the world (Hirano et al. 1967).

In the course of surveying the scope of other forms of neurologic disease, among the native population living on Guam another neurodegenerative disorder of high prevalence was noted among the native Chamorro population and this was referred to as parkinsonism-dementia complex of Guam (Hirano et al. 1961a). Parkinsonism-dementia complex of Guam is characterized by severe parkinsonism (that is, Parkinson's disease-like clinical features) with prominent muscular rigidity, bradykinesia (slowness of movement) and difficulties with gait. Although a resting tremor, rather typical of classic Parkinson's disease, may be seen in some patients, it is not a prominent or particularly disabling clinical feature among the patients on Guam. Accompanying the parkinsonian features is a profound loss of cognitive abilities or dementia that is reminiscent of Alzheimer's disease and is inevitably progressive. The patients show difficulties with recent memory, disorientation, difficulty with reasoning and inability to perform simple calculations. Cases of parkinsonism-dementia complex of Guam progress rapidly and patients with this disorder typically die within 5 years of the initial diagnosis being made. At autopsy, parkinsonism-dementia complex patients show evidence of severe degeneration of certain populations of nerve cells, including the dopaminergic neurons of the substantia nigra pars compacta, the neurons of the locus coeruleus and many other sites in the brain (Hirano et al. 1961b). Importantly, these cases show even more severe neurofibrillary tangle formation than is seen in the Guam amyotrophic lateral sclerosis cases, with severe involvement in the cerebral cortex, hippocampus, amygdala, substantia nigra pars compacta and other brainstem nuclei.

Guam amyotrophic lateral sclerosis/parkinsonism-dementia complex clearly represents a spectrum of environmentally determined neurodegenerative disorders in which neurofibrillary tangles represent a primary aspect of the cellular pathologic changes present in the brain. Based on electron microscopic studies, immunohistochemical and biochemical analysis, the neurofibrillary tangles encountered in Guam amyotrophic lateral sclerosis/parkinsonism-dementia complex cases are identical to the neurofibrillary tangles of cases of Alzheimer's disease (Hirano et al. 1968; Shankar et al. 1989; Buee-Scherrer et al. 1995). Clearly, gaining an understanding of the etiologic/pathogenetic mechanisms responsible for the unique disorders encountered on Guam will have important implications for an understanding of the relevant comparable age-related neurodegenerative disorders, as seen elsewhere in the world, namely, Alzheimer's disease, Parkinson's disease and amyotrophic lateral sclerosis. With this in mind, Perl and colleagues first studied Guam amyotrophic lateral sclerosis/parkinsonism-dementia complex cases using scanning electron microscopy with X-ray energy spectrometry (Perl et al. 1982). In this study we analyzed unstained frozen sections of brain tissue derived from Guam amyotrophic lateral sclerosis and parkinsonism-dementia complex cases with extensive neurofibrillary tangle formation and Guam controls that were tangle-free. This study documented evidence of prominent aluminum accumulation within the tangle-bearing neurons of the Guam amyotrophic lateral sclerosis/parkinsonism-dementia complex cases. The finding of prominent aluminum accumulation in the neurofibrillary tangle-bearing neurons of cases of Guam amyotrophic lateral sclerosis/parkinsonism-dementia complex have now been replicated employing five difference physical principles in studies conducted in three different laboratories (Garruto et al. 1984; Perl et al. 1986; Linton et al. 1987; Piccardo et al. 1988). Indeed, it has been estimated that within the neurofibrillary tangle-bearing neurons of Guam amyotrophic lateral sclerosis/parkinsonism-dementia complex cases the aluminum concentration can be 300 to 600 ppm, more than ten times that calculated for the tangle-bearing neurons of cases of Alzheimer's disease (Perl et al. 1986). The environmental source of these dramatic intraneuronal accumulations of aluminum remains unclear and their significance with respect to the etiology and pathogenesis of the Guam neurologic disorders is also unresolved.

Particular interest in the phenomenon of Guam amyotrophic lateral sclerosis/parkinsonism-dementia complex has centered on the identification of the underlying etiology of this high incidence focus occurring in such a remote island community. Zimmerman initially suggested that amyotrophic lateral sclerosis among the Chamorros of Guam likely represented an inherited disorder occurring in an isolated and therefore inbred community. Observations over the years have shown that while there are certain Chamorro families with a very high percentage of offspring affected by the disease, other families have been relatively spared from the affliction (Plato et al. 1967, 1969). Despite extensive efforts, studies on Guam to identify specific underlying genetically-based factors have been rather disappointing (Poorkaj et al. 2001; Perez-Tur et al. 1999). Although genetically-based susceptibility factors do likely play some role in the disorder, as will be discussed below, it is clear that Guam amyotrophic lateral sclerosis/parkinsonism-dementia complex appears to be an example of environmental/genetic interaction and that the ultimate etiology of the problem lies in long-term exposure to a putative environmental factor/s present on Guam.

Epidemiologic studies have provided strong evidence to support the concept that local environmental factors are primary in the etiology of Guam amyotrophic lateral sclerosis/parkinsonism-dementia complex. Studies performed in the 1960's and 1970's demonstrated that amyotrophic lateral sclerosis was the predominant form of neurodegeneration among the Chamorro population then living on Guam, although many cases of parkinsonism-dementia complex were also present. In addition, these studies showed that although amyotrophic lateral sclerosis and parkinsonism-dementia complex could be seen in any portion of the island, they occurred most commonly among certain villages on the southern part of Guam. Importantly, over the ensuing decades, the numbers of Chamorros suffering from amyotrophic lateral sclerosis has steadily dropped while parkinsonism-dementia complex still remains relatively common on the island (Garruto et al. 1985). Today, new cases of amyotrophic lateral sclerosis on Guam are occasionally seen but are rather rare (Galasko et al. 2002). Further, the ages of onset of both cases of amyotrophic lateral sclerosis and of parkinsonism-dementia complex have significantly changed over the past 30 years. During the interval of 1955 to 1965, the mean age of onset of amyotrophic lateral sclerosis among Chamorros was 47 years and of parkinsonism-dementia complex was about 52 years. However, the mean age of onset for cases diagnosed between 1985 and 1995 of amyotrophic lateral sclerosis cases was 53 years of age and of parkinsonism-dementia complex cases was 64 years. This represents an increase of approximately ten years in age of onset over this relatively short period of time. Such dramatic changes of the distribution and basic characteristics of a disease seen in a relatively stable population, all within an interval of less than one generation, do not occur in inherited disorders, which tend to remain rather stable within populations over many years of observation. Rather, it is much more likely that such dramatic changes reflect the changing impact of relevant environmental etiologic factors.

An additional and critical observation has been the identification of many cases of both amyotrophic lateral sclerosis and parkinsonism-dementia complex among Filipino migrants to Guam who have now had long-term residence on the island. Following the end of World War II and extending to the present time there has been a significant migration to Guam of individuals born and raised in the Philippines who now live permanently in Guam. Within the Filipino community living on Guam over an extended period, a considerable number of cases of amyotrophic lateral sclerosis have now been well documented (Garruto et al. 1981; Chen et al. 1982; Purohit et al. 1992). Virtually all of the cases have been Filipino men who moved to Guam in the late 1940's and early 1950's. A small number of autopsies have been performed on the amyotrophic lateral sclerosis cases among the Filipino migrants and neurofibrillary tangles have been documented in the brain specimens in about half of these cases. The extent of neurofibrillary tangle involvement has not been as dramatic as is typically encountered in the native Chamorro cases, but the finding of neurofibrillary tangles in the Filipino migrant cases clearly separates them from sporadic cases of amyotrophic lateral sclerosis (Hirano and Zimmerman 1962; Hirano et al. 1967), as seen elsewhere in the world. Neurofibrillary tangles are not encountered as a neuropathologic feature of amyotrophic lateral sclerosis cases other than in the high risk foci of the western Pacific. Cases of parkinsonism-dementia complex, a disease that is truly unique to Guam, have now also been documented both clinically and neuropathologically among this Filipino migrant community (Garruto et al. 1981; Chen et al. 1982; Purohit et al. 1992). Importantly, the Filipino migrants to Guam were born in many diverse districts of their native country which essentially rules out the possibility that they originate from a unique genetic isolate in the land of their birth.

Guam, especially the southern portion of the island, is volcanic and has an aluminum-rich bauxite soil. The northern part of the island is coral-based and thus is calcium-rich. It is of interest that although cases of amyotrophic lateral sclerosis and parkinsonism-dementia complex have been encountered in virtually every village on the island, the highest prevalence of amyotrophic lateral sclerosis/parkinsonism-dementia complex cases has traditionally been encountered among Chamorros living in the southern portions of Guam. Although several of the villages in the southern portion of the island have over the years shown the highest prevalence rates of neurodegenerative disease, the village of Umatac has had more cases than any other locale (see Fig. 6). Over the past 40 years in which neurologic disease has been most intensively studied, there have been more than 350 documented cases of either amyotrophic lateral sclerosis or parkinsonism-dementia complex among the inhabitants of this village of approximately 750 people. When one considers that at any time more that half of the residents of Umatac are less than 30 years old and therefore not within the at-risk age for these diseases, one appreciates the devastating problem amyotrophic lateral sclerosis/parkinsonism-dementia complex represents for this island population.

As mentioned before, Guam is part of an archipelago consisting of 15 islands. Some of these islands are volcanic, others based on coral, with a few having a mixture of coral and volcanic origins. Guam is an example of the latter. It is the largest, most populous and southern-most of the chain. The next nearest island, Rota, is only about 70 km north of Guam and has shown similar rates of amyotrophic lateral sclerosis/parkinsonism-dementia complex among its relatively small Chamorro population (total population, approximately 1500 individuals).

Figure 6. The appearance of the village of Umatac on the southwestern coast of Guam. Among the 750 inhabitants of this village, over the past 40 years there have been more than 350 documented cases of either amyotrophic lateral sclerosis or parkinsonism-dementia complex.

Interestingly, over many years of study, no cases of amyotrophic lateral sclerosis/parkinsonism-dementia complex have been documented among the Chamorro natives living on the next two islands, Tinian and Saipan (both coral-based islands) despite the fact that they are within 190 km of Guam. It would appear that amyotrophic lateral sclerosis/parkinsonism-dementia complex of Guam represent place-based disorders and these unique diseases are confined to individuals who live for long periods of time on these two particular islands (namely, Guam and Rota) but not on the two nearby islands (that is Tinian and Saipan). The people living on all four islands are derived from the same genetic stock (Yanagihara et al. 1983) and appear to have a similar diet, fish in the same waters, encounter the same plants, birds and insects, and share common cultural activities.

A large variety of possible etiologic agents have been considered for Guam amyotrophic lateral sclerosis/parkinsonism-dementia complex, ranging from constituents of water and soil to local plant toxins and the effects of infectious organisms. Importantly, on Guam there exists a wealth of archival death certificates records which date from the beginning of the 20th century. The island of Guam was awarded to the United States in 1898 as part of the Treaty of Paris which concluded the Spanish-American War. For many years after this the island was controlled and governed by the United States Navy. Dating from the United States takeover of administering the island, naval medical personnel had the duty to certify all deaths and their causes among the native population. These Naval death certificate documents have survived and provide clear reference to numerous deaths from amyotrophic lateral sclerosis among Chamorros living on Guam during the early years of the 20th century (see Fig. 7). This record represents irrefutable evidence that the putative environmental etiologic agent for the disease was already present on the island at that time. From this observation, it is clear that the cause of the problem has existed on Guam for over 100 years and can have nothing to do with the effects of the dramatic westernization of the island that has occurred in more recent years. Currently, Guam is entirely "westernized" and possesses every aspect of modern American "culture" currently available to

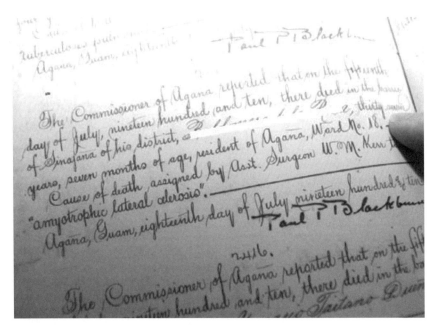

Figure 7. Death certificate entry from July, 1910 documenting the death of a 37 year old Guamanian man from amyotrophic lateral sclerosis. Many such entries are present in these documents.

those living in the mainland of the United States (numerous fast food restaurants, satellite TVs, pizza, shopping malls, golf courses, etc.). Prior to World War II, Guam could be characterized as a rural, self-sufficient society surviving on locally grown foodstuffs. Today virtually all foods are imported from elsewhere in the world and farming and livestock raising indigenous to Guam represents a miniscule portion of the native diet. Industry of any magnitude has not previously, nor is currently been practiced on the island. All this points to the existence of a putative naturally-occurring etiologic agent/s to which the population is exposed but which is not present on what appear to be remarkably similar neighboring islands.

Two additional observations must be considered in the interpretation of these findings, namely, migration studies and the identification of two additional foci of similar disorders. Prior to 1950, migration from Guam to the mainland United States was carefully controlled by the United States Navy and few Guamanians were permitted to leave the island and live elsewhere. In 1950, with the passage of the Organic Act of Guam, the inhabitants of Guam were provided with full United States citizenship and were allowed to migrate freely to the mainland. Since that time a significant number of Chamorros from Guam have taken the opportunity to migrate from the island and now live permanently in a number of communities on the west coast of the United States. In and around the cities of San Diego, San Jose, and Oakland, California and Bremerton, Washington, there are rather sizeable communities of Chamorros who were originally born and raised on Guam. Most of the inhabitants of these communities left Guam in their early 20's and now live permanently on the mainland United States.

It has now been clearly shown that despite being away from Guam for many years, these individuals, as they approach the age at risk seen among Chamorros living of Guam, are also at risk for developing amyotrophic lateral sclerosis/parkinsonism-dementia complex (Torres et al. 1957; Eldridge et al. 1969). Remarkably, it would appear that once having grown up on Guam, despite leaving the island and living on the United States mainland for over 20 years, they continue to carry the risk of developing the island's disease through their lifetime. It should be pointed out that despite leaving Guam, these individuals bring with them, to the extent that it is practical, their cultural practices and special dietary preferences. In addition, they have close ties to the island and tend to return to Guam, at least briefly, on a yearly basis. Nevertheless, the mechanism for this retention of risk within the migrant population remains unclear. Importantly, it is too soon to determine if their offspring, born and raised off Guam, will also continue to show an increased risk for developing amyotrophic lateral sclerosis/parkinsonism-dementia complex. This would appear unlikely, but it will be many years before the next generation of West coast living Guamanian Chamorros reaches their age at-risk for developing these disorders and will provide the answer to this important question.

The second important epidemiologic observation has been the identification of two other foci of similar neurodegenerative disorders, namely in the Kii Peninsula of Japan and in an isolated portion of southwestern New Guinea. The Kii Peninsula is an isolated mountainous region on the south eastern coast of the Japanese island of Honshu. The focus in the Kii Peninsula has been studied in much greater detail than that in New Guinea. The first account of a high incidence of amyotrophic lateral sclerosis in the Kii Peninsula of the Japanese island of Honshu was reported by Miura (1911). It was subsequently identified that there were actually two distinct and separate high incidence foci within this region, namely, in the Kozagawa township in the southern part of the peninsula and the isolated village of Hobara in the Mei Prefecture (Shiraki and Yase 1991). These two foci are located in a relatively remote mountainous region and are separated by approximately 200 km.

In the twenty years following World War II, the incidence rates for amyotrophic lateral sclerosis for these two locations in the Kii Peninsula were calculated at 55 per 100,000 population for Hobara and 14 per 100,000 for Kozagawa (as compared to 0.4 per 100,000 for the whole of Japan). Epidemiologic studies of the region indicated that the only additional cases of amyotrophic lateral sclerosis encountered in the region, in general, occurred among individuals who had grown up in one of these two villages and had subsequently moved elsewhere in the region. More recent clinical surveys have revealed the presence of additional cases of apparent parkinsonism in association with progressive dementia, a disorder that looks virtually identical to parkinsonism-dementia complex of Guam (Kokubo and Kuzuhara 2004; Kuzuhara and Kokubo 2005). Extensive neuropathologic studies by Shiraki and Yase (Shiraki and Yase 1991), plus the more recent publications by Kuzuhara and colleagues (Kuzuhara et al. 2001; Itoh et al. 2003; Kokubo and Kuzuhara 2004; Kuzuhara and Kokubo 2005) have revealed the presence of severe widespread neurofibrillary tangle formation in association with both the Kii Peninsula amyotrophic lateral sclerosis cases as well as those with parkinsonism and dementia. It would therefore appear that the phenomenon of endemic neurodegeneration seen in this isolated part of Japan is virtually identical to that encountered among the Chamorros living on Guam.

The southwestern New Guinea focus occurs among the extremely primitive and isolated Auyu and Jakai people living in a very remote district in southwestern New Guinea. These cases were first briefly reported by Gajdusek (1963) and then more detailed descriptions of the focus were provided by Gajdusek and Salazar in 1982 (Gajdusek 1982). In this focus, numerous cases of apparent amyotrophic lateral sclerosis and some cases of parkinsonism accompanied by dementia were documented clinically. In the New Guinea focus the age at which the natives became ill was younger than had been seen in either Guam or the Kii Peninsula, although documentation of the specific age of these primitive peoples has been poor. When first seen, the prevalence of amyotrophic lateral sclerosis within this remote focus was even higher than originally observed on Guam in the immediate post-World War II period.

In the 20 year period from Gajdusek's first observations to his second visit to the region, the distribution of disease had not spread to adjacent regions, despite the lack of any geographic barriers. To Gajdusek, this stability in distribution implied that the disease focus could not be secondary to any form of infectious organism. Further, no obvious cultural practices related to nutrition or other uses of plants, animal or mineral products were felt by Gajdusek to be responsible for the disorder. He also noted that flying insects, birds and most species of plants and animals appeared to be uniformly distributed across a large general region in which this focus occupied only a small portion. All this suggested to Gajdusek that some putative factor/s related to local soil and/or water supply was likely to relate to the etiology of the problem. This region is very difficult to reach and there has been little follow-up information. Recently, Spencer and colleagues have visited the region and reported a dramatic decline in the prevalence of amyotrophic lateral sclerosis with a lesser decline in those with parkinsonian features (Spencer et al. 2005). This parallels the changes in prevalence seen on Guam and again strongly suggests an environmental etiology with changing exposure to the putative cause of the disorders.

Unfortunately, no autopsied brain tissues have been available for study on individuals showing neurologic dysfunction from within the New Guinea focus. Accordingly, we do not have neuropathologic evidence regarding the disorder's similarity to those of Guam and the Kii peninsula. The clinical descriptions emerging from this focus appear consistent but neuropathologic confirmation would be of great help. If this is indeed a third focus of amyotrophic lateral sclerosis and parkinsonism-dementia complex-like disease, then the question remains as to what possible environmental factor/s link the three foci. Epidemiologic evidence from each site strongly indicates that these are "place" diseases, meaning that it affects individuals who live for long periods of time in that locality and leave unaffected, people of similar background and customs that live in nearby locations.

Could the three foci of neurodegeneration in the western Pacific represent the co-existence of two factors which when placed together, act synergistically? So far, consideration of such concepts has not yielded consistent answers. One possibility, suggested many years ago by Yase, is that they share common unique constituents of soil and/or drinking water (Yase

1972; Garruto and Yase 1986). Although there have been relatively few studies, to date, no unique characteristic of the water or soil of at least Guam and Kii have emerged other than the aluminum-rich bauxite soils of each site (McLachlan et al. 1989; Miller and Sanzolone 2003). However, many other populations live and thrive on bauxite soils without showing this kind of neurodegenerative disorder. One hypothesis that has been articulated is that a local deficiency of environmental sources of calcium and magnesium, both physiologically essential ionic constituents, combined with available sources of aluminum, could result in an increased uptake of aluminum, as an alternative potentially available dietary source of cations (Gajdusek 1984; Garruto and Yase 1986). There is data to support that some sources of water and soil in all three foci are calcium-deficient. However, evidence of a clinically detectible calcium deficiency among Guamanians has not been found (Ahlskog et al. 1995). Studies of water and soil in each region have not been rigorously performed and other explanations for the excess brain aluminum concentrations will likely need to be sought.

The geographic location of these three foci on the earth raises the following intriguing question: Is it merely coincidence that all three foci of amyotrophic lateral sclerosis/ parkinsonism-dementia complex lie essentially along the same longitude, yet are separated by thousands of miles and does this indicate a possible clue as to their underlying etiology? Is it just coincidence that the geographic location of the three known foci of amyotrophic lateral sclerosis/parkinsonism-dementia complex neurodegeneration consist of Guam, lying just beside the Marianas Trench and the Kii peninsula and New Guinea foci, are all strategically located on the earth near sites of the same tectonic plate interfaces? Could this offer a clue as to the underlying nature of three locations for these unique disease outbreaks? Those of us who work in the medical sciences clearly lack the expertise to consider this suggestion further. The knowledge and understanding of our colleagues in mineralogy, geology and geochemistry will be needed to further explore this novel hypothesis. Unraveling the mysteries of the three western Pacific foci of neurodegeneration will clearly provide important answers to questions relevant to the etiology and pathogenesis of amyotrophic lateral sclerosis and parkinsonism-dementia complex within the three high risk foci as well as the three analogous age-related neurodegenerative diseases seen elsewhere in the world, namely, Alzheimer's disease, Parkinson's disease and amyotrophic lateral sclerosis.

ACKNOWLEDGMENTS

Over the years, this work has been supported by grants from the NIH (AG-08812 and AG-14382) as well as the John Douglas French Foundation and the American Health Assistance Foundation.

REFERENCES

Ahlskog JE, Waring SC, Kurland LT, Petersen RC, Moyer TP, Harmsen WS, Maraganore DM, O'Brien PC, Esteban-Santillan C, Bush V (1995) Guamanian neurodegenerative disease: investigation of the calcium metabolism/heavy metal hypothesis. Neurology 45:1340-1344

Arnold A, Edgren DC, Palladino VS (1953) Amyotrophic lateral sclerosis. Fifty cases observed on Guam. J Nerv Ment Dis 117:135-139

Bomford A (2002) Genetics of haemochromatosis. Lancet 360:1673-1681

Bridelli MG, Tampellini D, Zecca L (1999) The structure of neuromelanin and its iron binding site studied by infrared spectroscopy. FEBS Lett 457:18-22

Buee-Scherrer V, Buee L, Hof PR, Leveugle B, Gilles C, Loerzel AJ, Perl DP, Delacourte A (1995) Neurofibrillary degeneration in amyotrophic lateral sclerosis/parkinsonism-dementia complex of Guam: immunochemical characterization of tau proteins. Am J Pathol 68:924-932

Candy JM, Oakley AE, Klinowski J, Carpenter TA, Perry RH, Atack JR, Perry EK, Blessed G, Fairbairn A, Edwardson JA (1986) Aluminosilicates and senile plaque formation in Alzheimer's disease. Lancet 1: 354-357

Castellani RJ, Perry G, Siedlak SL, Nunomura A, Shimohama S, Zhang J, Montine T, Sayre LM, Smith MA (2002) Hydroxynonenal adducts indicate a role for lipid peroxidation in neocortical and brainstem Lewy bodies in humans. Neurosci Lett 319:25-28

Chen KM, Makifuchi T, Garruto RM, Gajdusek DC (1982) Parkinsonism-dementia in a Filipino migrant: a clinicopathologic case report. Neurology 32:1221-1226

Crapper DR, Krishnan SS, Dalton AJ (1973) Brain aluminum distribution in Alzheimer's disease and experimental neurofibrillary degeneration. Science 180:511-513

Crapper DR, Krishnan SS, Quittkat S (1976) Aluminum, neurofibrillary degeneration and Alzheimer's disease. Brain 99:67-80

Datz C, Haas T, Rinner H, Sandhofer F, Patsch W, Paulweber B (1998) Heterozygosity for the C282Y mutation in the hemochromatosis gene is associated with increased serum iron, transferrin saturation, and hemoglobin in young women: a protective role against iron deficiency? Clin Chem 44:2429-2432

Dexter DT, Wells FR, Lees AJ, Agid F, Agid Y, Jenner P, Marsden CD (1989) Increased nigral iron content and alterations in other metal ions occurring in brain in Parkinson's disease. J Neurochem 52:1830-1836

Double KL, Gerlach M, Schunemann V, Trautwein AX, Zecca L, Gallorini M, Youdim MB, Riederer P, Ben-Shachar D (2003) Iron-binding characteristics of neuromelanin of the human substantia nigra. Biochem Pharmacol 66:489-494

Drayer BP (1989) Magnetic resonance imaging and extrapyramidal movement disorders. Eur Neurol 29 Suppl 1:9-12

Earle KM (1968) Studies on Parkinson's disease including x-ray fluorescent spectroscopy of formalin fixed brain tissue. J Neuropathol Exp Neurol 27:1-14

Eijkelkamp EJ, Yapp TR, Powell LW (2000) HFE-associated hereditary hemochromatosis. Can J Gastroenterol 14:121-125

Eldridge R, Ryan E, Rosario J, Brody JA (1969) Amyotrophic lateral sclerosis and parkinsonism-dementia in a migrant population from Guam. Neurology 19:1029-1037

Enochs WS, Sarna T, Zecca L, Riley PA, Swartz HM (1994) The roles of neuromelanin, binding of metal ions, and oxidative cytotoxicity in the pathogenesis of Parkinson's disease: a hypothesis. J Neural Transm Park Dis Dement Sect 7:83-100

Feder JN, Gnirke A, Thomas W, Tsuchihashi Z, Ruddy DA, Basava A, Dormishian F, Domingo R Jr., Ellis MC, Fullan A, Hinton LM, Jones NL, Kimmel BE, Kronmal GS, Lauer P, Lee VK, Loeb DB, Mapa FA, McClelland E, Meyer NC, Mintier GA, Moeller N, Moore T, Morikang E, Prass CE, Quintana L, Starnes SM, Schatzman RC, Brunke KJ, Drayna DT, Risch NJ, Bacon BR, Wolff RK (1996) A novel MHC class I-like gene is mutated in patients with hereditary haemochromatosis. Nat Genet 13:399-408

Fodinger M, Sunder-Plassmann G (1999) Inherited disorders of iron metabolism. Kidney Int Suppl 69:S22-S34

Gajdusek DC (1963) Motor neuron disease in natives of New Guinea. N Engl J Med 268:474-476

Gajdusek DC (1982) Foci of motor neuron disease in high incidence in isolated populations of East Asia and the Western Pacific. *In:* Human Motor Neuron Disease. Rowland LP (ed) Raven Press, p 363-393

Gajdusek DC (1984) Calcium deficiency induced secondary hyperparathyroidism and resultant CNS deposition of calcium and other metallic cations as the cause of ALS and PD in high incidence among the Auyu and Jakai people in west New Guinea. *In:* Amyotrophic Lateral Sclerosis in Asia and Oceania. Chen KM, Yase Y (eds) Shyan-Fu Chou National Taiwan University, Taipei, p 145-171

Galasko D, Salmon DP, Craig UK, Thal LJ, Schellenberg G, Wiederholt W (2002) Clinical features and changing patterns of neurodegenerative disorders on Guam, 1997-2000. Neurology 58:90-97

Garruto RM, Fukatsu R, Yanagihara R, Gajdusek DC, Hook G, Fiori C (1984) Imaging of calcium and aluminum in neurofibrillary tangle-bearing neurons in parkinsonism-dementia of Guam. Proc Natl Acad Sci USA 81:1875-1879

Garruto RM, Gajdusek DC, Chen KM (1981) Amyotrophic lateral sclerosis and parkinsonism-dementia among Filipino migrants to Guam. Ann Neurol 10:341-350

Garruto RM, Yanagihara R, Gajdusek DC (1985) Disappearance of high-incidence amyotrophic lateral sclerosis and parkinsonism-dementia on Guam. Neurology 35:193-198

Garruto RM, Yase Y (1986) Neurodegenerative disorders of the western Pacific: the search for mechanisms of pathogenesis. Trends Neurosci 9:368-374

Good PF, Hsu A, Werner P, Perl DP, Olanow CW (1998) Protein nitration in Parkinson's disease. J Neuropathol Exp Neurol 57:338-342

Good PF, Olanow CW, Perl DP (1992a) Neuromelanin-containing neurons of the substantia nigra accumulate iron and aluminum in Parkinson's disease: A LAMMA study. Brain Res 593:343-346

Good PF, Perl DP, Bierer LM, Schmeidler J (1992b) Selective accumulation of aluminum and iron in the neurofibrillary tangles of Alzheimer's disease: A laser microprobe (LAMMA) study. Ann Neurol 31:286-292

Gutteridge JM, Quinlan GJ, Clark I, Halliwell B (1985) Aluminum salts accelerate peroxidation of membrane lipids stimulated by iron salts. Biochim Biophys Acta 835:441-447

Hirano A, Arumugasamy N, Zimmerman HM (1967) Amyotrophic lateral sclerosis. A comparison of Guam and classical cases. Arch Neurol 16:357-363

Hirano A, Dembitzer HM, Kurland LT, Zimmerman HM (1968) The fine structure of some intraganglionic alterations. Neurofibrillary tangles, granulovacuolar bodies and "rod-like" structures as seen in Guam amyotrophic lateral sclerosis and parkinsonism-dementia complex. J Neuropathol Exp Neurol 27:167-182

Hirano A, Kurland LT, Krooth RS, Lessell S (1961a) Parkinsonism-dementia complex, an endemic disease on the island of Guam I. Clinical features. Brain 84:642-661

Hirano A, Malamud N, Kurland LT (1961b) Parkinsonism-dementia complex, an endemic disease on the island of Guam II. Pathological features. Brain 84:662-679

Hirano A, Zimmerman HM (1962) Alzheimer's neurofibrillary changes. A topographic study. Neurology 7: 227-242

Hirsch EC, Brandel JP, Galle P, Javoy-Agid F, Agid Y (1991) Iron and aluminum increase in the substantia nigra of patients with Parkinson's disease: an X-ray microanalysis. J Neurochem 56:446-451

Huang Y, Herman MM, Liu J, Katsetos CD, Wills MR, Savory J (1997) Neurofibrillary lesions in experimental aluminum-induced encephalopathy and Alzheimer's disease share immunoreactivity for amyloid precursor protein, A beta, alpha 1-antichymotrypsin and ubiquitin-protein conjugates. Brain Res 771: 213-220

Itoh N, Ishiguro K, Arai H, Kokubo Y, Sasaki R, Narita Y, Kuzuhara S (2003) Biochemical and ultrastructural study of neurofibrillary tangles in amyotrophic lateral sclerosis/parkinsonism-dementia complex in the Kii peninsula of Japan. J Neuropathol Exp Neurol 62:791-798

Jellinger K, Kienzl E, Rumpelmair G, Riederer P, Stachelberger H, Ben-Shachar D, Youdim MBH (1992) Iron-melanin complex in substantia nigra of parkinsonian brains: an x-ray microanalysis. J Neurochem 59:1168-1171

Jenner P, Olanow CW (1996) Oxidative stress and the pathogenesis of Parkinson's disease. Neurology 47: S161-S170

Klatzo I, Wisniewski H, Streicher E (1965) Experimental production of neurofibrillary pathology: I. Light microscopic observations. J Neuropathol Exp Neurol 24:187-199

Koerner DR (1952) Amyotrophic lateral sclerosis on Guam: a clinical study and review of the literature. Ann Intern Med 37:1204-1220

Kokubo Y, Kuzuhara S (2004) Neurofibrillary tangles in ALS and Parkinsonism-dementia complex focus in Kii, Japan. Neurology 63:2399-2401

Kuhn LC (1999) Iron overload: molecular clues to its cause. Trends Biochem Sci 24:164-166

Kurland LT (1958) Epidemiology: incidence, geographic distribution and genetic considerations. In: Pathogenesis and Treatment of Parkinsonism. Field W (ed) Charles C. Thomas Publisher, p 5-43

Kurland LT, Mulder DW (1954) Epidemiologic investigations of amyotrophic lateral sclerosis. 1. Preliminary report on geographic distribution, with special reference to the Mariana Islands, including clinical and pathologic observations. Neurology 4:355-378, 438-448

Kurland LT, Mulder DW, Sayre GP, Lambert E, Hutson W, Iriarte LG, Imus HA (1956) Amyotrophic lateral sclerosis in the Marianas Islands. Arch Neurol Psych 75:435-441

Kuzuhara S, Kokubo Y (2005) Atypical parkinsonism of Japan: amyotrophic lateral sclerosis-parkinsonism-dementia complex of the Kii peninsula of Japan (Muro disease): an update. Mov Disord 20 Suppl 12: S108-S113

Kuzuhara S, Kokubo Y, Sasaki R, Narita Y, Yabana T, Hasegawa M, Iwatsubo T (2001) Familial amyotrophic lateral sclerosis and parkinsonism-dementia complex of the Kii Peninsula of Japan: clinical and neuropathological study and tau analysis. Ann Neurol 49:501-511

Landsberg JP, McDonald B, Watt F (1992) Absence of aluminium in neuritic plaque cores in Alzheimer's disease. Nature 360:65-68

Linton RW, Bryan SR, Cox XB, Griffis DP, Shelburne JD, Fiori CE, Garruto RM (1987) Digital imaging studies of aluminum and calcium in neurofibrillary tangle-bearing neurons using SIMS (secondary ion mass spectrometry). Trace Elements Med 4:99-104

Malamud N, Hirano A, Kurland LT (1961) Pathoanatomic changes in amyotrophic lateral sclerosis on Guam. Neurology 5:401-414

Martin RB (1986) The chemistry of aluminum as related to biology and medicine. Clin Chem 32:1797-1806

Martyn CN, Coggon DN, Inskip H, Lacey RF, Young WF (1997) Aluminum concentrations in drinking water and risk of Alzheimer's disease. Epidemiology 8:281-286

McLachlan DR, McLachlan CD, Krishnan B, Krishnan SS, Dalton AJ, Steele JC (1989) Aluminum and calcium in soil and food from Guam, Palau and Jamaica: implications for amyotrophic lateral sclerosis and parkinsonism-dementia sysndromes of Guam. Env Geochem Health 11:45-53

Miller WR, Sanzolone RF (2003) Investigation of the possible connection of rock and soil geochemistry to the occurrence of high rates of neurodegenerative diseases on Guam and a hypothesis for the cause of the diseases. U.S. Department of the Interior, U.S. Geological Survey, Denver, CO, p 1-44

Miura K (1911) Amyotrophische lateralsklerose unter dem blide von sog. bulbarparalyse. Neurol Jap 10:366-369

Moalem S, Percy ME, Kruck TPA, Gelbart RR (2002) Epidemic pathogenic selection: an explanation for hereditary hemochromotosis? Med Hypotheses 59:325-329

Mura C, Le Gac G, Raguenes O, Mercier AY, Le Guen A, Ferec C (2000) Relation between HFE mutations and mild iron-overload expression. Mol Genet Metab 69:295-301

Murayama H, Shin RW, Higuchi J, Shibuya S, Muramoto T, Kitamoto T (1999) Interaction of aluminum with PHFtau in Alzheimer's disease neurofibrillary degeneration evidenced by desferrioxamine-assisted chelating autoclave method. Am J Pathol 155:877-885

Olanow CW, Drayer B (1987) Brain iron: MRI studies in Parkinson syndrome. *In:* Recent Developments in Parkinson's Disease, Volume 2. Fahn S, Marsden CD, Calne DB, Goldstein M (eds) MacMillan Healthcare, p 135-143

Olanow CW, Perl DP (1994) Free radicals and neurodegeneration. Trends Neurosci 17:193-194

Parkinson J (1955) An essay on the shaking palsy. London. 1817. *In:* Reproduced in: James Parkinson (1755-1824). Critchley M (ed) MacMillan, p 145-218

Perez-Tur J, Buee L, Morris HR, Waring SC, Onstead L, Wavrant-De Vrieze F, Crook R, Buee-Scherrer V, Hof PR, Petersen RC, McGeer PL, Delacourte A, Hutton M, Siddique T, Ashkog JE, Hardy J, Steele JC (1999) Neurodegenerative diseases of Guam: analysis of tau. Neurology 53:411-413

Perl DP, Brody AR (1980) Alzheimer's Disease: X-ray spectrographic evidence of aluminum accumulation in neurofibrillary tangle-bearing neurons. Science 208:297-299

Perl DP, Gajdusek DC, Garruto RM, Yanagihara RT, Gibbs CJ Jr. (1982) Intraneuronal aluminum accumulation in amyotrophic lateral sclerosis and parkinsonism-dementia of Guam. Science 217:1053-1055

Perl DP, Munoz-Garcia D, Good PF, Pendlebury WW (1986) Calculation of intracellular aluminum concentration in neurofibrillary tangle (NFT)-bearing and NFT-free hippocampal neurons of ALS/parkinsonism dementia of Guam using laser microprobe analysis. J Neuropathol Exp Neurol 45:379

Perry G, Sayre LM, Atwood CS, Castellani RJ, Cash AD, Rottkamp CA, Smith MA (2002) The role of iron and copper in the aetiology of neurodegenerative disorders: therapeutic implications. CNS Drugs 16:339-352

Piccardo P, Yanagihara R, Garruto RM, Gibbs CJ Jr., Gajdusek DC (1988) Histochemical and X-ray microanalytical localization of aluminum in amyotrophic lateral sclerosis and parkinsonism-dementia of Guam. Acta Neuropathol 77:1-4

Plato CC, Cruz MT, Kurland LT (1969) Amyotrophic lateral sclerosis/parkinsonism-dementia complex of Guam: further genetic investigations. Am J Hum Gen 21:133-141

Plato CC, Reed DM, Elizan TS, Kurland LT (1967) Amyotrophic lateral sclerosis/parkinsonism-dementia complex of Guam. IV. Familial and genetic investigations. Amer J Hum Gen 19:617-632

Poolsup N, Li Wan Po A, Knight TL (2000) Pharmacogenetics and psychopharmacotherapy. J Clin Pharm Ther 25:197-220

Poorkaj P, Tsuang D, Wijsman E, Steinbart E, Garruto RM, Craig UK, Chapman NH, Anderson L, Bird TD, Plato CC, Perl DP, Weiderholt W, Galasko D, Schellenberg GD (2001) TAU as a susceptibility gene for amyotropic lateral sclerosis-parkinsonism dementia complex of Guam. Arch Neurol 58:1871-8

Purohit DP, Perl DP, Steele JC (1992) ALS/parkinsonism dementia-complex among Filipino migrants to Guam: report of two cases with parkinsonian features. J Neuropathol Exp Neurol 51:323

Rao JK, Katsetos CD, Herman MM, Savory J (1998) Experimental aluminum encephalomyelopathy. Relationship to human neurodegenerative disease. Clin Lab Med 18:687-698

Roden DM (2001) Principles in pharmacogenetics. Epilepsia 42 Suppl 5:44-48

Rondeau V (2002) A review of epidemiologic studies on aluminum and silica in relation to Alzheimer's disease and associated disorders. Rev Environ Health 17:107-121

Savory J, Huang Y, Herman MM, Reyes MR, Wills MR (1995) Tau immunoreactivity associated with aluminum maltolate-induced neurofibrillary degeneration in rabbits. Brain Res 669:325-329

Shankar SK, Yanagihara R, Garruto RM, Grundke-Iqbal I, Kosik KS, Gajdusek DC (1989) Immunocytochemical characterization of neurofibrillary tangles in amyotrophic lateral sclerosis and parkinsonism-dementia of Guam. Ann Neurol 25:146-151

Shiraki H, Yase Y (1991) Amyotrophic lateral sclerosis and parkinsonism-dementia in the Kii Penisula: comparison with the same disorders in Guam and with Alzheimer's disease. *In:* Handbook of Clinical Neurology, Vol 15 (59), Diseases of the Motor System. deJung JMBV (ed) Elsevier Scientific Publishing, p 273-300

Sofic E, Paulus W, Jellinger K, Riederer P, Youdim MB (1991) Selective increase of iron in substantia nigra zona compacta of parkinsonian brains. J Neurochem 56:978-982

Spencer PS, Palmer VS, Ludolph AC (2005) On the decline and etiology of high-incidence motor system disease in West Papua (southwest New Guinea). Mov Disord 20 Suppl 12:S119-S126

Spillantini MG, Schmidt ML, Lee VM-Y, Trojanowski JQ, Jakes R, Goedert M (1997) Alpha synuclein in Lewy bodies. Nature 388:232-233

Stern AJ, Perl DP, Munoz-Garcia D, Good PF, Abraham C, Selkoe DJ (1986) Investigation of silicon and aluminum content in isolated plaque cores by laser microprobe mass analysis (LAMMA). 45:361

Terry RD, Pena C (1965) Experimental production of neurofibrillary pathology: Electron microscopy, phosphate histochemistry and electron probe analysis. J Neuropathol Exp Neurol 24:200-210

Thorburn D, Curry G, Spooner R, Spence E, Oien K, Halls D, Fox R, McCruden EA, MacSween RN, Mills PR (2002) The role of iron and haemochromatosis gene mutations in the progression of liver disease in chronic hepatitis C. Gut 50:248-252

Torres J, Iriarte LL, Kurland LT (1957) Amyotrophic lateral sclerosis among Guamanians in California. Calif Med 4:385-388

Wisniewski K, Jervis GA, Moretz RC, Wisniewski HM (1979) Alzheimer neurofibrillary tangles in diseases other than senile and presenile dementia. Ann Neurol 5:288-294

Worwood M (2005) Inherited iron loading: genetic testing in diagnosis and management. Blood Rev 19:69-88

Yanagihara RT, Garruto RM, Gajdusek DC (1983) Epidemiological surveillance of amyotrophic lateral sclerosis and parkinsonism-dementia in the Commonwealth of the Northern Mariana Islands. Ann Neurol 13:79-86

Yase Y (1972) The pathogenesis of amyotrophic lateral sclerosis. Lancet 2:292-296

Youdim MB, Ben Shachar D, Riederer P (1989) Is Parkinson's disease a progressive siderosis of substantia nigra resulting in iron and melanin induced neurodegeneration? Acta Neurol Scand Suppl 126:47-54

Zecca L, Tampellini D, Gatti A, Crippa R, Eisner M, Sulzer D, Ito S, Fariello R, Gallorini M (2002) The neuromelanin of human substantia nigra and its interaction with metals. J Neural Transm 109:663-672

Zimmerman HM (1945) Monthly report to Medical Officer in Command. US Naval Medical Research Unit No 2

Reviews in Mineralogy & Geochemistry
Vol. 64, pp. 135-152, 2006
Copyright © Mineralogical Society of America

5

Potential Role of Soil in the Transmission of Prion Disease

P. T. Schramm[1], C. J. Johnson[2,3], N. E. Mathews[4], D. McKenzie[3], J. M. Aiken[2,3] and J. A. Pedersen[1,4,5]

[1] *Molecular and Environmental Toxicology Center*
[2] *Program in Cellular and Molecular Biology*
[3] *Animal Health and Biomedical Sciences*
[4] *Nelson Institute for Environmental Studies*
[5] *Department of Soil Science*
University of Wisconsin
Madison, Wisconsin, 53706-1299, U.S.A.
e-mail: japedersen@soils.wisc.edu

BACKGROUND ON PRION DISEASES

Transmissible spongiform encephalopathies (TSE), or prion diseases, are a family of inevitably fatal neurodegenerative disorders affecting a variety of mammalian species. These diseases include scrapie in sheep and goats; bovine spongiform encephalopathy (BSE, "mad cow" disease) in cattle; chronic wasting disease (CWD) in North American deer, elk and moose; transmissible mink encephalopathy; and Creutzfeldt-Jakob disease (CJD, sporadic, familial and variant forms) and kuru in humans. These diseases are characterized by long incubation periods, spongiform degeneration of the brain and accumulation of an abnormally folded isoform of the prion protein, designated PrP^{Sc}, in brain tissue (Prusiner 1998).

CWD and scrapie are the only TSEs that appear to be environmentally transmitted. Therefore, this chapter focuses primarily on these two TSEs. Scrapie has been known in sheep for at least 250 years (McGowan 1922). Most clinically infected sheep exhibit the obvious feature of excessive rubbing and scratching of the skin; the term "scrapie" derives from this symptom. The peak incidence of scrapie occurs in sheep three to four years of age although the earliest cases are seen at 18 months and the latest in animals older than 10 years (Dickinson 1976). The origin of the scrapie agent is unknown, but a familial pattern exists in natural sheep scrapie suggesting that genetics and, possibly, vertical transmission are important. Scrapie has a world-wide distribution and has been documented wherever sheep are raised, with the exception of Australia and New Zealand.

CWD was first identified at a Colorado research facility in 1967 (Williams and Young 1980) and has since been identified in captive cervid populations in Wyoming, Wisconsin, Saskatchewan, South Dakota, Oklahoma, New York, Nebraska, Montana, Minnesota, Kansas and Alberta (Fig. 1). In the free-ranging cervid population, CWD has been found in Wyoming, Wisconsin, Utah, South Dakota, Saskatchewan, New Mexico, Nebraska, Illinois, New York, West Virginia, Alberta and Colorado (Fig. 1). The increase in the known range of CWD in free-ranging and captive cervids may be due in part to increased surveillance. Epidemiological data from expanding affected regions suggest that lateral transmission is the primary mode of disease spread (Miller et al. 2000).

Oral transmission of prion diseases is well established raising concern over the potential interspecies transmission of animal TSEs to humans. Interspecies transmission of BSE to sheep, felines and ungulates has occurred (Prusiner 1997; Horiuchi et al. 2000) and is

 DOI: 10.2138/rmg.2006.64.5

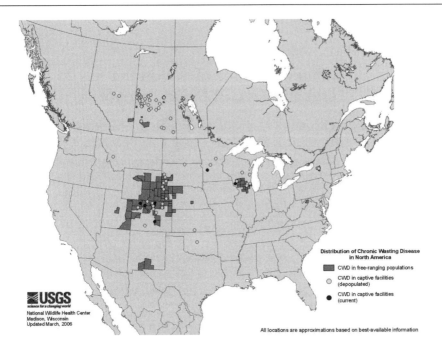

Figure 1. Distribution of cervid chronic wasting disease in North America. [Map produced by the U.S. Geological Survey: *www.nwhc.usgs.gov/disease_information/chronic_wasting_disease/north_america_ CWD_map.jsp* (accessed 12 June 2006).]

responsible for the emergence of variant CJD in humans. On the other hand, sheep scrapie does not appear to be orally transmissible to humans since the incidence of TSEs in populations consuming scrapie-infected sheep is not significantly higher than in other populations (Brown et al. 1987). Although CWD has been experimentally transmitted to squirrel monkeys (Marsh et al. 2005), no evidence exists for transmission of CWD to humans, and cell-free conversion assays suggest that the likelihood is low (Raymond et al. 2000). However, unlike BSE in cattle, infectivity is present in the muscle tissue of CWD-infected deer (Angers et al. 2006); thus, concerns remain about the possibility of CWD transmission into humans.

Etiology of prion diseases

Prion diseases were originally designated as "unconventional" or "slow" viruses based on the inability to identify a conventional virus and the long incubation periods associated with these infections (Sigurdsson 1954). The extreme resistance of these agents to ionizing and UV irradiation (Alper et al. 1978; Bellinger-Kawahara et al. 1987) combined with the inability to isolate a virus or scrapie-specific nucleic acid suggested that TSEs lacked a nucleic acid genome. These data support the hypothesis that a protein with self-replicating properties could be the TSE agent (Griffith 1967). The self-replicating protein hypothesis was refined and renamed as the prion hypothesis, which stated "prions are small proteinaceous particles which are resistant to inactivation by most procedures that modify nucleic acids" (Prusiner 1982). The discovery and characterization of the disease-associated prion protein (PrP^{Sc}) suggested that PrP^{Sc} may be a major component of the infections agent, if not the agent itself (Bolton et al. 1982). Subsequent studies demonstrated that PrP^{Sc}, rather than being a novel protein, was a conformational variant of a normal brain protein, PrP^C. Figure 2 shows electron micrographs of PrP^C and PrP^{Sc}. Circular dichroism and infrared spectroscopy indicate that, relative

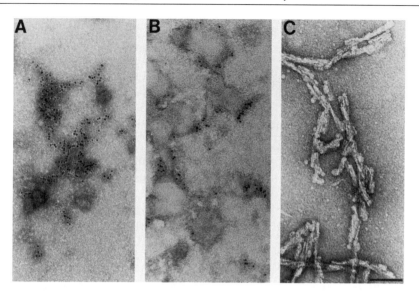

Figure 2. Transmission electron micrographs of prion proteins. (A) Cellular form of the prion protein, PrP^C. (B) Disease-associated form of the prion protein, PrP^{Sc}. (C) Negatively stained prion rods (aggregates of N-terminally truncated PrP^{Sc}). Bar = 100 nm. [Used by permission from Prusiner (1998) *Proc. Nat. Acad. Sci. USA*, Vol. 95, Fig. 3, p. 13367. Copyright (1998) National Academy of Sciences, U.S.A.]

to PrP^C, the disease-specific isoform has a higher β-sheet and lower α-helix content (Caughey et al. 1991). Although the three-dimensional structure of refolded, recombinant PrP^C has been elucidated by nuclear magnetic resonance (NMR) spectroscopy (Wüthrich and Riek 2001), difficulties in isolating pure PrP^{Sc} have thwarted attempts to determine its structure by NMR spectroscopy or X-ray crystallography. Recent electron crystallography data (Govaerts et al. 2004) suggest that PrP^{Sc} forms structured trimers with diameters of ~10 nm that aggregate into fibrils (typically 50 – 300 nm long) (Fig. 3).

In vitro cell culture studies have demonstrated that PrP^C is the precursor to PrP^{Sc} (Caughey et al. 1989; Borchelt et al. 1990). Inhibition of the migration of PrP^C to the cell surface blocks the formation of PrP^{Sc}, confirming that PrP^{Sc} arises via alternate processing or misprocessing of PrP^C (Caughey et al. 1989). Two different models for PrP^C conversion to PrP^{Sc} are currently being debated (Fig. 4). Both involve the direct interaction of PrP^C with PrP^{Sc} and consider PrP^{Sc} capable of conferring its abnormal conformation to the normal PrP^C molecule. The major difference between the two models is the nature of the PrP^{Sc}. In the template-assisted conversion model (Prusiner 1991), monomeric PrP^{Sc} is envisioned to interact with PrP^C, forming a heterodimer. PrP^C is then converted to PrP^{Sc} resulting in the formation of a homodimer, which then dissociates to form two PrP^{Sc} molecules (Fig. 4A, "Refolding" model). This cycle is then repeated resulting in the increase in both infectious titer and PrP^{Sc} accumulation. In the nucleation-dependent polymerization model (Jarrett and Lansbury 1993), however, the infectious PrP^{Sc} is present as a nucleant or seed, likely comprised of oligomers of PrP^{Sc} (Fig. 4B, "Seeding" model). Neither model, however, predicts how the conformation of PrP^{Sc} is conferred to the PrP^C molecule.

The function of the normal protein, PrP^C, has not been firmly established. PrP^C is normally expressed in mammalian neural tissue and may play a role in cellular resistance to oxidative stress and metal homeostasis in the brain (Brown 1999; Thackray et al. 2002). This host-encoded, 33-35 kDa sialoglycoprotein (Chesebro et al. 1985; Oesch et al. 1985) resides on the cell surface in lipid rafts (Taraboulos et al. 1995) and is tethered by a glycophosphotidylinositol

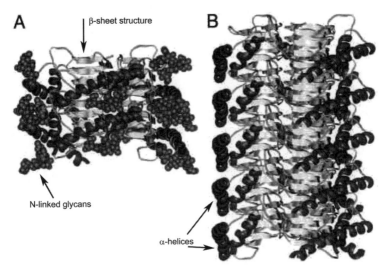

Figure 3. Proposed structures of pathological prion protein trimers and fibrils based on electron crystallography and molecular modeling. (A) Assembly of two structured trimers. (B) Fibrillization of structured trimers. Arrows indicate α-helices, β-sheet structure, and N-linked glycans (not shown in (B) for clarity). [Used by permission from Govaerts et al. (2004) *Proc. Nat. Acad. Sci. USA*, Vol. 101, Fig. 3, p. 8345. Copyright (1998) National Academy of Sciences, U.S.A.]

anchor (Stahl et al. 1987). PrP^C is concentrated at neuronal synapses (Sales et al. 1998; Herms et al. 1999) and axonally transported to nerve terminals (Borchelt et al. 1994), suggesting that PrP^C is important for neuronal activity. PrP^C also appears to be a metalloprotein binding copper *in vivo* (Brown et al. 1997), and perhaps an endocytic receptor for the uptake of extracellular copper (Pauly and Harris 1998).

The disease-associated form of the prion protein exists *in vivo* as an aggregate of PrP^{Sc} resulting in a build up of amyloid deposits, often surrounding regions of dead tissue or spongiosis (Fig. 5). The relationship of cell death to protein conversion is not clear, but loss of PrP^C or accumulation of PrP^{Sc} may be involved. Different strains of TSE agents exhibit different lesion patterns in the brain (i.e., distribution of vacuoles), physical properties, incubation times and host ranges (Bruce et al. 1976; Nonno et al. 2006). Protein aggregation may be responsible for toxic effects, but some evidence suggests that misfolded monomers or small oligomers may be responsible for toxicity (Stefani and Dobson 2003).

Resistance of prions to inactivation

Prions exhibit extraordinary resistance to conditions and treatments that inactivate conventional pathogens including exposure to ultraviolet, microwave and ionizing radiation, treatment with proteases and contact with most chemical disinfectants (Millison et al. 1976; Ernst and Race 1993; Taylor et al. 1995; Taylor 2000). Boiling does not to eliminate prion infectivity; high temperatures are more effective when combined with steam and pressure (autoclaving). A small fraction of infectivity in wet brain tissue can, however, withstand autoclaving at 134 °C for ≤60 min (Taylor et al. 1994). Dry heat sterilizes only at the extremely high temperatures (in excess of 600 °C) that can be achieved during incineration (Brown et al. 2000). Incineration at lower temperatures does not reliably eliminate prion infectivity. TSE agents withstand chemical decomposition retaining the ability to initiate infection after exposure to temperatures that should decompose or volatilize organic molecules (Brown et al. 2000, 2004). The effectiveness of incineration has been demonstrated only at the laboratory scale.

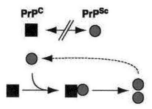

A "Refolding" model

B "Seeding" model

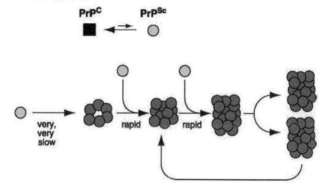

Figure 4. Models for conversion of PrP^C to PrP^{Sc}. (A) Template-assisted conversion or refolding model. (B) Nucleation-dependent or seeding model. [Used by permission from Weissmann (2002) *Proc. Nat. Acad. Sci. USA*, Vol. 99, Fig. 1, p. 16379. Copyright (1998) National Academy of Sciences, U.S.A.]

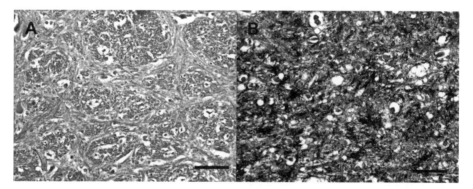

Figure 5. Obex region of the brainstem from white-tailed deer immunostained dark for PrP^{Sc}: (A) healthy tissue, (B) CWD-infected tissue. Note the extensive spongiform degeneration and dark reaction product marking PrP^{Sc} deposition in the CWD-infected tissue. Scale bars in each panel represent 20 μm. Stained tissue kindly provided by Chad Johnson.

ENVIRONMENTAL RESERVOIRS OF PRION INFECTIVITY

Epizoological and experimental evidence

Epizoological evidence. An environmental reservoir of prion infectivity appears to contribute to the maintenance of epizootics of CWD (Miller et al. 1998, 2004) and possibly scrapie (Pálsson 1979; Andréoletti et al. 2000). Evidence for environmental transmission of scrapie has long been noted but is largely anecdotal. Healthy sheep in Iceland contracted scrapie after grazing on fields previously occupied by infected animals, but not on fields where the disease had not been present (Greig 1940; Sigurdarson 1954; Pálsson 1979). These studies are difficult to interpret because they predate our understanding of the influence of sheep genetics on susceptibility to scrapie. More convincing are the observations of healthy elk (*Cervus elaphus*) contracting CWD after introduction into pens that previously housed infected animals (Miller et al. 1998; Williams et al. 2002).

Controlled field experiments. In a controlled field experiment (Fig. 6), Miller et al. (2004) demonstrated that the presence of decomposed infected carcasses as well as residual excreta from infected animals on the landscape were sufficient to transmit CWD to healthy mule deer (*Odocoileus hemionus*). This study employed three treatments (viz., CWD-infected carcass decomposed *in situ*, residual excreta from CWD-infected animals and live CWD-infected mule deer) to determine whether these exposures could transmit CWD to healthy deer. Healthy deer contracted CWD in all three types of paddocks.

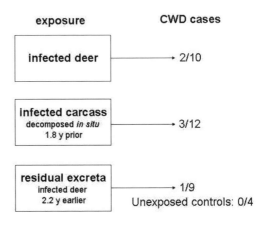

Figure 6. Controlled field experiment to examine environmental transmission of chronic wasting disease in mule deer. CWD cases given as number of infected per number of exposed animals at termination of experiment. [Based on Miller et al. (2004) *Emerg. Infect. Dis.* 10, 1003-1006.]

Hypothesized environmental reservoirs

The epidemiological and experimental studies cited above indicate that prion infectivity can be maintained in the environment for several years. Several hypotheses have been advanced on the nature of the putative environmental reservoir including soil, arthropod vectors and nematodes (Fitzsimmons and Pattison 1968; Brown and Gajdusek 1991; Post et al. 1999; Carp et al. 2000; Miller et al. 2004; Johnson et al. 2006d). As noted in a number of studies (Miller et al. 1998, 2004; Williams et al. 2002), arthropod vectors (e.g., hay mites, flesh flies) seem unlikely to account for the observed persistence of prion infectivity in the environment. Soils appear to represent a plausible environmental reservoir for prion infectivity. The remainder of this chapter focuses on the potential role of soil in the transmission of prion diseases.

Soil as an environmental reservoir of prion infectivity

For soil to serve as environmental reservoir, the following must hold: (1) prions must enter the environment; (2) prions must persist in soil; (3) prions in soil must retain infectivity; (4) prions must remain near the soil surface where they can be accessed by animals; (5) naïve animals must be exposed; and (6) the dose must be sufficient to cause infection. We discuss each of these factors below.

Introduction of prions into soil environments

Several plausible routes of prion introduction into soil environments can be envisioned (Fig. 7). Prions clearly enter soil environments when carcasses of infected animals decompose. Free-ranging CWD-infected animals die and decompose in the field. The remains of infected deer and elk dressed in the field by hunters (i.e., "gut piles") also represent a route of introduction of CWD agent into soils. Placentas of infected sheep carry high levels of infectivity (Race et al. 1998). Deposition of placenta to the soil surface may represent an important route of prion introduction.

The presence of PrP^{Sc} in gut-associated lymphatic tissue (e.g., tonsils, Peyer's patches, mesenteric lymph nodes) early in the disease course suggests that the agent could be shed through the alimentary system (Sigurdson et al. 1999; Miller and Williams 2002). Due to the lengthy incubation periods associated with TSE infection (long pre-clinical phase), infected animals may shed prions for extended periods. Prions have recently been detected in the saliva of CWD-infected mule deer by oral exposure of naïve white-tailed deer (*O. virginianus*), but in the feces or urine of the same animals (Mathiason et al. 2006). This latter finding may be due to the small sample size, the genotype of the recipient animals (Johnson et al. 2006a), and/or lower prion concentrations in urine and feces.

Urinary shedding of prions has been recently demonstrated in presymptomatic scrapie-infected animals having concurrent chronic kidney infection (Seeger et al. 2005). Earlier reports of urinary shedding in TSE-infected hamsters, cattle and humans (Shaked et al. 2001) have not been reproduced by others. The observations of Shaked et al. (2001) may have been due to a proteinase K-resistant enterobacterial outer membrane protein, of similar molecular mass as PrP, that exhibited nonspecific binding to a variety of anti-PrP antibodies (Furukawa et al. 2004).

Persistence of prions in soil

As discussed above, prions display remarkable resistance to a variety of conditions that inactivate conventional pathogens including heat, chemical disinfectants, proteases and

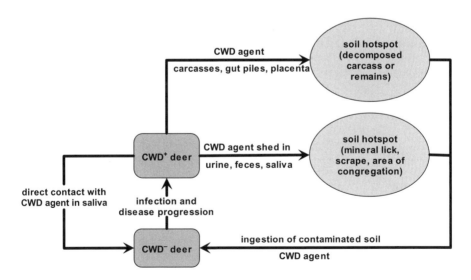

Figure 7. Potential routes of horizontal transmission of cervid chronic wasting disease including points of CWD agent entry into the environment. Direct contact between CWD⁺ and CWD⁻ deer may occur during sparring and grooming for bucks, during grooming in social groups for does, and between bucks and does during mating.

UV irradiation (Taylor 2000). Persistence of prions in soil environments requires that they withstand a variety of assaults including UV irradiation, freeze-thaw cycles, extracellular enzymes from bacteria and fungi, digestion by soil macrofauna, and abiotic transformation by reactive mineral phases. Furthermore, prions have the potential to interact with organic and inorganic soil components. Such interactions may influence the resistance of prions to various assaults and may alter their bioactivity.

The first study to examine the persistence of prions in soil involved interring scrapie-infected hamster brain material in garden soil for 3 years (Brown and Gajdusek 1991). Compared to a frozen control, infectivity in water washes of the soil was 1.8 to 2.6 log units less, but still represented $10^{5.6}$ to $10^{6.4}$ IU_{50}. The authors did not determine whether the apparent reduction in infectivity was attributable to attachment of PrP^{Sc} to soil particles or to degradation. This study was limited to a single, uncharacterized soil and did not include replicates for treatments or controls. Since soil represents a complex matrix of inorganic minerals and organic matter and soil properties vary considerably from location to location, these findings cannot be easily extrapolated to other settings.

Binding of prions to soils and soil components. The pathological form of the prion protein is extremely hydrophobic (Prusiner 1998) and is therefore expected to associate with soil components to a significant extent. Johnson et al. (2006b) examined the association of PrP^{Sc} with three common soil minerals (viz., montmorillonite, kaolinite and quartz) and four soils. In this study, PrP^{Sc} enriched from clinically infected hamster brains (Bolton et al. 1987; McKenzie et al. 1998) was exposed to pure minerals in aqueous suspension. Mineral-bound PrP^{Sc} was separated from unbound prion protein by density centrifugation through a sucrose cushion. Prion protein associated with mineral particles was recovered by 10-min extraction in a solution containing 10% sodium dodecyl sulfate (SDS; a denaturing anionic detergent) at 100 °C. The amounts of bound and unbound PrP^{Sc} were examined by SDS-polyacrylamide electrophoresis (SDS-PAGE) followed by immunoblotting (Western blotting). Of the soil minerals examined, montmorillonite (effective hydrodynamic diameter, d_h = 0.5-2 μm) exhibited the largest PrP^{Sc}-binding capacity followed by quartz microparticles (d_h = 1-5 μm) and kaolinite (d_h = 0.5-2) (Table 1). These results suggest that mineral surface properties contributed to the differences in PrP^{Sc} binding.

The binding of PrP^{Sc} to montmorillonite was avid; only extremely harsh conditions (viz., 10-min exposure to 10% SDS at 100 °C) were effective in desorbing the protein from the clay surface (Johnson et al. 2006b). Exposure to 10% SDS at lower temperatures was less effective in removing PrP^{Sc} from the montmorillonite. Conditions employed by previous investigators

Table 1. PrP^{Sc} Adsorption capacities for the minerals examined
[from Johnson et al. (2006b) *PLoS Pathogens*, Table 1.]

Mineral[a]	Binding Capacity (Sorbent Mass Basis) ($\mu g_{protein}\ mg_{mineral}^{-1}$)[a]	Binding Capacity (Sorbent Surface Area Basis) ($mg_{protein}\ m_{mineral}^{-2}$)[b]
Mte	87–174	2.8–5.7
Kte	1.7–2.6	0.15–0.22
Qtz	13.6–27.1	2.7–5.4

[a] Abbreviations: Kte, kaolinite; Mte, montmorillonite; Qtz, quartz microparticles. [b] Protein concentration determined by Bradford assay; PrP^{Sc} concentration was taken as 87% of total protein (Silveira et al. 2005). Reported adsorption capacities represent upper estimates, as the fraction of PrP^{Sc} in clarified preparations may have been lower. For Mte, binding capacity was based on the external (N_2-accessible) surface area.

(e.g., Morgan and Corke 1976; Docoslis et al. 2001) to desorb proteins from clay surfaces were ineffective in removing detectable PrP^{Sc} from montmorillonite (Johnson et al. 2006b). These conditions included increases in ionic strength, strong chaotropic agents (viz., 8 M guanidine HCl and 8 M urea) and pH extremes (pH 2.5 and 11.5). The attachment of proteins to clay surfaces is often maximal at the isoelectric point (pI) of the protein and decreases at pH > pI because of repulsive electrostatic interactions (Quiquampoix et al. 2002). Increasing suspension pH to above the protein pI often results in the detachment of proteins from clay surfaces (Armstrong and Chesters 1965; Quiquampoix et al. 1993). PrP^{Sc} aggregates exhibit an apparent pI of 4.6 (unpublished data), but were not desorbed in detectable amounts from the montmorillonite surface when pH was elevated to 11.5 (Johnson et al. 2006b).

Interestingly, PrP^{Sc} desorbed from montmorillonite, but not kaolinite or quartz, exhibited a reduction in molecular mass (Johnson et al. 2006b). Using antibodies directed against the *N*-terminus of the protein, Johnson et al. (2006b) determined that PrP^{Sc} desorbed from montmorillonite was cleaved at a site on the *N*-terminal flexible domain not required for infectivity (Supattapone et al. 1999). Cleavage of other proteins desorbed from Mte has not been reported.

The sorption of PrP^{Sc} to four soils differing in texture, organic carbon content and mineralogy was also examined (Johnson et al. 2006b). Although the sorption capacities of the soils were not quantified, the amount of sorbed PrP^{Sc} recovered varied among the soils suggesting differences in the strength of interaction. Clay content appeared to be an important determinant for the amount recovered. Leita et al. (2006) reported that PrP^{Sc} strongly interacted with two sandy loams and one clay loam. These investigators were unable to recover sorbed PrP^{Sc} from the soils, although the extractant used differed from that employed by Johnson et al. (2006b). Sorption of PrP^{Sc} to humic substances has not been investigated.

Several reports have appeared investigating the association of the non-pathogenic, nonglycosylated, recombinant PrP to clay minerals and soils (Revault et al. 2005; Vasina et al. 2005; Rigou et al. 2006) but relevance of these studies to the transmission of TSEs is limited because they did not employ the infectious agent in their experiments. As mentioned above, the tertiary structure (folding) of PrP^{C} differs dramatically from that of PrP^{Sc} resulting in the two isoforms exhibiting very different biophysical properties (Prusiner 1998; Riesner 2004).

Effect of sorption on preservation of prion infectivity. PrP^{Sc} is more resistant to proteases than most other proteins. Association of prions with mineral surfaces and humic substances may further protect the agent from degradation by microorganisms and extracellular proteases. This phenomenon has been observed for a variety of enzymes and *Bt* toxin, the insecticidal protein of *Bacillus thuringiensis* (Naidja et al. 2000; Stotzky 2000). The extent of protection depends on both the nature of the mineral surface and the protein in question.

Retention of bioactivity by prions in soils

For soil to contribute to lateral transmission of TSEs, prions attached to soil particles must retain infectivity. Adsorption of proteins to mineral surfaces can be accompanied by conformational changes that cause loss or diminution of function (e.g., Morgan and Corke 1976; Vettori et al. 1999; Naidja et al. 2000; Lecomte et al. 2001). Johnson et al. (2006b) examined whether attachment to clay surfaces would decrease infectivity and estimated that adsorption of prions enhanced infectivity by a factor of ~10 relative to unbound agent when inoculated intracerebrally (Table 2). This result was surprising, given the avidity of PrP^{Sc} attachment to the clay mineral, and suggested that association with montmorillonite reduced clearance from the brain (i.e., translocation out of the brain and degradation by proteases).

Bioavailability via the oral route of exposure could be diminished if detachment of PrP^{Sc} from mineral surfaces is necessary for uptake or enhanced if PrP^{Sc}-mineral complexes are

Table 2. Prions adsorbed to montmorillonite clay retain infectivity
[from Johnson et al. (2006b) *PLoS Pathogens*, Table 2.]

Inoculum	TSE-Positive Animals/ Total Animals	Onset of Clinical Symptoms (dpi)[a]
None	0/8	>200[b]
Mte (no PrP[Sc])	0/8	>200[b]
Mte-PrP[Sc] complex	10/10[c]	93 ± 4[d]
Mock supernatant[e] (no Mte)	8/8	103 ± 0[d]
Mock pellet[e] (no Mte)	8/8	178 ± 21[d]

[a] Mean days postinoculation (dpi) ± one standard deviation to the onset of clinical symptoms of TSE infection.
[b] None of the animals showed clinical symptoms of TSE infection or had protease-resistant PrP accumulation at the termination of the experiment at 200 dpi.
[c] Although 12 animals were inoculated, two non-TSE intercurrent deaths occurred at 8 dpi.
[d] Brains of infected animals were positive for protease-resistant PrP.
[e] Mock supernatant and mock pellet samples were generated by adding clarified PrP[Sc] (~0.2 µg) to buffer in the absence of soil minerals and processing identically to samples containing Mte.

directly taken up. The digestive tract secretes numerous surfactants (e.g., bile acids, lipid and protein surfactants) that may allow extraction of PrP[Sc] from soil or soil minerals. Association with mineral particles could, however, protect prions from proteolysis in the gastrointestinal tract, thereby enhancing infectivity. Detachment from mineral particles may not be required for uptake. Intestinal Peyer's patches sample microparticles similar in size to clay particles; in some cases such particles are translocated to lymph tissue (Desai et al. 1996; Hazzard et al. 1996). Both Peyer's patches and lymphatic tissues are peripheral sites of PrP[C]-to-PrP[Sc] conversion, involved in the early stages of CWD and scrapie infection (Sigurdson et al. 1999; Andréoletti et al. 2000; Miller and Williams 2002), and routing particle-associated prions to either of these sites could dramatically increase the chance of initiating TSE infection. Oral infectivity assays with soil- and soil-mineral-bound prions will be required to assess bioavailability via this exposure route.

Mobility of prions in soils

For prions in soils to contribute to the lateral transmission of TSEs, they must remain near the soil surface where animals can come in contact with them. Little published literature exists on the mobility of prions in soils. The strong attachment of PrP[Sc] to clay minerals (Johnson et al. 2006b) suggests that prions may be retained near the surface of fine textured soils. Attachment to clay particles may, however, facilitate transport through the soil column in some situations, as has been demonstrated for virus particles (Jin et al. 2000). Brown and Gajdusek (1991) observed a limited amount of prion leaching from a Petri dish with holes drilled through the bottom that contained infected hamster brain material and garden soil; no infectivity was observed more than 4 cm below the dish.

Inferences from virus transport literature. Because little information is available on the environmental behavior of prions, a brief consideration of research on virus transport may serve to frame our expectations of prion mobility in soils. Viruses are biological nanoparticles; like prions, non-enveloped viruses expose a protein surface to solution. While viruses are generally larger than monomers and small oligomers of PrP[Sc], the size range of larger PrP[Sc] aggregates overlaps that of small animal viruses (Silveira et al. 2005). The aggregation state of PrP[Sc] excreted from infected animals or released from decomposing carcasses is unknown but is likely smaller than those in PrP[Sc]-enriched fractions.

Factors influencing the transport of viruses include soil mineralogy and organic matter content (Lukasik et al. 1999; Ryan et al. 2002; Meschke and Sobsey 2003; Zhuang and Jin 2003). Iron and aluminum oxides appear to decrease virus movement (Lukasik et al. 1999; Zhuang and Jin 2003). Organic matter associated with soil particles increases retention of viruses in porous media, presumably by providing hydrophobic sites for attachment. In contrast, dissolved organic matter decreases virus attachment due to competition for surface sites (Schijven and Hassanizadeh 2000). Both electrostatic and hydrophobic interactions appear to be important in virus transport through porous media (Bales et al. 1991; Penrod et al. 1996; Redman et al. 1997; Dowd et al. 1998; Chattopadhyay and Puls 2000; Zhuang and Jin 2003). Not surprisingly, hydrophobic interactions were especially important for more hydrophobic viruses (Bales et al. 1991; Kinoshita et al. 1993; Chattopadhyay and Puls 1999, 2000). Both reversible and irreversible virus attachment have been reported in batch sorption and column leaching experiments (Bales et al. 1993; Loveland et al. 1996), depending upon the characteristics of the mineral surface. Loveland et al. (1996) demonstrated that viruses can be desorbed from mineral surfaces by raising the pH, decreasing ionic strength or adding a high ionic strength protein solution such as beef extract that can compete with virus for sorption sites (Bales et al. 1993). Surfactants reduce virus attachment to soil particles by competing for attachment sites, displacing attached viruses, or increasing virus solubility (Chattopadhyay et al. 2002). Although attachment to clay particles may limit virus (and PrP^{Sc}) mobility, association with colloidal clay minerals can also facilitate transport (Jin et al. 2000).

The analogy between prions and viruses is only partial. Prions exhibit surface charge heterogeneity, and aggregates are smaller and more hydrophobic than most viruses. Prion proteins exhibit a range of isoelectric points with at least eight charge isomers (Bolton et al. 1985). This charge heterogeneity is imparted by the number (0-2) and nature of N-linked glycans. Gel electrophoresis with immunoblotting yields three bands corresponding to the di-, mono- and unglycosylated protein. Mass spectrometry studies indicate that >160 PrP glycoforms exist in a single prion preparation (Baldwin 2001).

Animal exposure through soil ingestion

The early involvement of gut-associated lymphatic tissue in TSE infection argues for an oral route of exposure. Herbivores ingest soil both deliberately and incidentally during grazing and grooming (Beyer et al. 1994). Sites where infected carcasses, gut piles or placentas have decomposed may represent foci of TSE transmission via ingestion of prions associated with soils. The frequency that cervids [mule deer, white-tailed deer, elk, and moose (*Alces alces*)] and sheep visit such sites and the amount of soil sampled during visits has not been determined. The following discussion on deliberate soil ingestion focuses on mineral licks and scrapes because more is known about deer behavior at these sites than at those where carcasses, gut piles or placentas have decomposed. Given the much higher levels of infectivity in carcasses and gut piles, sites where these materials are deposited and decompose may harbor more infectivity and contribute more to TSE transmission than licks and scrapes.

In cervids, deliberate ingestion of soil occurs at mineral licks, artificial salt licks and scrapes (Atwood and Weeks 2003). Throughout the year, all members of the Cervidae family supplement their mineral intake by ingesting soil at locations that have higher levels of sodium and other cations (Atwood and Weeks 2002). Licks may be artificial (e.g., from mineral blocks used for livestock grazing) or natural, and the salts persist in clay soils for over 20 years, after diffusing into the soil (Thackston and Holbrook 1992). All natural licks used by white-tailed deer in one study (Weeks 1978) occurred in Stendal silt loams with subsoils having high clay contents. Licks formed in depressions, where runoff collected and evaporated, leaving dissolved minerals above the mostly impermeable subsoils (Weeks and Kirkpatrick 1976). Since minerals from licks are persistent and draw deer to their location on a regular basis for many years, they may serve as points of prion accumulation from deposited saliva, urine and feces.

Deliberate soil ingestion by deer also occurs at scrapes. Scrapes are created by heavy pawing of the soil to remove surface detritus, and then marked with deposits of glandular secretions and urine or feces (Hirth 1977; Kile and Marchinton 1977; Miller 1987). Primarily during the rut, male deer create these chemical signposts to communicate with other deer (Moore and Marchinton 1974; Hirth 1977). Both males and females are known to "scent mark" at established scrapes by urination (Moore and Marchinton 1971). Several studies suggest that females visit scrapes more frequently than males (Alexy et al. 2001). Multiple males visit each scrape with little indication of revisitation (Alexy et al. 2001). Soil sampling by deliberate licking the urine-saturated soil occurs by males as a means to determine the estrous condition of females and presence of other males (Moore and Marchington 1971). It is not known whether females sample soil at scrapes. Both males and females also mouth small branches above the scrapes (Kile and Marchinton 1977).

The most likely route of exposure to soil-associated CWD agent by deer appears to be via either of the above recurrent, deliberate behaviors (i.e., sampling soil at mineral licks and scrapes). Incidental ingestion of soil can also contribute to soil exposure. The annual average soil intake of sheep pastured all year was 4.5% of the dry matter intake (Fries 1996). Soil is estimated to comprise a minimum of 2% of a deer's diet (dry matter basis), annually, and may exceed 50% during the late spring and early summer at licks (Weeks and Kirkpatrick 1976).

Levels of infectivity in soils

For ingestion of soil to lead to TSE infection, sufficient quantities of prions (i.e., ≥ 1 oral IU_{50}) must be ingested. With intracerebral inoculation, 1 IU_{50} has been estimated to comprise approximately 10^5 PrP^{Sc} molecules in rodent models (Bolton et al. 1991). Oral exposure is approximately ~10^5-fold less efficient in mice and hamsters (Diringer et al. 1994, 1998). The number of molecules per IU_{50} has not been determined for scrapie in sheep or for CWD, nor has the efficiency of oral versus intracerebral dosing been determined for these TSEs. The infectious dose need not be acquired from a single exposure; repeated dosing can enhance transmission (Diringer et al. 1998). Frequent visitation of mineral licks and, perhaps, locations where carcasses or gut piles have decomposed, by cervids may enhance transmission. Since Johnson et al. (2006b) demonstrated that adsorption of PrP^{Sc} to the clay mineral montmorillonite enhanced infectivity by a factor of ~10 via the intracerebral route, sorption to particle surfaces may also enhance prion infectivity via the oral route by protecting the agent from degradation in the gastrointestinal tract (Martinsen et al. 2002). Such an effect has been demonstrated for bovine rotavirus and coronavirus (Clark et al 1998).

Levels of infectivity present in naturally prion–contaminated soils have not been measured. Until recently, methods to recover PrP^{Sc} from soils were unavailable. Johnson et al. (2006b) were able to recover PrP^{Sc} from soils and pure minerals experimentally spiked with the protein using SDS-PAGE sample buffer [10% SDS, 100 mM Tris pH 8.0, 7.5 mM EDTA, 100 mM dithiothreitol (DTT), 30% glycerol] at 100 °C. Rigou et al. (2006) described an electroelution method to extract recombinant PrP.

CONCLUSIONS

An environmental reservoir of prion infectivity appears to contribute to the transmission of cervid chronic wasting disease and, probably, sheep scrapie. Soil represents a likely candidate for an environmental reservoir because (1) oral exposure appears important in the natural transmission of scrapie and CWD; (2) accumulation of the infectious agent in gut-associated lymphatic tissue at early stages of infection argues for alimentary shedding [recently demonstrated for saliva (Mathiason et al. 2006)]; (3) sheep and deer ingest soil both deliberately and incidentally; and (4) prions persist in soil environments for ≥ 3 years.

Research conducted in our laboratories supports a role for soil in the transmission of prion diseases of sheep, deer and elk. Prions adsorb strongly to clay minerals and may serve to maintain infectivity close to the soil surface where they can be ingested by herbivores. Known behaviors of all members of the Cervidae family (deer, elk, moose) lead to deliberate ingestion of soils that have a high probability of contact with potentially infectious bodily secretions. Attachment to clay minerals enhances prion infectivity via the intracerebral route and may also do so via the oral route of exposure.

ACKNOWLEDGMENTS

The authors gratefully acknowledge financial support from the Department of Defense National Prion Research Program through grant DAMD17-03-1-0369 (JMA, DM and JAP) and the U.S. Environmental Protection Agency through grant 4C-R070-NAEX (JAP). PTS was supported by NIEHS training grant T32 ES07015-28.

REFERENCES

Abramov AY, Canevari L, Duchen MR (2004) Calcium signals induced by amyloid peptide and their consequences in neurons and astrocytes in culture. Biochim Biophys Acta 1742:81-87

Aiken JM, Williamson JL, Borchardt LM, Marsh RF (1990) Presence of mitochondrial D-loop DNA in scrapie-infected brain preparations enriched for the prion protein. J Virol 64:3265-3268

Alper T, Haig DA, Clarke MC (1978) The scrapie agent: evidence against its dependence for replication on intrinsic nucleic acid. J Gen Virol 41:503-516

Andréoletti O, Berthon P, Marc D, Sarradin P, Grosclaude J, van Keulen L, Schelcher F, Elsen, MJM, Lantier F (2000) Early accumulation of PrP^{Sc} in gut-associated lymphoid and nervous tissue of susceptible sheep from a Romanov flock with natural scrapie. J Gen Virol 81:3115-3126

Angers RC, Browning SR, Seward TS, Sigurdson CJ, Miller MW, Hoover EA, Telling GC (2006) Prions in skeletal muscles of deer with chronic wasting disease. Science 311:1117-1117

Armstrong DE, Chesters G (1964) Properties of protein-bentonite complexes as influenced by equilibration conditions. Soil Sci 98:39-52

Atwood TC, Weeks HP (2002) Sex- and age-specific patterns of mineral lick use by white-tailed Deer (*Odocoileus virginianus*). Am Midland Naturalist 148:289-296

Atwood TC, Weeks HP (2003) Sex-specific patterns of mineral lick preference in white-tailed deer. Northeastern Naturalist 10:409-414

Baldwin MA (2001) Mass spectrometric analysis of prion proteins. Adv Protein Chem 57:29-54

Bales RC, Hinkle SR, Kroeger TW, Stocking K, Gerba CP (1991) Bacteriophage adsorption during transport through porous media: Chemical perturbations and reversibility. Environ Sci Technol 25:2088-2095

Bales RC, Li S, Maguire KM, Yahya MT, Gerba CP (1993) MS–2 and poliovirus transport in porous medium: Hydrophobic effects and chemical perturbations. Water Resour Res 29:957–963

Baringer JR, Bowman KA, Prusiner SB (1983) Replication of the scrapie agent in hamster brain precedes neuronal vacuolation J Neuropathol Exp Neurol 42:539-547

Bellinger-Kawahara C, Cleaver JE, Diener TO, Prusiner SB (1987) Purified scrapie prions resist inactivation by UV irradiation. J Virol 61:159-166

Beyer WN, Connor EE, Gerould S (1994) Estimates of soil ingestion by wildlife. J Wildlife Manage 58:375-382

Bhattacharjee S, Ryan JN, Elimelech M (2002) Virus transport in physically and geochemically heterogeneous subsurface porous media. J Contam Hydroly 57:161-187

Bolton DC, McKinley MP, Prusiner SB (1982) Identification of a protein that purifies with the scrapie prion. Science 218:1309-1311

Bolton DC, Meter RK, Prusiner SB (1985) Scrapie PrP27-30 is a sialoglycoprotein. J Virol 53:596-606

Bolton DC, Bendheim PE, Marmorstein AD, Potempska A (1987) Isolation and structural studies of the intact scrapie agent protein. Archiv Biochem Biophys 258:579-590

Bolton DC, Rudelli RD, Currie JR, Bendheim PE (1991) Copurification of Sp33-37 and scrapie agent from hamster brain prior to detectable histopathology and clinical-disease. J Gen Virol 72:2905-2913

Borchelt DR, Scott M, Taraboulos A, Stahl N, Prusiner SB (1990) Scrapie and cellular prion proteins differ in their kinetics of synthesis and topology in cultured cells. J Cell Biol 110:743-752

Borchelt DR, Koliatsos VE, Guarnieri M, Pardo CA, Sisodia SS, Price DL (1994) Rapid anterograde axonal transport of the cellular prion glycoprotein in the peripheral and central nervous systems. J Biol Chem 269:14711-14

Bradford SA, Simunek J, Bettahar M, van Genuchten MT, Yates SR (2003) Modeling colloid attachment, straining and exclusion in saturated porous media. Environ Sci Technol 37:2242-2250

Brown DR, Qin K, Herms JW, Madlung A, Manson J, Strome R, Fraser PE, Kruck T, von Bohlen A, Schulz-Schaeffer W, Giese A, Westaway D, Kretzschmar H (1997) The cellular prion protein binds copper *in vivo.* Nature 390:684-687

Brown DR (1999) Prion protein expression aids cellular uptake and veratridine-induced release of copper. J Neurosci Res 58:717-725

Brown P, Cathala F, Raubertas RF, Gajdusek DC, Castaigne P (1987) The epidemiology of Creutzfeldt-Jakob disease: conclusion of a 15-year investigation in France and review of the world literature. Neurology 37(6):895-904

Brown P, Gajdusek DC (1991) Survival of scrapie virus after 3 years internment. Lancet 337:269-270

Brown P, Rau EH, Johnson BK, Bacote AE, Gibbs CJ, Gajdusek DC (2000) New studies on the heat resistance of hamster-adapted scrapie agent: Threshold survival after ashing at 600 °C suggests an inorganic template of replication. Proc Nat Acad Sci USA 97:3418-3421

Brown P, Rau EH, Lemieux P, Johnson BK, Bacote AE, Gajdusek C (2004) Infectivity studies of both ash and air emissions from simulated incineration of scrapie-contaminated tissues. Environ Sci Technol 38: 6155-6160

Bruce ME, Dickinson AG, Fraser H (1976) Cerebral amyloidosis in scrapie in the mouse: Effect of agent strain and mouse genotype. Neuropathol Appl Neurobiol 2:471-478

Butterfield AD, Drake J, Pocernich C, Castegna A (2001) Evidence of oxidative damage in Alzheimer's disease brain: central role for amyloid-peptide. Trends Molec Med 7:548-554

Carp RI, Meeker HC, Rubenstein R, Sigurdarson S, Papini M, Kascsak RJ, Kozlowski PB, Wisniewski HM (2000) Characteristics of scrapie isolates derived from hay mites. J Neurovirol 6:137-144

Caughey B, Race RE, Ernst D, Buchmeier MJ, Chesebro B. (1989) Prion protein biosynthesis in scrapie-infected and uninfected neuroblastoma cells J Virol 63:175-81

Caughey BW, Dong A, Bhat KS, Ernst D, Hayes SF, Caughey WS (1991) Secondary structure analysis of the scrapie-associated protein PrP 27-30 in water by infrared spectroscopy. Biochem 30:7672-7680

Caughey B (2001) Interactions between prion protein isoforms: The kiss of death? Trends Biochem Sci 26: 235-242

Chatelain J, Cathala F, Brown P, Raharison S, Court L, Gajdusek DC (1981) Epidemiologic comparisons between Creutzfeldt-Jakob disease and scrapie in France during the 12-year period 1968-1979. J Neurol Sci 51:329-37

Chattopadhyay S, Puls RW (1999) Adsorption of bacteriophages on clay minerals. Environ Sci Technol 33: 3609-3614

Chattopadhyay S, Puls RW (2000) Forces dictating colloidal interactions between viruses and soil. Chemosphere 8:1279-1286

Chattopadhyay D, Chattopadhyay S, Lyon WG, Wilson JT (2002) Effect of surfactants on the survival and sorption of viruses. Environ Sci Technol 36:4017-4024

Chesebro B, Race R, Wehrly K, Nishio J, Bloom M, Lechner D, Bergstrom S, Robbins K, Mayer L, Keith JM, Garron C, Haase A. (1985) Identification of scrapie prion protein-specific mRNA in scrapie-infected and uninfected brain. Nature 315:331-333

Cho HJ (1980) Requirement of a protein component for Scrapie infectivity. Intervirology 14:213-216

Chu Y, Jin Y, Yates MV (2000) Virus transport through saturated sand columns as affected by different buffer solutions. J Environ Qual 29:1103-1110

Clark KJ, Sarr AB, Grant PG, Phillips TD, Woode GN (1998) In vitro studies on the use of clay, clay minerals and charcoal to adsorb bovine rotavirus and bovine coronavirus. Vet Microbiol 63:137-146

Collee J, Bradley R (1997) BSE: A decade on – Part 1. Lancet 349:636-641

Derjaguin BV, Landau LD (1941) Theory of the stability of strongly charged lyophobic sols and of the adhesion of strongly charged particles in solutions of electrolytes. Acta Physicochim URSS 14:733-762

Desai MP, Labhasetwar V, Amidon G L, Levy RJ (1996) Gastrointestinal uptake of biodegradable microparticles: Effect of particle size. Pharm Res 13:1838-1845

Dickinson AG (1976) Scrapie in sheep and goats. *In*: Slow Virus Diseases of Animals and Man. Kimberlin RH (ed) North-Holland Publishing Co. p 209-241

Diringer H, Beekes M, Oberdieck U (1994) The nature of the scrapie agent - The virus theory. Ann New York Acad Sci 724:246-258

Diringer H, Roehmel J, Beekes M (1998) Effect of repeated oral infection of hamsters with scrapie. J Gen Virol 79:609-612

Docoslis A, Rusinski LA, Giese RF, van Oss CJ (2001) Kinetics and interaction constants of protein adsorption onto mineral microparticles - measurement of the constants at the onset of hysteresis. Colloids Surf B 22:267-283

Dowd SE, Pillai SD, Wang S, Corapcioglu MY (1998) Delineating the specific influence of virus isoelectric point and size on virus adsorption and transport through sandy soils. Appl Environ Microbiol 64:405-410

Eghiaian F, Grosclaude J, Lesceu S, Debey P, Doublet B, Treguer E, Rezaei H, Knossow M (2004) Insight into the PrP^C → PrP^{Sc} conversion from the structures of antibody-bound ovine prion scrapie-susceptibility variants. Proc Nat Acad Sci USA 101:10254-10259

Ernst DR, Race RE (1993) Comparative analysis of scrapie agent inactivation methods. J Virol Methods 41: 193-202

Fitzsimmons WM, Pattison IH (1968) Unsuccessful attempts to transmit scrapie by nematode parasites. Res Vet Sci 9:281-283

Fries GF (1996) Ingestion of sludge applied organic chemicals by animals. Sci Total Environ 185:93-108

Furukawa H, Doh-Ura K, Okuwaki R, Shirabe S, Yamamoto K, Udono H, Ito T, Katamine S, Niwa M (2004) A pitfall in diagnosis of human prion diseases using detection of protease-resistant prion protein in urine – Contamination with bacterial outer membrane proteins. J Biol Chem 279: 23661-23667

Greig JR (1940) Scrapie: Observations on the transmission of the disease by mediate contact. Vet J 96:203-206

Griffith JS (1967) Self-replication and scrapie. Nature 215:1043-1044

Govaerts C, Wille H, Prusiner SB, Cohen FE (2004) Evidence for assembly of prions with left-handed beta-helices into trimers. Proc Nat Acad Sci USA 101:8342-8347

Hadlow WJ, Kennedy RC, Race RE. (1982) Natural infection of Suffolk sheep with scrapie virus. J Infect Dis 146:657-664

Hazzard RA, Hodges GM, Scott JD, McGuinness CB, Carr KE (1996) Early intestinal microparticle uptake in the rat. J Anat 189:265-271

Herms J, Tings T, Gall S, Madlung A, Giese A, Siebert H, Schurmann P, Windl O, Brose N, Kretzschmar H (1999) Evidence of presynaptic localization and function of the prion protein. J Neurosci 19:8866-75

Hirth DH (1977) Social behavior of white-tailed deer in relation to habitat. Wildlife Monographs 53:1-55

Horiuchi M, Priola SA, Chabry J, Caughey B (2000) Interactions between heterologous forms of prion protein: binding, inhibition of conversion, and species barriers. Proc Natl Acad Sci USA 97:5836-5841

Israelachvili J (1991) Intermolecular and Surface Forces, 2nd ed. Academic Press

Jarrett JT, Lansbury,PT Jr. (1993) Seeding "one-dimensional crystallization" of amyloid: a pathogenic mechanism in Alzheimer's disease and scrapie? Cell 73:1055-58

Jin Y, Chu Y, Li Y (2000) Virus removal and transport in saturated and unsaturated sand columns. J Contam Hydrol 43:111-128

Jin Y, Pratt E, Yates MV (2000) Effect of mineral colloids on virus transport through saturated sand columns. J Environ Qual 29:532-539

Johnson C, Johnson J, Vanderloo JP, Keane D, Aiken JM, Mckenzie D (2006a) Prion protein polymorphisms in white-tailed deer influence susceptibility to chronic wasting disease. J Gen Virol 87:2109-2114

Johnson CJ, Phillips KE, Schramm PT, McKenzie DI, Aiken JM, Pedersen JA (2006b) Prions adhere to soil minerals and remain infectious. PLoS Pathogens 2:296-302

Kellings K, Meyer N, Mirenda C, Prusiner SB, Riesner D (1993) Analysis of nucleic acids in purified scrapie prion preparations. Arch Virol Suppl 7:215-225

Kile TL, Marchington, RL (1977) White-tailed deer rubs and scrapes: Spatial, temporal and physical characteristics and social role. Am Midland Naturalist 97:257-266

Kinoshita Y, Kuzuhara T, Kirigakubo M, Kobayashi M, Shimura K, Ikada Y (1993) Reduction in tumor formation on porous polyethylene by collagen immobilization. Biomaterials 14:546-550

Lecomte S, Hilleriteau C, Forgerit JP, Revault M, Baron MH, Hildebrandt P, Soulimane T (2001) Structural changes of cytochrome c(552) from *Thermus thermophilus* adsorbed on anionic and hydrophobic surfaces probed by FTIR and 2D-FTIR spectroscopy. Chembiochem 2:180-189

Leita L, Fornasier F, De Nobli M, Bertoli A, Genovesi S, Sequi P (2006) Interactions of prion proteins with soil. Soil Biol Biochem 38:1638-1644

Loveland JP, Ryan JN, Amy GL, Harvey RW (1996) The reversibility of virus attachment to mineral surfaces. Colloids Surf A 107:205-221

Lukasik J, Scott TM, Andryshak D, Farrah SR (2000) Influence of salts on virus adsorption to microporous filters. Appl Environ Microbiol 66:2914-20

Lupi O (2003) Could ectoparasites act as vectors for prion diseases? Int J Dermatol 42:425-429

Marsh RF, Kincaid AE, Bessen RA, Bartz JC (2005) Interspecies transmission of chronic wasting disease prions to squirrel monkeys (*Saimiri sciureus*). J Virol 79:13794-13796

Martinsen TC, Taylor DM, Johnsen R, Waldum HL (2002) Gastric acidity protects mice against prion infection? Scand J Gastroenterol 37:497-500

Mathiason CK, Powers JG, Dahmes SJ, Osborn DA, Miller KV, Warren RJ, Mason GL, Hays SA, Hayes-Klug J, Seelig DM, Wild MA, Wolfe LL, Spraker TR, Miller MW, Sigurdson CJ, Telling GC, Hoover, EA. (2006) Infectious prions in the saliva and blood of deer with chronic wasting disease. Science 314: 133-136

Mattson MP (1999) Impairment of membrane transport and signal transduction systems by amyloidogenic proteins. Methods Enzymol 309:733-768

Mattson MP, Liu D (2002) Energetics and oxidative stress in synaptic plasticity and neurodegenerative disorders. Neuromol Med 2:215-231

McGowan JP (1922) Scrapie in sheep. Scott J Agric 5:365-375

McKenzie D, Bartz J, Mirwald J, Olander D, Marsh R, Aiken J (1998) Reversibility of scrapie inactivation is enhanced by copper. J Biol Chem 273:25545-25547

Mertz PA, Somerville RA, Wisniewski HM, Iqbal K (1981) Abnormal fibrils from scrapie-infected brain. Acta Neuropathol 54:63-74

Meschke JS, Sobsey MD (2003) Comparative reduction of Norwalk virus, poliovirus type 1, F+ RNA coliphage MS2 and Escherichia coli in miniature soil columns. Water Sci Technol 47(3):85-90

Miller KV, Marchinton RL, Forand KJ, Johansen KL (1987) Dominance, testosterone levels, and scraping activity in a captive herd of white-tailed deer. J Mammalogy 68:812-817

Miller MW, Wild MA, Williams ES (1998) Epidemiology of chronic wasting disease in captive Rocky Mountain elk. J Wildlife Dis 34:532-538

Miller MW, Williams ES (2003) Horizontal prion transmission in mule deer. Nature 425:35-36

Miller MW, Williams ES, Hobbs NT, Wolfe LL (2004) Environmental sources of prion transmission in mule deer. Emerg Infect Dis 10:1003-1006

Miller MW, Wild MA, Williams ES (1998) Epidemiology of chronic wasting disease in captive Rocky Mountain elk. J Wildlife Dis 34:532-538

Millison GC, Hunter GD, Kimberlin RH (1976) The physico-chemical nature of the scrapie agent. *In*: Slow Virus Diseases of Animals and Man. Kimberlin RH (ed) North Holland, p 243-266

Moore WG, Marchington RL (1971) Marking behavior and its social function in white-tailed deer. *In:* The behavior of ungulates and its relation to management. Vol 1. Geist V, Walther F (eds) IUCN, Publication No. 24, p 447-456

Morgan HW, Corke CT (1976) Adsorption, desorption, and activity of glucose oxidase on selected clay species. Can J Microbiol 22:684-693

Mori I, Nishiyama Y, Yokochi T, Kimura Y (2005) Olfactory transmission of neurotropic viruses. J Neurovirol 11:129-137

Naidja A, Huang PM, Bollag JM (2000) Enzyme-clay interactions and their impact on transformations of natural and anthropogenic organic compounds in soil. J Environ Qual 29:677-691

Nonno R, Di Bari MA, Cardone F, Vaccari G, Fazzi P, Dell'Omo G, Cartoni C, Ingrosso L, Boyle A, Galeno R, Sbriccoli M, Lipp H-P, Bruce M, Pocchiari M, Agrimi U (2006) Efficient transmission and characterization of Creutzfeldt–Jakob Disease strains in bank voles. PLoS Pathogens 2:112-120

Oesch B, Westaway D, Wälchl M, McKinley MP, Kent SBH, Aebersold R, Barr, RA, Tempst P, Teplow D, Hood LE, Prusiner SB, Weissmann C (1985) A cellular gene encodes scrapie PrP 27-30 protein. Cell 40: 735-746

Pálsson, PA (1979) Rida (scrapie) in Iceland and its epidemiology. *In*: Slow Transmissible Diseases of the Nervous System. 1st ed. Prusiner SB, Hadlow WJ (ed) Academic Press, p 357-366

Pauly PC, Harris DA (1998) Copper stimulates endocytosis of the prion protein. J Biol Chem 273:33107-33110

Penrod SL, Olson TM, Grant SB (1996) Deposition kinetics of two viruses in packed beds of quartz granular media. Langmuir 12:5576-5587

Post K, Riesner D, Walldorf V, Mehlhorn H (1999) Fly larvae and pupae as vectors for scrapie. Lancet 354: 1969-1970

Prusiner SB (1982) Novel proteinaceous infectious particles cause scrapie. Science 216:136-144

Prusiner SB (1991) Molecular biology of prion diseases. Science 252:1515-22

Prusiner SB (1997) Prion diseases and the BSE crisis. Science 278:245-251

Prusiner SB (1998) The prion diseases. Brain Pathol 8:499–513

Prusiner SB, McKinley MP, Groth DF, Bowman KA, Mock NI, Cochran SP, Masiarz FR (1981) Scrapie contains a hydrophobic protein. Proc Natl Acad Sci 78:6675-6679

Quiquampoix H, Servagent-Noinville S, Baron M-H (2002) Enzyme adsorption on soil mineral surfaces and consequences for the catalytic activity. *In*: Enzymes in the Environment: Activity, Ecology and Applications. Burns RG, Dick RP (eds) Marcell_Dekker, p 285-306

Quiquampoix H, Staunton S, Baron MH, Ratcliffe RG (1993) Interpretation of the pH dependence of protein adsorption on clay mineral surfaces and its relevance to the understanding of extracellular enzyme activity in soil. Colloids Surf A 75: 85-93

Race R, Jenny A, Sutton D (1998) Scrapie infectivity and proteinase K-resistant prion protein in sheep placenta, brain, spleen, and lymph node: Implications for transmission and antemortem diagnosis. J Infect Dis 178:949-953

Ranville JF, Schmiermund RL (1998) An overview of environmental colloids. *In* Perspectives in Environmental Chemistry. Macalady DL (ed), Oxford University Press, p 25-56

Raymond GJ, Bossers A, Raymond LD, O'Rourke KI, McHolland LE, Bryant III PK, Miller MW, Williams ES, Smits M, Caughey B (2000) Evidence of a molecular barrier limiting susceptibility of humans, cattle and sheep to chronic wasting disease. EMBO J 19:4425–4430

Redman JA, Grant SB, Olson TM, Hardy ME, Estes MK (1997) Filtration of recombinant Norwalk virus particles and bacteriophage MS2 in quartz sand: Importance of electrostatic interactions. Environ Sci Technol 31:3378-3383

Revault M, Quiquampoix H, Baron MH, Noinville S (2005) Fate of prions in soil: Trapped conformation of full-length ovine prion protein induced by adsorption on clays. Biochim Biophys Acta - Gen Subjects 1724:367-374

Riesner D (2004) Biochemistry and structure of PrP^C and PrP^{Sc}. Br Med Bull 66:21-33

Rigou P, Rezaei H, Grosclaude J, Staunton S, Quiquampoix H (2006) Fate of prions in soil: Adsorption and extraction by electroelution of recombinant ovine prion protein from montmorillonite and natural soils. Environ Sci Technol 40:1497-1503

Sales N, Rodolfo K, Hassig R, Faucheux B, Giamberardino LD, Moya KL (1998) Cellular prion protein localization in rodent and primate brain. Eur J Neurosci 10:2464-71.

Schijven JF, Hassanizadeh SM (2000) Removal of viruses by soil passage: overview of modeling, processes, and parameters. Critical Rev Environ Sci and Technol 30:49-127

Scott MRD, Butler DA, Bredesen DE, Walchli M, Hsiao KK, Prusiner SB (1988) Prion protein gene expression in cultured cells. Protein Eng 2:69-76

Seeger H, Heikenwalder M, Zeller N, Kranich J, Schwarz P, Gaspert A, Seifert B, Miele G, Aguzzi A (2005) Coincident scrapie infection and nephritis lead to urinary prion excretion. Science 310:324-326

Shaked GM, Shaked Y, Kariv-Inbal Z, Halimi M, Avraham I, Gabizon R (2001) A protease-resistant prion protein isoform is present in urine of animals and humans affected with prion diseases. J Biol Chem 276: 31479-31482

Sigurdarson B (1954) RIDA, a chronic encephalitis of sheep (with general remarks on infectious which develop slowly and some of their special characteristics). Br Vet J 110:341-354

Sigurdson CJ, Williams ES, Miller MW, Spraker TR, O'Rourke KI, Hoover EA (1999) Oral transmission and early lymphoid tropism of chronic wasting disease PrP^{res} in mule deer fawns (*Odocoileus hemionus*). J Gen Virol 80:2757-2764

Silveira JR, Raymond GJ, Hughson AG, Race RE, Sim VL, Hayes SF, Caughey B (2005) The most infectious prion protein particles. Nature 437:257-261

Sposito G (1984) The Surface Chemistry of Soils. Oxford University Press

Stahl N, Borchelt DR, Hsaio K, Prusiner SB (1987) Scrapie prion protein contains a phosphatidylinsitol glycolipid. Cell 51:229-240

Stahl N, Baldwin MA, Teplow DB, Prusiner SB (1993) Structural studies of the scrapie prion protein using mass spectrometry and amino acid sequencing. Biochem 32:1991-2002

Stefani M, Dobson CM (2003) Protein aggregation and aggregate toxicity: New insights into protein folding, misfolding diseases and biological evolution. J Mol Med 81:678-699

Stotzky G (2000) Persistence and biological activity in soil of insecticidal proteins from *Bacillus thuringiensis* and of bacterial DNA bound on clays and humic acids. J Environ Qual 29:691-705

Supattapone S, Bosque P, Muramoto T, Wille H, Aagaard C, Peretz D, Nguyen HOB, Heinrich C, Torchia M, Safar J, Cohen FE, DeArmond SJ, Prusiner SB, Scott M (1999) Prion protein of 106 residues creates an artificial transmission barrier for prion replication in transgenic mice. Cell 96:869-878

Taraboulos A, Scott M, Semenov A, Avrahami D, Laszlo L, Prusiner SB (1995) Cholesterol depletion and modification of COOH-terminal targeting sequence of the prion protein inhibit formation of the scrapie isoform. J Cell Biol 129:121-132

Taylor DM, Fraser H, McConnell I, Brown DA, Brown KL, Lamza KA, Smith GRA (1994) Decontamination studies with the agents of bovine spongiform encephalopathy and scrapie. Arch Virol 139:313-326

Taylor DM (1995) Survival of mouse-passaged bovine spongiform encephalopathy agent after exposure to paraformaldehyde-lysine-periodate and formic acid. Vet Microbiol 44:111-112

Taylor DM (2000) Inactivation of transmissible degenerative encephalopathy agents: a review. Vet J 159:10-17

Thackray AM, Knight R, Haswell SJ, Bujdoso R, Brown DR (2002) Metal imbalance and compromised antioxidant function are early changes in prion disease. Biochem J 362:253-258

Thackston RE, Holbrook HT (1992) Artificial mineral licks: Longevity, use and attitudes. Proc Southeastern Assoc Fish Wildlife Assoc 46:188-193

Towne EG (2000) Prairie vegetation and soil nutrient responses to ungulate carcasses. Oecologia 122:232-239

Vasina EN, Dejardin P, Rezaei H, Grosclaude J, Quiquampoix H (2005) Fate of prions in soil: Adsorption kinetics of recombinant unglycosylated ovine prion protein onto mica in laminar flow conditions and subsequent desorption. Biomacromolecules 6:3425-3432

Verwey EJ, Overbeek JTG (1948) Theory of the Stability of Lyophobic Colloids. Elsevier

Vettori C, Calamai L, Yoderc M, Stotzky G, Gallori E (1999) Adsorption and binding of AmpliTaq DNA polymerase on the clay minerals, montmorillonite and kaolinite. Soil Biol Biochem 31:587-593

Vettori C, Gallori E, Stotzky G. (2000) Clay minerals protect bacteriophage PBS1 of *Bacillus subtilis* against inactivation and loss of transducing ability by UV radiation. Can J Microbiol 46:770-773

Weeks HP, Kirkpatrick CM (1976) Adaptations of white-tailed deer to naturally occurring sodium deficiencies. J Wildlife Manage 40:610-625

Will RG (1993) Epidemiology of Creutzfeldt-Jakob disease. Br Med Bull 49:960-970

Williams ES, Miller MW (2002) Chronic wasting disease in deer and elk in North America. *In*: Infectious diseases of wildlife: Detection, diagnosis, and management, part two. Bengis RG (ed) Rev Sci Tech Off Int Epiz 21:305-316

Wisniewski HM, Sigurdarson S, Rubenstein R, Kascsak RJ, Carp RI (1996) Mites as vectors for scrapie. Lancet 347:1114

Wüthrich K, Riek R (2001) Three-dimensional structures of prion proteins. Adv Protein Chem 57:55-82

You Y, Vance GF, Sparks DL, Zhuang J, Jin Y (2003) Sorption of MS2 bacteriophage to layered double hydroxides: effects of reaction time, pH, and competing anions. J Environ Qual 32:2046-53

Zhuang J, Jin Y (2003) Virus retention and transport as influenced by different forms of soil organic matter. J Environ Qual 32:816-823

Reviews in Mineralogy & Geochemistry
Vol. 64, pp. 153-178, 2006
Copyright © Mineralogical Society of America

6

Interaction of Iron and Calcium Minerals in Coals and their Roles in Coal Dust-Induced Health and Environmental Problems

Xi Huang*, Terry Gordon, William N. Rom

[1]*Department of Environmental Medicine*
New York University School of Medicine
550 First Avenue, PHL Room 802, New York, New York, 10016, U.S.A.

Robert B. Finkelman

[2]*U.S. Geological Survey*
12201 Sunrise Valley Drive; Mail Stop 956
Reston, Virginia, 20192-0002, U.S.A.

*e-mail: huangx02@med.nyu.edu

ABSTRACT

Epidemiological studies using pollutant gases (e.g., SO_2) and particle characteristics (e.g., elemental carbon) indicate that products of fossil fuel combustion are important contributors to particulate matter (PM)-associated hospital admissions and mortality. Coal is one of the world's most important fossil fuels, providing 40% of electricity worldwide. Besides individuals exposed to PM in ambient air, coal mining can cause adverse health effects in workers exposed to coal dusts at the workplace. Among the respiratory diseases, coal workers' pneumoconiosis (CWP) has received the most attention because of its clear occupational association. The field of CWP research is one of the few areas in occupational health in which considerable epidemiological data are available. This offers a good opportunity to focus on the relationship between epidemiological data and physico-chemical and/or biological characteristics of coals. The objective of this review is to assess whether some physico-chemical parameters play a role in the observed regional differences in the prevalence of CWP among various coalmine regions. We mainly concentrate on the chemical interaction of two minerals, pyrite (FeS_2) and calcite ($CaCO_3$) in the coals and their role in causing occupational lung diseases (e.g., pneumoconiosis) and other environmental problems (e.g., acid mine drainage). Therefore, understanding the chemical interaction of the two minerals in the coal may lead to the identification of the causal components in coal dusts as well as in PM. Examples from U.S.A. coals are used to illustrate the chemical interaction and geological distribution of iron and calcium minerals in various coalmine regions and how the differences in levels of these types of minerals contribute to the observed regional differences in the prevalence of CWP. Molecular mechanisms leading to the CWP development are also discussed, particular in the aspects of oxidative stress and inflammation.

INTRODUCTION

A diversity of airborne dusts, gases, fumes, and vapors can cause adverse health effects in individuals exposed in the workplace or indoor environment or ambient air. For example, long-term exposure to airborne particulate matter (PM) for years or decades is associated

 DOI: 10.2138/rmg.2006.64.6

with elevated total, cardiovascular, and infant mortality (Anonymous 1996; Brunekreef and Forsberg 2005; Delfino et al. 2005; Lacasana et al. 2005). With respect to morbidity, respiratory symptoms, lung growth, and function of the immune system are affected (Kappos et al. 2004). Short-term studies show consistent associations of exposure to daily concentrations of PM with mortality and morbidity on the same day or the subsequent days (Schwela 2000). Patients with asthma, chronic obstructive pulmonary diseases (COPD), pneumonia, and other respiratory diseases as well as patients with cardiovascular diseases and diabetes are especially affected (Seaton et al. 1995; Pope 1996). Increasing epidemiologic evidence points to underlying components linked to fossil fuel combustion (Delfino et al. 2005).

Coal is a combustible, sedimentary, organic rock, which is composed mainly of carbon, hydrogen, and oxygen (Levine et al. 1982; Meyers 1982). A fossil fuel, it is formed from vegetation which has been consolidated between rock strata and altered by the combined effects of pressure and heat over million years to form coal. Coal is an aggregate of heterogeneous substances composed of organic and inorganic materials (Meyers 1982). The degree of change undergone by a coal as it matures from peat to anthracite—known as coalification—has an important bearing on its physical and chemical properties and is referred to as the "rank" of the coal. Coal rank is defined as the extent to which the organic materials have matured during geological time ongoing from peat to anthracite. Coal rank can be roughly estimated by the carbon content in the coal, molar ratio of carbon /hydrogen (C/H), heat value, volatile materials, or moisture. The four major coal types ranked in order of increasing heat value are:

lignite < sub-bituminous < bituminous < anthracite

Low rank coals, such as lignite and sub-bituminous coals are typically softer, friable materials with a dull, earthly appearance. They are characterized by high moisture levels and relatively low carbon content, and therefore a lower energy content. Higher rank coals are generally harder and stronger and often have a black, vitreous luster. They contain more carbon, have lower moisture content, and produce more energy. Anthracite is at the top of the rank scale and has a correspondingly higher carbon and energy content and a lower level of moisture.

The inorganic portion of coal can range from a few percent to more than 50% (by weight) and is composed of phyllosilicates (kaolinite, illite, etc.), quartz, carbonates, sulfides, sulfates, and other minerals (Meyers 1982). In general, Al and Fe are the main metals in the coals and As, Ni, Zn, Cd, Co, Cu are trace metals that represent only a very small fraction of the inorganic constituents (Finkelman 1995).

Coal is one of the world's most important sources of energy, fueling almost 40% of electricity worldwide (for more information, visit *http://www.worldcoal.org*). For example, Poland relies on coal for over 94% of its electricity, China for 77%, and Australia for 76%. In the U.S.A., 50% of electricity is generated in coal-fired plants. With the recent energy crisis in California, coal is looking more appealing than it has been in years. Recent debates in the U.S.A. have focused on increasing coal use. In fact, energy costs from a new coal power plant are relatively low, between 3.5 and 4 cents per kilowatt-hour. However, health and environmental costs such as occupational lung disease compensation can bring the total costs from 3.5-4 to as high as 5.5-8.3 cents per kilowatt-hour (Jacobson and Masters 2001). According to the Work-related Lung Disease Surveillance Report (NIOSH 2003), Federal "Black Lung" Program payments totaled more than $1.5 billion for nearly 190,000 beneficiaries in 1999. Coal workers' pneumoconiosis (CWP) deaths accounted for half of the pneumoconiosis deaths during the 10-year period from 1990 to 1999, clearly outnumbering deaths associated with other types of pneumoconiosis, such as asbestosis and silicosis. Among the occupations listed by the Census Industry Code, coal mining is the highest risk job associated with asthma and COPD deaths with a proportionate mortality ratio of 1.98 [95% confidence interval (CI) 1.84-2.12, adjusted for age, sex and race], as compared to the second highest risk job of trucking service of 1.29 (95% CI 1.22-1.37) (NIOSH 2003).

In this book chapter, we will mainly focus on the chemical interaction of iron and calcium minerals in the coals and their role in causing occupational lung diseases (e.g., pneumoconiosis) and other environmental problems (e.g., acid mine drainage). Epidemiological studies using pollutant gases (e.g., SO_2) and particle characteristics (e.g., elemental carbon) have provided certain evidence that products of fossil fuel combustion are important contributors to PM-associated hospital admissions and mortality (Delfino et al. 2005). Therefore, understanding the chemical interaction of minerals in the coal may link the sources of PM and also identify some of the causal components in PM. Moreover, the health risks to the general population due to PM exposure in ambient air as well as the cost to the federal government in "black lung" disability benefits in the coalmine industry highlight the need for surveillance and screening programs carefully monitoring early adverse effects of airborne particles. Knowing the active component(s) in the particles and their pathological interactions with biological molecules will certainly help to develop such programs. Examples from U.S.A. coals will be used to illustrate the geological distribution of iron and calcium minerals in various coalmine regions as well as how the differences in levels of these types of minerals contribute to the observed regional differences in the prevalence of CWP. Molecular mechanisms leading to the CWP development are also discussed, particularly in the aspects of oxidative stress and inflammation.

PATHOLOGIES AND CLINICAL DIAGNOSIS OF CWP

The term "pneumoconiosis" embraces non-neoplastic reactions of the lungs to inhaled mineral or organic dusts, and the resultant alteration in their structure, excluding asthma, bronchitis and emphysema (Gross 1962; Honma et al. 2004). It can be defined as a dust overload disease in the lungs and the tissue's reaction to its presence (Castranova and Vallyathan 2000). Unlike silicosis, being the fibrotic disease of the lungs caused by inhalation of dust containing silicon dioxide, primarily in the free crystalline form, CWP results from inhalation of coal dust, a complex mixture usually containing relatively small amounts of free crystalline silica (Honma et al. 2004). Mineralogical analysis of tissue specimens can be performed in a number of different ways in order to distinguish various types of pneumoconiosis due to exposure to mixed dust (Funahashi et al. 1977; Mastin et al. 1988). One approach is the ashing of the tissue and analysis of recovered dust gravimetrically and by X-ray diffraction. To obtain accurate identification of the mineral species and to quantify their abundance, X-ray diffraction analysis is required. Another approach involves *in situ* analysis of tissue sections by analytical electron microscopy (scanning, transmission, or scanning transmission electron microscopy). A convenient and frequently employed technique is the examination of 5-μm-thick sections in a scanning electron microscope equipped with a backscattered electron detector and an energy dispersive spectrometer. An advantage of this technique is that it can be performed on a section serial to the one used to make the pathological diagnosis. Depending upon the relative content of silica and coal in the lung, the terms anthracosis, silicosis, and anthraco-silicosis have been used (Murthy 1952; Akazaki and Inagaki 1959; Razemon and Ribet 1961; Huang et al. 1999).

Several diagnostic techniques allow a comprehensive approach to coal workers' occupational lung diseases (Remy-Jardin et al. 1992; Balaan et al. 1993; Wagner et al. 1993; Honma et al. 2004). These techniques include chest radiograph, high resolution computed tomography, bronchoalveolar lavage, and measurements of lung function. Radiological findings of CWP on chest radiographs include a mixture of small rounded and irregular opacities as defined by the 1980 International Labor Office (ILO) International Classification of Radiographs of Pneumoconiosis (Gurney 1993; Bauer et al. 2006). This variety has been further subdivided into categories 1, 2, and 3 according to the extent and distribution of the nodular opacities. Pathological studies have shown that these densities represent small aggregates of coal dust with some surrounding fibrosis (Vallyathan et al. 1996). Respiratory impairment is found in CWP patients with category 2 and 3 (Yeoh and Yang 2002; Naidoo et al. 2004).

Certain subjects with CWP can develop progressive massive fibrosis (PMF). PMF appears on a chest film as a dense opacity or opacities that initially may be relatively small, but may grow to occupy the entire lung. Chest x-ray films of a normal lung from a male patient never exposed to coal dusts can be viewed at *http://www.radiologyresource.org* and category II CWP as well as PMF showing diffuse, small light areas (3 to 5 mm) on both sides of the lungs can be viewed at *http://www.nlm.nih.gov/medlineplus/ency/imagepages*. Recently, imaging from CT and chest radiograph as well as pathological findings obtained with lung biopsy (hematoxylin-eosin stain) have been compared among different types of pneumoconiosis including silicosis, CWP, asbestosis, berylliosis, talcosis, and siderosis (Chong et al. 2006). Histological examination reveals large areas of collagen deposition with frequent cavitation and there is marked respiratory impairment in subjects with PMF (Seal et al. 1986; Yeoh and Yang 2002; Bauer et al. 2006).

EPIDEMIOLOGICAL STUDIES ON CWP

Mortality and morbidity studies of coal workers have identified three pathologies that are related to coal dust inhalation: pneumoconiosis, including simple pneumoconiosis and PMF, chronic obstructive bronchopneumopathy, such as emphysema and/or chronic bronchitis, and stomach cancer (Morgan 1971; Rockette 1980; Swaen et al. 1995; Meijers et al. 1997). These studies have generally shown increased standard mortality rates for accidents, respiratory diseases, and stomach cancer (Miller and Jacobsen 1985). Among the respiratory diseases, CWP has received the most attention because of its clear occupational association. Asthma, emphysema, and chronic bronchitis in coal workers have a significant positive correlation with (a) the number of years worked at the coal mine independent of age at death; and (b) the severity of pneumoconiosis (Leigh et al. 1983, 1994). While mining exposures contribute significantly to lung disease, smoking is a major factor in the development of lung cancer and COPD, thus necessitating a comprehensive approach for prevention and control of mining-related occupational lung disease (Jacobsen et al. 1977; Ross and Murray 2004).

During the last thirty years, the prevalence of CWP has fallen consistently in the U.S.A.. (Ross and Murray 2004). For example, in active underground mines included in the National Institute for Occupational Safety and Health (NIOSH) Coal Workers' Surveillance Program, rates fell from >10% in the early 1970s to <2% in the late 1990s (NIOSH 2003). Similar trends have been noted in Europe as well as in South Africa (Ross and Murray 2004). The decrease in CWP prevalence may be partly due to lowered dust levels in coal mines. In the U.S.A., the permissible exposure limit for respirable coal dust is 2 mg/m^3. In contrast to CWP, asthma, emphysema, and chronic bronchitis presently appear to be more relevant in contributing to the morbidity and mortality of coal workers (Meyer et al. 2001; NIOSH 2003; Baur and Latza 2005). Moreover, pneumoconiosis can become progressive after cessation of dust exposure. Therefore, extensive at-risk populations consist of retired or ex-coal miners and those currently active in the workplace (Soutar et al. 1986; Gautrin et al. 1994; Attfield and Seixas 1995).

It has been shown in the U.S.A., Great Britain, France, and Germany that the prevalence and severity of CWP differs markedly between different coalmines despite comparable exposures to respirable dust (Reisner and Robock 1977; Hurley et al. 1982; Amoudru 1987; Attfield and Castellan 1992). For example, the first round of the U.S. National Study of CWP (NSCWP), which was completed in 1971, examined a total of 9,076 miners from 29 underground bituminous and 2 anthracite mines (Morgan et al. 1973). The average exposure concentration in these mines during that period was 3 mg/m^3 (Attfield and Morring 1992a). It was found that 41% of the eastern Pennsylvania anthracite miners had simple pneumoconiosis and a further 14% had PMF, but the comparable figures for bituminous miners in Colorado and Utah were 4% and 0.4% (Morgan et al. 1973; Attfield and Morring 1992a). Figure 1 shows that for the same number of years spent underground or for the same job category, the regional difference in the prevalence of CWP remains. The follow-up studies at the same mines (in 1972-1975, 1977-

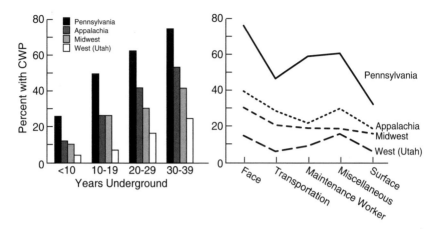

Figure 1. Relationship of CWP prevalence to years of underground exposure and to principal job according to the geographic region. [Reprinted from Morgan et al. (1973) in *Arch Environ Health* with permission of Heldref Publications, a division of Helen Dwight Reid Educational Foundation, *www.heldref.org*.].

1981, 1985-1988 and 1996-2002) have shown that the overall prevalence of CWP has decreased in the U.S.A. but the regional differences persists with a greater risk in Eastern coal miners (Pennsylvania and West Virginia) than in Western coal miners (Utah and Colorado) (Attfield and Seixas 1995; Goodwin and Attfield 1998; Pon et al. 2003b; Antao et al. 2005).

Figure 1 also clearly shows that the prevalence of CWP in the four coalmine regions is related to years of underground exposure. It is apparent that overall CWP prevalence for miners who worked less than 10 years is lower than those who worked underground for 30 years or longer regardless of coalmine regions. It is noteworthy that, as a result of increasing emphasis on exposure-response characterization for occupational hazards, quantitative estimation of exposure has been developed in occupational epidemiology (Seixas et al. 1991, 1992; Seixas and Checkoway 1995). Cumulative dust exposure estimates can be calculated for each miner by multiplying the coal dust concentration and the time worked underground. It is commonly expressed as units of gram-years per cubic meter (g-yr/m^3) (Attfield and Morring 1992a). Application of these exposures to exposure-response has led to detection of a relationship between dust exposure and prevalence of CWP and between dust exposure and ventilatory function (Attfield and Hodous 1992; Attfield and Morring 1992b; Naidoo et al. 2004; Soutar et al. 2004). These relationships were stronger than those obtained by using surrogate measure of exposure, such as tenure underground.

The relationship of CWP to the miner's principal job is also shown in Figure 1. It is evident that the prevalence of CWP is highest at the mine face and least on the surface. The apparent anomaly of a high prevalence of the disease among miscellaneous underground workers may be explained by the fact that many of these workers have duties which, at times, involve face work, and also have duties at coal transfer points on conveyer systems where dust concentrations may be high (Morgan et al. 1973). Therefore, marked regional differences in the prevalence of CWP exist despite comparable levels of dust exposure and years spent underground. The disease is more common in Pennsylvania than in the rest of Appalachia. It occurs less frequently in the Mid-West and much less so in the West. Such large regional differences in the prevalence of CWP cannot be explained by the slight differences present in exposure concentration or mining techniques (Morgan et al. 1973; Attfield and Seixas 1995). Since the same techniques and technicians were used in each round of the NSCWP, factors such as the X-ray reader variation or changes in X-ray standard should not have contributed to the regional differences in CWP (Attfield and Morring 1992a). In France, coal miners of Provence have not had CWP (0%) whereas the prevalence of CWP in coal miners of Nord Pas de Calais was 24% (Amoudru 1987). In Great Britain, the proportional mortality ratios for CWP varied from 135 (95% CI 16-488) in Leicestershire county to 3825 (95% CI 1538-7881) in South Glamorgan county (Coggon et al. 1995). The observed regional differences in the prevalence of CWP indicate that physico-chemical characteristics of coal mine dust responsible for CWP development may vary from region to region. Quartz, coal rank, carbon-centered free radicals, transition metals, as well as bioavailable iron (BAI) have been considered important components in the coal dusts that may contribute to the CWP development. Therefore, differences in levels of these causative factors may be associated with the observed regional differences in the prevalence of CWP.

PHYSICO-CHEMICAL CHARACTERISTICS THOUGHT TO BE IMPORTANT IN COAL DUST-INDUCED CWP

Quartz

Epidemiological studies of the relationship between the prevalence of CWP and environmental measurements have consistently revealed that the predominant adverse exposure factor is respirable mixed coal dusts (Attfield and Wagner 1993; Attfield et al. 2004). Because silica is an important component of the mixed coal dust and free crystalline silica (quartz) is a well-known mineral in inducing silicosis, a fibrotic lung disease, CWP was originally thought to be a variant of silicosis (Borm and Tran 2002). Indeed, quartz is more biologically active than the mixed coal dust (Albrecht et al. 2002; Ernst et al. 2002) and the high biological reactivity of quartz may be due to the unique properties of the silanol groups (-SiOH) (Castranova and Vallyathan 2000; Fubini et al. 2001). For example, it was shown that quartz as cumulative exposure was a highly statistically significant predictor of polymorphonuclear leukocytes (PMNs) counts in bronchoalveolar lavage fluid, a marker for lung injury. PMNs are small, actively motile white blood cells containing many lysosomes (a vesicle containing hydrolytic enzymes) and specializing in phagocytosis, a process of ingestion and digestion of particles within cells. Cumulative coal dust exposure did not significantly add to the prediction of PMNs above that predicted by cumulative quartz exposure (Kuempel et al. 2003). However, CWP also results from the inhalation of coalmine dust that usually contains relatively small amounts of quartz. It has been found that quartz is only a minor contributor to CWP development in general (Walton et al. 1977; Borm and Tran 2002). Moreover, pneumoconiosis was also found in coal trimmers who shoveled coal that contained little or no silica and in graphite and carbon electrode workers who were exposed to carbonaceous materials free of silica (Miller and Ramsden 1961; Lister and Wimborne 1972). Since the presence of silica or silicates is found to be unnecessary for the development of pneumoconiosis, the question arises what other factor(s) in the coal dust is responsible for the disease.

Coal rank

Coal rank was found to play a role in CWP, since risk increases with coal rank (Miller and Jacobsen 1985; Attfield and Morring 1992b; Kuempel et al. 1995). Laboratory coal breakage studies have shown a positive correlation with the amount of respirable-size particles found in the product increasing with coal rank (Moore and Bise 1984). It is suggested that higher rank coals with a higher electrostatic charge on breakage may contribute to the increased incidence of CWP in high rank coal regions (Page 2000; Page and Organiscak 2000). As mentioned earlier, coal rank is only an indicator of the carbonization of organic materials in the coals and is not an active component, chemically speaking, which can induce lung injury. Investigation on the cytotoxic characteristics of coal mine dust yielded opposite results: coal dust from low rank coal causes a higher cytotoxicity than coal dust from high rank coal. Leaching of coal dust by dichloromethane and physiological fluids (aqueous solution in the presence of lecithin, a lung surfactant) shows that high concentrations of phenolic compounds are extracted from low rank coals (Schulz 1997). Values of phenol index decrease with increasing coal rank. The amounts of phenolic compounds leached by the fluids correlate with high cytotoxicity of the coal dusts. Using two coals of Pennsylvania and Utah with high and low prevalence of CWP, respectively, the leachate from the Pennsylvania coal inhibited cell growth while chemicals leached from the same size particles of the Utah coal stimulated cell growth (Christian and Nelson 1978; Christian et al. 1979). Results of mineralogical analyses of coal dust samples and cytotoxicity tests showed that the mineral content and cytotoxic potential of dusts collected from the same mine, and even from the same underground site, at different times, varied considerably. Using Affymetrix GeneChip arrays, it has been shown that the coal from Pennsylvania coalmine region with a high rank and a high prevalence of CWP altered 908 gene expressions in primary human bronchial epithelial cells (Hu et al. 2003). The coal from Utah with a low rank and a low CWP prevalence modified 356 genes out of 12,000 genes examined. It was suggested by Hu et al. (2003) that differences in levels of metals, particularly bioavailable iron but not coal rank of the two coals, may contribute to the differences in the gene expression profile. In conclusion, a relationship between coal rank and CWP prevalence was established by epidemiological studies. However, a correlation between coal rank and cell cytotoxicity has not yet been established in biological studies (Reisner and Robock 1977; Gormley et al. 1979).

Carbon-centered free radicals

Since the discovery of the paramagnetic property of coal-like materials, coal and carbon are known as a "tank" of carbon-centered free radicals, also termed as "fossil radicals" (Ubersfeld et al. 1954). Electron spin resonance (ESR) has demonstrated that two types of carbon-centered free radicals exist in coals, which might respectively be attributed to the macromolecular phase and molecular phase of coal (Huang et al. 1999). It has been suggested that coal consists of a macromolecular phase cross-linked to form a three-dimensional network and a molecular phase situated in the pores of the macromolecular phase (Duber and Wiekowski 1982; Retcofsky 1982). It was postulated that the types of free radicals in coals might play a role in the fibrogenicity of coal dust (Artemov and Reznik 1980; Dalal et al. 1989, 1990). It has been shown that the free radicals in anthracite as well as low rank coal are very sensitive to air (Huang et al. 1999). Grinding produced more free radicals in anthracite coals than in bituminous coals, which mimics the breakage of coal during mining processes (Dalal et al. 1989). Exposure of freshly ground bituminous coals to air induced a slight increase of free radicals and a slight decrease after long term exposure (e.g., months). The lung tissue samples of coal workers deceased due to CWP showed similar ESR spectra and free radical intensities as those of coal samples, indicating that these free radicals are very stable, even after 20 years of staying in the pulmonary medium (Dalal et al. 1989; Huang et al. 1999). Treatments of coal samples with O_2, glutathione (GSH), and H_2O_2 showed that the carbon-centered free radicals had little chemical reactivity in high rank coals, adding further evidence that these types of free radicals are very inert (Huang et al. 1999).

Transition metals

Asbestos, silica, and coal dusts have long been known to produce various types of lung diseases. Due to the mixed nature of various chemical compositions in these dusts, it is extremely difficult to identify the active compound(s) in the dusts that is responsible for the disease (Costa and Dreher 1997). Recently, progress has been made in metal-induced parenchymal lung disorders. For example, lung disorders can arise from exposure to aluminum, beryllium, cadmium, cobalt, copper, iron, mercury, and nickel (Kelleher et al. 2000). Metal fume fever, an inhalation fever syndrome, is a common condition among welders, with onset of symptoms typically 4-12 h after the inhalation of high levels of respirable metal fumes (Antonini et al. 2003). Metalworking fluids, along with microbial changes that occur in fluid composition, during use and storage in the workplace, were shown to be responsible for the pulmonary effects reported for workers exposed to metalworking fluid aerosols (Gordon 2004). Beryllium may act as an antigen and chronic beryllium disease is a hypersensitivity disorder caused by beryllium exposure in the workplace and is characterized by the accumulation of beryllium-specific CD4$^+$ T cells in the lung and granulomatous inflammation (Fontenot and Maier 2005). CD4$^+$ T cells are T helper cells, a sub-group of leukocytes, expressing CD4 protein, a glycoprotein. These cells play an important role in establishing and maximizing the capabilities of the body immune system. Other metals such as iron, cadmium, and mercury induce nonspecific damage, probably by initiating production of reactive oxygen species (Kelleher et al. 2000; Valko et al. 2005).

Based on the observation that transition metals, especially iron, are concentrated in the lungs of miners with pneumoconiosis (Guest 1978), Ghio and Quigley (1994) postulated that the body's endogenous iron (Fe^{3+}) sequestered by humic-like substances in certain coals may play a role in pneumoconiosis (Ghio and Quigley 1994). It was shown that oxidants generated by coal dusts after exposure to exogenous FeCl$_3$ increased with the concentration of humic-like substances. By studying the ability of different coals to catalyze the generation of $^{\cdot}$OH from H$_2$O$_2$, Dalal et al. (1995) found that the potential to induce $^{\cdot}$OH formation and lipid peroxidation by the coal dusts exhibits a good correlation with the available surface iron (Dalal et al. 1995). The leachable Ni^{2+} content of coal dust was proposed to be an additional cytotoxic parameter (Carlberg et al. 1971; Sichletidis et al. 2004). Using a system mimicking phagolysosomes, it was found that beside Fe^{2+}, Fe^{3+}, and Ca^{2+}, trace amounts of Ni, Cu, and As were released in proportion to the CWP prevalence among three coal mine regions of Pennsylvania, West Virginia, and Utah (Zhang et al. 2002). However, it has not yet been shown that the levels of humic-like substances or the available surface iron in the coals correlate with the prevalence of CWP from various coalmine regions. Moreover, it is questionable that humic-like acid can sequester the endogenous iron *in vivo*, since the body iron is tightly bound to iron proteins such as ferritin. Ferritin is an iron storage protein with a high binding affinity for iron (>10^{36}) and a huge capacity of binding up to 4,500 atoms of iron per molecule of ferritin (Harrison and Arosio 1996; Torti and Torti 2002). As we have shown below, the surface iron in certain coals is leachable under acidic conditions, suggesting that these coals might release a much larger amount of bioavailable iron (BAI) in cells.

Bioavailable iron (BAI)

Iron is the best-known transition metal capable of producing oxidants through Fenton, Haber-Weiss or autoxidation reactions (Kamp et al. 1992; Morris et al. 1995; Meneghini 1997; Huang 2003). However, not all iron compounds in coal are bioavailable for oxidant formation and subsequent adverse health effects. Because ferric ion (Fe^{3+}) is a weak oxidant, it was first postulated that acid soluble Fe^{2+} is the active compound in the coal in inducing CWP (Huang et al. 1993). Later on, BAI was defined as iron (both Fe^{2+} and Fe^{3+}) released by coal in 10 mM phosphate solution, pH 4.5, which mimics conditions in the phagolysosomes of cells (Zhang et al. 2002).

Acid solubilization of iron and buffering capacity of coals. Particles deposited on the alveolar epithelium are phagocytosed by alveolar macrophages (AMs) (Stuart and Ezekowitz 2005). Particles are entrapped in a membrane-bound phagosomal vacuole, which subsequently fuses with lysosomes to form secondary lysosomes or phagolysosomes. AMs dissolve many inorganic particles at faster rate than simulated lung fluid or saline. The low phagolysosomal pH is suggested to be one mechanism behind the ability of AMs to efficiently dissolve inorganic particles (Lundborg et al. 1985; Kreyling et al. 1991; Beletskii et al. 2005).

Using a pH 4.5 phosphate solution (10 mM), which mimics the phagolysosomal pH, Table 1 shows that average levels of bioavailable Fe^{2+}, Fe^{3+}, Ni, and Cu are high in Pennsylvania coal samples, lower in West Virginia coals, and lowest in Utah coals (Huang et al. 1998; Zhang et al. 2002). Levels of bioavailable Cr were below detection limit (<0.1 mg/l) in all of the coal

Table 1. Average levels of bioavailable iron, calcium, and other metals in three coal mine regions with different prevalence of CWP. [a]

Region	CWP [b]	Fe^{2+}	Fe^{3+}	Ca^{2+}
Utah ($n = 10$)	4%	1.95 ±1.93 (Range 0-19.3)	7.62 ± 2.21 (0-25.8)	717.5 ± 84.64 (286.4-1042)
West Virginia ($n = 10$)	10%	1963.14 ± 886.90 (0-7535.9)	2695.7 ± 1192.7 (0-9897.8)	498.83 ± 131.5 (15.6-1249.9)
Pennsylvania ($n = 8$)	26%	7288.2 ± 2710.7 (543.8-20045.7)	4860.8 ± 1376.1 (672-10401.2)	662.9 ± 70.5 (531.1-1127.5)

Region	Ni	Cu	As	Cr
Utah ($n = 10$)	0 [c]	0	0	0
West Virginia ($n = 10$)	0.11 ± 0.062	0.17 ± 0.076	0.30 ± 0.20	0
Pennsylvania ($n = 8$)	0.4 ± 0.06	0.32 ± 0.074	0.13 ± 0.12	0

a. Metal concentrations were parts per million (ppm) of coal (w/w). 2.7 g of coal samples were suspended in 30 mL phosphate solution (10 mM, pH 4.5) for 3 days, which mimics the phagolysosomal conditions of cells. Metals released under such conditions were considered as bioavailable. After filtration, levels of Fe^{2+}, Fe^{3+}, and Ca^{2+} in the coals were measured by spectrophotometer using 2,2'-dipyridyl, DFO, and Arsenazo III as their color-forming complexes, respectively. Levels of Ni, Cu, As, and Cr were determined by atomic absorption. Data are presented as Mean ± Standard Error (SE) of the coal samples from each region.

b. Prevalence of CWP was from Attfield and Morning (1992a).

c. Below detection limit. Reprinted from Zhang et al. (2002).

samples. Because of the heterogeneity of the coal samples, the standard deviations in each region are large as reflected by the wide range of each metal measured (Table 1). Ca^{2+} is present in all coals, though the Utah coals have highest levels of Ca^{2+} released. Epidemiological studies have shown that CWP is frequent in Pennsylvania coal workers (Morgan et al. 1973; Attfield et al. 1984; Attfield and Althouse 1992). In contrast, coals from Utah with low prevalence of CWP had little acid soluble Fe^{2+} and Fe^{3+} (Table 1). The coals from West Virginia with an intermediate prevalence of CWP, released a moderate level of acid soluble Fe^{2+} and Fe^{3+}. It was further found that the pH of aqueous coal suspensions from Pennsylvania were acidic and those from Utah were basic or neutral (Huang et al. 1998). By titrating the Utah coals with sulfuric acid (H_2SO_4), these Utah coals consumed large amounts of acid in order to reduce the pH of the aqueous coal suspensions to 4.5. These results suggest that buffering capacity of the Utah coals is one of the physico-chemical parameters controlling the dissolution of a particle. Interestingly, the buffering capacities of Utah coals were shown to correlate positively with the calcium oxide (CaO) content in the high temperature ash of each coal (Huang et al. 1998). These results indicate that high buffering capacities of Utah coals are related to the presence of calcite ($CaCO_3$) in the coals. Calcite can easily react with sulfuric acid to produce $CaSO_4$. It was observed that coals associated with a high prevalence of CWP contain much less calcite (Huang et al. 1993, 1998; Zhang et al. 2002). Thus, these coals have low buffering capacity, and Fe^{2+} can be readily released in alveolar macrophages. In contrast, coals which were shown to induce a low prevalence of CWP contained more calcite. For example, the coals from Gardanne of Provence, France, which contains a high percentage of calcite (>10% w/w), had a very high buffering capacity (Huang et al. 1993). This coal did not even release any detectable Fe^{2+} in 50 mM HCl and no CWP has been reported in the miners who worked in these Gardanne coalmines (Amoudru 1987; Huang et al. 1993). As mentioned earlier, the coals from Western U.S.A., such as Utah and Colorado also contain higher levels of calcite than those from Pennsylvania and West Virginia coal mines. As mentioned previously, epidemiological studies have shown that coals from Utah and Colorado are less hazardous to coal workers than the coals from Pennsylvania and West Virginia. These results indicate that calcite may play an antagonistic role in CWP development by preventing acid solubilization of iron and, thus, makes iron non-bioavailable for oxidant formation and adverse health effects.

Pyrite oxidation and factors controlling the stability of BAI. It is well known that pyrite (FeS_2, including its polymorph marcasite, the same chemistry as pyrite but a different structure) is a typical constituent of coal (Meyers 1982). The surface oxidation in air and air-saturated aqueous solutions of the iron subsulfide has been studied in detail (Lowson 1982; Usher et al. 2004, 2005). Previous studies have shown that pyrite-rich coals tend to undergo rapid low-temperature oxidation (Huggins et al. 1983). This oxidation may give rise to mining complications including acid mine drainage from abandoned mine faces and coal-refuse piles, roof weakness, pillar collapses in underground mines, and spontaneous combustion of stock piles (Zodrow and McCandlish 1978; Huang et al. 1994; Younger 1997). Multiple physico-chemical factors, such as oxygen partial pressure, temperature, particle size, crystallinity, and relative humidity play a role in pyrite oxidation, and these conditions are likely to vary in different coal mines (Lowson 1982; Huggins et al. 1983; Huang et al. 1994). Iron present in coal can become bioavailable by pyrite oxidation, which produces ferrous sulfate and sulfuric acid as follows:

$$2\ FeS_2 + 7\ O_2 + 2\ H_2O \rightarrow 2\ FeSO_4 + 2\ H_2SO_4 \tag{1}$$

$FeSO_4$ is water soluble (150 g/l at 20 °C) and, thus, readily bioavailable for oxidant formation. However, the subsequent human exposure to coal dust containing $FeSO_4$ depends upon the stability of the formed $FeSO_4$. If $FeSO_4$ is oxidized prior to inhalation by coal workers, there is no iron available for oxidant formation and lung injury. It was found that pH of the coal played the most important role in stabilizing $FeSO_4$, such that a final pH < 4.5 after oxidation of pyrite stabilized $FeSO_4$, whereas at high pH the conversion of Fe^{2+} to Fe^{3+} was immediate (Huang et al. 1994). Increasing the pH would facilitate ferrous ion oxidation to goethite (FeOOH), which is water-insoluble and, thus, not bioavailable for redox reactions. Recent studies have shown that by monitoring the reaction *in situ* with horizontal attenuated total reflectance infrared spectroscopy, water was the primary source of oxygen in the sulfate product, while the oxygen atoms in the iron oxyhydroxide product was from dissolved molecular oxygen (Usher et al. 2004, 2005). Moreover, it has been shown that the rate of oxidation of Fe^{2+} by oxygen in abiotic system is a function of pH. It has been estimated that the reaction half life of Fe^{2+} at pH 3.5 and 7.0 are 1,000 days and 8 min, respectively (Singer and Stumm 1969; Stumm and Lee 1961). Therefore, there is not much toxicity in the gastric system (pH 1-2) caused by oral uptake of a normal dose of ferrous sulfate because Fe^{2+} oxidation is very slow at pH < 3.5. However, in lung medium in which the pH is usually near 7, oxidation of Fe^{2+} proceeds quickly and oxidants resulting from the interaction of Fe^{2+} and O_2 may damage lung cells and cause cytotoxicity. This suggests that the inhalation of iron may be more hazardous than ingestion of iron. Indeed, pulmonary injury after aspiration of $FeSO_4$ has been reported in a patient showing acute bronchial damage and early histological change in the biopsy specimens after the exposure (Godden et al. 1991). A delayed occurrence of bronchial stenosis after inhalation of iron has also been described (Tarkka et al. 1988; Mizuki et al. 1989).

The pH of water in contact with a particular coal depends not only on its origin but also on the various factors that, over geological time, influence the formation of sulfide and carbonate minerals in the coal. Coals rich in carbonates such as calcite have high pHs, whereas coals rich in sulfides such as pyrite have rather low pHs. The functional chemical groups (carboxyl, phenol, or quinone) on the coal's surface may also influence the pH of a particular coal. The acidity of a particular coal is not constant and can vary at different coalmines over time due to the vertical and lateral variations in coal composition. This can be explained by the ongoing oxidation of coal and its accompanying minerals, such as pyrite.

Calcite ($CaCO_3$) is the major component in coal that consumes acid and neutralizes the pH as follows:

$$CaCO_3 + H_2SO_4 \rightarrow CaSO_4 + H_2O + CO_2 \qquad (2)$$

Therefore, little BAI will accumulate when calcite is present, such as in the coals from the Utah coal mine region (Huang et al. 1998; Zhang et al. 2002). If calcite is absent in the coals, sulfuric acid produced from reaction (1) would solubilize other iron compounds (e.g., $FeCO_3$) and release more BAI. To demonstrate the protective role of calcite in inhibiting BAI, predetermined proportions of calcite were added into three Pennsylvania coals. After suspending the mixtures in the aqueous phosphate solution (10 mM, pH 4.5), it was shown that level of pH as well as Ca^{2+} in the coals were increased in a calcite concentration-dependent manner (Zhang and Huang 2005). In contrast, levels of BAI (both Fe^{2+} and Fe^{3+}) were decreased. Interestingly, calcite also showed a significant inhibitory effect on Pennsylvania coal-induced ferritin formation in human lung epithelial A549 cells. These results indicate that the addition of calcite can inhibit the bioavailability of iron in the iron-containing Pennsylvania coals.

Besides pH and calcite, it has been shown that oxidizing activities and iron levels of coal samples decrease as a function of the duration of exposure to air (Huang et al. 1994). It has been shown that BAI content, even in low pH coals (e.g., pH 4.5), could decrease to half after 2 months of exposure to air. These results indicate that the formation of $FeSO_4$ is a dynamic process and that the oxidizing activity generated by coal samples can vary from the time when the coal dust is extracted to the time when it is tested in biological studies. Coal dust exposed to air for too long (days to months) will lose their oxidizing activity, hence biological activity. This could explain why no significant effects of coal dusts were observed in previous *in vitro* and *in vivo* studies (Castranova et al. 1985; Castranova and Vallyathan 2000), because the coal dusts that were recovered from respirators or sample collectors were exposed to air for months if not years. Oxidizing activity of these coal dusts had already disappeared before any cell or animal treatments were performed.

In conclusion, three factors are important in controlling levels of BAI in the coals. (1) pH of coals: The initial low pH can favor the stabilization of $FeSO_4$, but is not a determinant factor. $FeSO_4$ may be originally present in the coals. The majority of $FeSO_4$ is from oxidation of pyrite, which can yield BAI and simultaneously stabilize BAI by lowering pH. (2) The presence of calcite in the coals: The buffering capacity of the coal dusts must be low so that H_2SO_4 produced by pyrite oxidation can sufficiently lower the pH to 4.5 and stabilize $FeSO_4$. Excessive H_2SO_4 can further release BAI from the coal dusts and enhance their oxidizing potential. If the levels of calcite in the coals are high, calcite neutralizes sulfuric acid and thus, increases the pH of the coals. At a relatively high pH (> 4.5), BAI can be readily oxidized to biologically inactive goethite. (3) Duration of coal's exposure to air prior to inhalation: Levels of BAI as well as oxidizing activities of coals decrease as a function of time of exposure to air. Therefore, the longer the duration of coal's exposure to air is, the less the cytotoxicity of coal will be. In summary, BAI may be the active compound in coal dust-induced lung disease, but BAI is a transient compound in coal dusts. The formation and stabilization of BAI may vary from one place or time to another due to the many factors related to the physico-chemical properties of the coals and the heterogeneity of coal samples taken from different areas of certain mines.

Mapping and prediction of coals' pneumoconiotic potencies with BAI content in the coals. Based on the hypothesis that BAI is the active component in the coals inducing CWP, the differences in the levels of BAI in the coals may be responsible for the observed regional differences in the prevalence of CWP. CWP prevalence data from the first National Study of CWP (NSCWP) as well as physico-chemical data from US Geological Survey (USGS) coal quality database were used for the correlation studies. Based on the names of the coal mines, counties, and states, several thousand analyses of coal samples contained in the USGS coal quality database were searched. Ninety-four coal samples from twenty-four coalmines within seven states matched the locations in the NSCWP. These are bituminous coals obtained from mines within the same state, county, and coal seam as those samples used in the first NSCWP (Morgan et al. 1973). Most of the samples in the USGS database were collected in the period of 1975-1985 (Finkelman and Gross 1999).

As mentioned above, BAI mainly consists of water-soluble iron, such as ferrous and ferric sulfate, which can be originally present in the coals or can be obtained by the oxidation of pyrite (Huang et al. 1998; Zhang et al. 2002). Another possible source of BAI is acid solubilization of siderite or ferrous silicate ($FeSiO_3$). Using the USGS coal database, levels of BAI in each coal have been calculated. It can be seen from reaction (1) that one mole of pyrite will produce one mole of BAI as ferrous sulfate and one mole of sulfuric acid. However, levels of pyrite in the USGS coal database were not measured directly. Since only pyritic sulfur content is available in the database, reaction (1) shows that one mole of pyritic sulfur will produce half a mole of BAI and half a mole of sulfuric acid.

Previous studies have shown that BAI is stable only in an acidic environment (Stumm and Lee 1961; Singer and Stumm 1969; Huang et al. 1994). If there is calcite in the coal, calcite will consume the acid and neutralize the pH as shown in reaction (2). If calcite is absent in the coals, sulfuric acid produced from reaction (1) would solubilize other iron compounds (e.g., $FeCO_3$) and release more BAI as follows:

$$FeCO_3 + H_2SO_4 \rightarrow FeSO_4 + H_2O + CO_2 \tag{3}$$

According to the chemical reactions (1-3), it was concluded that the total available molar amounts of sulfuric acid in a given amount of coal would be $[H_2SO_4]$ = [½ pyritic sulfur (S_{py}) + sulfate – calcite]. If $[H_2SO_4] \leq 0$, this indicates that acid is completely consumed by calcite and, concomitantly, iron will be oxidized. Therefore, there will be no BAI in that coal. If $[H_2SO_4] > 0$, the excess acid would stabilize BAI and, possibly, also leach out other iron compounds, such as siderite, thus releasing additional BAI.

Based on the USGS coal database (*http://energy.er.usgs.gov/products/databases/CoalQual*), pyritic sulfur and sulfates as percents of coal are available for calculations. Calcite and siderite in the coals were not measured. However, levels of calcium oxide (CaO) and the total amount of iron (shown as Fe_2O_3) were measured in high temperature ashes of the coals and the ash yield in the coal is also available from the USGS database. It can be assumed that the calcium oxide was all derived from calcite and the iron was derived from pyrite, two of the most common minerals in coal, thus maximizing the calcite and pyrite estimates. Since one mole of CaO is formed by the decomposition of one mole of calcite at high temperature, the same molar amounts of CaO can be used as measures of calcite. Therefore, the millimolar amounts of pyritic sulfur and sulfates per 100 g dry coal in each individual coal can be calculated. The millimolar amounts of CaO and total iron (Fe_2O_3) per 100 g dry coal can also be calculated after taking into consideration the ash yield in each individual coal.

The average levels of total sulfuric acid (½ S_{py} + sulfate), amount of acid available for solubilization of other iron compounds (½ S_{py} + sulfate – calcium oxide), total iron, and BAI in each coalmine region with known CWP prevalence have also been calculated (Huang et al. 2005). To determine BAI, it has been discovered that the amount of BAI in the coal should be equal to the lesser value between the amount of available acid (½ S_{py} + SO_4^{2-} – CaO) and Fe_2O_3, because: 1) if the coal has an excessive amount of acid and a limited amount of iron, BAI will be limited by the amount of iron present; 2) if the coal has less acid but more iron present, BAI will then be limited by the amount of acid since excess iron cannot be solubilized and, therefore, cannot become bioavailable. Table 2 shows the average levels of BAI (millimoles/100 g dry coal) from 7 states with corresponding CWP prevalence reported in the first NSCWP.

It has been shown that CWP prevalence correlated well with BAI (correlation coefficient r = 0.94, 95% CI 0.66-0.999, $p < 0.0015$), as well as with pyritic sulfur (r = 0.91, 95% CI 0.35-

Table 2. Average levels of total sulfuric acid ($\frac{1}{2}$ S_{py} + SO_4), available amount of acid ($\frac{1}{2}$ S_{py} + SO_4 – CaO), total iron, and predicted BAI.[a]

State	# of Mines	CWP (%)	$S_{py}/2$ + SO_4	$S_{py}/2$ + SO_4 – CaO	Fe_2O_3	BAI
Pennsylvania	9	45.35	18.61	14.63	12.48	11.82
Ohio	6	31.80	19.91	14.69	12.86	9.07
Kentucky	13	29.00	13.17	7.49	9.78	6.25
West Virginia	8	28.25	9.15	4.57	7.27	4.77
Alabama	13	16.70	9.65	6.77	8.85	5.29
Utah	4	13.10	4.14	–3.19	2.69	1.09
Colorado	41	4.60	1.92	–2.69	3.68	0.15

a. Units for the chemicals are millimoles per 100 gram dry coal. Levels of pyritic sulfur, sulfate, CaO, and Fe_2O_3 were obtained from the USGS database for each coalmine. Values in the individual coal samples were calculated first and then averaged for the coalmine region for each of the physico-chemical parameters listed in the Table. [From Huang et al. (2005) with permission of Taylor and Francis *http://www.tandf.co.uk/journals.*]

0.99, $p < 0.0048$), and total iron ($r = 0.85$, 95% CI 0.20-0.97, $p < 0.016$), but not significantly with coal rank ($r = 0.59$, 95% CI –0.26-0.91, $p < 0.16$) or silica ($r = 0.28$, 95% CI –0.55-0.82, $p < 0.54$). No association of CWP with CaO itself was observed ($r = –0.18$, 95% CI –0.78-0.60, $p < 0.69$).

The relationship between CWP and BAI is well described by a linear model (Huang et al. 2005). Based on the levels of BAI in each coal that was calculated with the method mentioned above, each coal's pneumoconiotic potency has been derived in seven thousand coal samples collected and analyzed by the USGS. Figure 2 shows that there is a geographic distribution of coals with different levels of BAI and, therefore, possibly different pneumoconiotic potencies. For example, in the Western States, most coals do not have BAI, which may pose less risk for CWP to coalminers, as shown in the light color. In the Eastern States, there is a trend for possibly high CWP risk coals (dark color), ranging from Pennsylvania to Ohio to West Virginia and Kentucky. There is also an apparent trend of low CWP risk coal (light color) from West Virginia to Tennessee to Alabama. Since CWP prevalence was much higher at the first round of the NSCWP than the current epidemiological data, the prevalence of CWP in the map is probably overestimated, in part due to reduced dust exposure. However, the indication of the relative risk of CWP in coal mining in various coalmine regions may still be valid and useful for CWP prediction.

OTHER ENVIRONMENTAL PROBLEMS RELATED TO COAL USE

Coal fly ash (CFA)-induced lung diseases

CFA are complex particles of a variable composition, which is mainly dependent on the source of coal and the combustion process. Although electric utilities generally are required to remove more than 99% of the noncombustible mineral material, fine and ultrafine particles produced during the combustion process can be emitted into the environment and remain airborne for a long time. Toxic constituents in the CFA are considered to be metals, polycyclic aromatic hydrocarbons, and possibly silica (Borm 1997). Treatment of human lung epithelial A549 cells with CFA from Utah resulted in an increase in mRNA as well as protein levels of interleukin-8 (IL-8) (Smith et al. 2000). IL-8 is a basic, proinflammatory cytokine produced by monocytes, vascular endothelium, and other cell types, which acts on neutrophils as a chemoattractant, activator, and modulator of endothelial adhesion and transmigration. It was

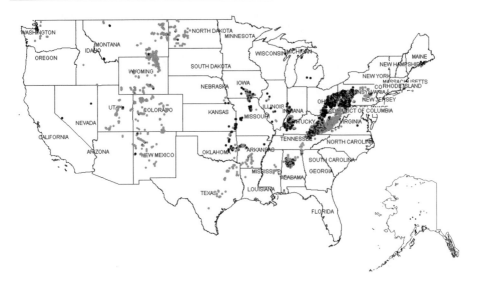

Figure 2. Mapping and prediction of coal's pneumoconiotic potency by the level of BAI in the coal (Bottom right: Alaska coalmine region). Reprinted from Huang et al. (2005).

suggested that BAI was the likely component that induced IL-8 and further shown that iron present in CFA could be mobilized in A549 cells, which leads to ferritin induction (Smith et al. 1998). The particle size of CFA as well as the source of coal played important roles in determining levels of BAI (Veranth et al. 2000a,b). In guinea pigs exposed to CFA from the Illinois coalmine region, total lung capacity, vital capacity, and diffusing capacity for carbon monoxide were significantly reduced below the control values (Chen et al. 1990), whereas exposure to Montana lignite fly ash at comparable concentration and particle size did not show alterations in diffusing capacity. It was suggested that part of the sulfate (either sulfuric acid or iron sulfate) in the Illinois fly ash was responsible for the adverse effects observed in the exposed animals. In fact, decreased pulmonary host defense has been found in mice exposed to carbonaceous particles coated with sulfuric acid (Clarke et al. 2000). Recent studies have shown that airway inflammation and hyperresponsiveness may be caused by a mixture of its major metal components (e.g., Ni, V, Fe, Cu, Zn) but not always by any individual metal alone in the particles (Hamada et al. 2002; Schaumann et al. 2004). Generally speaking, most studies on fly ash toxicity were not designed to elucidate effects of its BAI content nor did they include coalmine dust as a reference (for more details, see review Borm 1997). Therefore, it is unknown whether CFA from regions with high prevalence of CWP (e.g., Pennsylvania and West Virginia) are more potent in inducing lung injury than those from regions with low prevalence of CWP (e.g., Utah). A systematic study with a large number of coals and CFA should provide valuable information on the toxicological profile of the two types of dusts. Because these two types of dusts are from the same source, interpretation and comparison to prove the role of BAI should be relatively straightforward due to the known physico-chemical compositions of the dusts and few confounding factors.

Role of pyrite oxidation in acid mine drainage

Acid mine drainage (AMD) caused by coal and metal mining is typically highly acidic with elevated levels of dissolved metals. AMD is formed by a series of complex geo-chemical and possibly microbial reactions that occur when water comes in contact with pyrite in coal (Anonymous 1999). The dissolved metals, mainly iron, remain in solution until the pH rises to a level where precipitation occurs. As shown in reaction (1), pyritic sulfur is oxidized to sulfate and ferrous ion is released. The "rate determining step" in the overall acid generating processes is the conversion of ferrous to ferric ions as follows:

$$4 \, Fe^{2+} + O_2 + 4H^+ \rightarrow 4 \, Fe^{3+} + 2 \, H_2O \tag{4}$$

This reaction rate is pH dependent with the reaction proceeding slowly under acidic conditions (pH 2-3) and several orders of magnitude faster at pH values greater than 5 (Stumm and Lee 1961; Singer and Stumm 1969). Certain bacteria can also increase the rate of oxidation from ferrous to ferric ion. Once a ferric ion is formed, it can undergo hydrolysis. Hydrolysis is a reaction that splits the water molecule as follows:

$$4 \, Fe^{3+} + 12 \, H_2O \rightarrow 4 \, Fe(OH)_3 + 12 \, H^+ \tag{5}$$

As shown in reaction (5), three moles of acidity are generated per mole of iron as a byproduct. The formation of ferric hydroxide precipitate (known as "yellowboy") is pH dependent. Solids form if the pH is above about 3.5 but below 3.5 therefore, little or no solids will precipitate. Thus, as the acidic waters containing dissolved ions encounters near neutral surface waters the pH is increased and iron oxides (and to a lesser extent, manganese and aluminum oxides) precipitate forming the ubiquitous brown and orange stains that characterize thousands of miles of streams in the Appalachian region alone. Data compiled in 2000 indicated that drainage from abandoned coalmines in Appalachia affected more than 9,500 miles of streams (G.E. Conrad, personal communication, 2006).

It is important to note that ferric ion (not oxygen) can oxidize additional pyrite as follows:

$$FeS_2 + 14 \, Fe^{3+} + 8 \, H_2O \rightarrow 15 \, Fe^{2+} + 2 \, SO_4^{2-} + 16 \, H^+ \tag{6}$$

This reaction is cyclic and self propagating and moves very rapidly and continues until either ferric ion or pyrite is depleted (Anonymous 1999b).

The benefit of calcite and other alkaline agents, such as calcium oxide (lime), calcium hydroxide, anhydrous ammonia (NH_3), and soda ash (Na_2CO_3), have long been recognized in treating AMD (Cravotta 2003; Hossner and Doolittle 2003; Aziz et al. 2004). Open limestone channels may be the simplest passive treatment method (e.g., buried beds of limestone). Principles of this treatment are based on the fact that dissolution of the limestone adds alkalinity to the water and raises the pH. However, armoring or the coating of the calcite by $FeCO_3$ and $Fe(OH)_3$ produced by neutralization reduces the generation of alkalinity, so large quantities of calcite are needed to ensure the long term success (*http://www.dep.state.pa.us/dep/deputate/minres/bamr/amd/science_of_amd.htm*).

The unsightly staining of the streambeds is only the most visible manifestation of AMD. The acidic, metal laden water renders the stream water undrinkable and affects the ecosystem for miles downstream from the source. The cost for chemical treatment of AMD has been estimated as more than $1 million per day (Clark 1995).

Role of pyritic sulfur in acid rain

The term "acid rain" is commonly used to refer to the deposition of acidic components in rain, snow, fog, dew, or dry particles. The evidence is strong that most of the acidity in the rain is caused by sulfur dioxide (SO_2) released from the smokestacks of coal-burning power plants and other industrial sources (Srivastava and Jozewicz 2001; Carmichael et al. 2002). However, nitrogen oxides also contribute to the acidity. Locally, emissions from motor vehicles can be a major contributor to atmospheric acidity.

In the atmosphere, the sulfur dioxide is converted into sulfuric acid. This may be carried to the ground in rain or snow, but often particles containing sulfuric acid settle out of dry air. So the problem of acid rain is really one of acid deposition in dry as well as wet weather. Acid rain can affect the environment in several ways. The acid precipitation can leach cations from the soil that are essential for healthy plant growth and increase the amount of aluminum that can hinder the uptake of water and essential nutrients by plants. Evidence suggests that acid precipitation has stressed trees and has caused the decline of the red spruce and the sugar maple in the U.S. Northeast. Direct precipitation and runoff has caused acidification of lakes and streams and increased dissolved aluminum adversely affecting the health of ecosystems (*http://www.hbrook.sr.unh.edu/hbfound/report.pdf*).

One method of preventing the release of SO_2 from power plants, is to use calcite as a lime-based desulfurizing agent. For example, pulverized limestone is pneumatically injected into the upper part of the boiler near the superheater where it absorbs 75-85% of the SO_2 in the boiler flue gas (Ekman et al. 2003). Because limestone is added into the boiler rather than into the humidification reactor to produce more effective SO_2 capture, the addition of calcite in relatively small amounts should not decrease British thermal units of the coals, a factor of concern by the coal industries. Industry efforts to reduce acid rain have been effective. Recent reports indicate that acid deposition has been reduced by 10 to 25% in the Eastern U.S.A. (*http://bqs.usgs.gov/acidrain/Program.pdf*) and results from nationwide monitoring stations indicate that downward trends in sulfate deposition throughout the U.S.A. (Anonymous 1999a).

PATHOGENESIS OF COAL DUST-INDUCED LUNG INJURY

Oxidative stress

CWP is considered as one of the human lung pathologies related to oxidative stress. Oxidative stress is a disturbance in the oxidant/antioxidant steady state in favor of oxidants, which leads to cellular damage. Accumulating evidence suggests that release of reactive oxygen species (ROS) can play an important role in diverse pathologies including inflammation, tissue aging, cardiac ischemia, arthritis, cancer, and fibrosis (Vallyathan and Shi 1997; Schins and Borm 1999; Langen et al. 2003; Land and Wilson 2005). The presumed mechanism is through the ability of ROS to induce biochemical alterations in macromolecules such as DNA, lipids, and proteins. It has been known that ROS, produced by leukocytes and macrophages as a bactericidal mechanism of host defense, can also damage surrounding tissue. Recent evidence indicates that ROS either from endogenous (e.g., macrophages) or exogenous (e.g., coal dusts) sources also serve as signaling molecules, which can activate transcription factors such as nuclear factor-κB (NF-κB) and activator protein-1 (AP-1) and alter mitogenic and fibrogenic signals such as interleukin-6 (IL-6) (Chatterjee and Fisher 2004; Frey and Malik 2004). Figure 3 depicts a simplified mechanism of coal dust-induced CWP development.

Oxidative property of coal dusts

The oxidative property of coal dusts is primarily attributed to its transition metal constituents, which typically include Fe, Cr, Co, Ni, Mn, As, Zn, and V (Finkelman 1999; Zhang et al. 2002). Using ESR, it was found that aqueous coal filtrates containing Fe, Zn, Cu, Ni, and Co were able to produce ROS such as ˙OH or ferryl ($Fe^{IV}=O$) (Huang et al. 1993). These species are very electrophilic, hence very toxic. It was further shown that ROS resulting from ROS-producing coal dusts are able to inactivate alpha-1-antitrypsin (α_1-AT), which is a major serum inhibitor of elastases, enzymes contributing to emphysema development (Huang et al. 1993). Inherited deficiency in α_1-AT and inactivation of α_1-AT by cigarette smoke show increased susceptibility to emphysema (Janoff 1985; Snider 1992). Average levels of BAI, an important fraction of total iron, which is capable of catalyzing ROS formation, can reach as high as 0.7% (w/w) in the Pennsylvania coalmine region (Huang et al. 1998; Zhang et al. 2002).

Oxidants, cytokines, and growth factors in CWP

It is well known that dusts deposited on the alveolar epithelium are phagocytized by alveolar macrophages (AMs), which can lead to ROS formation. It has been shown that release

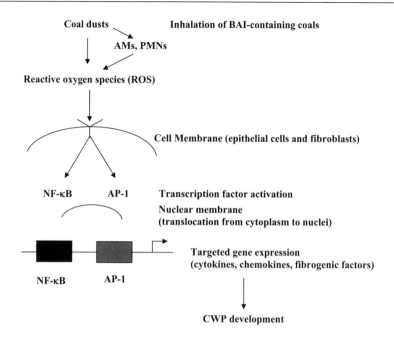

Figure 3. Proposed mechanism involved in the CWP development. ROS are produced directly from BAI-containing (exogenous) as well as a result of phagocytosis of coal dust by alveolar macrophages (AMs) and polymorphonucelar leukocytes (PMNs) (endogenous). These ROS activate transcription factors (NF-κB, AP-1, etc.) through membrane receptors. Following activation, transcription factors are translocated from cytoplasm to nuclei and bind to the promoter regions of target genes. This increases the production of inflammatory cytokines, chemokines, and fibrogenic factors, leading to CWP development.

of H_2O_2 and $O_2^{-\bullet}$ is increased in AMs from coal workers with emphysema and pneumoconiosis (Rom 1990, 1991; Wallaert et al. 1990). Concurrently, glutathione (GSH), a major cellular antioxidant known to protect the lung from oxidants, was significantly decreased. In contrast, glutathione peroxidase and Cu/Zn superoxide dismutase were increased in the red blood cells of miners with pneumoconiosis (Borm et al. 1990; Perrin-Nadif et al. 1996).

Interaction between AMs and epithelial cells may promote inflammatory responses to air pollution particles. Lung epithelial Type II pneumocytes are well known to grow and divide to epithelialize the alveolar surface through differentiation. Co-culture of AMs and type II rat lung epithelial cells synergistically enhanced basal levels of pro-inflammatory cytokine tumor necrosis factor-α (TNF-α) (Tao and Kobzik 2002). The extent of dust-induced lung injury may also be related to the amount of time a particle is present in the interstitial compartment (Brody 1986; Davis 1986). Particles on the alveolar surface, which are not cleared by AMs, can be transported rapidly across the epithelium to the interstitium (Brody et al. 1982; Brody and Overby 1989). Type II cells express proteases, which are likely playing a role in clearing fibrin deposits from the alveolar space in a number of forms of acute and chronic lung injury, including the inflammatory phases of hypersensitivity pneumonitis, toxin-, and asbestos-mediated injury, idiopathic pulmonary fibrosis, and acute lung injury (Sitrin et al. 1987; Lee et al. 1994; Leikauf et al. 2002). It has been suggested that, although inflammatory events immediately following lung injury initiate the fibrotic process, the response of the pulmonary epithelium can be a major determinant of the pattern of tissue repair, and thus, of the ultimate degree of a permanent lung damage (Crouch 1990). There is strong evidence that fibroblasts are the predominant source of fibrous proteins of the extracellular matrix accumulated in the lung interstitium during the development of fibrosis (Olman 2003). The interaction of AMs and fibroblasts is important in contributing to the accumulation of collagen in affected lungs (Crouch 1990; Castranova 2004).

Because cytokines can generate ROS or deplete antioxidants such as glutathione, activation of cytokines is an important mechanism for CWP development (Jardine et al. 2002; Zhang and Huang 2003). IL-6, IL-8, granulocyte-macrophage-colony stimulating factor (GM-CSF), TNF-α, and transforming growth factor (TGF) are the most important cytokines in inflammation, emphysema, and fibrosis (Kelley 1990; Piguet et al. 1990; Borm and Schins 2001; Yucesoy et al. 2002; Kelly et al. 2003; Morris et al. 2003). Cytokines/growth factors, which are responsible for the development of inflammation and fibrosis can be divided into two groups, i.e., mitogenic (those that increase cell proliferation) and fibrogenic (those that enhance extracellular matrix synthesis). Examples of mitogenic cytokines are TGF-α, IL-1, TNF-α, and insulin-like growth factor, while TGF-β is a fibrogenic cytokine and IL-6 can be both mitogenic and fibrogenic (Hirano 1998; Pittet et al. 2001). It has been shown that soluble IL-1β bioactivity and IL-1β-dependent IL-6 up-regulation are critical mediators of fibroblast activation and proliferation in acute lung injury (Olman et al. 2004). A single instillation of silica in mice leads to a marked increase in the level of lung TNF-α production. Most interestingly, silica-induced lung fibrosis is almost completely prevented by anti-TNF-α antibodies (Piguet et al. 1990). It has been shown that serum levels of TNF receptors and IL-6 may be associated with the fibrotic process occurring in CWP, while serum cytokine levels may be correlated with the severity of CWP (Zhai et al. 2002). It was found that a spontaneous release of TNF-α by AMs was significantly higher in active coal miners than in retired miners (Lassalle et al. 1990). Studies on the release of monocyte TNF-α after *in vitro* stimulation with coal dusts have shown that TNF-α was increased in miners, especially in the early stages of pneumoconiosis. Miners who showed an abnormally high dust-stimulated release of TNF-α had an increased risk of progression to fibrosis (Schins and Borm 1995; Kim et al. 1999).

Fibronectins and collagens are major constituents of extracellular matrix, which bind to other matrix components to promote adhesion, spreading, and migration of various cell types. Airway wall remodeling on an inflammatory basis is believed to be fundamental to the development of COPD in workers exposed to particles (Churg and Wright 2002). AMs from miners with simple pneumoconiosis have been shown to spontaneously release increased amounts of fibronectin (Rom et al. 1987; Vallyathan et al. 2000). Antioxidant enzymes, such as catalase, glutathione peroxidase, and superoxide dismutase, showed a significant increase above control, respectively, in coal miners with category 2/2 CWP. Significant increases in the secretion of IL-1, IL-6, TNF-α, TGF-β, fibronectin, and α_1-AT also were evident in coal miners with disease. This up-regulation of antioxidant defenses and cytokines was not evident in coal miners in the absence of clinically evident radiographic disease (Vallyathan et al. 2000).

Oxidant-signaling in cytokine formation and CWP

AP-1, NF-κB, and nuclear factor of activated T cells (NFAT) are important transcription factors sensitive to oxidative stress and have been shown to play key roles in gene expression of many inflammatory cytokines (Rahman 2002; Castranova 2004). It has been shown that coal from the Pennsylvania coalmine region with a high prevalence of CWP transactivates both AP-1 and NFAT (Huang et al. 2002). In contrast, coal from the Utah coalmine region, which has a low prevalence of CWP, has no such effects. The Pennsylvania coals stimulate the mitogen-activated protein kinase (MAPK) family members of extracellular signal-regulated kinases (ERKs) and p38 MAPK but not c-Jun-NH$_2$-terminal kinase (JNKs), as determined by the phosphorylation assay (Fig. 4).

Increasing evidence demonstrates that IL-6 plays a central role in the acute phase reaction and tissue inflammation, and therefore, may contribute to the initiation and progression of

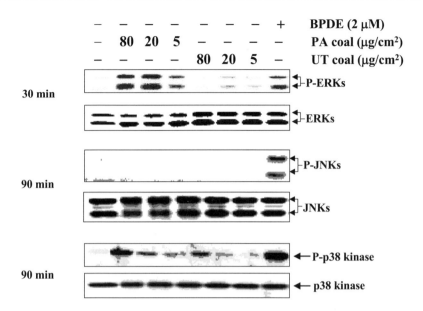

Figure 4. Effects of the Pennsylvania and Utah coals on the phosphorylation of ERKs, JNKs, and p38 MAPK. Cells were seeded into each well of 96-well plates. After being cultured at 37 °C overnight, the cells were starved for 12 h by replacing medium with 0.1% FBS MEM. Then, the cells were treated with the Pennsylvania or Utah coals for different periods of time as indicated, or various amounts of coals. After lysis, phosphorylated and non-phosphorylated ERKs, JNKs, and p38 kinase proteins were assayed using the corresponding specific antibodies. The phosphorylated and non-phosphorylated proteins were analyzed using the same transferred membrane blots. BPDE was used as a positive control for JNKs. [Reproduced with permission of the American Thoracic Society from Huang et al. (2002), *American Journal of Respiratory Cell and Molecular Biology*, Vol. 27, p. 571, Fig. 3.]

CWP (Vallyathan et al. 2000; Ishihara and Hirano 2002). It is known that human IL-6 gene promoter and enhancer region contains several binding sites of *cis*-activating transcription factors such as NF-kB, AP-1, NF-IL-6, and CREB (cAMP-responsive element binding protein) (Hirano 1998; Mann et al. 2002). It has been shown that levels of IL-6 in cells treated with coals from three coalmine regions correlated well with CWP prevalence from that region (Huang and Zhang 2003). It has been further shown that the increase in IL-6 protein and mRNA by the Pennsylvania coal was completely eliminated by the pretreatment of cells with PD98059, a specific ERKs pathway inhibitor, and SB202190, a p38 MAPK inhibitor. Considering the observed effects of Pennsylvania coals on lipid peroxidation, ferritin induction, and its prevention by deferoxamine, a specific iron chelator (Zhang and Huang 2002, 2003), it is reasonable to conclude that BAI in the Pennsylvania coals is most likely the active compound that activates AP-1 and induces IL-6. These results suggest that activation of AP-1 by the iron-containing Pennsylvania coals may contribute to the high prevalence of CWP and COPD in workers from that coalmine region.

In summary, interaction of pyrite and calcite, two important minerals in the coals, may contribute to the CWP development though oxidative stress mechanisms. Pyrite is present in almost every coal while calcite exists in abundance only in certain coalmine regions. In the absence of calcite, oxidation of pyrite results in the formation of BAI, which can lead to the ROS formation in the lung. Calcite can play a protective role in coal dust-induced lung injury by: 1) increasing pH resulting in rapid oxidation of BAI prior to coal workers' inhalation; and 2) inhibiting acid solubilization of iron compounds and preventing iron from becoming bioavailable. Lung injury caused by exposure to coal dust may be due to direct production of ROS by BAI in the coal dust and/or indirect stimulation of phagocytic cells (e.g., AMs, PMNs) by BAI to release ROS (Fig. 3). The formed ROS can directly cause lung damage and stimulate AMs and Type II cells through signaling pathways to secrete inflammatory cytokines, mediators, and growth factors, which induce fibroblast proliferation and enhance collagen synthesis. Therefore, the differences in levels of BAI may contribute to the observed regional differences in the prevalence of CWP among various coalmine regions.

ACKNOWLEDGMENT

This research was supported in part by a grant from NIOSH OH03561 and NIH grants ES010344, ES000260, and CA016087.

REFERENCES

Akazaki K, Inagaki Y (1959) On the experimental anthracosis, anthracosilicosis and the relationship of these to tuberculosis in complication. Tohoku J Exp Med 71:195-207

Albrecht C, Borm PJ, Adolf B, Timblin CR, Mossman BT (2002) *In vitro* and *in vivo* activation of extracellular signal-regulated kinases by coal dusts and quartz silica. Toxicol Appl Pharmacol 184:37-45

Amoudru C (1987) Les pneumoconioses - dimension actuelle des problemes. Colloque INSERM 155:3-40

Anonymous (1996) Health effects of outdoor air pollution. Committee of the Environmental and Occupational Health Assembly of the American Thoracic Society. Am J Respir Crit Care Med 153:3-50

Anonymous (1999) The Science of Acid Mine Drainage and Passive Treatment. Pennsylvania Department of Environmental Protection Report. *http://www.dep.state.pa.us/dep/deputate/minres/bamr/amd/science_of_amd.htm*

Antao VC, Petsonk EL, Sokolow LZ, Wolfe AL, Pinheiro GA, Hale JM, Attfield MD (2005) Rapidly progressive coal workers' pneumoconiosis in the United States: geographic clustering and other factors. Occup Environ Med 62:670-674

Antonini JM, Lewis AB, Roberts JR, Whaley DA (2003) Pulmonary effects of welding fumes: review of worker and experimental animal studies. Am J Ind Med 43:350-360

Artemov AV, Reznik LA (1980) Questions on the methodology of biological investigation of fibrogenicity of coal dust. Gig Truda Prof Zabol 2:51-55

Attfield M, Reger R, Glenn R (1984) The incidence and progression of pneumoconiosis over nine years in U.S. coal miners: II. Relationship with dust exposure and other potential causative factors. Am J Ind Med 6:417-425

Attfield M, Wagner G (1993) Respiratory disease in coal miners. *In:* Environmental and Occupational Medicine. Rom WN (ed) Little, Brown, and Company, p 325-344

Attfield MD, Althouse RB (1992) Surveillance data on US coal miners' pneumoconiosis, 1970 to 1986. Am J Public Health 82:971-977

Attfield MD, Castellan RM (1992) Epidemiological data on US coal miners' pneumoconiosis, 1960 to 1988. Am J Public Health 82:964-970

Attfield MD, Hodous TK (1992) Pulmonary function of US coal miners related to dust exposure estimates. Am Rev Respir Dis 145:605-609

Attfield MD, Morring K (1992a) The derivation of estimated dust exposures for US coal miners working before 1970. Am Ind Hyg Assoc J 53:248-255

Attfield MD, Morring K (1992b) An investigation into the relationship between coal workers' pneumoconiosis and dust exposure in U.S. coal miners. Am Ind Hyg Assoc J 53:486-492

Attfield MD, Seixas NS (1995) Prevalence of pneumoconiosis and its relationship to dust exposure in a cohort of U.S. bituminous coal miners and ex-miners. Am J Ind Med 27:137-151

Attfield MD, Wood JM, Antao VC, Pinheiro GA (2004) Changing patterns of pneumoconiosis mortality--United States, 1968-2000. MMWR Morb Mortal Wkly Rep 53:627-632

Aziz HA, Yusoff MS, Adlan MN, Adnan NH, Alias S (2004) Physico-chemical removal of iron from semi-aerobic landfill leachate by limestone filter. Waste Manag 24:353-358

Balaan MR, Weber SL, Banks DE (1993) Clinical aspects of coal workers' pneumoconiosis and silicosis. Occup Med 8:19-34

Bauer TT, Heyer CM, Duchna HW, Andreas K, Weber A, Schmidt EW, Ammenwerth W, Schultze-Werninghaus G (2006) Radiological Findings, Pulmonary Function and Dyspnea in Underground Coal Miners. Respiration, doi: 10.1159/000090200 (*in press*)

Baur X, Latza U (2005) Non-malignant occupational respiratory diseases in Germany in comparison with those of other countries. Int Arch Occup Environ Health 78:593-602

Beletskii A, Cooper M, Sriraman P, Chiriac C, Zhao L, Abbot S, Yu L (2005) High-throughput phagocytosis assay utilizing a pH-sensitive fluorescent dye. Biotechniques 39:894-897

Borm PJ (1997) Toxicity and occupational health hazards of coal fly ash (CFA). A review of data and comparison to coal mine dust. Ann Occup Hyg 41:659-676

Borm PJ, Meijers JM, Swaen GM (1990) Molecular epidemiology of coal worker's pneumoconiosis: application to risk assessment of oxidant and monokine generation by mineral dusts. Exp Lung Res 16:57-71

Borm PJ, Schins RP (2001) Genotype and phenotype in susceptibility to coal workers' pneumoconiosis. the use of cytokines in perspective. Eur Respir J Suppl 32:127s-133s

Borm PJ, Tran L (2002) From quartz hazard to quartz risk: the coal mines revisited. Ann Occup Hyg 46:25-32

Brody AR (1986) Pulmonary cell interactions with asbestos fibers in vivo and in vitro. Chest 89:155S-159S.

Brody AR, Overby LH (1989) Incorporation of tritiated thymidine by epithelial and interstitial cells in bronchiolar-alveolar regions of asbestos-exposed rats. Am J Pathol 134:133-140

Brody AR, Roe MW, Evans JN, Davis GS (1982) Deposition and translocation of inhaled silica in rats. Quantification of particle distribution, macrophage participation, and function. Lab Invest 47:533-542.

Brunekreef B, Forsberg B (2005) Epidemiological evidence of effects of coarse airborne particles on health. Eur Respir J 26:309-318

Carlberg JR, Crable JV, Limtiaca LP, Norris HB, Holtz JL, Mauer P, Wolowicz FR (1971) Total dust, coal, free silica, and trace metal concentrations in bituminous coal miners' lungs. Am Ind Hyg Assoc J 32:432-440

Carmichael GR, Streets DG, Calori G, Amann M, Jacobson MZ, Hansen J, Ueda H (2002) Changing trends in sulfur emissions in Asia: implications for acid deposition, air pollution, and climate. Environ Sci Technol 36:4707-4713

Castranova V (2004) Signaling pathways controlling the production of inflammatory mediators in response to crystalline silica exposure: role of reactive oxygen/nitrogen species. Free Radic Biol Med 37:916-925

Castranova V, Bowman L, Reasor MJ, Lewis T, Tucker J, Miles PR (1985) The response of rat alveolar macrophages to chronic inhalation of coal dust and/or diesel exhaust. Environ Res 36:405-419.

Castranova V, Vallyathan V (2000) Silicosis and coal workers' pneumoconiosis. Environ Health Perspect 108 Suppl 4:675-684

Chatterjee S, Fisher AB (2004) ROS to the rescue. Am J Physiol Lung Cell Mol Physiol 287:L704-705

Chen LC, Lam HF, Kim EJ, Guty J, Amdur MO (1990) Pulmonary effects of ultrafine coal fly ash inhaled by guinea pigs. J Toxicol Environ Health 29:169-184

Chong S, Lee KS, Chung MJ, Han J, Kwon OJ, Kim TS (2006) Pneumoconiosis: comparison of imaging and pathologic findings. Radiographics 26:59-77

Christian RT, Nelson J (1978) Coal: response of cultured mammalian cells corresponds to the prevalence of coal workers pneumoconiosis. Environ Res 15:232-241.

Christian RT, Nelson JB, Cody TE, Larson E, Bingham E (1979) Coal workers' pneumoconiosis: in vitro study of the chemical composition and particle size as causes of the toxic effects of coal. Environ Res 20:358-365

Churg A, Wright JL (2002) Airway wall remodeling induced by occupational mineral dusts and air pollutant particles. Chest 122:306S-309S

Clark LB (1995) Coal Mining and Water Quality. International Energy Agency (IEA) Coal Research, IEACR/80

Clarke RW, Antonini JM, Hemenway DR, Frank R, Kleeberger SR, Jakab GJ (2000) Inhaled particle-bound sulfate: effects on pulmonary inflammatory responses and alveolar macrophage function. Inhal Toxicol 12:169-186

Coggon D, Inskip H, Winter P, Pannett B (1995) Contrasting geographical distribution of mortality from pneumoconiosis and chronic bronchitis and emphysema in British coal miners. Occup Environ Med 52: 554-555

Costa DL, Dreher KL (1997) Bioavailable transition metals in particulate matter mediate cardiopulmonary injury in healthy and compromised animal models. Environ Health Perspect 105 Suppl 5:1053-1060

Cravotta CA, 3rd (2003) Size and performance of anoxic limestone drains to neutralize acidic mine drainage. J Environ Qual 32:1277-1289

Crouch E (1990) Pathobiology of pulmonary fibrosis. Am J Physiol 259:L159-184

Dalal NS, Newman J, Pack D, Leonard S, Vallyathan V (1995) Hydroxyl radical generation by coal mine dust: possible implication to coal workers' pneumoconiosis (CWP). Free Radic Biol Med 18:11-20

Dalal NS, Shi XL, Vallyathan V (1990) Role of free radicals in the mechanisms of hemolysis and lipid peroxidation by silica: comparative ESR and cytotoxicity studies. J Toxicol Environ Health 29:307-316

Dalal NS, Suryan MM, Vallyathan V, Green FH, Jafari B, Wheeler R (1989) Detection of reactive free radicals in fresh coal mine dust and their implication for pulmonary injury. Ann Occup Hyg 33:79-84

Davis GS (1986) Pathogenesis of silicosis: current concepts and hypotheses. Lung 164:139-154

Delfino RJ, Sioutas C, Malik S (2005) Potential role of ultrafine particles in associations between airborne particle mass and cardiovascular health. Environ Health Perspect 113:934-946

Duber S, Wiekowski AB (1982) E. P. R. study of molecular phases in coal. Fuel 61:433-436

Ekmen I, Der VK, Sarkus TA (2003) LIFAC Sorbent Injection Desulfurization Demonstration Project. U.S. Department of Energy Project Fact Sheet 3-20–3-23

Ernst H, Rittinghausen S, Bartsch W, Creutzenberg O, Dasenbrock C, Gorlitz BD, Hecht M, Kairies U, Muhle H, Muller M, Heinrich U, Pott F (2002) Pulmonary inflammation in rats after intratracheal instillation of quartz, amorphous SiO_2, carbon black, and coal dust and the influence of poly-2-vinylpyridine-N-oxide (PVNO). Exp Toxicol Pathol 54:109-126

Finkelman RB (1995) Modes of occurrence of environmentally-sensitive trace elements in coal. *In:* Environmental Aspects of Trace Elements. Swaine DJ, Goodarzi F (eds) Kluwer Academic Publishers, p 24-50

Finkelman RB (1999) Trace elements in coal: environmental and health significance. Biol Trace Elem Res 67: 197-204.

Finkelman RB, Gross PMK (1999) The types of data needed for assessing the environmental and human health impacts of coal. Int J Coal Geol 40:91-101

Fontenot AP, Maier LA (2005) Genetic susceptibility and immune-mediated destruction in beryllium-induced disease. Trends Immunol 26:543-549

Frey RS, Malik AB (2004) Oxidant signaling in lung cells. Am J Physiol Lung Cell Mol Physiol 286:L1-3

Fubini B, Fenoglio I, Elias Z, Poirot O (2001) Variability of biological responses to silicas: effect of origin, crystallinity, and state of surface on generation of reactive oxygen species and morphological transformation of mammalian cells. J Environ Pathol Toxicol Oncol 20 Suppl 1:95-108

Funahashi A, Siegesmund KA, Dragen RF, Pintar K (1977) Energy dispersive x-ray analysis in the study of pneumoconiosis. Br J Ind Med 34:95-101

Gautrin D, Auburtin G, Alluin F, Brice FM, Chouraki B, Francois P, Marquet M, Poure C, Senecot B, Szmacinski R, et al. (1994) Recognition and progression of coal workers' pneumoconiosis in the collieries of northern France. Exp Lung Res 20:395-410

Ghio AJ, Quigley DR (1994) Complexation of iron by humic-like substances in lung tissue: role in coal workers' pneumoconiosis. Am J Physiol 267:L173-179.

Godden DJ, Kerr KM, Watt SJ, Legge JS (1991) Iron lung: bronchoscopic and pathological consequences of aspiration of ferrous sulphate. Thorax 46:142-143

Goodwin S, Attfield M (1998) Temporal trends in coal workers' pneumoconiosis prevalence. Validating the National Coal Study results. J Occup Environ Med 40:1065-1071

Gordon T (2004) Metalworking fluid—the toxicity of a complex mixture. J Toxicol Environ Health A 67:209-219

Gormley IP, Collings P, Davis JM, Ottery J (1979) An investigation into the cytotoxicity of respirable dusts from British collieries. Br J Exp Pathol 60:523-536

Gross P (1962) Pneumoconiosis. The dilemma of definition and classification. Arch Environ Health 5:269-270

Guest L (1978) The endogenous iron content, by Mossbauer spectroscopy, of human lungs--II. Lungs from various occupational groups. Ann Occup Hyg 21:151-157.

Gurney JW (1993) Pneumoconiosis assistant: a hypermedia-based classification of the pneumoconioses. J Thorac Imaging 8:143-151

Hamada K, Goldsmith CA, Suzaki Y, Goldman A, Kobzik L (2002) Airway hyperresponsiveness caused by aerosol exposure to residual oil fly ash leachate in mice. J Toxicol Environ Health A 65:1351-1365

Harrison PM, Arosio P (1996) The ferritins: molecular properties, iron storage function and cellular regulation. Biochim Biophys Acta 1275:161-203

Hirano T (1998) Interleukin-6. *In:* The Cytokine Handbook, 3rd edition. Thomson A (ed) Academic Press, p 197-228

Honma K, Abraham JL, Chiyotani K, De Vuyst P, Dumortier P, Gibbs AR, Green FH, Hosoda Y, Iwai K, Williams WJ, Kohyama N, Ostiguy G, Roggli VL, Shida H, Taguchi O, Vallyathan V (2004) Proposed criteria for mixed-dust pneumoconiosis: definition, descriptions, and guidelines for pathologic diagnosis and clinical correlation. Hum Pathol 35:1515-1523

Hossner LR, Doolittle JJ (2003) Iron sulfide oxidation as influenced by calcium carbonate application. J Environ Qual 32:773-780

Hu W, Zhang Q, Su WC, Feng Z, Rom W, Chen LC, Tang M, Huang X (2003) Gene expression of primary human bronchial epithelial cells in response to coal dusts with different prevalence of coal workers' pneumoconiosis. J Toxicol Environ Health A 66:1249-1265

Huang C, Li J, Zhang Q, Huang X (2002) Role of bioavailable iron in coal dust-induced activation of activator protein-1 and nuclear factor of activated T cells: difference between Pennsylvania and Utah coal dusts. Am J Respir Cell Mol Biol 27:568-574

Huang X (2003) Iron overload and its association with cancer risk in humans: evidence for iron as a carcinogenic metal. Mutat Res 533:153-171

Huang X, Fournier J, Koenig K, Chen LC (1998) Buffering capacity of coal and its acid-soluble Fe^{2+} content: possible role in coal workers' pneumoconiosis. Chem Res Toxicol 11:722-729

Huang X, Laurent PA, Zalma R, Pezerat H (1993) Inactivation of alpha 1-antitrypsin by aqueous coal solutions: possible relation to the emphysema of coal workers. Chem Res Toxicol 6:452-458

Huang X, Li W, Attfield MD, Nadas A, Frenkel K, Finkelman RB (2005) Mapping and prediction of coal workers' pneumoconiosis with bioavailable iron content in the bituminous coals. Environ Health Perspect 113:964-968

Huang X, Zalma R, Pezerat H (1994) Factors that influence the formation and stability of hydrated ferrous sulfate in coal dusts. Possible relation to the emphysema of coal miners. Chem Res Toxicol 7:451-457

Huang X, Zalma R, Pezerat H (1999) Chemical reactivity of the carbon-centered free radicals and ferrous iron in coals: role of bioavailable Fe^{2+} in coal workers pneumoconiosis. Free Radical Res 30:439-451

Huang X, Zhang Q (2003) Coal-induced interleukin-6 gene expression is mediated through ERKs and p38 MAPK pathways. Toxicol Appl Pharmacol 191:40-47

Huggins FE, Huffman GP, Lin MC (1983) Observations on low-temperature oxidation of minerals in bituminous coals. Int J Coal Geol 3:157-182

Hurley JF, Burns J, Copland L, Dodgson J, Jacobsen M (1982) Coalworkers' simple pneumoconiosis and exposure to dust at 10 British coalmines. Br J Ind Med 39:120-127

Ishihara K, Hirano T (2002) IL-6 in autoimmune disease and chronic inflammatory proliferative disease. Cytokine Growth Factor Rev 13:357

Jacobsen M, Burns J, Attfield MD (1977) Smoking and coal workers' pneumoconiosis. *In:* Inhaled Particles IV. Walton WH (ed) Pergamon Press, p 750-771

Jacobson MZ, Masters GM (2001) Energy. Exploiting wind versus coal. Science 293:1438

Janoff A (1985) Elastases and emphysema. Current assessment of the protease-antiprotease hypothesis. Am Rev Respir Dis 132:417-433

Jardine H, MacNee W, Donaldson K, Rahman I (2002) Molecular mechanism of transforming growth factor (TGF)-beta1-induced glutathione depletion in alveolar epithelial cells. Involvement of AP-1/ARE and Fra-1. J Biol Chem 277:21158-21166

Kamp DW, Graceffa P, Pryor WA, Weitzman SA (1992) The role of free radicals in asbestos-induced diseases. Free Radic Biol Med 12:293-315

Kappos AD, Bruckmann P, Eikmann T, Englert N, Heinrich U, Hoppe P, Koch E, Krause GH, Kreyling WG, Rauchfuss K, Rombout P, Schulz-Klemp V, Thiel WR, Wichmann HE (2004) Health effects of particles in ambient air. Int J Hyg Environ Health 207:399-407

Kelleher P, Pacheco K, Newman LS (2000) Inorganic dust pneumonias: the metal-related parenchymal disorders. Environ Health Perspect 108 Suppl 4:685-696

Kelley J (1990) Cytokines of the lung. Am Rev Respir Dis 141:765-788

Kelly M, Kolb M, Bonniaud P, Gauldie J (2003) Re-evaluation of fibrogenic cytokines in lung fibrosis. Curr Pharm Des 9:39-49

Kim KA, Lim Y, Kim JH, Kim EK, Chang HS, Park YM, Ahn BY (1999) Potential biomarker of coal workers' pneumoconiosis. Toxicol Lett 108:297-302

Kreyling WG, Nyberg K, Nolibe D, Collier CG, Camner P, Heilmann P, Lirsac N, Lundborg M, Matejkova E (1991) Interspecies comparison of phagolysosomal pH in alveolar macrophages. Inhalation Toxicol 3: 91-100

Kuempel ED, Attfield MD, Vallyathan V, Lapp NL, Hale JM, Smith RJ, Castranova V (2003) Pulmonary inflammation and crystalline silica in respirable coal mine dust: dose-response. J Biosci 28:61-69

Kuempel ED, Stayner LT, Attfield MD, Buncher CR (1995) Exposure-response analysis of mortality among coal miners in the United States. Am J Ind Med 28:167-184

Lacasana M, Esplugues A, Ballester F (2005) Exposure to ambient air pollution and prenatal and early childhood health effects. Eur J Epidemiol 20:183-199

Land SC, Wilson SM (2005) Redox regulation of lung development and perinatal lung epithelial function. Antioxid Redox Signal 7:92-107

Langen RC, Korn SH, Wouters EF (2003) ROS in the local and systemic pathogenesis of COPD. Free Radic Biol Med 35:226-235

Lassalle P, Gosset P, Aerts C, Fournier E, Lafitte JJ, Degreef JM, Wallaert B, Tonnel AB, Voisin C (1990) Abnormal secretion of interleukin-1 and tumor necrosis factor alpha by alveolar macrophages in coal worker's pneumoconiosis: comparison between simple pneumoconiosis and progressive massive fibrosis. Exp Lung Res 16:73-80

Lee YC, Hogg R, Rannels DE (1994) Extracellular matrix synthesis by coal dust-exposed type II epithelial cells. Am J Physiol 267:L365-374

Leigh J, Driscoll TR, Cole BD, Beck RW, Hull BP, Yang J (1994) Quantitative relation between emphysema and lung mineral content in coalworkers. Occup Environ Med 51:400-407

Leigh J, Outhred KG, McKenzie HI, Glick M, Wiles AN (1983) Quantified pathology of emphysema, pneumoconiosis, and chronic bronchitis in coal workers. Br J Ind Med 40:258-263

Leikauf GD, McDowell SA, Wesselkamper SC, Hardie WD, Leikauf JE, Korfhagen TR, Prows DR (2002) Acute lung injury: functional genomics and genetic susceptibility. Chest 121:70S-75S

Levine DG, Schlosberg RH, Silbernagel BG (1982) Understanding the chemistry and physics of coal structure (A Review). Proc Natl Acad Sci USA 79:3365-3370

Lister WB, Wimborne D (1972) Carbon pneumoconiosis in a synthetic graphite worker. Br J Ind Med 29:108-110

Lowson RT (1982) Aqueous oxidation of pyrite by molecule of oxygen. Chem Rev 82:461-497

Lundborg M, Eklund A, Lind B, Camner P (1985) Dissolution of metals by human and rabbit alveolar macrophages. Br J Ind Med 42:642-645

Mann J, Oakley F, Johnson PW, Mann DA (2002) CD40 induces interleukin-6 gene transcription in dendritic cells: regulation by TRAF2, AP-1, NF-kappa B, AND CBF1. J Biol Chem 277:17125-17138

Mastin JP, Stettler LE, Shelburne JD (1988) Quantitative analysis of particulate burden in lung tissue. Scanning Microsc 2:1613-1629

Meijers JM, Swaen GM, Slangen JJ (1997) Mortality of Dutch coal miners in relation to pneumoconiosis, chronic obstructive pulmonary disease, and lung function. Occup Environ Med 54:708-713

Meneghini R (1997) Iron homeostasis, oxidative stress, and DNA damage. Free Radical Biol Med 23:783-792

Meyer JD, Holt DL, Chen Y, Cherry NM, McDonald JC (2001) SWORD '99: surveillance of work-related and occupational respiratory disease in the UK. Occup Med (Lond) 51:204-208

Meyers RA (1982) Coal Structure. Academic Press

Miller AA, Ramsden FR (1961) Carbon pneumoconiosis. Br J Ind Med 18:103-113

Miller BG, Jacobsen M (1985) Dust exposure, pneumoconiosis, and mortality of coalminers. Br J Ind Med 42:723-733

Mizuki M, Onizuka O, Aoki T, Tsuda T (1989) A case of remarkable bronchial stenosis due to aspiration of delayed-release iron tablet. Nihon Kyobu Shikkan Gakkai Zasshi 27:234-239

Moore MP, Bise CJ (1984) The relationship between the Hardgrove grindability index and the potential for respirable dust generation. *In:* Proceedings of the Coal Mine Dust Conference, Generic Mineral Technology Center for Respirable Dust, October 8-10. Morgantown, West Virginia, p 250-255

Morgan WK (1971) Coal worker's pneumoconiosis. Am Ind Hyg Assoc J 32:29-34

Morgan WKC, Burgess DB, Jacobson G, O'Brien RJ, Pendergrass E, Reger RB, Shoub EP (1973) The prevalence of coal workers' pneumoconiosis in US coal miners. Arch Environ Health 27:221-230

Morris CJ, Earl JR, Trenam CW, Blake DR (1995) Reactive oxygen species and iron--a dangerous partnership in inflammation. Int J Biochem Cell Biol 27:109-122

Morris DG, Huang X, Kaminski N, Wang Y, Shapiro SD, Dolganov G, Glick A, Sheppard D (2003) Loss of integrin alpha(v)beta6-mediated TGF-beta activation causes Mmp12-dependent emphysema. Nature 422:169-173

Murthy BS (1952) Silicosis, anthracosis and predisposition to pulmonary tuberculosis in cement industry. Antiseptic 49:297-299

Naidoo RN, Robins TG, Solomon A, White N, Franzblau A (2004) Radiographic outcomes among South African coal miners. Int Arch Occup Environ Health 77:471-481

NIOSH (2003) Work-related Lung Disease Surveillance Report 2002. US Government Printing Office

Olman MA (2003) Epithelial cell modulation of airway fibrosis in asthma. Am J Respir Cell Mol Biol 28:125-128

Olman MA, White KE, Ware LB, Simmons WL, Benveniste EN, Zhu S, Pugin J, Matthay MA (2004) Pulmonary edema fluid from patients with early lung injury stimulates fibroblast proliferation through IL-1 beta-induced IL-6 expression. J Immunol 172:2668-2677

Page SJ (2000) Relationship between electrostatic charging characteristics, moisture content, and airborne dust generation for subbituminous and bituminous coals. Aerosol Sci Technol 32:249-267

Page SJ, Organiscak JA (2000) Suggestion of a cause-and-effect relationship among coal rank, airborne dust, and incidence of workers' pneumoconiosis. Aihaj 61:785-787

Perrin-Nadif R, Auburtin G, Dusch M, Porcher JM, Mur JM (1996) Blood antioxidant enzymes as markers of exposure or effect in coal miners. Occup Environ Med 53:41-45

Piguet PF, Collart MA, Grau GE, Sappino AP, Vassalli P (1990) Requirement of tumour necrosis factor for development of silica-induced pulmonary fibrosis. Nature 344:245-247.

Pittet JF, Griffiths MJ, Geiser T, Kaminski N, Dalton SL, Huang X, Brown LA, Gotwals PJ, Koteliansky VE, Matthay MA, Sheppard D (2001) TGF-beta is a critical mediator of acute lung injury. J Clin Invest 107:1537-1544

Pon MRL, Roper RA, Petsonk EL, Wang ML, Castellan RM, Attfield MD, Wagner GR (2003) Pneumoconiosis prevalence among working coal miners examined in federal chest radiograph surveillance programs--United States, 1996-2002. MMWR Morb Mortal Wkly Rep 52:336-340

Pope CA, 3rd (1996) Adverse health effects of air pollutants in a nonsmoking population. Toxicology 111:149-155

Rahman I (2002) Oxidative stress, transcription factors and chromatin remodelling in lung inflammation. Biochem Pharmacol 64:935-942

Razemon P, Ribet M (1961) Surgical treatment of pulmonary tuberculosis associated with anthracosis and silicosis in coal miners. J Thorac Cardiovasc Surg 41:281-290

Reisner MTR, Robock K (1977) Results of Epidemiological, Mineralogical and Cytotoxicological Studies on the Pathogenicity of Coal Mine Dusts. Pergamon Press

Remy-Jardin M, Remy J, Farre I, Marquette CH (1992) Computed tomographic evaluation of silicosis and coal workers' pneumoconiosis. Radiol Clin North Am 30:1155-1176

Retcofsky HL (1982) Magnetic resonance studies of coal. *In:* Coal Science. Dryden IG (ed) Academic Press, p 43-82

Rockette HE (1980) Mortality Patterns of Coal Miners. Ann Arbor Sciences

Rom WN (1990) Basic mechanisms leading to focal emphysema in coal workers' pneumoconiosis. Environ Res 53:16-28

Rom WN (1991) Relationship of inflammatory cell cytokines to disease severity in individuals with occupational inorganic dust exposure. Am J Ind Med 19:15-27

Rom WN, Bitterman PB, Rennard SI, Cantin A, Crystal RG (1987) Characterization of the lower respiratory tract inflammation of nonsmoking individuals with interstitial lung disease associated with chronic inhalation of inorganic dusts. Am Rev Respir Dis 136:1429-1434

Ross MH, Murray J (2004) Occupational respiratory disease in mining. Occup Med (Lond) 54:304-310

Schaumann F, Borm PJ, Herbrich A, Knoch J, Pitz M, Schins RP, Luettig B, Hohlfeld JM, Heinrich J, Krug N (2004) Metal-rich ambient particles (particulate matter 2.5) cause airway inflammation in healthy subjects. Am J Respir Crit Care Med 170:898-903

Schins RP, Borm PJ (1995) Epidemiological evaluation of release of monocyte TNF-alpha as an exposure and effect marker in pneumoconiosis: a five year follow up study of coal workers. Occup Environ Med 52: 441-450.

Schins RP, Borm PJ (1999) Mechanisms and mediators in coal dust induced toxicity: a review. Ann Occup Hyg 43:7-33

Schulz HM (1997) Coal mine workers' pneumoconiosis (CWP): *in vitro* study of the release of organic compounds from coal mine dust in the presence of physiological fluids. Environ Res 74:74-83

Schwela D (2000) Air pollution and health in urban areas. Rev Environ Health 15:13-42

Seal RM, Cockcroft A, Kung I, Wagner JC (1986) Central lymph node changes and progressive massive fibrosis in coalworkers. Thorax 41:531-537

Seaton A, MacNee W, Donaldson K, Godden D (1995) Particulate air pollution and acute health effects. Lancet 345:176-178.

Seixas NS, Checkoway H (1995) Exposure assessment in industry specific retrospective occupational epidemiology studies. Occup Environ Med 52:625-633.

Seixas NS, Moulton LH, Robins TG, Rice CH, Attfield MD, Zellers ET (1991) Estimation of cumulative exposures for the nation study of coal workers' pneumoconiosis. Appl Occup Environ Hyg 6:1032-1041

Seixas NS, Robins TG, Attfield MD, Moulton LH (1992) Exposure-response relationships for coal mine dust and obstructive lung disease following enactment of the Federal Coal Mine Health and Safety Act of 1969. Am J Ind Med 21:715-734

Sichletidis L, Tsiotsios I, Chloros D, Daskalopoulou E, Ziomas I, Michailidis K, Kottakis I, Konstantinidis TH, Palladas P (2004) The effect of environmental pollution on the respiratory system of lignite miners: a diachronic study. Med Lav 95:452-464

Singer P, Stumm W (1969) Acidic mine drainage: the rate-determining step. Science 167:1121-1123

Sitrin RG, Brubaker PG, Fantone JC (1987) Tissue fibrin deposition during acute lung injury in rabbits and its relationship to local expression of procoagulant and fibrinolytic activities. Am Rev Respir Dis 135:930-936

Smith KR, Veranth JM, Hu AA, Lighty JS, Aust AE (2000) Interleukin-8 levels in human lung epithelial cells are increased in response to coal fly ash and vary with the bioavailability of iron, as a function of particle size and source of coal. Chem Res Toxicol 13:118-125

Smith KR, Veranth JM, Lighty JS, Aust AE (1998) Mobilization of iron from coal fly ash was dependent upon the particle size and the source of coal. Chem Res Toxicol 11:1494-1500

Snider GL (1992) Emphysema: the first two centuries--and beyond. A historical overview, with suggestions for future research: Part 1. Am Rev Respir Dis 146:1334-1344

Soutar CA, Hurley JF, Miller BG, Cowie HA, Buchanan D (2004) Dust concentrations and respiratory risks in coalminers: key risk estimates from the British Pneumoconiosis Field Research. Occup Environ Med 61: 477-481

Soutar CA, Maclaren WM, Annis R, Melville AW (1986) Quantitative relations between exposure to respirable coalmine dust and coalworkers' simple pneumoconiosis in men who have worked as miners but have left the coal industry. Br J Ind Med 43:29-36

Srivastava RK, Jozewicz W (2001) Flue gas desulfurization: the state of the art. J Air Waste Manag Assoc 51: 1676-1688

Stuart LM, Ezekowitz RA (2005) Phagocytosis: elegant complexity. Immunity 22:539-550

Stumm W, Lee GF (1961) Oxygenation of ferrous iron. Indust Engineer Chem 53:143-146

Swaen GM, Meijers JM, Slangen JJ (1995) Risk of gastric cancer in pneumoconiotic coal miners and the effect of respiratory impairment. Occup Environ Med 52:606-610

Tao F, Kobzik L (2002) Lung macrophage-epithelial cell interactions amplify particle-mediated cytokine release. Am J Respir Cell Mol Biol 26:499-505

Tarkka M, Anttila S, Sutinen S (1988) Bronchial stenosis after aspiration of an iron tablet. Chest 93:439-441

Torti FM, Torti SV (2002) Regulation of ferritin genes and protein. Blood 99:3505-3516

Ubersfeld J, Etienne A, Combrisson J (1954) Paramagnetic resonance, a new property of coal-like materials. Nature 174:614

Usher CR, Cleveland CA Jr., Strongin DR, Schoonen MA (2004) Origin of oxygen in sulfate during pyrite oxidation with water and dissolved oxygen: an in situ horizontal attenuated total reflectance infrared spectroscopy isotope study. Environ Sci Technol 38:5604-5606

Usher CR, Paul KW, Narayanasamy J, Kubicki JD, Sparks DL, Schoonen MA, Strongin DR (2005) Mechanistic aspects of pyrite oxidation in an oxidizing gaseous environment: an in situ HATR-IR isotope study. Environ Sci Technol 39:7576-7584

Valko M, Morris H, Cronin MT (2005) Metals, toxicity and oxidative stress. Curr Med Chem 12:1161-1208

Vallyathan V, Brower PS, Green FH, Attfield MD (1996) Radiographic and pathologic correlation of coal workers' pneumoconiosis. Am J Respir Crit Care Med 154:741-748

Vallyathan V, Goins M, Lapp LN, Pack D, Leonard S, Shi X, Castranova V (2000) Changes in bronchoalveolar lavage indices associated with radiographic classification in coal miners. Am J Respir Crit Care Med 162: 958-965

Vallyathan V, Shi X (1997) The role of oxygen free radicals in occupational and environmental lung diseases. Environ Health Perspect 105 Suppl 1:165-177

Veranth JM, Smith KR, Hu AA, Lighty JS, Aust AE (2000a) Mobilization of iron from coal fly ash was dependent upon the particle size and source of coal: analysis of rates and mechanisms. Chem Res Toxicol 13:382-389

Veranth JM, Smith KR, Huggins F, Hu AA, Lighty JS, Aust AE (2000b) Mossbauer spectroscopy indicates that iron in an aluminosilicate glass phase is the source of the bioavailable iron from coal fly ash. Chem Res Toxicol 13:161-164

Wagner GR, Attfield MD, Parker JE (1993) Chest radiography in dust-exposed miners: promise and problems, potential and imperfections. Occup Med 8:127-141

Wallaert B, Lassalle P, Fortin F, Aerts C, Bart F, Fournier E, Voisin C (1990) Superoxide anion generation by alveolar inflammatory cells in simple pneumoconiosis and in progressive massive fibrosis of nonsmoking coal workers. Am Rev Respir Dis 141:129-133

Walton WH, Dodgson J, Hadden GG, Jacobsen M (1977) The effect of quartz and other non-coal dusts in coal workers' pneumoconiosis. In: Inhaled Particles IV. Walton WH (ed) Pergamon Press, p 669-689

Yeoh CI, Yang SC (2002) Pulmonary function impairment in pneumoconiotic patients with progressive massive fibrosis. Chang Gung Med J 25:72-80

Younger PL (1997) The longevity of minewater pollution: a basis for decision-making. Sci Total Environ 194-195:457-466

Yucesoy B, Vallyathan V, Landsittel DP, Simeonova P, Luster MI (2002) Cytokine polymorphisms in silicosis and other pneumoconioses. Mol Cell Biochem 234-235:219-224

Zhai R, Liu G, Ge X, Bao W, Wu C, Yang C, Liang D (2002) Serum levels of tumor necrosis factor-alpha (TNF-alpha), interleukin 6 (IL-6), and their soluble receptors in coal workers' pneumoconiosis. Respir Med 96:829-834

Zhang Q, Dai J, Ali A, Chen L, Huang X (2002) Roles of bioavailable iron and calcium in coal dust-induced oxidative stress: possible implications in coal workers' lung disease. Free Radic Res 36:285-294

Zhang Q, Huang X (2002) Induction of ferritin and lipid peroxidation by coal samples with different prevalence of coal workers' pneumoconiosis: role of iron in the coals. Am J Ind Med 42:171-179

Zhang Q, Huang X (2003) Induction of interleukin-6 by coal containing bioavailable iron is through both hydroxyl radical and ferryl species. J Biosci 28:95-100

Zhang Q, Huang X (2005) Addition of calcite reduces iron's bioavailability in the Pennsylvania coals--potential use of calcite for the prevention of coal workers' lung diseases. J Toxicol Environ Health A 68: 1663-1679

Zodrow E, McCandlish K (1978) Hydrated sulfates in the Sydney coalfield, Cape Breton Nova Scotia. Can Mineral:17-22

Reviews in Mineralogy & Geochemistry
Vol. 64, pp. 179-221, 2006
Copyright © Mineralogical Society of America

7

Mineral-Induced Formation of Reactive Oxygen Species

Martin A. A. Schoonen*[1,2,3], Corey A. Cohn[1,2], Elizabeth Roemer[3,4], Richard Laffers[1,2], Sanford R. Simon[3,4,5], Thomas O'Riordan[3,6]

[1]*Department of Geosciences*
[2]*Center for Environmental Molecular Science*
[3]*Minerals, Metals, Metalloid, and Toxicity (3MT) Graduate Training Program*
[4]*Department of Pathology*
[5]*Department of Biochemistry and Cell Biology*
[6]*Department of Medicine, Division of Pulmonary and Critical Care*
Stony Brook University
Stony Brook, New York, 11794, U.S.A.
*e-mail: martin.schoonen@stonybrook.edu

INTRODUCTION

The term reactive oxygen species, ROS, is defined by the US National Library of Medicine (NIH 2006) as:

"Molecules or ions formed by the incomplete one-electron reduction of oxygen. These reactive oxygen intermediates include singlet oxygen; superoxides; peroxides; hydroxyl radical; and hypochlorous acid. They contribute to the microbicidal activity of phagocytes, regulation of signal transduction and gene expression, and the oxidative damage to nucleic acids; proteins; and lipids."

This chapter explores the role of minerals in the formation of reactive oxygen species. Five different mechanisms by which minerals may promote the formation and transformation of ROS species are explored (Fig. 1). These are:

1. *Mineral release of metal ions:* Metals that are released into body fluids via congruent or incongruent mineral dissolution can act as catalysts. In this mechanism minerals are a source of metals, but are not directly involved in any of the reactions.

2. *Surface-bound metal-promoted reactions:* Insoluble metal-containing minerals can catalyze formation of ROS from molecular oxygen. In this mechanism the conversion of molecular oxygen takes place on the mineral surface, with the mineral surface itself, or adsorbed species, acting as an electron donor.

3. *Intrinsic or mechanically-induced surface defects:* Defects on the mineral surface, either intrinsic to the mineral structure or generated by crushing, can react to form ROS. In this mechanism, highly reactive defects combine with water, molecular oxygen, or carbon dioxide to form ROS.

4. *Inflammatory cell/mineral interactions:* Insoluble particles that deposit in the airways and alveoli of the lung may activate airway epithelial cells or macrophages as a result of binding to the cell surface or engulfment mechanisms that bring the particles into the cell interior; these processes can lead to production of secondary cellular ROS, including hydroxyl radicals and other very reactive entities. Hence, minerals that do not promote ROS formation via the first three mechanisms *in vitro* may still show upregulation of ROS *in vivo*.

1529-6466/06/0064-0007$05.00 DOI: 10.2138/rmg.2006.64.7

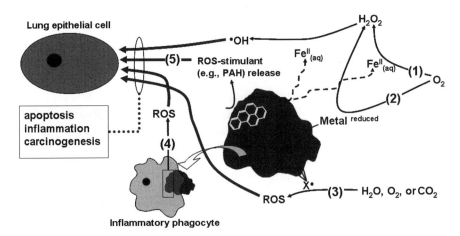

Figure 1. Mineral-induced ROS formation and transformations. Numbers refer to mechanisms listed in text. [Used by permission of Corey Cohn and Martin Schoonen, Stony Brook University, 2006]

Table 1. Abbreviations used in this chapter.

8-oxodG	8-oxodeoxyguanosine
Amplex UltraRed	10-acetyl-3,7-dihydroxyphenoxazine, hydrogen peroxide indicator
APF	3'-(p-aminophenyl) fluorescein, $C_{26}H_{17}NO_5$, fluorogenic spintrap
BMPO	5-*tert*-butoxycarbonyl-5-methyl-1-pyrroline N-oxide, spin trap
DCFH	2',7'-dichlorofluorescin, reactive oxygen species indicator
DFO	Desferrioxamine, ligand
DMPO	5,5-dimethyl-1-pyrroline N-oxide, spintrap
EDTA	ethylenediaminetetraacetic acid, $C_{10}H_{16}N_2O_8$, ligand
EPR	Electron paramagnetic resonance spectroscopy
ESR	Electron spin resonance spectroscopy
HRP	Horseradish Peroxidase
MPO	myeloperoxidase
NADPH	reduced nicotinamide adenine dinucleotide phosphate
NHE	Normal Hydrogen Electrode
PAH	Polycyclic aromatic hydrocarbon
PBN	N-*tert*-Butyl-α-phenylnitrone; $C_{11}H_{15}NO$, spintrap
PF	Proxyl fluorescamine; 5-(2-carboxyphenyl)-5-hydroxy-1-((2,2,5,5-tetramethyl-1-oxypyrrolidin-3-yl)methyl)-3-phenyl-2-pyrrolin-4-one; $C_{26}H_{28}N_2O_5^-$
POBN	N-*tert*-Butyl-α-(4-pyridyl)nitrone N-oxide
RCS	Reactive Chlorine Species
Resorufin	$C_{12}H_6NNaO_3$
RNS	Reactive Nitrogen Species
ROS	Reactive Oxygen Species
SOD	superoxide dismutase
Tempo-9-ac	(4-((9-acridinecarbonyl)amino)-2,2,6,6-tetramethylpiperidin-1-oxyl); $C_{23}H_{26}N_3O_2$, fluorogenic spintrap

5. *Mineral surface-sorbed compounds:* Minerals may act as "carriers" of ROS-inducing compounds (e.g., PAH). Sorption onto or aggregation of non-mineral ROS-inducing compounds with minerals can transform inhaled minerals into carriers. Dissolution of the carrier species into the alveolar fluid may induce formation of ROS. Hence, in this mechanism the mineral acts as a substrate for a ROS stimulant (e.g., PAH). Mineral surface charge, surface functionalities, and hydrophobicity/hydrophilicity are important factors that dictate the interaction between a ROS stimulant and the mineral.

In reality a mineral may promote ROS formation or transformation via a combination of the mechanisms listed above. In this chapter, we use the term "mineral-induced" without implying a specific mechanism. In addition, the term "Earth material" will be used for all natural particles including minerals, glasses, and composite materials such as coal, soil and sediments.

The second part of the definition provided by the US National Library of Medicine provides a powerful motivation for the study of the role of minerals in ROS formation or transformations. While it is well documented that exposure to a range of particles of various composition can lead to ROS formation, the factors that contribute to mineral-induced ROS formation are far from completely understood. ROS and related oxidants are integral components of the normal function of the human body and important in the human immune system, signal transduction and gene expression. However, the concentrations of ROS are tightly regulated in the body by a balance between pro-oxidants and anti-oxidants (Fujimoto et al. 1996; Heiser and Elstner 1998; Babior 2000). ROS can cause cell death by disrupting cell membranes (necrosis) or initiating programmed cell death (apoptosis or "cellular suicide"). In sublethal concentrations, ROS can upregulate production of cytokines and other inflammatory mediators and can promote mutagenesis and carcinogenesis. It should be remembered that ROS generated by inhaled particulate materials can be active either extracellularly or intracellularly. Within the tightly regulated intracellular microenvironment even small and transient increases in ROS levels can markedly alter cell metabolism. Particle characteristics (e.g., metallic content, surface area, particle size and morphology) as well as cellular characteristics (e,g, embryologic origin [epithelial cells (lung-surface cells) vs. phagocytic mesenchymal cells (innate immune system cells)], state of differentiation, and pre-existing levels of activation) combine to produce a broad spectrum of responses.

Chronic inflammation associated with mineral-generated ROS production has been linked to a number of diseases including cardiovascular disease (Brook et al. 2003), lung diseases (i.e., silicosis, asbestosis) (Kamp et al. 1992), and lung cancer (Knaapen et al. 2004). Oxidative stress has been linked to memory loss and learning impairment (Fujimoto et al. 1996). Hence, it is important to understand which minerals can induce formation of primary ROS or transformations to secondary ROS. Furthermore, from a biological standpoint, it is important to study the cellular responses to mineral exposures, including the upregulation of ROS or impairment of antioxidant defense systems.

This chapter is divided into four parts. The first part provides background. It consists of a short description of ROS species, thermodynamic data related to ROS species, a summary of common analytical methods and assays for the determination of ROS in cell-free (acellular) systems and within cellular systems. Part II is a detailed discussion of the five mechanisms listed above. Part III discusses silicosis in detail to illustrate the complexities involved in unraveling the relation between a mineral exposure and a disease. Closing remarks are presented in Part IV.

PART I: BACKGROUND

The objective of this part of the chapter is to briefly provide background information that is of importance in understanding the mechanisms discussed in Part II of the chapter. This section is written with the geochemist in mind. Hence, we have made an attempt to place the

Table 2. List of selected terms used in this chapter.

Term	Explanation
apoptosis	Programmed cell death (PCD) of an unwanted cell in a multicellular organism. The apoptotic cell usually undergoes phagocytosis
chemokine	Chemokines are small, secreted protein signals. Chemokines are chemoattractants for leukocytes, recruiting monocytes and neutrophils from the blood to sites of infection or damage
coal worker's pneumoconiosis	CPW, Coal worker's pneumoconiosis, is a lung condition caused by the inhalation of dust, characterized by formation of fibrotic changes in lungs
cytokine	Cytokines are proteins produced by inflammatory and other cell types that are involved in immune responses
deoxyribonucleic acid	Deoxyribonucleic acid (DNA) is a nucleic acid—usually in the form of a double helix—that contains the genetic instructions specifying the biological development of all cellular forms of life
endocytosis	Internalization of particles or fluid by cells without forming a vesicle around the engulfed material
endogonic reaction	Energetically unfavorable chemical reactions
exogonic reaction	Energetically favorable chemical reactions
granulocyte	A leucocyte that has small granules in the cytoplasm
inflammation	First response of the immune system to infection or irritation
leukocytes	Leukocytes or white blood cells are a component of blood. They defend the body against infectious agents and foreign materials as part of the immune system
macrophage	A type of white blood cell that surrounds and kills microorganisms, removes dead cells, and stimulates the action of other immune system cells
monocyte	A large white blood cell that is formed in the bone marrow, enters the blood, and migrates into the connective tissue where it differentiates into a macrophage
necrosis	Unprogrammed death of cells and results in production of extracellular debris which can in turn be proinflammatory
neutrophil	A granulocyte leucocyte that is the chief phagocytic white blood cell.
phagocytes	An immune system cell that can surround and kill microorganisms and remove dead cells. Phagocytes include macrophages
phagocytosis	Phagocytosis is a form of endocytosis wherein large particles are enveloped by the cell membrane of a cell and internalized to form a phagosome. Phagocytosis is performed by macrophages and granulocytes
point mutation	type of genetic mutation that causes the replacement of a single base nucleotide with another nucleotide
Ribonucleic acid	Ribonucleic acid, (RNA), consists of nucleotide monomers.. It is transcribed from DNA by enzymes called RNA polymerases. RNA serves as the template for translation of genes into proteins, transferring amino acids to the ribosome to form proteins, and also translating the transcript into proteins
Silicosis	Lung disease caused by inhalation of crystalline silica dust, and is marked by inflammation and scarring

various forms of ROS in the context of concepts familiar to most geochemists. The role of ROS in human health is an active area of biomedical research with broad implications. It is far beyond the scope of this chapter to even attempt to summarize the biomedical literature on this topic. Instead the focus here is to provide the necessary background that will allow geochemists to appreciate how minerals may form, transform, or interact with ROS. Literature sources, such as Gilbert and Colton (1999), provide extensive background information on the role of ROS in human physiology and disease.

Formation, function and reactivity of ROS species and related oxidants

In this section, formation, function and chemical reactivity of ROS and related oxidants are briefly discussed. The emphasis in this chapter will be on mineral-induced reactions that lead to formation of primary ROS or to transformation to secondary ROS; however, it is useful to start with a short overview of the formation of ROS and related oxidants within the human body and their function to provide the necessary context. Besides molecular oxygen and its intermediate reduction products (superoxide, hydrogen peroxide, and hydroxyl radical), the oxidants nitric oxide, peroxynitrite and hypochlorite are also introduced. These latter species are often referred to as reactive nitrogen species (RNS) and reactive chlorine species (RCS). The RNS and RCS are included in this overview because there are several important reactions in which ROS are transformed into RNS or RCS or react with RNS or RCS to yield secondary species.

Enzymatic formation of oxidants

White blood cells or leukocytes are the first line of defense against invading pathogens. Specialized leukocytes participate in a robust, nonspecific, defense system known as the innate immune response that deploys enzymatically-generated oxidants to kill invading pathogens. Stimulated either by the presence of an invader or via cell signaling, four enzymes [NADPH oxidase, superoxide dimutase (SOD), nitric oxide synthase (NOS), and myeloperoxidase (MPO)] produce four primary oxidants: $O_2^{\bullet-}$, H_2O_2, $NO^{\bullet}$, HOCl (Babior 2000). These oxidants may be formed in response to internalization of pathogens by phagocytic cells (monocytes, macrophages, and neutrophils) or to binding of pathogens to the cell surface. The primary oxidants can either directly detoxify the pathogens or can react to form other oxidants and radicals that inactivate pathogens. Figure 2 shows some of the key transformation reactions involving the four primary oxidants. A deficiency in any of the four enzymes that produce the primary oxidants leads to weakening of the innate immune system.

It is beyond the scope of this chapter to discuss the biochemistry involved in the regulation of the enzyme activity and the enzymatic mechanisms in any level of detail. The reviews by Babior (2000) and by Heiser and Elstner (1998) provide excellent introductions to this topic. In the remainder of this section the origin and function of each of the primary oxidants are discussed and the roles of molecular oxygen and the hydroxyl radical and peroxynitrite, two of the most reactive secondary oxidants are also considered.

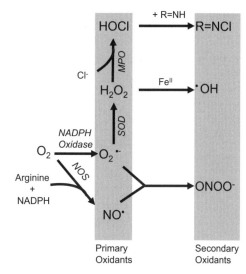

Figure 2. Formation pathways of ROS within human body. Primary oxidants are formed enzymaticaly, secondary oxidants are formed via subsequent reactions.

It should be noted that the generation of primary oxidants leads to free radical chain reactions with organic molecules that will produce a myriad of radicals. For example, as illustrated in Equation (1) below, hydroxyl radical, $^{\bullet}OH$, may react via hydrogen abstraction and create a new radical carbon species.

$$^{\bullet}OH + R\text{-}C\text{-}H \rightarrow H_2O + R\text{-}C^{\bullet} \tag{1}$$

Secondary, carbon-based radicals may be longer lived and act as important chain propogators.

Molecular oxygen. Molecular oxygen, O_2, has two unpaired electrons in its $\pi*$ orbitals. Despite the fact that it is a bi-radical species, molecular oxygen is relatively stable. Molecular oxygen can accept four electrons and the overall reaction (Eqn. 2) leads to water.

$$O_2(aq) + 4H^+ + 4e^- \rightarrow 2H_2O \tag{2}$$

However, the reduction of oxygen, especially when catalyzed by metal-containing enzymes, proceeds via intermediate steps in which electrons are transferred singly in a stepwise fashion.

Provoked by the presence of an invader or stimulated by cellular signaling mediators, phagocytes will convert molecular oxygen into superoxide, see below. This is the first of a series of reactions outlined in Figure 2 that lead to the formation of the four primary oxidants. Hence, as the innate immune system is engaged, molecular oxygen is consumed.

Superoxide. The first step in the reduction of molecular oxygen leads to superoxide, $O_2^{\bullet-}$ (Eqn. 3), a radical.

$$O_2(aq) + e^- \rightarrow O_2^{\bullet-}(aq) \tag{3}$$

The superoxide anion has an unpaired electron and is the conjugate base of the weak acid HO_2 (Eqn. 4).

$$HO_2^{\bullet}(aq) = H^+ + O_2^{\bullet-}(aq) \qquad pK = -4.88 \tag{4}$$

Given the value of the acid dissociation constant for $HO_2^{\bullet}$, it is expected that at pH values in excess of 4.88 $O_2^{\bullet-}$ will be the dominant form. Hence in lung fluid where the pH is buffered at a value of 7.4, less than 1% of the total superoxide concentration will be present as $HO_2^{\bullet}$. Superoxide can be stabilized as a potassium salt KO_2, which is often used as a source of superoxide *in vitro*.

As pointed out by Sawyer and Valentine (1981), the name superoxide may lead one to believe that this is a very reactive species. In fact, superoxide is rather stable and does not possess the reactivity one may associate with a radical. Superoxide shows little reactivity toward organic matter; it is a single-electron reductant of iron (Eqn. 5).

$$Fe^{3+}(aq) + O_2^{\bullet-}(aq) \rightarrow Fe^{2+} + O_2(aq) \tag{5}$$

In the lungs, ferric iron may be complexed with organic ligands, which increases the rate of this rate and facilitate redox cycling of iron within cells and intracellular fluids (Rose and Waite 2005).

Superoxide does play an important role in the immune system. It is produced by the enzyme NADPH oxidase as the first step in the immune system's response against pathogens, Figure 2. The enzyme superoxide dismutase (SOD) facilitates a reaction between two superoxide molecules yielding hydrogen peroxide and molecular oxygen (Eqn. 6):

$$2\,O_2^{\bullet-} + 2H^+ \rightarrow O_2 + H_2O_2 \tag{6}$$

SOD is an extremely efficient biocatalyst, it has the highest turnover number of any known enzyme (Babior 2000). Hydrogen peroxide, one of the products of this reaction, is not a radical, but is nevertheless a crucial component of pathways of ROS generation as it can undergo subsequent reactions leading to several radicals (Fig. 2).

Hydrogen peroxide and peroxide anion. The acid dissociation equilibrium of hydrogen peroxide favors the protonated form under most environmental and physiological pH conditions (Eqn. 7).

$$H_2O_2 = HO_2^- + H^+ \qquad pK = 11.65 \tag{7}$$

H_2O_2 is not a radical like superoxide and it is also rather unreactive toward biomolecules. For example, RNA is stable in dilute hydrogen peroxide solutions (Cohn et al. 2004). Most free radicals of biological relevance react within a few molecular diameters from where they are formed; however, hydrogen peroxide can be expected to be transported over much longer distances before it reacts. For example, most natural waters have nanomolar concentrations of hydrogen peroxide (Cooper and Zika 1983); higher concentrations are found in waters exposed to sunlight (Cooper et al. 1988; Emmenegger 1998; Wilson et al. 2000b). Hydrogen peroxide is injected into groundwater and added to wastewater to promote the degradation of the organic pollutants by forming hydroxyl radicals via reaction with ferrous iron (EPA 1998; Watts et al. 1999a; Kwan and Voelker 2003; Ma et al. 2005; Watts and Teel 2005). The reactivity of hydroxyl radicals is discussed below.

The SOD-catalyzed conversion of superoxide to hydrogen peroxide and molecular oxygen is an especially prominent reaction in mononuclear phagocytes (monocytes and macrophages). Given that hydrogen peroxide is a key reactant that leads to very reactive oxidants, such as hydroxyl radicals and hypochlorous acid (Fig. 3), its intracellular concentration is tightly controlled. The enzyme catalase is used to protect cells from hydrogen peroxide-induced damage by facilitating the conversion of hydrogen peroxide to water and oxygen.

Hydroxyl radical. With a bond dissociation energy of 146 kJ/mole, the O-O bond in hydrogen peroxide is relatively weak compared to other covalent bonds (e.g., C-C 357 kJ/mole, S-S 231 kJ/mole). Homolytic cleavage of this bond leads to the formation of hydroxyl radical, •OH (Eqn. 8).

$$H_2O_2 \rightarrow 2\text{•OH} \tag{8}$$

The hydroxyl radical is one of the most reactive species in nature. It reacts nonspecifically with most organic molecules. Reactions take place within nanoseconds after •OH formation.

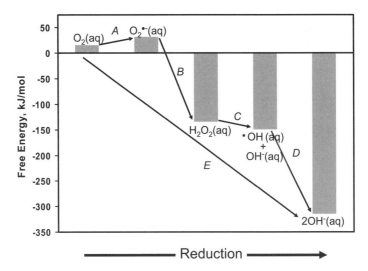

Figure 3. Free Energy diagram for (step-wise) O_2 reduction.
See Table 3 for reactions and thermodynamic data.

Hence, reactions involving 'OH take place within a few molecular diffusion distances from the sites of formation (Pryor 1986).

The homolytic cleavage of hydrogen peroxide requires either thermal activation, exposure to UV, or the presence of a catalyst. At a temperature of 160 °C, half of the hydrogen peroxide present will undergo homolysis in one hour. By contrast, capture of a UV photon leads to the slow dissociation of hydrogen peroxide at room temperature (this is the reason H_2O_2 solutions are typically stored in the dark). The addition of a reduced transition metal, such as ferrous iron and cupric copper leads to a heterolytic decomposition with OH^- and 'OH as products. This is exploited in the so called Fenton reaction (Eqn. 9) in which hydrogen peroxide is mixed with a solution of a ferrous iron salt (Fenton 1894; Walling 1975).

$$H_2O_2(aq) + Fe^{2+} \rightarrow {}^{\bullet}OH(aq) + OH^-(aq) + Fe^{3+} \tag{9}$$

Given its reactivity, the formation of hydroxyl radical in biological systems is tightly controlled. This is accomplished in cellular systems by limiting the availability of labile ferrous iron and controlling the concentration of hydrogen peroxide, the two Fenton reactants (Valko et al. 2005).

Nitric oxide and peroxynitrite. Nitric oxide, $NO^{\bullet}$, is a radical species which can further react with superoxide to form peroxynitrite, $ONOO^-$ (Vásquez-Vivar et al. 1996; Burney et al. 1999). Nitric oxide is present in the environment as an air pollutant emitted by combustion engines (Berner and Berner 1996). Nitric oxide, although it is a free radical, is nevertheless relatively unreactive toward biomolecules and can, therefore, be transported through lipid membranes in and out of cells. Because of its relatively low chemical reactivity and its capacity to pass from cell to cell, nitric oxide is now recognized as an especially important signaling molecule (Burney et al. 1999). For example, nitric oxide levels in the inner lining of blood vessel signal the smooth muscle around the vessel to contract or dilate, thereby controlling blood flow. The formation of nitric oxide is synthesized by at least three different nitric oxide synthase (NOS) enzymes (Vásquez-Vivar et al. 1998) associated with endothelial cells (eNOS), neurons (nNOS), and inflammatory leukocytes (inducible or iNOS). Given its relatively low reactivity, it has been suggested that the conversion of nitric oxide to the much more highly reactive peroxynitrite anion is a critical step in the defense mechanism against invading bacteria (Pou et al. 1995; Vásquez-Vivar et al. 1996; Burney et al. 1999). The peroxynitrite species is a strong oxidant, capable of oxidizing nucleic acids and lipids among other biomolecules (Vásquez-Vivar et al. 1996).

Peroxynitrite is a weak acid ($pK_1 = 6.8$). Hence, at pH 7.4, the anionic form will be slightly more abundant than the acid form. Below pH 6.8, the acid form will dominate. The reaction between superoxide ions and nitric oxide that results in formation of peroxynitrite proceeds at a near diffusion-limited rate ($k = 6.7 \ 10^9 \ M^{-1} \ s^{-1}$) and may compete effectively with the SOD catalyzed dismutase of superoxide (Burney et al. 1999).

Hypochlorous acid. Hypochlorous acid, HOCl, and its conjugate base hypochlorite, OCl^-, is used as household bleach and is the most widely used mild disinfectant in water treatment (Stumm and Morgan 1995). At physiological pH (7.4), about 50% of the acid will be present as OCl^-. Hypochlorous acid reacts with a broad spectrum of biomolecules and causes cell death (hence its effectiveness as a disinfectant) (Whiteman et al. 2002). The broad-spectrum reactivity of hypochlorous acid is exploited in the immune system. Hypochlorous acid is synthesized within the neutrophil, a type of phagocytic leukocyte that is especially active in engulfing and killing bacteria. The enzyme myeloperoxidase (MPO) catalyzes the oxidation of halides by hydrogen peroxide (Whiteman et al. 2002; Fig. 2). Given that chloride is the most abundant halide in cell fluid, the predominant product is HOCl (Babior 2000):

$$H_2O_2 + Cl^- \rightarrow HOCl + OH^- \tag{10}$$

Thermodynamics of ROS

The thermodynamics of the stepwise reduction of molecular oxygen have been studied extensively. The most important thermodynamic data are summarized in this section. This provides the necessary background for discussing the kinetics and catalysis of ROS formation and transformation reactions.

The relative thermodynamic stability of ROS acts as an important constraint on their reactivity. In Figure 3, the stepwise reduction of molecular oxygen is presented in a free energy diagram. In this diagram, all one-electron transfer reactions are written so that only electrons and protons are consumed in the reactions. Since, by convention, the free energy of formation for both the proton and electron equal zero, the free energy differences among the different ROS can be considered to account fully for the difference in free energy for the reactions in which these species are transformed. The complete reactions and their equilibrium constants are summarized in Table 3. There are several important observations to be made here.

The first step in the stepwise reduction of dissolved molecular oxygen is thermodynamically unfavorable (endergonic), with a positive ΔG_r^o of 15.52 kJ/mole (Table 3). In essence, adding the first electron to molecular oxygen is an activation step along the stepwise reduction path. This activation barrier explains in part the relative inertness of molecular oxygen. All other steps are thermodynamically favorable (exergonic); with the reduction of superoxide to hydrogen peroxide as well as the hydroxyl radical to hydroxide steps as the most energetically favorable reactions. Hence, these two steps provide a very strong oxidative driving force. In other words, superoxide and hydroxyl radical have the thermodynamic potential to oxidize a wide range of compounds (whether these species react rapidly depends on the reaction mechanism and cannot be predicted on the basis of the thermodynamics). The reduction of hydrogen peroxide to hydroxyl radical and hydroxide is exergonic but requires the dissociation of an O-O bond. This bond has a dissociation energy of 146 kJ/mole (Atkins and de Paula 2002), which represents a formidable activation barrier. It is this activation barrier that explains the relative stability of hydrogen peroxide.

The free energy diagram presented in Figure 3 also suggests that there are two steps within the cascade that could be catalyzed. Catalysis that allows two electrons to be added to molecular oxygen in one step would lead to hydrogen peroxide and avoids the endergonic first step in the cascade. For example, oxidation of pyrite, FeS_2, is known to produce hydrogen peroxide (Ahlberg and Broo 1997; Cohn et al. 2005c). As a semiconductor mineral, pyrite can transfer two electrons in concert and avoid the formation of a superoxide intermediate. The second reaction that benefits from catalysis is the dissociation of hydrogen peroxide. For example, the addition of ferrous or ferric iron is known to facilitate the decomposition of hydrogen peroxide [the Fenton reaction referred to above; Fenton (1894)]. The exact mechanism has been a matter

Table 3. Thermodynamic data for selected reactions involving ROS[a].

	Reaction	ΔG_r^o (kJ mol^{-1})	log K_{25}	E^o (pH = 7) (mV)
A	$O_2(aq) + e^- = O_2^{\bullet-}(aq)$	15.52	−2.7	−161
B	$O_2^{\bullet-}(aq) + e^- + 2H^+ = H_2O_2(aq)$	−165.9	29.1	892
C	$H_2O_2(aq) + e^- = OH^-(aq) + {}^\bullet OH(aq)$	−15.42	2.71	575
D	${}^\bullet OH(aq) + e^- = OH^-(aq)$	−165.0	28.91	2126
E	$O_2(aq) + 4 e^- + 2H^+ = 2 OH^-(aq)$	−330.9	57.97	3432

[a] Thermodynamic data taken from Stumm and Morgan (1995)

of debate for close to a century, but it is thought to involve an activated reaction complex that involves a ferrous iron cation and a hydrogen peroxide. The ferrous iron transfers an electron to hydrogen peroxide, forcing the dissociation of the O-O bond into a hydroxyl radical and a hydroxide ion. It has been suggested that as an intermediate a ferryl [Fe(IV)] species forms (Pierre and Fontecave 1999; Alegría et al. 2003). Ferrous iron acts as a catalyst in the Fenton reaction (Eqn. 9) if the ferric iron product is reduced. Some ferric iron is reduced by hydrogen peroxide back to ferrous iron (Eqn. 11) (Alegría et al. 2003).

$$Fe^{3+}(aq) + H_2O_2 \ (aq) = Fe^{2+} + HO_2^\bullet + H^+ \tag{11}$$

This reaction is slow, but leads to regeneration of ferrous iron which can then further promote the Fenton reaction. For example, it has been shown that adding ferric iron to a hydrogen peroxide solution leads to an induction period in which no hydrogen peroxide is decomposed. After the induction period the reaction proceeds at the same rate as when the reaction is initiated with ferrous iron (Alegría et al. 2003). In many systems the reduction of ferric iron takes place with organic reductant. Other reducible transition state metals (e.g., Cu, Co, V, Cr) can also catalyze the Fenton reaction (Pierre and Fontecave 1999; Valko et al. 2005). The deleterious effects of ROS are predominantly related to reactions involving hydroxyl radicals, which are more reactive than superoxide. Hence, controlling the concentration of metals that can catalyze the Fenton reaction is crucial. Chelation with a ligand that prevents the formation of a ternary complex between hydrogen peroxide, metal ion, and the ligand is a key requirement (Graf and Goldsmith 1955). It has been shown that ternary ligand-metal-H_2O_2 complexes will actually accelerate the Fenton reaction (Graf and Goldsmith 1955; Alegría et al. 2003) whereas desferrioxamine (DFO) and phytate can complex iron and inhibit the formation of the ternary ligand-metal-H_2O_2 complex (Graf et al. 1984).

It is common to analyze the thermodynamics of redox reactions involving ROS on the basis of a comparison of standard potentials for the redox couples involved. While this approach is convenient, it can lead to some serious misconceptions, as pointed out by Pierre and Fontecave (1999). To illustrate the problems with this approach consider the reduction of hydrogen peroxide to hydroxyl radical (Eqn. 9), the molecular oxygen/superoxide couple (Eqn. 6), and several other relevant reactions. It is customary to construct a redox ladder, which is nothing more than the standard potentials for the reactions of interest at a common set of conditions, e.g., pH 7 (Stumm and Morgan 1995). The vertical scale of the ladder can be either in mV or Volts with respect to the NHE or pe. It is common to use the relative position of two half reactions on a redox ladder to evaluate in which direction an overall reaction between the two half cells will proceed spontaneously (the higher redox couple will accept electrons; whilst the lower couple will donate electrons). For example, the standard potential of the $O_2(aq)/O_2^{\bullet-}$ redox couple suggests that any redox couple involving ferric and ferrous iron with a standard potential in excess of -160 mV can lead to reduction of the ferric iron to ferrous iron (Fig. 4). Hence, one would predict that $Fe(H_2O)_6^{3+}$ as well as Fe(III)-EDTA will be reduced to ferrous iron. However, this only holds if the superoxide activity (or effective concentration) is equal to that of dissolved molecular oxygen. As pointed out by Pierre and Fontecave (1999), the concentration of superoxide in a normal cell is around 10^{-11} M and the concentration of molecular oxygen will be around 3.5×10^{-5} M. Applying the Nernst Equation, the actual $O_2(aq)/O_2^{\bullet-}$ ratio leads to a redox potential of +230 mV, an upward shift of 390 mV. As a result, a 50/50 Fe(II)-EDTA/Fe(III)-EDTA solution is no longer expected to be reduced by the $O_2(aq)/O_2^{\bullet-}$ couple. Conversely, upregulation of superoxide production at the expense of molecular oxygen will shift the actual potential provided by the $O_2(aq)/O_2^{\bullet-}$ activity ratio down. When interpreting which oxidation reactions the $H_2O_2/^\bullet OH + OH^-$ couple can drive, it should be kept in mind that the activity of $^\bullet OH$ will be many orders of magnitude smaller than that for H_2O_2, effectively raising the position of this couple upward. For example, if the activity of the radical is a factor of 10^{10} lower than that for hydrogen peroxide, the position of

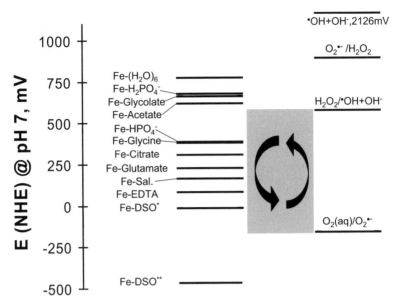

Figure 4. Standard redox potentials for couples involving dissolved iron species and ROS.

the effective position of the redox couple will be raised by 590 mV. As a result this couple can drive almost any oxidation reaction. It is also important to keep in mind that reactions that are feasible energetically may not proceed at significant rates due to kinetic inhibitions.

Analytical methods to determine ROS

The reactivity of ROS, particularly hydroxyl radical, presents a considerable analytical challenge. To meet this challenge several strategies have been developed for the analysis of ROS in acellular and cellular studies. For very reactive species (e.g., hydroxyl radical), there are two different approaches. One can add a scavenger to the system of interest and analyze the conversion of the scavenger. In this strategy the scavenger is added in sufficiently high concentration to effectively out compete all other radical-consuming reactions. An example of a common scavenger is benzoic acid, which is converted to salicylic acid as it reacts with hydroxyl radical (Winterbourn 1987). An alternative strategy is to use a radical probe. Probes, such as nitrobenzene and anisole, are in essence radical scavengers but they are deployed in low concentrations (typically less than 5 μM) and their reaction does not affect the steady state concentration of radicals. The radical concentration in the experimental system is determined on the basis of loss or conversion of probe molecules. This strategy is often deployed in photochemical studies in which the interaction of photons with the system of interest produces radicals and the objective is to determine the steady state radical concentration under illumination (Zepp et al. 1992; Kwan and Voelker 2002).

While scavenging techniques relying on the conversion of organic molecules is widely used it is also possible to use structural changes to biomolecules to demonstrate the presence of radicals. For example, Cohn and coworkers have developed a protocol that uses strand breakage in yeast RNA as an assay that is suitable for the study of mineral-induced hydroxyl radical formation in mineral slurries and coal slurries (Cohn et al. 2005b). In brief, this method makes use of a molecular probe specific to RNA (Ribogreen from Invitrogen). The probe will fluoresce when bound to RNA. However, if the yeast RNA strand is shortened due to interaction with hydroxyl radical, the molecular probe will no longer fluoresce. Thus, the loss of Ribogreen fluorescence serves as a measure of hydroxyl radical-induced strand breakage. In addition to the method developed by Cohn, there are a number of other standard assays that are all based on radical-induced strand breakage of nucleic acids. In most cases, such methods make use of DNA, rather than RNA. The strand breakage is often detected by studying the mobility of the nucleic acid in a gel electrophoresis experiment. The mobility of a nucleic acid in an agarose gel depends on its size and structure. A tightly coiled DNA can unwind as it reacts with a hydroxyl radical. This (partial) unwinding as a result of the interaction with a radical will result in increased hydrodynamic radius of the linear polymer and a consequent lowered mobility. Conversely, fragmentation of nucleic acids will lead to shorter pieces that will have a higher mobility. Several examples of the use of gel electrophoresis will be given later in the chapter. In addition to introducing nucleic acid strand breaks, the hydroxyl radical may also oxidize the bases. Guanine is particularly susceptible and its main oxidized product, 8-oxodeoxyguanosine (8-oxodG) is often used as a marker for hydroxyl radical-induced oxidation.

Radicals can also be detected by forming a stable adduct that can be detected. This strategy is often referred to as spin trapping. The concentration of the stable adduct is often determined using electron spin resonance spectroscopy (ESR) or in some cases by fluorescence measurements. For cellular systems it is also useful to analyze for biomarkers that are indicative of the upregulation of ROS. In this section, we briefly present some of the most commonly used strategies and techniques. In addition we discuss in more detail assays our group have developed or adapted for studies involving mineral suspension ROS detection. The first part of this section is devoted to ESR and spin trapping, followed by a section on the challenges related to ROS detection in cellular systems. Table 4 summarizes techniques that are used commonly for ROS detections.

Table 4. Common ROS detection methods*.

Probe	Detects	Equipment	Sensitivity	Literature[#]
DMPO	$^{\bullet}OH$, $(O_2^{\bullet})^-$	ESR	nM range	1
DCFH	H_2O_2, $^{\bullet}OH$, $ONOO^-$, $NO^{\bullet}$	fluorometer	nM range	2
APF	$^{\bullet}OH$, ^-OCl	fluorometer	nM range	3
PF	$^{\bullet}OH$, $(O_2^{\bullet})^-$	fluorometer	nM range	4
Tempo-9-ac	$^{\bullet}OH$, $(O_2^{\bullet})^-$	fluorometer	nM range	5
Amplex Red	H_2O_2	fluorometer	10 nM	6
LCV	H_2O_2	UV-Vis	µM range	7
Cu-DMP	H_2O_2	fluorometer	1.0 µM	8
Scopoletin	H_2O_2	fluorometer	25 nM	9
Benzoic acid	$^{\bullet}OH$	GC	nM range	10
Luminol and Lucigenin chemiluminescence	H_2O_2, $^{\bullet}OH$, $(O_2^{\bullet})^-$	luminometer	nM range	11

*The ROS detection methods listed here have been used by our group; however, there are many other methods available. The literature citations are examples of papers that use the given method. #*Literature Sources*: 1: Fubini et al. (1995); Shi et al. (2004); 2: LeBel et al. (1992); Rota et al. (1999); 3: Setsukinai et al. (2003); 4:Pou et al. (1993, 1995); Li et al. (1997); 5: Borisenko et al. (2004); 6: Zhou et al. (1997); 7: Mottola et al. (1970); Zhang and Wong (1994); Cohn et al. (2005c); 8: Kosaka et al. (1998); 9: Cooper et al. (1988); Holm et al. (1987); Wilson et al. (2000a); 10: Winterbourn (1987); 11: Casadevall et al. (1984); Yıldız and Demiryürek (1998); Ohyama et al. (2001); Fach et al. (2002); Iwata and Yano (2003)

Electron spin resonance spectroscopy (ESR). Electron spin resonance spectroscopy, often also referred to as electron paramagnetic resonance spectroscopy, is frequently used to determine the products of a spin trap experiment. In ESR spectroscopy, electronic spins are excited by placing the sample in a magnetic field. Unpaired electrons will resonate at specific magnetic field strengths (termed hyperfine splitting-constants) between a state in which their spin is aligned parallel and antiparallel to the external magnetic field. The energy difference associated with the conversion between the two spin states is measured in an ESR experiment and detected as an energy absorbance as a function of magnetic field strength. The absorbance spectrum yields information on the local environment around the unpaired electron. ESR spectra are typically presented as the first derivative of the absorbance vs. magnetic field strength as this conversion improves the apparent spectral resolution and facilitates the interpretation of subtle features in the absorbance spectrum (the same strategy is often used in FTIR spectroscopy; Smith 1996). The ESR technique is both quantitative and qualitative. The signal intensity is related to the number of excited unpaired electrons and identification of radical adducts is determined based on the hyperfine splitting constants. ESR is used to probe unpaired electrons in the solid, liquid and gaseous state. Liquid samples are used in spin trapping experiments. ESR is a well-established technique; the text by Gordy (1980) provides further detail into the theory and practice of the technique. Geochemists may find the introduction by Calas (1988) in one of the earlier Reviews in Mineralogy and Geochemistry volumes useful, although that review is exclusively focused on solid state ESR. An excellent on-line tutorial of the technique is also available (Bruker Biospin 2006).

Spin trapping. In a spin trap experiment, a spin probe molecule is reacted with the sample and the reaction product is analyzed. Typically the reaction products are analyzed using ESR, but some spin probes have been developed that will alter in fluorescence intensity upon reaction with a target radical. Hence, for those probes, the fluorescence intensity of the reaction products is determined. Spin trapping of transient free radicals has been widely used in the biomedical research community, but has seen limited application in the geochemical research community. The purpose of this brief section is to introduce spin trapping techniques as well as potential problems in applying the technique to mineral slurries.

As mentioned above, the goal of a spin trap experiment is to form an adduct that is more stable than the free radical and then measure the concentration of the adduct. A suite of nitrone and nitroso spin traps (e.g., PNB, POBN, DMPO) are routinely used in biomedical research (Khan et al. 2003). These spin traps can be added to cells and have even been injected into mice to detect the presence of superoxide *in vivo* (Dikalova et al. 2001; Takeshita et al. 2004). A successful spin trap experiment requires that the spin trap reacts rapidly and efficiently with the target free radical. A second requirement is that the adduct has a sufficiently long half-life that it remains stable over the course of the ESR measurements. The half-life of the DMPO-OH adduct in aqueous solutions is only on the order of 3 min; the BMPO-OH adduct has a half life approaching 20 min in aqueous solution (Khan et al. 2003). The practical implication is that experiments have to be conducted where there is immediate access to an ESR facility.

One of the challenges faced in spin trap experiments is the level of specificity that can be achieved. One of the most commonly used spin traps, DMPO, reacts with hydroxyl radical, superoxide, and dissolved ferric iron. The reaction of DMPO with superoxide and hydroxyl radical leads to two different adducts, DMPO-OOH and DMPO-OH, respectively. However, the DMPO-OOH adduct spontaneously decays to DMPO-OH on a timescale that makes it impossible to measure the DMPO-OOH adduct. This limitation can be circumvented by a standard method to determine if the signal is derived from superoxide or hydroxyl radical. This involves conducting a series of ESR experiments as shown in Figure 5 taken from (Shi et al. 2005). The top spectrum of Figure 5 is a characteristic 4-line ESR spectrum of the DMPO-OH adduct. This spectrum could have originated from either the interaction of the superoxide

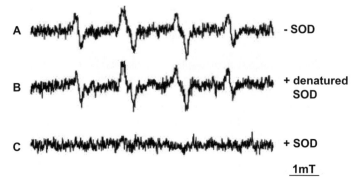

Figure 5. 1,6-Benzo(a)pyrene quinone (BPQ) induced superoxide radical formation in human mammary epithelial cell line, MFC-10A. Spectrum A: BMPO (50 mM) was mixed with 1E6 cells, 100 μL cell medium, and 10 μM BPQ was added to start reaction. Spectrum B: same as spectrum A but with pre-added heat denaturated SOD (400 U/mL). Spectrum C: same as spectrum A with pre-added SOD (400 U/mL). [Reprinted from Shi et al., *Archives of Biochemistry and Biophysics*, Vol. 437, p. 59-68. Copyright (2005),with permission from Elsevier.]

radical or the hydroxyl radical with the probe. However, addition of the enzyme superoxide dismutase (SOD) shows that superoxide is responsible for the 4-line DMPO-OH spectrum and not hydroxyl radical. A control experiment with denatured SOD is necessary to rule out that the addition of SOD itself would alter the spectrum. SOD is inactivated (denatured) by heating.

In the case of studies involving iron-bearing minerals, it is important to note that the presence of ferric iron can induce the formation of a DMPO-OH adduct in the absence •OH (Makino et al. 1990). The interaction between ferric iron and DMPO presents a problem for studies with iron-containing minerals. However, there is a standard protocol to verify the presence of radicals. In this protocol another reactant, which acts as a *radical scavenger* (e.g., ethanol, mannitol, dimethyl sulfoxide), is added to compete with DMPO for reaction with •OH. Addition of radical scavengers is often used to verify the presence of a particular radical species and to determine the contribution of the radical species when non-radical reactions will lead to the same spectrum (Rice-Evans et al. 1991). When ethanol is used as a radical scavenger, its reaction with •OH forms the α-hydroxyethyl radical that adds to DMPO resulting in the DMPO-CH(CH$_3$)OH adduct (Makino et al. 1990). In the absence of •OH, but presence of ferric iron (or other nucleophiles), ethanol will add to DMPO with its oxygen and not carbon moiety. This then results in the formation of a DMPO-OCH$_2$CH$_3$ adduct (Makino et al. 1990). The DMPO-CH(CH$_3$)OH and DMPO-OCH$_2$CH$_3$ adducts have different ESR spectra, see Figure 6.

Fluorogenic spin trap methods are relatively new (Blough and Simpson 1988; Pou et al. 1993; Li et al. 1997; Zhou et al. 1997; Setsukinai et al. 2003; Borisenko et al. 2004) and have only recently been applied to the study of soil slurries by Blough and coworkers (Petigara et al. 2002). These reagents are relatively stable free radicals themselves, which form spin-paired adducts with more highly reactive radicals. The emission from the fluorophore in these reagents is quenched by the delocalized unpaired spin in the unreacted probe, but in the spin-paired products the quantum yield becomes significant. The major advantage of these spin traps is that they do not require an ESR facility to quantify the adduct. A relatively simple and inexpensive fluorometer can be used to determine the adduct concentration. This simplicity also has the advantage that fluorogenic spin trap experiments can be conducted in anaerobic chambers and in the field. Fluorogenic spin traps can also be used to image the spatial distribution of the radical formation within cells or tissue (Pou et al. 1995). It should be pointed out, however, that in many cases, ESR spectra are specific to certain radicals, while fluorogenic spin traps do not

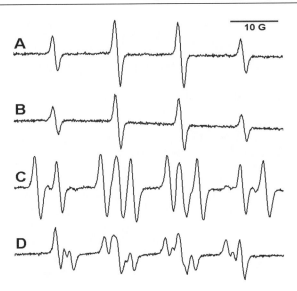

Figure 6. ESR spin-trapping with 100 mM 5,5-dimethyl-1-pyrroline N-oxide (DMPO): (A) Fenton reaction (120 µM H_2O_2 + 25 µM Fe(II)), (B) 100 µM Fe(III), (C) Fenton + 33% EtOH, (D) 500 µM Fe(III) + 33% EtOH. The experiments were conducted by mixing the reactants and immediately injection into a liquid flow-through cell at room temperature (24 ± 2 °C) with the following ESR settings: magnetic field of 3470 ± 100 G, microwave power of 20 mW, modulation frequency of 100 kHz and amplitude of 1 G, receiver gain of 2 × 10^5, time constant of 0.64 s, and scan time of 2 min 47 s.

provide a spectrum—only a change in fluorescence. A fluorogenic spin trap for hydroxyl radical will show an increase in fluorescence as it reacts with the radical, but it will not generate the diagnostic ESR spectrum. The use of some fluorogenic spin traps is complicated by the fact that they can react with a number of radicals with a considerable range of efficiency in formation of the spin-paired fluorescent product; hence, it is important to develop a panel of assays.

Our group has been using fluorogenic spin traps to study the formation of radicals within mineral slurries and coal slurries as well as in cell cultures exposed to mineral particles, see Table 4. All the methods listed in Table 4 were designed to be used in biomedical studies and not to detect radicals in mineral or coal slurries. We have, however, found that the methods listed in Table 4 can be used to determine radicals in mineral/coal slurries and or cell cultures exposed to minerals. In all these experiments, a reactant probe is added to the slurry or culture at the start of the experiment. In experiments with mineral slurries, aliquots are withdrawn periodically and the fluorescence of the spin trap is determined after filtration. Calibration curves for hydroxyl radical can be created using the Fenton reaction or reaction of horseradish peroxidase (HRP) with hydrogen peroxide. For some probes, the addition of HRP to H_2O_2 in the presence of a fluorogenic spin trap results in conversion of H_2O_2 to water and oxidation of the spin trap (Eqn. 12).

$$\text{probe} + H_2O_2 \rightarrow \text{oxidized probe} + H_2O \tag{12}$$

Thus, measured fluorescence can be related to a known equivalent hydrogen peroxide concentration. In experiments performed to evaluate mineral particle generation of ROS in cell cultures, the fluorescence of the probe in the presence of cells exposed to minerals is compared to the fluorescence of the probe in the presence of cells that are untreated. These measurements can be efficiently performed directly on the reaction mixtures of particles, cells, and probe using a well plate reader.

By combining fluorogenic spin traps with fluorogenic assays for hydrogen peroxide, it is possible to rapidly determine hydrogen peroxide and hydroxyl radical formation in mineral slurries. The most commonly used fluorogenic probe is DCFH which will react with a broad spectrum of ROS (Rota et al. 1999). DCFH is widely used in ROS detection in cell cultures. A critical step in its use in cell cultures is the esterification of DCFH, which generates a product which is highly cell-permeable and ensures uptake of DCFH into the cells. Within the cell, nonspecific esterases in the cytosol of viable cells regenerates the DCFH. Most commercially available DCFH reagent preparations are designed for use in cell cultures and, therefore, in the ester form. Hence, when DCFH is used with cell-free mineral slurries, it is important to revert the ester form to DCFH (LeBel et al. 1992). The DCHF protocol does require daily preparation of the reagent and its storage under an inert gas, which makes it cumbersome. Amplex UltraRed, a H_2O_2 specific assay, is a better suited for work with mineral slurries. In brief, the Amplex UltraRed reagent (10-acetyl-3,7-dihydroxyphenoxazine) reacts with H_2O_2 in a 1:1 stoichiometry in the presence of horseradish peroxidase (HRP) to produce highly fluorescent Resorufin ($C_{12}H_6NNaO_3$). Compared to the most widely used probe for H_2O_2, DCFH, Amplex UltraRed is 5 to 20 times as sensitive, and Resorufin is more stable to chemical and photo-induced oxidation (Zhou et al. 1997). Amplex UltraRed has been used to measure H_2O_2 production in cells exposed to asbestos (Xu et al. 2002), cancer cells (Wagner et al. 2005), plant cells (Janisch and Schempp 2004), and brain mitochondia (Gyulkhandanyan and Pennefather 2004).

ROS analysis in cell cultures. ROS analysis in cell cultures presents additional challenges. The initial products generated by reactions of minerals with water and oxygen—observed in studies of mineral slurries—may themselves serve as substrates for further reactions catalyzed by enzymes present in the cellular environment. The products formed by such reactions may themselves be highly reactive and longer lived than such highly reactive species as hydroxyl radicals. These secondary species may then go on to modify proteins, lipids, and nucleic acids within cells, triggering a range of pathologic consequences. An appreciation of the potential consequences of reactivity of various mineral dusts and particulates therefore requires a panel of assays to detect formation of initial mineral-derived reactive species as well as secondary reactive species, which may be generated by enzymatic conversions of the initial reactive species within live cells. For example, hydroxyl radicals, which may be formed directly at the mineral surface, may trigger formation of secondary species such as superoxide ions, hypochlorite ions, and hydrogen peroxide by serving as substrates for secondary reactions mediated by cellular components. Once such secondary species are formed, further reactions initiated by the response of specialized inflammatory cells to modifications of their constituents produced by primary and secondary reactive oxygen species may lead to generation of yet additional highly chemically reactive species, such as hypochlorite ions, which are typically formed only when such cells have been stimulated by exogenous "foreign" materials. Therefore, it would be desirable to use a range of assays that are not only sensitive and specific, but, which can also detect secondary and tertiary products and be compatible with the cells and enzymes involved in generating such species.

In addition to probes that detect hydroxyl radicals via direct reaction, such as APF, other fluorogenic spin traps are available that are capable of detecting a number of free radical species, including hydroxyl radicals and can be used as long as a suitable molecule is also present to serve as a mediator. These fluorogenic spin traps are far more sensitive detectors of free radicals than the "parent" spin traps themselves because of the inherently lower limits of detection by fluorescence spectroscopy than electron spin resonance spectroscopy. Proxyl fluorescamine (PF) was developed by Blough's laboratory (Blough and Simpson 1988) to serve as a fluorogenic derivative of the proxyl spin trap: the fluorophore is quenched in the presence of the nitroxide with its delocalized unpaired electron but emits with a high quantum yield when the nitroxide is converted to the corresponding spin-paired hydroxylamine by alkyl free radicals. In the presence of hydroxyl radicals, however, PF fails to undergo direct reaction. Blough and coworkers (Pou et al. 1993; Li et al. 1997) have overcome this lack of reactivity by adding dimethylsulfoxide (DMSO) (< 1 M, which is compatible with normal cell functions; 5% DMSO = 705 mM), to a mixture of a ferrous salt and a source of hydrogen peroxide, components which generate hydroxyl radicals via the Fenton reaction. Under these conditions, the hydroxyl radicals react with the DMSO to form methyl radicals, which then react with the PF to yield the fluorescent hydroxylamine. Blough et al. have gone on to show that PF can also be used as a spin trap for hydroxyl radicals generated intracellularly by metabolism of a quinone-based drug by a mouse epidermal cell line. As an alternative reagent, Borisenko et al. (2004) have employed TEMPO-acridine (TEMPO-Ac) as a fluorogenic spin trap. Conversion of this nitroxide to a fluorescent product, acridine-piperidine, cannot be effectively achieved except by thiyl radicals, formed under physiologically relevant conditions within cells from the free radical scavenger, glutathione, under oxidizing conditions. Borisenko et al. added phenol and hydrogen peroxide to cells to "commandeer" their intracellular myeloperoxidase. TEMPO-Ac in the presence of exogenous glutathione and phenol can also be used to detect hydroxyl radicals formed from mineral suspensions. Furthermore, TEMPO-Ac along with low concentrations of phenol can be used as an intracellular probe to detect secondary formation of thiyl free radicals from the endogenous glutathione within MonoMac 6 cells and human neutrophils which have been allowed to phagocytose (i.e., engulf) mineral particles.

Using chemiluminescence with luminol or lucigenin, particle-induced ROS have been detected in cell cultures. The probes become luminescent when valence electrons are promoted to higher energy states by an oxidation reaction. This luminescence is detected with a luminometer. Lucigenin and luminol have been used to detect particle-induced superoxide and hydrogen peroxide (Casadevall et al. 1984; Fach et al. 2002; Iwata and Yano 2003; Ohyama et al. 2001) and luminol has been shown to react with hydroxyl radicals (Yıldız and Demiryürek 1998).

PART II: MECHANISMS OF MINERAL-INDUCED ROS FORMATION

In Part II each of the five mechanisms listed in the beginning of the chapter will be discussed in detail. Specific examples from studies—mostly *in vitro*—conducted by our group as well as by others are used to illustrate these mechanisms.

Homogeneous catalysis of ROS (trans)formation by dissolved metal ions (mechanism 1)

Minerals may promote the formation or transformation of ROS by (partially) dissolving into body fluids. The dissolution of minerals is covered in the chapter by Plumlee et al. (2006) in this volume. Important concepts such as biodurability and bioavailability are introduced and discussed in depth in that chapter. In addition the chapter by Reeder et al. (2006) discusses the importance of metal speciation and the techniques that can be used to determine metal composition and speciation in Earth materials and the changes these materials undergo when exposed to body fluids. There is also a vast body of literature on the thermodynamics and kinetics of the dissolution of minerals in aqueous fluids ranging from distilled water, to seawater, to hydrothermal water in the geochemical literature (Lasaga 1998). However, in the context of ROS formation, it is important to consider the release of metals and metalloids as minerals are reacting with biofluids. For this chapter, we will only consider the interaction with lung fluid and the more acid (pH ~ 4.5) environment encountered in phagocytes. Inhalation of mineral dust may provoke lung cells and can lead to an array of lung diseases as well as exacerbating other diseases as mentioned in the introduction. ROS species play an important role in the development of these diseases.

The role of metals in the formation and transformation of ROS has been extensively studied in the biomedical community (see review by Halliwell and Gutteridge 1990). Figure 7 adapted from Kawanishi et al. (2002) shows, for example, how metals and metalloids may

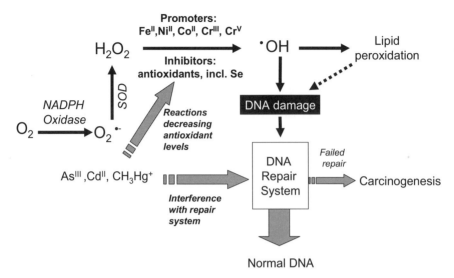

Figure 7. Role of metals and metalloids in promotion of carcinogenesis.
Figure adapted from Kawanishi et al. (2002).

play a role in DNA damage, which can lead to cancer. It is important to stress that dissolved metals/metalloids may either promote DNA damage by catalyzing the formation of hydroxyl radical or by interfering with DNA repair, which is an important defense strategy against mutations. It has been estimated that mammalian cells experience as many as 10,000 "hits" by ROS per day (Ames 1987). Hence, partial impairment of the repair system can shift the balance between DNA damage and repair, thereby increasing the risk for cancer.

Iron is the most abundant transition metal in the body and its role in ROS transformation reactions has received the most attention (Toyokuni 1996; Welch et al. 2002; Valko et al. 2005), although other metals, such as copper, nickel, cobalt, chromium, and vanadium, have also been implicated in the formation and transformation of ROS (Shi and Dalal 1992; Shi et al. 1992; Stohs and Bagchi 1995; Lloyd and Phillips 1999; Strli et al. 2003; Valko et al. 2005). Below we will focus mainly on reactions involving dissolved iron. The effect of chelation on ROS formation is also discussed as this is an important factor in ROS formation and DNA damage (Lloyd and Phillips 1999). Iron is an essential element; however, iron can also promote deleterious reactions involving ROS (Morris et al. 1995; Welch et al. 2002). Within the body, iron speciation is tightly controlled (Harrison and Arosio 1996). Exposure to iron-bearing minerals can lead to an increase in labile or free iron, which can catalyze several of the one-electron steps in the molecular oxygen reduction chain. The kinetics of the reduction of molecular oxygen in stepwise reactions with dissolved iron has been studied not only because of the biomedical relevance (Graf et al. 1984), but also because these reactions are important in redox cycling in natural waters (Emmenegger 1998) and in remediation processes that rely on the formation of ROS species to degrade organic pollutants, such as solvents (Watts et al. 1997; Watts and Teel 2005) and dyes (Ma et al. 2005). The iron-catalyzed reduction of hydrogen peroxide to hydroxyl radical and hydroxide is perhaps of greatest physiological significance. As discussed in Part I, free ferrous iron as well as some ferrous iron complexes can promote this reaction. However, complexing ferrous iron with DFO inhibits the Fenton reaction.

While the overall reaction rate between ferrous iron and oxygen has received consider-able attention in the geochemical community, the role and formation of superoxide, hydrogen peroxide, and hydroxyl radical as reaction intermediates has received less attention in this community. However, recent studies that have shown the presence of ROS in natural waters (Cooper and Zika 1983; Holm et al. 1987; Willey et al. 1996; Herut et al. 1998; Price et al. 1998; Goldstone and Voelker 2000; Voelker et al. 2000; Wilson et al. 2000a,b; Kieber et al. 2001; Goldstone et al. 2002; Scott et al. 2003; Hakkinen et al. 2004; Peake and Mosley 2004; Vione et al. 2006) and their role in the degradation of organic pollutants (EPA 1998; Watts et al. 1999b; Petigara et al. 2002; Kwan and Voelker 2002, 2003, 2004) have drawn attention the importance of ROS in the environment. Figure 8 shows the formation of hydroxyl radical in a ferrous solution when exposed to air, while the radical is not formed if the solution is kept strictly anaerobic. The aerobic solution was saturated with air, while the anaerobic solution was purged with nitrogen and kept in an anaerobic chamber with an atmosphere of 5% H_2 and 95% N_2. While Figure 8 shows the formation of hydroxyl radical, Figure 9 shows the effect on the stability of RNA. The experiments summarized in Figure 9

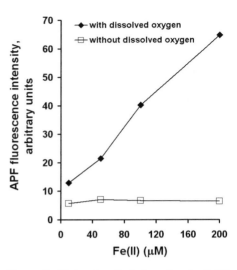

Figure 8. Hydroxyl radical formation in Fe(II) solutions in the presence and absence of dissolved oxygen. Fe(II) solutions were incubated with 10 μM APF at pH 7.4 for 24 h in the dark at room temperature (24 ± 2 °C) either at atmospheric levels of O_2 or in an anaerobic glove box followed by fluorescence readings.

were conducted with aerated solutions. It is thought that the presence of dissolved iron leads to the formation of hydroxyl radical, which is capable of degrading RNA. Hydroxyl radical will rapidly degrade nucleic acids by hydrogen abstraction or addition to bases (Tullius and Dombroski 1986; Heilek and Noller 1996; Brenowitz et al. 2002; Tullius and Greenbaum 2005).

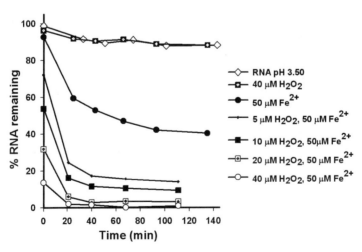

Figure 9. Yeast RNA degradation in the presence of H_2O_2, Fe(II), and Fenton reagents. 1.5 mg/L RNA was exposed to the various reagents at room temperature (24 ± 2 °C) in the presence of atmospheric levels of dissolve oxygen. RNA was quantified using the fluorescent dye RiboGreen (Invitrogen).

Other reducible transition metals may also promote the Fenton reaction, although there are fewer studies that have addressed the kinetics and mechanisms of the non-iron Fenton metals. In a recent comprehensive review of the role of metal ions, iron, cobalt, copper, and vanadium are listed as "Fenton" metals (Valko et al. 2005). In addition, there is ongoing discussion whether nickel is a Fenton metal. Shi et al. (1993a) have shown that both trivalent and pentavalent chromium ion form hydroxyl radical when exposed to hydrogen peroxide. The mechanism involving Cr(III) is uncertain. There is good evidence, however, that hexavalent Cr is reduced in the cell to Cr(III) via an intermediate Cr(V) state. ESR studies indicate that the Cr(V) species is oxidized to Cr(VI) while hydrogen peroxide is reduced to hydroxyl radical and hydroxide. A comparative study by Strli et al. (2003) suggests that Cr(III) is a more efficient Fenton catalyst than ferrous iron is. There is no consensus on whether Ni(II) is a Fenton metal. Strli et al. (2003) also found that Ni(II) will only promote the Fenton reaction above pH 7.5. Complexation of Ni(II) may enhance its reactivity. Similar to iron, complexation of Ni(II) with a ligand that forms a strong complex with Ni(III) stabilizes Ni(III) within the water stability field and makes it possible for nickel to cycle between the two oxidation states (Shi et al. 1993b; Van den Broeke et al. 1998). In other words, Ni^{3+}(aq) is not stable in water, but chelation is expected to stabilize this unusual valence state and allow the Ni(II) complex to promote the Fenton reaction. A similar argument has been made for the reactivity of cobalt (Leonard et al. 2004). Chromium and nickel are of great interest because both metals are known or suspected carcinogens, yet they are common in metal contaminants and some ultramafic rock types and soils developed on these types of rocks are naturally rich in these metals (Som and Joshi 2002).

Mineral surface-promoted formation of ROS by catalysis of O_2 reduction (mechanism 2)

While in the preceding mechanism the mineral played a rather passive role, reactions that take place at the mineral surface in the mechanism are considered in this section. Within the geochemical community there is a growing appreciation for the catalytic role mineral surfaces can play in electron transfer (redox) reactions (Schoonen and Strongin 2005). In fact, one of the best studied redox reactions is the surface-mediated oxygenation of transition metals, such as iron. It has been shown that the reaction between ferrous iron and molecular oxygen is much faster when the ferrous iron is sorbed onto a substrate, particularly a ferric-oxide surface (Wehrli et al. 1989; Wehrli 1990). An in-depth review of this topic has been presented recently in the context of environmental chemistry (Schoonen and Strongin 2005). A second type of mineral-promoted O_2 reduction reactions are reactions in which the mineral itself is oxidized. In this section, these two types of mechanisms are briefly reviewed.

Reactions with sorbates. The rate of the reaction between molecular oxygen and dissolved ferrous iron $[Fe(H_2O)_6^{2+}]$ is about six orders of magnitude slower than the same reaction with the ferrous iron sorbed onto goethite (FeOOH). The reason for this increase in rate is that coordination of ferrous iron with hydroxyl groups derived from the surface effectively shifts the redox potential of the ferrous iron couple downward, making it a better electron donor. For example, the redox potential for a bidentate surface complex of Fe(II) onto goethite has been estimated to lie at 360 mV (NHE) (Wehrli 1990), while the standard potential for the $Fe(H_2O)_6^{2+}/Fe(H_2O)_6^{3+}$ redox couple lies at 770 mV (NHE) (Bard et al. 1985). Similarly, ferrous iron sorbed on silica has an estimated redox potential of between 230 mV (Strathmann and Stone 2003) and 550 mV (Buerge and Hug 1999). By comparison, ferrous iron as a structural component of silicate minerals has an estimated redox potential ranging from 330 to 520 mV (NHE). It should be emphasized that the redox couples involving sorbed or structural ferrous iron are difficult to measure. However, there is a growing body of literature that supports the notion that sorbed or structural ferrous iron are far better electron donors than free ferrous iron. In essence, the formation of bonds with a mineral surface or the coordination to other components within a mineral tunes the ferrous/ferric couple to different redox levels, not unlike ligands can tune this redox couple, see Figure 10.

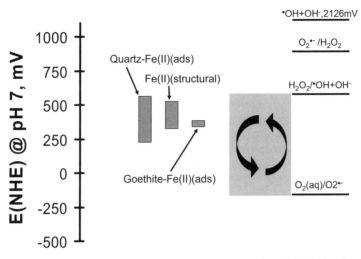

Figure 10. Estimated redox couples for adsorbed and structural Fe(II)/Fe(III) in mineral systems. Diagram based on compilation by Schoonen and Strongin (2005).

While most of the geochemical literature and environmental science literature has been focused on sorbed or structural ferrous iron, there are some reports that the oxidation reaction of other metals is similarly affected. Oxidation of free dissolved Mn(II) with molecular oxygen as electron acceptor is extremely slow (Davies and Morgan 1989; Martin 2005). However, Mn(II) sorbed onto alumina, quartz, and goethite reacts rapidly (Davies and Morgan 1989). The implication is that manganese taken up in a cell as free dissolved Mn(II) may not promote ROS formation, but Mn(II) sorbed onto a mineral might be able to promote ROS formation or transformation. The effect of sorption has also been studied on the reduction of Cr(VI) (Deng and Stone 1996). Hexavalent chromium is reduced by low molecular weight carboxylic acids. The presence of goethite, alumina, and titanium dioxide catalyze these reduction reactions. Hence, it is possible that inhalation of dust containing hexavalent chromium may promote the reduction of this carcinogenic metal and make it available for the (heterogeneous) catalysis of the Fenton reaction.

Recent, unpublished work by Levangie in our group shows that iron sorbed onto quartz promotes the decomposition of RNA. In this study, the rate of RNA decomposition was measured for (1) a pure, hydrothermally-synthesized quartz; (2) a RNA solution without any solids or trace elements; (3) ferrous iron sorbed onto quartz, and (4) an equivalent amount of dissolved ferrous iron or more without quartz. The methods used in the work by Levangie have been reported elsewhere (Cohn et al. 2005b). As seen in Figure 11, RNA decomposes most rapidly in the experiment with ferrous iron sorbed onto quartz compared to uncomplexed ferrous iron. The batch experiments with RNA solution alone and transition metal-free quartz + RNA are identical; hence, pure synthetic quartz that has not undergone any mechanical stress does not induce RNA degradation. This simple study corroborates the notion that redox metals sorbed onto inhaled particles may be important reactants that may lead to lung malignancies.

Oxidation of ferrous-iron containing minerals. The oxidation of ferrous-iron containing minerals by oxygenated water has been extensively studied by geochemists and environmental scientists. Perhaps the best studied reaction is the oxidation of pyrite, FeS_2. Pyrite is the most abundant metal sulfide on Earth and it is found in almost all anaerobic environments (Rickard et al. 1995; Schoonen 2004). It is a common mineral component of coal and accounts, together with its dimorph marcasite, for half or more of the sulfur content in coal (Calkins 1994). The

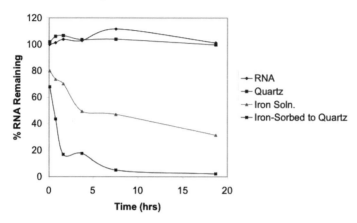

Figure 11. Degradation of yeast RNA exposed to quartz with or without adsorbed iron. RNA degradation under aerobic conditions in the presence of pure synthetic quartz, iron sorbed to the surface of pure synthetic quartz, and dissolved iron (779 μM). Note that the iron-sorbed quartz was prepared by exposing 1 g of pure synthetic quartz to a solution of 40mL of water with 50 mg of Mohr's salt added under anaerobic conditions. The dissolved iron concentration in the slurry with iron-sorbed quartz was 251 μM. Bath experiments were carried out buy mixing 2.64 mg/L RNA with 1.56g/L quartz and iron-sorbed quartz. Samples were filtered and the RNA was quantified using the fluorescent dye, RiboGreen.

oxidation of pyrite by molecular oxygen is a key reaction in the development of acid mine drainage, which is a widespread environmental problem in coal mining and in base-metal mining districts around the world (Banks et al. 1997). The reaction with molecular oxygen leads to the formation of hydrogen peroxide (Ahlberg and Broo 1996; Cohn et al. 2005c) and hydroxyl radical in solution (Cohn et al. 2006). A critical step is the interaction of molecular oxygen with the pyrite surface (Moses et al. 1987; Borda et al. 2004; Usher et al. 2004). Theoretical calculations as well as electrochemical studies have shown that molecular oxygen binds to an exposed ferrous iron (Biegler et al. 1975; Rosso et al. 1999). Hence, the interaction between molecular oxygen and the pyrite surface may be represented by the following reactions.

$$Fe^{II}(pyrite) + O_2 \rightarrow Fe^{III}(pyrite) + (O_2^{\bullet})^- \tag{12}$$

$$Fe^{II}(pyrite) + (O_2^{\bullet})^- + 2H^+ \rightarrow Fe^{III}(pyrite) + H_2O_2 \tag{13}$$

On the pyrite surface these two one-electron transfer reactions may be effectively accomplished by one two-electron transfer reaction. A simultaneous two-electron step bypasses the thermodynamically unfavorable initial one-electron step from molecular oxygen to superoxide, see Figure 3. In addition, the ferric iron species formed on pyrite may readily be reduced by the pyrite substrate, which is a semiconductor (Xu and Schoonen 2000). This reduction would make it possible to rapidly cycle the surface-bound iron between the two oxidation states with molecular oxygen as the terminal electron acceptor (Moses et al. 1987; Eggleston et al. 1996). Considering that hydroxyl radical-induced degradation of nucleic acids is well-documented, we propose that the mechanism that leads to nucleic acid degradation in the presence of pyrite is by Fenton-generated hydroxyl radicals.

The reaction mechanism for the formation of hydrogen peroxide in pyrite suspensions is clearly a surface mediated process. Ahlberg and Broo (1996), using electrochemical techniques, showed that pretreatment of the surface changes the concentration of hydrogen peroxide found in solution. Cohn et al. (2005c) showed that addition of an excess of EDTA stabilizes hydrogen peroxide formed in pyrite slurries and that the concentration is proportional to the amount of pyrite surface area in the slurry. Cohn speculated that the excess EDTA interfered with the Fenton reaction, which decomposes hydrogen peroxide. This notion is corroborated by an additional study in which the formation of hydroxyl radical was demonstrated by ESR spin trapping (Cohn et al. 2005a).

The effect of the formation of ROS in pyrite slurries on nucleic acids was investigated using yeast RNA, ribosomal RNA, plasmid DNA, and adenine. The results with yeast RNA, ribosomal RNA, and plasmid DNA have been reported in a recent publication (Cohn et al. 2006). Figure 12 shows the effect of the presence of pyrite on plasmid DNA and ribosomal RNA. In these experiments the progress of the nucleic acid through the agarose gel is greatly influenced by changes to the nucleic acid strand lengths. It is clear from the gel that the presence of pyrite leads to strand breakage. It has been proposed that the reaction between hydroxyl radical and the bases in nucleic acids constitutes an important pathway in the degradation of the nucleic acids. To evaluate this we conducted several proof-of-concept experiments with adenine. As shown in Figure 13, adenine is slowly decomposed in pyrite slurries. The decomposition is inhibited by either adding EDTA, ethanol (a radical scavenger), or the enzyme catalase. Catalase decomposes hydrogen peroxide and prevents the transformation of this ROS to hydroxyl radical.

Within the biomedical community the reaction of iron associated with asbestos has received considerable attention. It has been proposed that the presence and coordination of iron is an important cofactor in the toxicity of asbestos (Fubini et al. 2001a; Martra et al. 2003). Similarly, it has been suggested that structural iron in asbestos exposed to the surface is important in the production of asbestos-induced ROS (Hardy and Aust 1995; Gazzano et

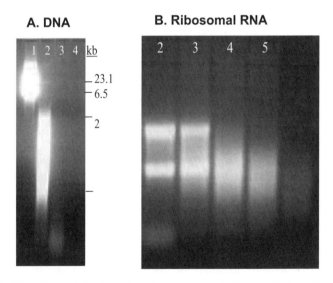

Figure 12. Nucleic acid degradation in pyrite-water suspensions. (A): 100 mg/L plasmid DNA exposed to pyrite (~1-50 mm in size, ~1 m²/g BET surface area) at different particle loadings: lane 1) nothing added, 2) 6 g/L pyrite, 3) 64 g/L pyrite, 4) 640 g/L pyrite. The suspensions were vortexed for 4 h and filtered samples were observed on a 3% agarose gel. The negatively-charged RNA samples were loaded at the top of the agarose gels. When a voltage is applied across the gel, the RNA strands travel downward with the smaller strands moving faster than larger strands. DNA strand-lengths are given on the right side of the gel. (B) 100 mg/L ribosomal RNA exposed to 100 g/L pyrite. Samples were taken at various times after mixing the RNA with pyrite: 1) nothing added, 2) pyrite mixed with RNA for 0.5 min, 3) 10 min, 4) 30 min, 5) 1.5 h. The two bright fluorescent bands from the RNA are due to 28S and 18S ribosomal moieties.

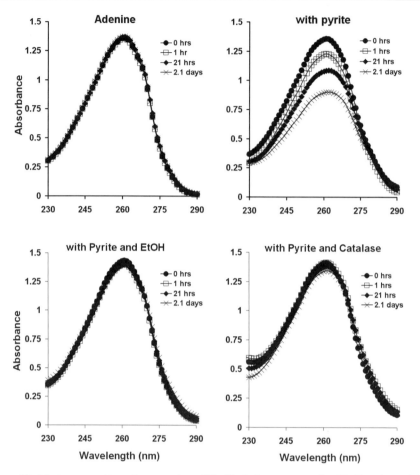

Figure 13. Adenine exposed to pyrite suspensions. 100 μM adenine was exposed to 10 g/L pyrite particles (~1-50 μm in size, ~1 m²/g BET surface area) with nothing added, 50% EtOH, or 64 kUnits catalase. Samples were filtered and wavelength scans were recorded.

al. 2005). This last notion is corroborated by studies on the ROS formation and cytotoxicity of asbestos fibers that have been exposed to siderophores or have undergone iron leaching (Martino et al. 2003, 2004; Martra et al. 2003; Daghino et al. 2005; Favero-Longo et al. 2005) (Fig. 14). One important difference between pyrite and asbestos-form minerals is that the latter are insulators. Hence, their reactive capacity is limited to the surface-exposed iron (and possibly other redox metals).

Hetland et al. (2001) evaluated whether the inflammatory response after exposure to different quarry stones was induced by the soluble fraction or the mineral solids. They showed that the upregulation of a cytokine (IL-8) in A549 lung cells is more pronounced when exposed to a slurry of solid quarry stone particles as opposed to just the leachate of these solids. In their system, both metal sorbates and structural metals may have contributed to the formation of ROS. One of the quarry stones that Hetland and coworkers studied was a basalt. Our group has shown that a principal component of basalt, forsterite (Mg olivine), forms ROS and degrades yeast RNA (Cohn et al. 2005b).

In summary, this second mechanism has only been investigated on a limited basis. It is to be expected that many minerals may promote ROS formation either via reaction of the molecular oxygen with sorbed metals or by a reaction with structural metals. In a recent study we report on a first set of minerals we have evaluated for their ability to form ROS and degrade yeast RNA (Cohn et al. 2005b). The results are summarized in Figure 15.

Mineral defect-driven formation of ROS (mechanism 3)

Stress induces defects in minerals, adding to intrinsic defects that any mineral contains (Borg and Dienes 1992). In the context of human health, the most common form of stress-induced defects is related to grinding of minerals. Grinding is a common step in the processing of

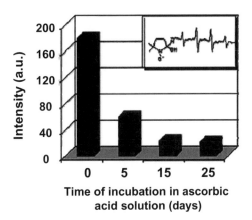

Figure 14. Free radical release from aqueous suspensions of crocidolite fibers. Integrated intensity of the EPR spectra of the [DMPO-OH] adduct produced by suspending in the aqueous solution of H_2O_2 and DMPO the original crocidolite fibers and fibers incubated for 5, 10, and 25 days in aqueous solution of ascorbic acid. Inset: scheme of the [DMPO-OH] adduct and the corresponding EPR spectrum. [Used by permission Amercian Chemical Society, from Martra et al. (2003) *Chem. Res. Tox.* 16, Figure 4, 328-335].

Earth materials and frequently leads to an occupational exposure. Exposure to olivine dust by workers in the glass blowing industry prompted Victor M. Goldschmidt, often referred to as the father of modern Geochemistry (Mason 1992), to contribute to an *in vivo* study on the effect of olivine dust in rats (King et al. 1945). While the objective of grinding in mining operations is typically to decrease grain size and break waste minerals from ore minerals or coal, grinding

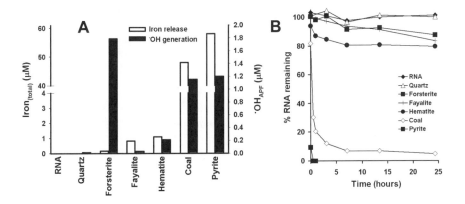

Figure 15. Iron release, hydroxyl radical generation, and RNA degradation induced by several common minerals. Minerals are all from Wards except for the coal (no. 2685b, ~2-3% pyrite), which is from the National Institute for Standards and Technology. The minerals were crushed (63-38 μm) in an agate mill ~50 days before their use in experiments. (A) Iron release measured with ferrozine and hydroxyl radicals quantified using APF after 24 h exposures in opaque centrifuge tubes. (B) 1.5 mg/L yeast RNA exposed to 0.5 m²/L mineral suspensions at room temperature (24 ± 2 °C). RNA was quantified from filtered samples using RiboGreen (Invitrogen). Data is from Cohn et al. (2006).

does more than simply increase the specific surface area of an Earth material. Grinding can change the physical and chemical properties of minerals, including the mineral composition, defect density, lattice strain, and transition-metal spin state (Balaz 2003). Not unexpectedly, there are active research programs in hydrometallurgy and mining engineering to exploit these so called *mechanochemical* reactions to improve mineral separation (Vandeventer et al. 1993; Kirillova et al. 2000; Peng et al. 2003; Yigit and Ozkan 2004) and benefaction (Achimovicova and Balaz 2005), leaching (Hu et al. 2003; Hu et al. 2004; Achimovicova and Balaz 2005) and bioleaching (Balaz et al. 1994; Balaz 2003), as well as to synthesize nanomaterials (Balaz et al. 2003, 2004).

Within the medical community, quartz has received the most attention. In particular Bice Fubini's group in Turin has been leading the way in advancing our understanding of the role of surface defects in the formation of ROS (Fenoglio et al. 2000a,b; 2003; Fubini 1998; Fubini et al. 1987, 1990, 1999, 2001a,b, 2004; Fubini and Hubbard 2003; Cakmak et al. 2004;). As shown in Figure 16, grinding of quartz leads to several types of surface defects resulting from homolytic or heterolytic cleavage of Si-O bonds. These defects can react with water and oxygen to form ROS (Fubini and Hubbard 2003). Recent theoretical calculations by Narayanasamy and Kubicki (2005) suggest that there will always be a small pool of surface radicals on a quartz surface. This work also indicates that the reaction between water and a surface Si-O$^\bullet$ radical, producing a hydroxyl radical, is thermodynamically favorable. The presence and reactivity of the surface defects (in essence, surface-bound radicals) has been studied extensively by Fubini's group using solid state ESR and spin trapping among other techniques. These surface-bound radicals then react with target molecules to produce an array of soluble radicals. As may be expected, the reactivity of freshly ground quartz drops off as the material is exposed to humid air. Hence, not surprisingly the toxicity of freshly ground dust in animal studies is higher than for aged materials for a review, see Fubini and Hubbard (2003). Other Earth materials that have been investigated in the context of grinding-induced radical formation are metal oxides (Costa et al. 1989b; Fubini et al. 1995; Lison et al. 1997), sulfides (Costa et al. 1989a), asbestos (Fubini et al. 1995), and zeolites (Fubini et al. 1995).

A fundamental issue that has not been resolved is the distinction between contributions of surface defects and bulk defects to the overall reactivity. It is expected that the contribution of bulk defects is minimal for insulator materials. In other words, the reactivity of insulator materials is expected to be dominated, if not solely determined, by the surface defect density. By contrast, it is expected that grinding of metallic materials and semiconducting materials leads to defects in the bulk that can migrate to the surface and contribute to their stress-induced reactivity. The difference in stress-induced reactivity between these two endmember models is

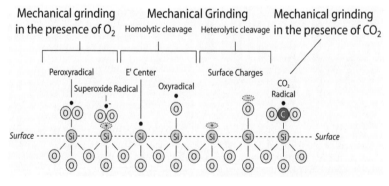

Figure 16. Reactive functionalities at the quartz surface and their formation mechanisms. After Fubini et al. (2003), courtesy of Joel Hurowitz, Stony Brook University.

schematically illustrated in Figure 17. The shape of curve b is notional; the key attribute is that the surface normalized reactivity of the material is dependent on the grinding time. Building on our earlier work on the use of yeast RNA as a probe of mineral-induced hydroxyl formation, we have used the rate of RNA decomposition to evaluate how grinding affects the reactivity of minerals. The rates of RNA decomposition are normalized with respect to surface-area based on BET measurement. For pyrite these experiments show that the surface normalized rate data are dependent on grinding time (Fig. 18), which is indicative of a major contribution of bulk defect. By contrast, the results with quartz indicate that the reactivity appears to be dominated by contributions from the surface (Fig. 18). Note also that quartz is far less reactive than pyrite (the rates of quartz were multiplied by a factor of 1000 in Fig. 18).

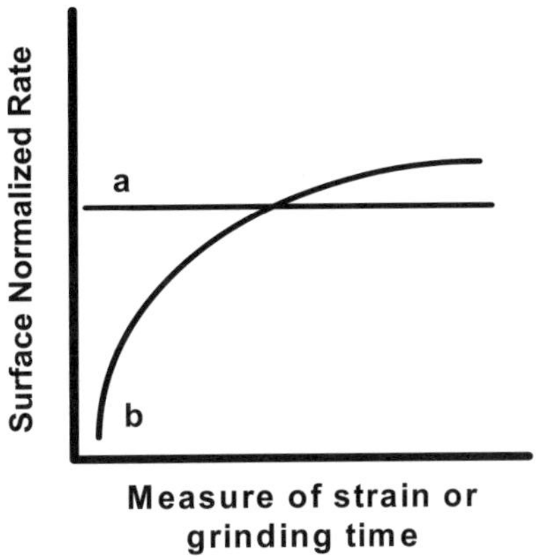

Figure 17. Schematic diagram illustrating two endmember conditions. Curve a represents a material in which only surface defects contribute to the reactivity, curve b is a material with a considerable contribution of the bulk to the reactivity.

Cellular particle-induced formation of ROS (mechanism 4)

The fourth mechanism to be considered is the cellular response to the presence of particles. We purposefully use the term particle because we consider here the non-specific reaction of the immune system to the presence of an otherwise inert solid. In reality, particulate matter may very well promote ROS formation or transformation directly, but those mechanisms are discussed separately (see text on mechanism 2). The cellular response to a provocation brought on by inhaling particulates has been extensively studied (Driscoll et al. 1997; Schins 2002) and it is beyond the scope of this chapter to review the subject. For the purpose of this chapter, we will outline some of the key cellular processes that can contribute to ROS production triggered

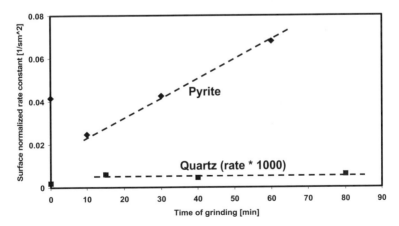

Figure 18. Surface-normalized rates of yeast RNA decomposition by ground pyrite and ground quartz . Note that the rates for quartz were multiplied by a factor of 1000. Data points for unground materials are not taken into account for regression as these points are subject to large uncertainty in BET measurements. Data suggest that bulk defects migrate to the surface of pyrite, while reactivity of quartz is limited to surface defects.

by the mere fact that particles have been inhaled. A general introduction to the topic of pulmonary physiology and some of the processes briefly described in this section is provided by Levitzky (2003)

In the mammalian lung (Fig. 19), gas exchange is facilitated by an alveolar complex that combines a huge surface area with a thin alveolar capillary membrane. Such a delicate structure would be vulnerable to injury from inhaled particulates (organic and inorganic) which gain entry to lungs as a consequence of the critical need to exchange air continuously between the alveolar space and the external atmosphere. There are, however, mechanisms of defense against this inadvertent deposition of particles in the alveolar space. First, the angular configuration of the upper extrapulmonary airway and the branching structure of lower airways (the "tracheobronchial tree") cause particles to deposit on the walls of these airways through inertial impaction. After deposition in airways proximal to the terminal airspaces, the particles become engulfed in mucus and are in turn transported proximally by the movement of cilia to the pharynx and then swallowed, a process called mucociliary clearance. In addition, some particles, especially those that cause irritation, can be expelled by cough clearance. In contrast, if an insoluble particle deposits in the airspaces distal to the ciliated airways, it will be ingested by alveolar macrophages, which will in due course migrate into the ciliated tracheobronchial tree or deliver the particle to regional lymph nodes. This process is called alveolar clearance.

The process of phagocytosis is designed to protect the delicate alveolar tissue from noxious particles, including infectious agents. Sometimes, however, the process of phagocytosis can itself be more injurious to the lung than the continued presence of free exogenous particles in the alveolar space; many investigators have studied this phenomenon. There is evidence that if a macrophage is overloaded with an inert particle, it will secrete inflammatory mediators. In addition if the particles are very small (ultrafine with a diameter less than 0.1 μm) more inflammation is generated compared to an equivalent mass of larger particles, possibly because the ultrafine particles have more surface area. Surface activity, surface charge and surface smoothness are other potential variables. The presence of transition metals is also believed to contribute to inflammatory potential of ingested particles. Inflammatory cells such as macrophages may exhibit greater inflammatory responses to ingested particulates if they

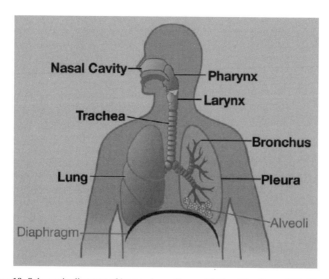

Figure 19. Schematic diagram of human lung. Source National Institute of Health. *http:// www.nlm.nih.gov/medlineplus/lungsandbreathing.html.*

have been primed by an earlier different proinflammatory stimulus such as lipopolysaccharide, a bacterial product. While this is well recognized *in vitro* there are some intriguing clinical and population health correlations. For example the acute death rates in urban areas increase in response to subtle increase in airborne particulate material with most of the additional deaths occurring in people with chronic cardiopulmonary disease. Similar observations of increased mortality and morbidity associated with certain forms of industrial exposure to inhaled particulates, as may be encountered in mining, combined with chronic exposure to cigarette smoke are also consistent with the concept of a "two-hit" process of activation of the inflammatory response with attendant negative impact on health.

At the molecular level, phagocytosis is a complex process, phagocytes, such as macrophages and neutrophils, engulf and destroy invading bacteria. A coating process *with biomolecules*, opsonization, is required for the attack on cellular invaders, which have a negative surface, just like the phagocytes. In addition to alleviating electrostatic repulsion, opsonization leaves the surface covered with molecular "ligands" that can be engaged by the engulfing cell in a process of receptor-mediated phagocytosis. As shown in a study with albumin-coated fluorescent latex particles, reaction of the albumin-coated particles with an anti-albumin antibody followed by deposition of these opsonized particles in the lung caused pulmonary inflammation in hamsters, but the uncoated particles showed little or no response (Kobzik et al. 1993). The implication is that opsonization followed by binding of antibody-coated particles to immunoglobulin receptors in the lung can transform an inert material into a stimulant which is recognized by the lung as pro-inflammatory.

Once the alveolar phagocytes become activated, they start to release signaling compounds (e.g., cytokines, transcription factors, and enzymes) as well as ROS and RNS to counteract the provocation. The production of one subclass of signaling compounds, exemplified by the so-called chemokine Interleukin-8 (IL-8), has the effect of recruiting additional inflammatory cells. While pathogens are typically efficiently engulfed and decomposed by the battery of ROS, RNS, acids, and enzymes that are unleashed within the phagocytes, particles, particularly fibrous particulates, can lead to frustrated attempts at engulfment and disposal. Macrophages are 12 to 18 μm in size. Hence, any particles that approach this length may thwart total engulfment (Fig. 20). This process of "frustrated phagocytosis," a phenomenon seen in both macrophages and neutrophils, is known to induce the production of ROS (Vallyathan et al. 1998). As a result, exposure to even an inert particle may lead to chronic upregulation of various signaling compounds and elevated levels of ROS and RNS (Fubini and Hubbard 2003). Conversely, some particles that show ROS formation *in vitro* may be readily cleared by destruction within the phagosome or migration of the cell

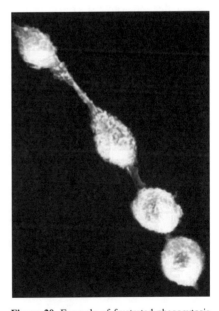

Figure 20. Example of frustrated phagocytosis induced in a Rat AM cell culture by exposure to glass fibers with a length of 17 micron or more. [Used with permission by Dr. Vincent Castranova, National Institute for Occupational Health and Safety, Morgantown, WV. URL: *http://www.epa.gov/oswer/asbestos_ws/docs/castranova.pdf.*]

containing the internalized particles towards the gastrointestinal tract where it is digested; such mechanisms of clearance do not necessarily present a health problem.

The cellular response to the inhalation of "inert" particles is confounded by the state of the cells. It has been shown that pre-existing pulmonary inflammation unrelated to an acute particle exposure event may lead to a stronger proinflammatory response (Stringer and Kobzik 1998; Imrich et al. 2000). In order words, one or more previous stimuli to inflammatory cells in the lung can amplify the ROS formation associated with a subsequent exposure of those cells to particulates. The modulation of the response to a provocation involves a complex system of receptors and cell signaling molecules (Kobzik et al. 1993; Stringer and Kobzik 1998; Imrich et al. 2000). The details of the interactions between minerals and various cells components, cell signaling molecules, and receptors are not well understood.

The effect of inhalation of inert particles has been extensively studied to unravel the cellular response without the confounding factors related to particle reactivity. In many of these studies, cells are exposed to inert polystyrene beads (Palecanda and Kobzik 2000) or TiO_2 (Driscoll et al. 1991; Stringer and Kobzik 1998). In Figure 21, we show the results of a comparative study conducted by our group in which the formation of ROS was detected as conversion of DCFH to its fluorescent product DCF in cells exposed to polystyrene beads, TiO_2, SiO_2, and pyrite. Out of these particles, pyrite shows the highest DCF production. All other particles show no difference from exposure to beads. While TiO_2 is typically thought of to be inert, a recent *in vitro* study in which co-cultures of alveolar macrophages and lung epithelial cells were used showed that TiO_2 does trigger a cellular response in the co-culture, but not in separate alveolar macrophages or lung epithelial cells (Tao and Kobzik 2002). It is thought that the synergistic effect in the co-cultures is the result of cell signaling between the two types of cells (Driscoll et al. 1996; Tao and Kobzik 2002) and may be a more realistic *in*

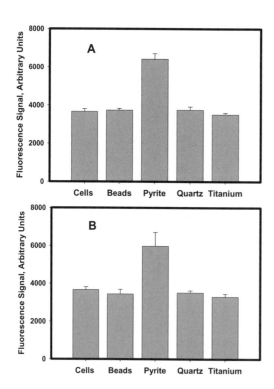

Figure 21. Formation of secondary ROS in MonoMac cells, based on conversion of DCFH to DCF. Panel A (top) MonoMac cells exposed to 400 microgram of mineral per mL of cell culture volume, except for column labeled cells, which contained no minerals. Panel B (bottom), same as Panel A except cells are pretreated and provoked by exposure to Lipid Poly Saccharide (LPS). [Figure based on a presentation by Daccueil F, Fijman A, Roemer EJ, Simon SR (2005) Measures of production of oxidant species in vitro and in cells: assessment for evaluation of possible health risks. Meeting of the Society for In Vitro Biology, Baltimore, MD.]

vitro model of *in vivo* conditions. Huang and coworkers provide more background on mineral-induced cell signaling and cytokine upregulation (Huang et al. 2006).

Minerals as carriers of ROS stimulants (mechanism 5)

The final mechanism in this chapter considers the role of minerals as carriers for compounds that stimulate ROS formation or transformation. In this mechanism the mineral provides a substrate for a stimulant and may promote its transport and residence in the lung. The co-exposures of cigarette smoke and minerals (mostly asbestos) (Valavanidis et al. 1996), diesel exhaust and aerosols (Nikula et al. 1997; Murphy et al. 1998; Garcon et al. 2001; Dybdahl et al. 2003, 2004; Li et al. 2003; Sorensen et al. 2003; Muller et al. 2004; Avogbe et al. 2005), and endotoxins (Imrich et al. 2000) and aerosols (Monn and Becker 1999; Heinrich et al. 2003; Kleinman et al. 2003; Schins et al. 2004) have received considerable attention. The interactions between the mineral substrate and the ROS stimulant often lead to ROS production over and above those seen in the exposures to the isolated components. These augmented responses are partly a reflection of the simultaneous activation of inflammatory cells by two discrete stimuli but are also considered to arise from the physical association of the two stimuli with each other as they trigger the cells.

As an example of augmented stimulation by two activating materials which are in direct association, the response to co-exposure to cigarette smoke and asbestos may be considered. It is well-documented that the co-exposure of tobacco smoke and asbestos leads to an increase in free radicals (Valavanidis et al. 1996; Kamp et al. 1998; Jung et al. 2000; Churg 2003) and other measures of inflammation and cell injury over either smoke or asbestos alone. The synergistic effect between cigarette smoke and asbestos may be caused by several factors. First, the sorption of smoke-derived compounds onto the mineral surface changes its surface properties. It has been shown that the sorption of smoke-derived compounds onto asbestos leads to an increased adhesion onto epithelial cells, thereby increasing the net uptake. The morphology of the mineral particle may lead to low clearance rates, which increases the residence time of the stimulant in the sites where inflammatory cells are most abundant. *In vivo* studies show an increased penetration of asbestos fibers into the walls of the airways. The higher uptake, penetration, and longer residence time of the ROS stimulant *in situ*ations of co-exposure to these two materials eventually leads to an exaggerated fibrotic response, which impairs lung function. Churg (2003) showed that addition of ROS scavengers can lead to a decrease of adhesion and uptake of these materials. This observation suggests that redox reactions may modify the chemical functionality of the adsorbate-substrate system. An ESR study showed that iron derived from asbestos reacts with hydrogen peroxide derived from tobacco-derived tar (Valavanidis et al. 1996).

The interaction of exogenous ROS stimulants and mineral surfaces is an area in which geochemists are poised to make a contribution. Interaction of PAH and other suspected ROS stimulants with mineral surfaces is an active research area in the geosciences and environmental sciences (Gustafsson et al. 1996; Arey 2000; Krauss et al. 2000; Kubicki 2000; Van Metre et al. 2000; Zhu et al. 2004). However, the fate of these composite particles has not been explored by these communities in the presence of inflammatory cells or even cell-free lung fluids or proxies that mimic some of the complexity encountered as composite particles are inhaled.

PART III: SILICOSIS AS AN EXAMPLE OF A MINERAL-INDUCED DISEASE

Silicosis is a form of lung disease resulting from occupational exposure to silica dust over a period of years. Silicosis causes slowly progressive fibrosis of the lungs but the degree of impairment of lung function in workers exposed to silica varies from minimal to respiratory failure. This range of disease severity reflects variations in the composition of silica dust as well as differences in the inflammatory responses of individuals exposed to similar levels of silica.

Silicosis is often associated with occupations, such as mining, masonary, construction, glass and ceramic fabrication, and stone carving, in which workers are exposed to silica dust over long periods of time. Silicosis is, in some cases, an irreversible disease. Even after the exposure ceases the disease may progress. Deposition of silica particles in the lungs leads to chronic inflammation and induces the growth of nodules, which may coalesce into fibrotic growths. Ultimately, such growths in the lungs contribute to respiratory stress and in some cases respiratory failure (Rimal et al. 2005). It has been estimated that about 250 deaths/year are related to silica exposure (Castranova and Vallyathan 2000). While occupational exposures have been well documented, silicosis may also occur in household activity as shown in an example from Saudi Arabia where an Afghani woman was diagnosed with the disease. In that case the exposure was related to cleaning duties as a child that included the daily cleaning of the walls and floor of a mud hut with a hard brush (Safa and Machado 2003). Whether smoking and silicosis are positively correlated remains uncertain (Hessel et al. 2003); however, smokers diagnosed with silicosis have a higher risk of developing lung cancer (Kurihara and Wada 2004). Silicosis may also contribute to immune system deficits. For example, mine workers are often at increased risk for tuberculosis (Davies 2001; Rimal et al. 2005).

The mechanism by which silica exposure leads to silicosis remains an area of active research(Castranova and Vallyathan 2000; Elias et al. 2000; Donaldson et al. 2001; Rimal et al. 2005). As summarized by Castranova and Vallyathan (2000), it is not clear what causes the disease. It has been suggested that silanol groups on silica surfaces lead to a strong interaction with biological membranes, possibly leading to cell damage. Another explanation is that crystalline silica interacts strongly with scavenger receptors on macrophages, thereby stimulating a cellular release of ROS/RNS. Finally, it has been noted that radicals on the silica surface react with water to yield ROS, which may induce a cascade of cell signaling events leading to inflammation. The ability of the solid to generate ROS and trigger cell signaling to induce inflammation has perhaps received most attention, although a combination of mechanisms cannot be ruled out. In the remainder of this section, factors that are of interest to geochemists in relation to silicosis will be briefly reviewed.

Mineralogy

The mineralogical form of the silica exposure is thought to be an important factor in the development of silicosis. In the context of silicosis, the word silica is used to mean SiO_2 polymorphs as well as silicates. One of the most widely used test materials is a quartz sand from a Tertiary formation in Germany (DQ-12 quartz). This sand has been described as 87 to 89% crystalline quartz, some amorphous silica, and a small fraction of kaolinite (Adamis et al. 2000; Donaldson et al. 2001). A chemical analysis of the DQ12 quartz sand indicates that the sand is a very pure quartz sand (Donaldson et al. 2001), with 98.51 wt% SiO_2, less than 0.05 wt% Fe_2O_3, and 0.1 wt% Al_2O_3. In studies that compare the toxicity of various forms of silica, crystalline quartz often emerges as one of the most toxic forms of silica. For example, an *in vivo* and *in vitro* comparison of quartz, diatomaceaous Earth, mordenite and clinoptilolite showed marked difference in toxicity (Adamis et al. 2000). As part of this study the materials were instilled in rats and the response to the instillation was studied over a period of a year. Quartz induced acute, subacute and chronic inflammation and fibrosis in rats. Diatomaceous Earth showed acute and subacute inflammation, but the effects moderated after 60 days. Mordenite showed signs of inflammation, which has been attributed to the fibrous morphology of this zeolite. By contrast, clinoptilolite was inert. Other studies have compared different types of crystalline quartz. Fubini and coworkers showed that different commercial quartz powders have very different biological reactivity *in vivo* and *in vitro* (Bruch et al. 2004; Fubini et al. 2004; Seiler et al. 2004). Donaldson et al. (2001) pointed out that DQ-12 quartz is far more inflammatory than quartz collected at actual worksites. Cazmak et al. (2004) on the other hand found that DQ-12 and several commercial quartz powders elicited about the same response in A549 epithelial cells. Schwarze et al. (2002) studied the inflammatory response of various quarry stones by intratracheal instillation of ground rock samples in rats. They found that quartz content and metal release to solution were poor predictors of inflammatory response. A mylonite quarry stone generated the most pronounced response.

Surface chemistry

The role of surface chemistry in the toxicity of quartz has been well established (Fubini 1998). There are two lines of evidence that corroborate the notion that the surface chemistry of quartz is an important factor in its toxicity. On one hand there are several studies that show that treatment of the surface with a sorbate that also scavenges radicals depresses the toxicity of quartz in *in vivo* studies (Duffin et al. 2001; Albrecht et al. 2004, 2005; Stone et al. 2004). Figure 22 shows that rats instilled with uncoated quartz have elevated hydrogen peroxide concentrations in their alveolar solution and significantly higher levels of MPO, the enzyme that produces HClO (Fig. 23). Fubini attributed the difference in biological activity of four different commercial quartz flowers to their difference in surface composition (Fubini et al. 2004). The other line of evidence is based on studies that evaluated the state of the surface in relation to its capacity to produce ROS and RNS, which is an important step in the development of the disease. Fubini and coworkers showed, using solid state ESR, that freshly

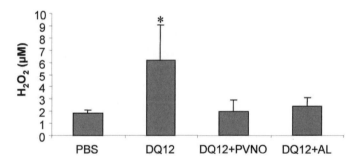

Figure 22. H_2O_2 generation in the rat lung. H_2O_2 levels were analyzed spectrophotometrically in the BALF obtained from rats exposed to non-coated DQ12 or DQ12 coated with AL or PVNO (7 days after i.t. instillation). Data are expressed as mean ± SD (n = 5). *p < 0.01 vs. PBS (ANOVA, Tukey). [Permission under Creative Commons Attribution License. Original citation Albrecht et al. (2005) Respiratory Research 6:129. *http://respiratory-research.com/content/6/1/129*]

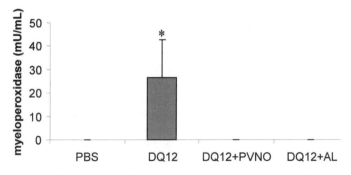

Figure 23. Myeloperoxidase activity in BALF of rat lungs 7 days following i.t. instillation of 2 mg DQ12 or DQ12 coated with either AL or PVNO. Data are expressed as mean ± SD (n = 5). *p < 0.01 vs. PBS (ANOVA, Tukey). [Permission under Creative Commons Attribution License. Original citation Albrecht et al. (2005) Respiratory Research 6:129. *http://respiratory-research.com/content/6/1/129*]

fractured quartz has a higher surface concentration of radicals. These surface radicals react with water to produce hydroxyl radicals (Vallyathan et al. 1998). Reactions with NO may yield peroxynitrite. The reactivity imparted by the mechanical stress dissipates over the course of days when the material is left to react with ambient air (Fubini et al. 1990). Vallyathan et al. (1995) instilled freshly ground quartz and aged quartz powder in rats and found that the freshly ground quartz induced a far more pronounced inflammatory response. Heating of a fibrous crystobalite to 1300 °C inactivated an otherwise cytoxic material by removing surface radicals and increasing its hydrophobicity (Fubini et al. 1999).

Metal impurities and Co-exposures

While pure quartz has been shown to induce an inflammatory reaction in rats, the effect is enhanced by the presence of iron (Castranova et al. 1997) and coal dust (Donaldson et al. 2001; Stone et al. 2004). Iron is a common natural trace constituent in quartz. In addition, grinding of quartz is likely to introduce iron, which increases its reactivity (Fenoglio et al. 2001, 2003). The augmentation of the inflammatory response, although not uniform (Donaldson et al. 2001), is likely due to the fact that ferrous/ferric iron can act as a Fenton element. In our group we have shown that iron sorbed onto pure quartz can drastically accelerate the degradation of RNA, which presumably requires the formation of hydroxyl radical (Fig. 11). Coal contains quartz and silicates as mineral components. The prevalence of silicosis among coal miners has been linked to the co-exposure to these minerals and coal dust (Borm 2002). Pyrite is another common mineral component in coal. Its presence in coal may be a factor in silicosis as pyrite is know to spontaneously generate hydroxyl radicals (Cohn et al. 2006).

PART IV: CONCLUDING REMARKS

There is growing *in vitro* and *in vivo* evidence that minerals such as asbestos, quartz, pyrite, and forsterite can induce the formation of ROS. The development and adaptation of ROS assays for use in mineral suspensions allows for broad testing of Earth materials. By further developing acellular assays it may become possible to rapidly assess the ROS-formation potential for Earth materials. These results can then be combined with assays of further ROS generation in the presence of cells as well as other biomedical *in vitro* tests. In aggregate these *in vitro* results can provide useful guidance for subsequent *in vivo* tests (Fubini 1996), which are the preferred basis for a toxicological assessment (Bernstein et al. 2005).

Although some of the mechanisms by which minerals induce ROS in aqueous solutions have been resolved, much remains largely unknown. However, surface chemistry appears to play an important role as it is thought to influence adhesion, interaction with receptors, sorption of ROS stimulants, and the direct formation of ROS. Geochemists are in a unique position to contribute to a better fundamental understanding of the role of minerals in diseases such as silicosis and asbestosis. Experimental geochemists can contribute by synthesizing materials for proof-of-concept studies, characterize synthetic as well as natural materials and surfaces, and induce mechanical stress in controlled ways. Theoretical geochemists can team up with their biochemical colleagues to explore the interaction of mineral surfaces with receptor molecules as well as ROS stimulants, such as PAH and components of cigarette smoke, to better understand cellular ROS formation. The first step is to form close collaborations among the relevant disciplines, start cross-disciplinary discussions, and develop common objectives. It is our hope that this chapter is helpful in starting this process.

ACKNOWLEDGMENTS

The authors are grateful for seed funds provided by the Office of the VP for Research at Stony Brook and support from the Center for Environmental Molecular Science at Stony

Brook (CEMS). CEMS is supported by NSF-Chemistry. Research reported in this chapter has also been funded in part by US-DOE and NSF-EAR. The 3MT graduate training and research program at Stony Brook is supported by an NSF-IGERT award. Nita Sahai is thanked for handling the review of this contribution and three reviewers are thanked for their insights.

REFERENCES

Achimovicova M, Balaz P (2005) Influence of mechanical activation on selectivity of acid leaching of arsenopyrite. Hydrometallurgy 77:3-7

Adamis Z, Tátrai E, Honma K, Six E, Ungváry G (2000) *In vitro* and *in vivo* tests for determination of the pathogenicity of quartz, diatomaceous Earth, mordenite and clinoptilolite. Ann Occup Hyg 44:67-74

Ahlberg E, Broo AE (1997) Electrochemical reaction mechanisms at pyrite in acidic perchlorate solutions. J Electrochem Soc 144:1281-1286

Ahlberg E, Broo AE (1996) Oxygen reduction at sulphide minerals. 3. The effect of surface pre-treatment on the oxygen reduction at pyrite. Int J Miner Process 47:49-60

Albrecht C, Knaapen AM, Becker A, Höhr D, Haberzettl P, van Schooten FJ, Borm PJ, Schins RP (2005) The crucial role of particle surface reactivity in respirable quartz-induced reactive oxygen/nitrogen species formation and APE/Ref-1 induction in rat lung. Respir Res 6:129

Albrecht C, Schins RPF, Hohr D, Becker A, Shi T, Knaapen AM, Borm PJA (2004) Inflammatory time course after quartz instillation. Am J Respir Cell Biol 31:292-301

Alegría Y, Liendo F, Núñeza O (2003) On the Fenton degradation mechanism. The role of oxalic acid. ARKIVOC 2003(x):538-549

Ames BN (1987) Oxidative DNA damage, cancer, and aging. Ann Intern Med 107:526–545

Arey J (2000) Urban air--causes and consequences of urban air pollution. *In:* Environmental Medicine. Moller L (ed) Joint Industrial Safety Council, p 52-71

Atkins P, de Paula J (2002) Physical Chemistry, 7th ed. Freeman

Avogbe PH, Ayi-Fanou L, Autrup H, Loft S, Fayomi B, Sanni A, Vinzents P, Moller P (2005) Ultrafine particulate matter and high-level benzene urban air pollution in relation to oxidative DNA damage. Carcinogenesis 26:613-620

Babior BM (2000) Phagocytes and oxidative stress. Am J Med 109:33-44

Balaz P (2003) Mechanical activation in hydrometallurgy. Int J Miner Process 72:341-354

Balaz P, Boldizarova E, Godocikova E, Briancin J (2003) Mechanochemical route for sulphide nanoparticles preparation. Mater Lett 57:1585-1589

Balaz P, Godocikova E, Kril'ova L, Lobotka P, Gock E (2004) Preparation of nanocrystalline materials by high-energy milling. Mater Sci Eng A 386:442-446

Balaz P, Kusnierova M, Varencova VI, Misura B (1994) Mineral properties and bacterial leaching of intensively ground sphalerite and sphalerite pyrite mixture. Int J Miner Process 40:273-285

Banks D, Younger PL, Arnesen R-T, Iversen ER, Banks SB (1997) Mine-water chemistry: the good, the bad and the ugly. Environ Geol 32:157-174

Bard AJ, Parsons R, Jordan J (1985) Standard Potentials in Aqueous Solution. Marcel Dekker

Berner EK, Berner RA (1996) Global Environment. Prentice-Hall

Bernstein D, Castranova V, Donaldson K, Fubini B, Hadley J, Hesterberg T, Kane A, Lai D, McConnell EE, Muhle H, Oberdorster G, Olin S, Warheit DB, Group IRSIW (2005) Testing of fibrous particles: short-term assays and strategies. Inhal Toxicol 17:497-537

Biegler T, Rand DAJ, Woods R (1975) Oxygen reduction on sulphide minerals. Part I. Kinetics and mechanism at rotated pyrite electrodes. J Electroanal Chem 60:151-162

Blough NV, Simpson DJ (1988) Chemically mediated fluorescence yield switching in nitroxide-fluorophore adducts: iotical sensors of radical/redox reactions. J Am Chem Soc 110:1915-1917

Borda MJ, Strongin DR, Schoonen MA (2004) A vibrational spectroscopic study of the oxidation of pyrite by molecular oxygen. Geochim Cosmochim Acta 68:1807-1813

Borg RJ, Dienes GJ (1992) The Physical Chemistry of Solids. Academic Press

Borisenko GG, Martin I, Zhao Q, Amoscato AA, Tyrunia YY, Kagan VE (2004) Glutathione propagates oxidative stress triggered by myeloperoxidase in HL-60 cells. J Biol Chem 279:23453-23462

Borm PJA (2002) Particle toxicology: From coal mining to nanotechnology. Inhal Toxicol 14:311-324

Brenowitz M, Chance MR, Dhavan G, Takamoto K (2002) Probing the structural dynamics of nucleic acids by quantitative time-resolved and equilibruim hydroxyl radical 'footprinting'. Curr Opin Struct Biol 12:648-653

Brook RD, Brook JR, Rajagopalan S (2003) Air pollution: the "heart" of the problem. Curr Hypertension Rep 5:32-39

Bruch J, Rehn S, Rehn B, Borm PJ, Fubini B (2004) Variation of biological responses to different respirable quartz flours determined by a vector model. Int J Hyg Environ Health 207:203-216

Bruker Biospin (2006) What is EPR? URL: *http://www.bruker-biospin.com/brukerepr/whatiseprintroduction.h tml* (accessed October 2006)

Buerge IJ, Hug SJ (1999) Influence of mineral surfaces on chromium(VI) reduction by iron(II). Environ Sci Technol 33:4285-4291

Burney S, Caulfield JL, Niles JC, Wishnok JS, Tannenbaum SR (1999) The chemistry of DNA damage from nitric oxide and peroxynitrite. Mutat Res 424:37-49

Cakmak GD, Schins RP, Shi T, Fenoglio I, Fubini B, Borm PJ (2004) *In vitro* genotoxity assessment of commercial quartz flours in comparison to standard DQ12 quartz. Int J Hyg Environ Health 207:105-113

Calas G (1988) Electron paramagnetic resonance. Rev Mineral 18:513-572

Calkins WH (1994) The chemical forms of sulfur in coal - a review. Fuel 73:475-484

Casadevall T, Cruz-Reyna S, Rose W, Bagley S, Finnegan D, Zoller W (1984) Crater lake and post-eruption hydrothermal activity, El Chichon volcano, Mexico. J Volcanol Geotherm Res 23:169-191

Castranova V, Vallyathan V (2000) Silicosis and coal workers' pneumoconiosis. Environ Health Perspect 108: 675-684

Castranova V, Vallyathan V, Ramsey D, McLaurin J, Pack D, Leonard S, Barger M, Ma J, Dalal N, Teass A (1997) Augmentation of pulmonary reactions to quartz inhalation by trace amounts of iron-containing particles. Environ Health Perspect 105 Suppl 5:1319-1324

Churg A (2003) Interactions of exogenous or evoked agents and particles: the role of reactive oxygen species. Free Radic Biol Med 34:1230-1235

Cohn C, Mueller S, Wimmer E, Leifer N, Greenbaum S, Strongin DR, Schoonen M (2006) Pyrite-induced hydroxyl radical formation and its effect on nucleic acids. Geochem Trans 7:3 (4 April 2006)

Cohn C, Mueller S, Wimmer E, Leifer ND, Greenbaum SG, Schoonen M (2005a) Mineral-generated reactive oxygen species. *In:* Abstracts of papers of the American Chemical Society 229:U884-U884 017-GEOC Part 1, March 13, 2005

Cohn CA, Borda MJ, Schoonen MA (2004) RNA decomposition by pyrite-induced radicals and possible role of lipids during the emergence of life. Earth Planet Sci Lett 225:271-278

Cohn CA, Laffers R, Schoonen MA (2005b) Using yeast RNA as a probe for generation of hydroxyl radicals by Earth materials. Environ Sci Technol 40(8):2838-2843

Cohn CA, Pak A, Schoonen MAA, Strongin DR (2005c) Quantifying hydrogen peroxide in iron-containing solutions using leuco crystal violet. Geochem Trans 6:47-52

Cooper W, Zika R (1983) Photochemical formation of hydrogen peroxide in surface and ground waters. Science 220:711-712

Cooper W, Zika R, Petanse R, Plane J (1988) Photochemical formation of H_2O_2 in natural waters exposed to sunlight. Environ Sci Technol 22:1156-1160

Costa D, Guignard J, Pezerat H (1989a) Production of free-radicals arising from the surface-activity of minerals and oxygen. 2. Arsenides, sulfides, and sulfoarsenides of iron, nickel, and copper. Toxicol Ind Health 5: 1079-1097

Costa D, Guignard J, Zalma R, Pezerat H (1989b) Production of free radicals arising from the surface activity of minerals and oxygen. Part I. Iron mine ores. Toxicol Ind Health 5:1061-1078

Daghino S, Martino E, Fenoglio I, Tomatis M, Perotto S, Fubini B (2005) Inorganic materials and living organisms: surface modifications and fungal responses to various asbestos forms. Chemistry 11:5611-5618

Davies JC (2001) Silicosis and tuberculosis among South African goldminers--an overview of recent studies and current issues. S Afr Med J 91:562-566

Davies SHR, Morgan JJ (1989) Manganese(II) oxidation kinetics on oxide surfaces. Colloids Surf A Physicochem Eng Asp 129:63-77

Deng B, Stone AT (1996) Surface-catalyzed chromium(IV) reduction: reactivity comparisons of different organic reductants and different oxide surfaces. Environ Sci Technol 30:2484-2494

Dikalova AE, Kadiiska MB, Mason RP (2001) An *in vivo* ESR spin-trapping study: Free radical generation in rats from formate intoxication— role of the Fenton reaction. PNAS 98:13549-13553

Donaldson K, Stone V, Duffin R, Clouter A, Schins R, Borm P (2001) The quartz hazard: effects of surface and matrix on inflammogenic activity. J Environ Pathol Toxicol Oncol 20 Suppl 1:109-118

Driscoll K, Carter J, Hassenbein D, Howard B (1997) Cytokines and particle-induced inflammatory cell recruitment. Environ Health Perspect 105 Suppl 5:1159-1164

Driscoll KE, Howard BW, Carter JM, Asquith T, Johnston C, Detilleux P, Kunkel SL, Isfort RJ (1996) Alpha-quartz-induced chemokine expression by rat lung epithelial cells: effects of *in vivo* and *in vitro* particle exposure. Am J Pathol 149:1627-1637

Driscoll KE, Lindenschmidt RC, Maurer JK, Perkins L, Perkins M, Higgins J (1991) Pulmonary response to inhaled silica or titanium dioxide. Toxicol Appl Pharmacol 111:201-210

Duffin R, Gilmour PS, Schins RP, Clouter A, Guy K, Brown DM, MacNee W, Borm PJ, Donaldson K, Stone V (2001) Aluminium lactate treatment of DQ12 quartz inhibits its ability to cause inflammation, chemokine expression, and nuclear factor-kappaB activation. Toxicol Appl Pharmacol 176:10-17

Dybdahl M, Risom L, Bornholdt J, Autrup H, Loft S, Wallin H (2004) Inflammatory and genotoxic effects of diesel particles *in vitro* and *in vivo*. Mutat Res 562:119-131

Dybdahl M, Risom L, Moller P, Autrup H, Wallin H, Vogel U, Bornholdt J, Daneshvar B, Dragsted LO, Weimann A, Poulsen HE, Loft S (2003) DNA adduct formation and oxidative stress in colon and liver of Big Blue (R) rats after dietary exposure to diesel particles. Carcinogenesis 24:1759-1766

Eggleston CM, Ehrhardt J-J, Stumm W (1996) Surface structural controls on pyrite oxidation kinetics: An XPS-UPS, STM, and modeling study. Am Mineral 81:1036-1056

Elias Z, Poirot O, Daniere MC, Terzetti F, Marande AM, Dzwigaj S, Pezerat H, Fenoglio I, Fubini B (2000) Cytotoxic and transforming effects of silica particles with different surface properties in Syrian hamster embryo (SHE) cells. Toxicol *in vitro* 14:409-422

Emmenegger L (1998) Oxidation kinetics of Fe(II) in a eutrophic swiss lake. Environ Sci Technol 32:2990-2996

EPA US (1998) Field Applications of *In situ* Remediation Technologies: Chemical Oxidation. Environmental Protection Agency, Washington D.C., Report # EPA 542-R-98-008

Fach E, Waldman WJ, Williams M, Long J, Meister RK, Dutta PK (2002) Analysis of the biological and chemical reactivity of zeolite-based aluminosilicate fibers and particulates. Environ Health Perspect 110: 1087-1096

Favero-Longo SE, Turci F, Tomatis M, Castelli D, Bonfante P, Hochella MF, Piervittori R, Fubini B (2005) Chrysotile asbestos is progressively converted into a non-fibrous amorphous material by the chelating action of lichen metabolites. J Environ Monit 7:764-766

Fenoglio I, Croce A, Di Renzo F, Tiozzo R, Fubini B (2000a) Pure-silica zeolites (Porosils) as model solids for the evaluation of the physicochemical features determining silica toxicity to macrophages. Chem Res Toxicol 13:489-500

Fenoglio I, Fonsato S, Fubini B (2003) Reaction of cysteine and glutathione (GSH) at the freshly fractured quartz surface: a possible role in silica-related diseases? Free Radic Biol Med 35:752-762

Fenoglio I, Martra G, Coluccia S, Fubini B (2000b) Possible role of ascorbic acid in the oxidative damage induced by inhaled crystalline silica particles. Chem Res Toxicol 13:971-975

Fenoglio I, Prandi L, Tomatis M, Fubini B (2001) Free radical generation in the toxicity of inhaled mineral particles: the role of iron speciation at the surface of asbestos and silica. Redox Report 6:235-241

Fenton HJH (1894) Oxidation of tartaric acid in the presence of iron. J Chem Soc 65:899-910

Fubini B (1998) Surface chemistry and quartz hazard. Ann Occup Hyg 42:521-530

Fubini B (1996) Use of physico-chemical and cell-free assays to evaluate the potential carcinogenicity of fibres. IARC Sci Publ 140:35-54

Fubini B, Bolis V, Giamello E (1987) The surface chemistry of crushed quartz dust in relation to its pathogenicity. Inorg Chim Acta 138:193-197

Fubini B, Fenoglio I, Ceschino R, Ghiazza M, Martra G, Tomatis M, Borm P, Schins R, Bruch J (2004) Relationship between the state of the surface of four commercial quartz flours and their biological activity *in vitro* and *in vivo*. Int J Hyg Environ Health 207:89-104

Fubini B, Fenoglio I, Elias Z, Poirot O (2001a) On the variability of the biological responses to silicas: effect of origin, ctrystallinity and state of the surface on the generation of reactive oxygen species and consequent morphological transformations in cells. J Environ Pathol Toxicol Oncol 20:87-100

Fubini B, Fenoglio I, Elias Z, Poirot O (2001b) Variability of biological responses to silicas: effect of origin, crystallinity, and state of surface on generation of reactive oxygen species and morphological transformation of mammalian cells. J Environ Pathol Toxicol Oncol 20 Suppl 1:95-108

Fubini B, Giamello E, Volante M, Bolis V (1990) Chemical functionalities at the silica surface determining its reactivity when inhaled. Formation and reactivities of surface radicals. Toxicol Ind Health 6:571-598

Fubini B, Hubbard A (2003) Reactive oxygen species (ROS) and reactive nitrogen species (RNS) generation by silica in inflammation and fibrosis. Free Radic Biol Med 34:1507-1516

Fubini B, Mollo L, Giamello E (1995) Free radical generation at the solid/liquid interface in iron containing minerals. Free Radic Res 23:593-614

Fubini B, Zanetti G, Altilia S, Tiozzo R, Lison D, Saffiotti U (1999) Relationship between surface properties and cellular responses to crystalline silica: studies with heat-treated cristobalite. Chem Res Toxicol 12: 737-745

Fujimoto H, Sakata K, Fukui K (1996) Transient bonds and chemical reactivity of molecules. Int J Quantum Chem 60:401-408

Garcon G, Zerimech F, Hannothiaux MH, Gosset P, Martin A, Marez T, Shirali P (2001) Antioxidant defense disruption by polycyclic aromatic hydrocarbons-coated onto Fe_2O_3 particles in human lung cells (A549). Toxicology 166:129-137

Gazzano E, Foresti E, Lesci IG, Tomatis M, Riganti C, Fubini B, Roveri N, Ghigo D (2005) Different cellular responses evoked by natural and stoichiometric synthetic chrysotile asbestos. Toxicol Appl Pharmacol 206:356-364

Gilbert DL, Colton CA (1999) Reactive Oxygen Species in Biological Systems An Interdisciplinary Approach. Kluwer/Plenum

Goldstone JV, Pullin MJ, Bertilsson S, Voelker BM (2002) Reactions of hydroxyl radical with humic substances: Bleaching, mineralization, and production of bioavailable carbon substrates. Environ Sci Technol 36:364-372

Goldstone JV, Voelker BM (2000) Chemistry of superoxide radical in seawater: CDOM associated sink of superoxide in coastal waters. Environ Sci Technol 34:1043-1048

Gordy W (1980) Theory and Applications of Electron Spin Resonance. Wiley

Graf DL, Goldsmith JR (1955) Dolomite-magnesian calcite relations at elevated temperatures and CO_2 pressures. Geochim Cosmochim Acta 7:109-128

Graf E, Mahoney JR, Bryant RG, Eaton JW (1984) Iron-catalyzed hydroxyl radical formation. Stringent requirement for free iron coordination site. J Biol Chem 259:3620-3624

Gustafsson O, Haghseta F, Chan C, MacFarlane J, Gschwend PM (1996) Quantification of the dilute sedimentary soot phase: implications for PAH speciation and bioavailability. Environ Sci Technol 31:203-209

Gyulkhandanyan AV, Pennefather PS (2004) Shift in the localization of sites of hydrogen peroxide production in brain mitochondria by mitochondrial stress. J Neurochem 90:405-421

Hakkinen PJ, Anesio AM, Graneli W (2004) Hydrogen peroxide distribution, production, and decay in boreal lakes. Can J Fish Aquat Sci 61:1520-1527

Halliwell B, Gutteridge JMC (1990) Role of free radicals and catalytic metal ions in human disease: an overview. Methods Enzymol 189:1-85

Hardy JA, Aust AE (1995) Iron in asbestos chemistry and carcinogenicity. Chem Rev 95:97-118

Harrison PM, Arosio P (1996) The ferritins: molecular properties, iron storage function and cellular regulation. Biochim Biophys Acta 1275:161-203

Heilek GM, Noller HF (1996) Directed hydroxyl radical probing of the rRNA neighborhood of ribosomal protein S13 using tethered Fe(II). RNA 2:597-602

Heinrich J, Pitz M, Bischof W, Krug N, Borm PJA (2003) Endotoxin in fine (PM2.5) and coarse (PM2.5-10) particle mass of ambient aerosols. A temporo-spatial analysis. Atmos Environ 37:3659-3667

Heiser I, Elstner EF (1998) The biochemistry of plant stress and disease: oxygen activation as a basic principle. Ann NY Acad Sci 851:224-232

Herut B, Shoham-Frider E, Kress N, Fiedler U, Angel DL (1998) Hydrogen peroxide production rates in clean and polluted coastal marine waters of the Mediterranean, Red and Baltic Seas. Mar Pollut Bull 36:994-1003

Hessel PA, Gamble JF, Nicolich M (2003) Relationship between silicosis and smoking. Scand J Work Environ Health 29:329-336

Hetland RB, Myhre O, Lag M, Hongve D, Schwarze PE, Refsnes M (2001) Importance of soluble metals and reactive oxygen species for cytokine release induced by mineral particles. Toxicology 165:133-144

Holm T, George G, Barcelona M (1987) Fluorometric determination of hydrogen peroxide in groundwater. Anal Chem 59:582-586

Hu HP, Chen QY, Yin ZL, Zhang PM, Wang GF (2004) Effect of grinding atmosphere on the leaching of mechanically activated pyrite and sphalerite. Hydrometallurgy 72:79-86

Hu HP, Chen QY, Yin ZL, Zhang PM, Ye LS (2003) Effect of aging conditions on the leaching of mechanically activated pyrite and sphalerite. Metall Mater Trans B 34:639-645

Huang X, Gordon T, Rom WN, Finkelman RB (2006) Interaction of iron and calcium minerals in coals and their roles in coal dust-induced health and environmental problems. Rev Mineral Geochem 64:153-178

Imrich A, Ning YY, Kobzik L (2000) Insoluble components of concentrated air particles mediate aveolar macrophage responses *in vitro*. Toxicol Appl Pharmacol 167:140-150

Iwata T, Yano E (2003) Reactive oxygen metabolite production induced by asbestos and glass fibers: Effect of fiber milling. Ind Health 41:32-38

Janisch K, Schempp H (2004) Evaluation of the oxidative burst in suspension cell culture of Phaseolus vulgaris. Z Naturforsch C 59:849-855

Jung M, Davis WP, Taatjes DJ, Churg A, Mossman BT (2000) Asbestos and cigarette smoke cause increased DNA strand breaks and necrosis in bronchiolar epithelial cells *in vivo*. Free Radic Biol Med 28:1295-1299

Kamp DW, Graceffa P, Pryor WA, Weitzman SA (1992) The role of free radicals in asbestos-induced diseases. Free Radic Biol Med 12:293-315

Kamp DW, Greenberger MJ, Sbalchierro JS, Preusen SE, Weitzman SA (1998) Cigarette smoke augments asbestos-induced alveolar epithelial cell injury: role of free radicals. Free Radic Biol Med 25:728-739

Kawanishi S, Hiraku Y, Murata M, Oikawa S (2002) The role of metals in site-specific DNA damage with reference to carcinogenesis. Free Radic Biol Med 32:822–832

Khan N, Wilmot CM, Rosen GM, Demidenko E, Sun J, Joseph J, rsquo, Hara J, Kalyanaraman B, Swartz HM (2003) Spin traps: *in vitro* toxicity and stability of radical adducts. Free Radic Biol Med 34:1473-1481

Kieber RJ, Cooper WJ, Willey JD, Avery GB (2001) Hydrogen peroxide at the Bermuda Atlantic Time Series Station. Part 1: Temporal variability of atmospheric hydrogen peroxide and its influence on seawater concentrations. J Atmos Chem 39:1-13

King EJ, Rogers N, Gilchrist M, Goldschmidt VM, Nagelschmidt G (1945) The effect of olivine on the lungs of rats. J Pathol Bacteriol 57:488-491

Kirillova YA, Yusupov TS, Shumskaya LG, Asanov IP (2000) Flotation behavior of sulfides on mechanical activation. J Mining Sci 36:87-90

Kleinman MT, Sioutas C, Chang MC, Boere AJF, Cassee FR (2003) Ambient fine and coarse particle suppression of alveolar macrophage functions. Toxicol Lett 137:151-158

Knaapen AM, Borm PJA, Albrecht C, Schins RPF (2004) Inhaled particles and lung cancer. Part A: Mechanisms. Int J Cancer 109:799-809

Kobzik L, Huang S, Paulauskis JD, Godleski JJ (1993) Particle opsonization and lung macrophage cytokine response. *In vitro* and *in vivo* analysis. J Immunol 151:2753-2759

Kosaka K, Yamada H, Matsui S, Echigo S, Shishida K (1998) Comparison among the methods for hydrogen peroxide measurements to evaluate advanced oxidation processes: Application of a spectrophotometric method using copper(II) ion and 2,9-dimethyl-1,10-phenanthroline. Environ Sci Technol 32:3821-3824

Krauss M, Wilcke W, Zech W (2000) Polycyclic aromatic hydrocarbons and polychlorinated biphenyls in forest soils: depth distribution as indicator of different fate. Environ Pollut 110:79 - 88

Kubicki JD (2000) Molecular mechanics and quantum mechanical modeling of hexane soot structure and interactions with pyrene. Geochem Trans 1:41-46

Kurihara N, Wada O (2004) Silicosis and smoking strongly increase lung cancer risk in silica-exposed workers. Ind Health 42:303-314

Kwan WP, Voelker BM (2004) Influence of electrostatics on the oxidation rates of organic compounds in heterogeneous Fenton systems. Environ Sci Technol 38:3425-3431

Kwan WP, Voelker BM (2003) Rates of hydroxyl radical generation and organic compound oxidation in mineral-catalyzed Fenton-like systems. Environ Sci Technol 37:1150-1158

Kwan WP, Voelker BM (2002) Decomposition of hydrogen peroxide and organic compounds in the presence of dissolved iron and ferrihydrite. Environ Sci Technol 36:1467-1476

Lasaga AC (1998) Kinetic Theory in the Earth Sciences. Princeton University Press

LeBel CP, Ischiropoulos H, Bondy SC (1992) Evaluation of the probe 2',7'-dichlorofluorescin as an indicator of reactive oxygen species formation and oxidative stress. Chem Res Toxicol 5:227-231

Leonard SS, Harris GK, Shi XL (2004) Metal-induced oxidative stress and signal transduction. Free Radic Biol Med 37:1921-1942

Levitzky MG (2003) Pulmonary Physiology, 6th Edition. McGraw-Hill

Li B, Gutierrez PL, Blough NV (1997) Trace determination of hydroxyl radical in biological systems. Anal Chem 69:4295-4302

Li N, Sioutas C, Cho A, Schmitz D, Misra C, Sempf J, Wang M, Oberley T, Froines J, Nel A (2003) Ultrafine particulate pollutants induce oxidative stress and mitochondrial damage. Environ Health Perspect 111: 455-460

Lison D, Lardot C, Huaux F, Zanetti G, Fubini B (1997) Influence of particle surface area on the toxicity of insoluble manganese dioxide dusts. Arch Toxicol 71:725-729

Lloyd DR, Phillips DH (1999) Oxidative DNA damage mediated by copper(II), iron(II) and nickel(II) Fenton reactions: evidence for site-specific mechanisms in the formation of double-strand breaks, 8-hydroxydeoxyguanosine and putative intrastrand cross-links. Mutat Res 424:23-36

Ma J, Song W, Chen C, Ma W, Zhao J, Tang Y (2005) Fenton degradation of organic compounds promoted by dyes under visible irradiation. Environ Sci Technol 39:5810-5815

Makino K, Hagiwara T, Hagi A, Nishi M, Murakami A (1990) Cautionary note for DMPO spin trapping in the presence of iron ion. Biochem Biophys Res Commun 172:1073-1080

Martin ST (2005) Precipitation and dissolution of iron and manganese oxides. *In:* Environmental Catalysis. Grassian V (ed) CRC Press, p 61-81

Martino E, Cerminara S, Prandi L, Fubini B, Perotto S (2004) Physical and biochemical interactions of soil fungi with asbestos fibers. Environ Toxicol Chem 23:938-944

Martino E, Prandi L, Fenoglio I, Bonfante P, Perotto S, Fubini B (2003) Soil fungal hyphae bind and attack asbestos fibers. Angew Chem Int Ed Engl 42:219-222

Martra G, Tomatis M, Fenoglio I, Coluccia S, Fubini B (2003) Ascorbic acid modifies the surface of asbestos: possible implications in the molecular mechanisms of toxicity. Chem Res Toxicol 16:328-335

Mason B (1992) Victor Moritz Goldschmidt: Father of Modern Geochemistry. Special Publication No. 4. The Geochemical Society

Monn C, Becker S (1999) Cytotoxicity and induction of proinflammatory cytokines from human monocytes exposed to fine (PM2.5) and coarse particles (PM10-2.5) in outdoor and indoor air. Toxicol Appl Pharmacol 155:245-252

Morris CJ, Earl JR, Trenam CW, Blake DR (1995) Reactive oxygen species and iron - a dangerous partnership in inflammation. Int J Biochem Cell Biol 27:109-122

Moses CO, Nordstrom DK, Herman JS, Mills AL (1987) Aqueous pyrite oxidation by dissolved oxygen and by ferric iron. Geochim Cosmochim Acta 51:161-1572

Mottola HA, Simpson BE, Gorin G (1970) Absorptiometric determination of hydrogen peroxide in submicrogram amounts with leuco crystal violet and peroxidase as catalyst. Anal Chem 42:410-411

Muller AK, Farombi EO, Moller P, Autrup HN, Vogel U, Wallin H, Dragsted LO, Loft S, Binderup ML (2004) DNA damage in lung after oral exposure to diesel exhaust particles in Big Blue (R) rats. Mutat Res 550: 123-132

Murphy SA, BéruBé KA, Pooley FD, Richards RJ (1998) The response of lung epithelium to well characterised fine particles. Life Sci 62:1789-1799

Narayanasamy J, Kubicki JD (2005) Mechanism of hydroxyl radical generation from a silica surface: molecular crbital calculations. J Phys Chem B 109: 21796-21807

NIH (2006) Genetics home reference. U.S. National Library of Medicine, *http://ghr.nlm.nih.gov/*

Nikula KJ, Avila KJ, Griffith WC, Mauderly JL (1997) lung tissue responses and sites of particle retention differ between rats and Cynomolgus monkeys exposed chronically to diesel exhaust and coal dust*1. Fundam Appl Toxicol 37:37-53

Ohyama M, Otake T, Morinaga K (2001) Effect of size of man-made and natural mineral fibers on chemiluminescent response in human monocyte-derived macrophages. Environ Health Perspect 109: 1033-1038

Palecanda A, Kobzik L (2000) Alveolar macrophage-environmental particle interaction: analysis by flow cytometry. Methods 21:241-247

Peake BM, Mosley LM (2004) Hydrogen peroxide concentrations in relation to optical properties in a fiord (Doubtful Sound, New Zealand). N Z J Mar Freshwater Res 38:729-741

Peng YJ, Grano S, Fornasiero D, Ralston J (2003) Control of grinding conditions in the flotation of galena and its separation from pyrite. Int J Miner Process 70:67-82

Petigara BR, Blough NV, Mignerey AC (2002) Mechanisms of hydrogen peroxide decomposition in soils. Environ Sci Technol 36:639-645

Pierre JL, Fontecave M (1999) Iron and activated oxygen species in biology: the basic chemistry. Biometals 12: 195-199

Plumlee GS, Morman SA, Ziegler TL (2006) The toxicological geochemistry of earth materials: an overview of processes and the interdisciplinary methods used to understand them. Rev Mineral Geochem 64:5-57

Pou S, Bhan A, Bhadti VS, Wu SY, Hosmane RS, Rosen GM (1995) The use of fluorophore-containing spin traps as potential probes to localize free radicals in cells with fluorescence imaging methods. FASEB J 9: 1085 - 1090

Pou S, Huang YI, Bhan A, Bhadti VS, Hosmane RS, Wu SY, Cao GL, Rosen GM (1993) A fluorophore-containing nitroxide as a probe to detect superoxide and hydroxyl radical generated by stimulatd neutrophils. Anal Biochem 212:85-90

Price D, Fauzi R, Mantoura C, Worsfold PJ (1998) Shipboard determination of hydrogen peroxide in the western Mediterranean sea using flow injection with chemiluminescence detection. Anal Chim Acta 377:145-155

Pryor WA (1986) Oxy-radicals and related species: their formation, lifetimes, and reactions. Annu Rev Physiol 48:657-663

Reeder RJ, Schoonen MAA, Lanzirotti A (2006) Metal speciation and its role in bioaccessibility and bioavailability. Rev Mineral Geochem 64:59-113

Rice-Evans CA, Diplock AT, Symons MCR (1991) Techniques in free radical research. *In*: Laboratory Techniques in Biochemistry and Molecular Biology. Burdon RH, Knippenberg PH (eds) Elsevier, p 292

Rickard D, Schoonen MAA, Luther GW (1995) Chemistry of iron sulfides in sedimentary environments. *In*: Geochemical Transformations of Sedimentary Sulfur. Vairavamurthy MA, Schoonen MAA (eds) American Chemical Society, Symposium Series 612:168-193

Rimal B, Greenberg AK, Rom WN (2005) Basic pathogenetic mechanisms in silicosis: current understanding. Curr Opin Pulm Med 11:169-173

Rose AL, Waite TD (2005) Reduction of organically complexed ferric iron by superoxide in a simulated natural water. Environ Sci Technol 39:2645-2650

Rosso KM, Becker U, Hochella MFJ (1999) The interaction of pyrite {100} surfaces with O_2 and H_2O: Fundamental oxidation mechanisms. Am Mineral 84:1549-1561

Rota C, Chignell CF, Mason RP (1999) Evidence for free radical formation during the oxidation of 2'-7'-dichlorofluorescin to the fluorescent dye 2'-7'-dichlorofluorescein by horseradish peroxidase: possible implications for oxidative stress measurements. Free Radic Biol Med 27:873-881

Safa WF, Machado JL (2003) Silicosis in a housewife. Saudi Med J 24:101-103

Sawyer DT, Valentine JS (1981) How super is superoxide? Acc Chem Res 14:393-400

Schins RPF (2002) Mechanisms of genotoxicity of particles and fibers. Inhal Toxicol 14:57-78

Schins RPF, Lightbody JH, Borm PJA, Shi T, Donaldson K, Stone V (2004) Inflammatory effects of coarse and fine particulate matter in relation to chemical and biological constituents. Toxicol Appl Pharmacol 195: 1-11

Schoonen MAA (2004) Mechanisms of sedimentary pyrite formation. *In:* Sulfur Biogeochemistry—Past and Present. GSA Special Paper 379. Amend JP, Edwards KJ, Lyons TW (eds) Geological Society of America, p 117–134

Schoonen MAA, Strongin DR (2005) Catalysis of electron transfer reactions at mineral surfaces. *In:* Environmental Catalysis. Grassian V (ed) CRC Press, p 37-60

Schwarze P, Hetland R, Refsnes M, Låg M, Becher R (2002) Mineral composition other than quartz is a critical determinant of the particle inflammatory potential. Int J Hyg Environ Health 204:327 - 331

Scott DT, Runkel RL, McKnight DM, Voelker BM, Kimball BA, Carraway ER (2003) Transport and cycling of iron and hydrogen peroxide in a freshwater stream: Influence of organic acids. Water Resour Res 39(11): 1308, doi:10.1029/2002WR001768

Seiler F, Rehn B, Rehn S, Bruch J (2004) Different toxic, fibrogenic and mutagenic effects of four commercial quartz flours in the rat lung. Int J Hyg Environ Health 207:115-124

Setsukinai K, Urano Y, Kakinuma K, Majima HJ, Nagano T (2003) Development of novel fluorescence probes that can reliably detect reactive oxygen species and distinguish different species. J Biol Chem 278:3170-3175

Shi H, Hudson LG, Ding W, Wang S, Cooper KL, Liu S, Chen Y, Shi X, Liu KJ (2004) Arsenite causes DNA damage in keratinocytes via generation of hydroxyl radicals. Chem Res Toxicol 17:871-878

Shi H, Timmins G, Monske M, Burdick A, Kalyanaraman B, Liu Y, Clément JL, Burchiel S, Liu KJ (2005) Evaluation of spin trapping agents and trapping conditions for detection of cell-generated reactive oxygen species. Arch Biochem Biophys 437:59-68

Shi X, Dalal NS (1992) The role of superoxide radical in chromium (VI)-generated hydroxyl radical: The Cr(VI) Haber-Weiss cycle. Arch Biochem Biophys 291:323-327

Shi X, Dalal NS, Kasprzak KS (1993a) Generation of free radicals from hydrogen peroxide and lipid hydroperoxides in the presence of Cr(III). Arch Biochem Biophys 302:294-299

Shi X, Dalal NS, Kasprzak KS (1993b) Generation of free radicals in reactions of Ni(II)-thiol complexes with molecular oxygen and model lipid hydroperoxides. J Inorg Biochem 15:211-225

Shi X, Dalal NS, Kasprzak KS (1992) Generation of free radicals from lipid hydroperoxides by Ni^{2+} in the presence of oligopeptides. Arch Biochem Biophys 299:154-162

Smith BC (1996) Fundamentals of Fourier Transform Infrared Spectroscopy. CRC Press

Som SK, Joshi R (2002) Chemical weathering of serpentinite and Ni enrichment in Fe oxide at Sukinda Area, Jajpur District, Orissa, India. Econ Geol 97:165-172

Sorensen M, Autrup H, Moller P, Hertel O, Jensen SS, Vinzents P, Knudsen LE, Loft S (2003) Linking exposure to environmental pollutants with biological effects. Mutat Res 544:255-271

Stohs SJ, Bagchi D (1995) Oxidative mechanisms in the toxicity of metal-ions. Free Radic Biol Med 18:321-336

Stone V, Jones R, Rollo K, Duffin R, Donaldson K, Brown DM (2004) Effect of coal mine dust and clay extracts on the biological activity of the quartz surface. Toxicol Lett 149:255 - 259

Strathmann TJ, Stone AT (2003) Mineral surface catalysis of reactions between FeII and oxime carbamate pesticides. Geochim Cosmochim Acta 67:2775-2791

Stringer B, Kobzik L (1998) Environmental particulate-mediated cytokine production in lung epithelial cells (A549): role of preexisting inflammation and oxidant stress. J Toxicol Environ Health A 55:31-44

Strli M, Kolar J, Šelih V-S, Ko ar D, Pihlara B (2003) A comparative study of several transition metals in Fenton-like reaction systems at circum-neutral pH. Acta Chim Slov 50:619–632

Stumm W, Morgan JJ (1995) Aquatic Chemistry: Chemical Equilibria and Rates in Natural Waters, 3rd Edition. Wiley-Interscience

Takeshita K, Fujii K, Anzai K, Ozawa T (2004) *In vivo* monitoring of hydroxyl radical generation caused by x-ray irradiation of rats using the spin trapping/epr technique. Free Radic Biol Med 36:1134-1143

Tao F, Kobzik L (2002) Lung macrophage-epithelial cell interactions amplify particle-mediated cytokine release. Am J Respir Cell Mol Biol 26:499-505

Toyokuni S (1996) Iron-induced carcinogenesis: the role of redox regulation. Free Radic Biol Med 20:553-566

Tullius TD, Dombroski BA (1986) Hydroxyl radical "footprinting": High-resolution information about DNA-protein contacts and application to λ repressor and Cro protein. Proc Natl Acad Sci USA 83:5469-5473

Tullius TD, Greenbaum JA (2005) Mapping nucleic acid structure by hydroxyl radical cleavage. Curr Opin Chem Biol 9:127-134

Usher CR, Cleveland CA, Strongin DR, Schoonen MAA (2004) Origin of oxygen in sulfate during pyrite oxidation with water and dissolved oxygen: An *in situ* horizontal attenuated total reflectance infrared spectroscopy isotope study. Environ Sci Technol 38:5604-5606

Valavanidis A, Balomenou H, Macropoulou I, Zarodimos I (1996) A study of the synergistic interaction of asbestos fibers with cigarette tar extracts for the generation of hydroxyl radicals in aqueous buffer solution. Free Radic Biol Med 20:853-858

Valko M, Morris H, Cronin MTD (2005) Metals, toxicity and oxidative stress. Curr Med Chem 12:1161-1208

Vallyathan V, Castranova V, Pack D, Leonard S, Shumaker J, Hubbs A, Shoemaker D, Ramsey D, Pretty J, McLaurin J (1995) Freshly fractured quartz inhalation leads to enhanced lung injury and inflammation. Potential role of free radicals. Am J Respir Crit Care Med 152:1003-1009

Vallyathan V, Shi X, Castranova V (1998) Reactive oxygen species: Their relation to pneumoconiosis and carcinogenesis. Environ Health Perspect 106:1151-1156

Van den Broeke LT, Graslund A, Nilsson JL, Wahlberg JE, Scheynius A, Karlberg AT (1998) Free radicals as potential mediators of metal-allergy: Ni^{2+}- and Co^{2+}-mediated free radical generation. Eur J Pharm Sci 6: 279-286

Van Metre PC, Mahler BJ, Furlong ET (2000) Urban spraw lleaves its PAH signature. Environ Sci Technol 34: 4064 -4070

Vandeventer JSJ, Ross VE, Dunne RC (1993) The effect of galvanic interaction on the behavior of the froth phase during the flotation of a complex sulfide ore. Minerals Eng 6:1217-1229

Vásquez-Vivar J, Kalyanaraman B, Martásek P, Hogg N, Masters BS, Karoui H, Tordo P, Pritchard KA (1998) Superoxide generation by endothelial nitric oxide synthase: the influence of cofactors. Proc Natl Acad Sci USA 95:9220-9225

Vásquez-Vivar J, Santos AM, Junqueira VB, Augusto O (1996) Peroxynitrite-mediated formation of free radicals in human plasma: EPR detection of ascorbyl, albumin-thiyl and uric acid-derived free radicals. Biochem J 314(3):869-876

Vione D, Falletti G, Maurino V, Minero C, Pelizzetti E, Malandrino M, Ajassa R, Olariu RI, Arsene C (2006) Sources and sinks of hydroxyl radicals upon irradiation of natural water samples. Environ Sci Technol 40: 3775-3781

Voelker BM, Sedlak DL, Zafiriou OC (2000) Chemistry of superoxide radical in seawater: Reactions with organic Cu complexes. Environ Sci Technol 34:1036-1042

Wagner BA, Evig CB, Reszka KJ, Buettner GR, Burns CP (2005) Doxorubicin increases intracellular hydrogen peroxide in PC3 prostate cancer cells. Arch Biochem Biophys 440:181-190

Walling C (1975) Fenton's reagent revisited. Acc Chem Res 8:125-131

Watts RJ, Foget MK, Kong SH, Teel AL (1999a) Hydrogen peroxide decomposition in model subsurface systems. J Hazard Mater 69:229-243

Watts RJ, Jones AP, Chen PH, Kenny A (1997) Mineral-catalyzed Fenton-like oxidation of sorbed chlorobenzenes. Water Environ Res 69:269-275

Watts RJ, Teel AL (2005) Chemistry of modified Fenton's reagent (catalyzed H$_2$O$_2$ propagations-CHP) for *in situ* soil and groundwater remediation. J Environ Eng 131:612-622

Watts RJ, Udell MD, Kong SH, Leung SW (1999b) Fenton-like soil remediation catalyzed by naturally occurring iron minerals. Environ Eng Sci 16:93-103

Wehrli B (1990) Redox reactions of metal ions at mineral surfaces. *In*: Aquatic Chemical Kinetics. Stumm W (ed) Wiley, p 311-336

Wehrli B, Sulzberger B, Stumm W (1989) Redox processes catalyzed by hydrous oxide surfaces. Chem Geol 78:167-179

Welch KD, Davis TZ, van Eden ME, Aust SD (2002) Deleterious iron-mediated oxidation of biomolecules. Free Radic Biol Med 32:577-583

Whiteman M, Hooper DC, Scott GS, Koprowski H, Halliwell B (2002) Inhibition of hypochlorous acid-induced cellular toxicity by nitrite. Proc Natl Acad Sci USA 99:12061-12066

Willey JD, Kieber RJ, Lancaster RD (1996) Coastal rainwater hydrogen peroxide: Concentration and deposition. J Atmos Chem 25:149-165

Wilson C, Hinman N, Sheridan R (2000a) Hydrogen peroxide formation and decay in iron-rich geothermal waters: The relative roles of abiotic and biotic mechanisms. Photochem Photobiol 71:691-699

Wilson CL, Hinman NW, Cooper WJ, Brown CF (2000b) Hydrogen peroxide cycling in surface geothermal waters of Yellowstone National Park. Environ Sci Technol 34:2655-2662

Winterbourn CC (1987) The ability of scavengers to distinguish OH· production in the iron-catalyzed Haber-Weiss reaction: Comparison of four assays for OH. Free Radic Biol Med 3:33-39

Xu A, Zhou HN, Yu DZL, Hei TK (2002) Mechanisms of the genotoxicity of crocidolite asbestos in mammalian cells: Implications from mutation patterns induced by reactive oxygen species. Environ Health Perspect 110:1003-1008

Xu Y, Schoonen MAA (2000) The absolute energy position of conduction and valence bands of selected semiconducting minerals. Am Mineral 85:543-556

Yigit E, Ozkan SG (2004) Effects of surface alteration phenomena on flotability of copper minerals of Kure-Bakibaba ore. Colloids Surf A Physicochem Eng Asp 248:105-109

Yıldız G, Demiryürek AT (1998) Ferrous iron-induced luminol chemiluminescence: A method for hydroxyl radical study. J Pharmacol Toxicol Methods 39:179-184

Zepp RG, Faust BC, Hoigne J (1992) Hydroxyl radical formation in aqueous reactions (pH 3-8) of iron(II) with hydrogen peroxide: the photo-Fenton reaction. Environ Sci Technol 26:313-319

Zhang LS, Wong GTF (1994) Spectrophotometric determination of H_2O_2 in marine waters with leuco crystal violet. Talanta 41:2137-3145

Zhou M, Dimu Z, Panchuk-Voloshina N, Haugland RP (1997) A stable nonfluorescent derivative of resorufin for the fluorometric determination of trace hydrogen peroxide: Applications in detecting the activity of phagocyte NADPF oxidase and other oxidases. Anal Biochem 253:162-168

Zhu D, Herbert BE, Schlautman MA, Carraway ER, Hur J (2004) Cation-pi bonding: a new perspective on the sorption of polycyclic aromatic hydrocarbons to mineral surfaces. J Environ Qual 33:1322-1330

Reviews in Mineralogy & Geochemistry
Vol. 64, pp. 223-282, 2006
Copyright © Mineralogical Society of America

8

Bone: Nature of the Calcium Phosphate Crystals and Cellular, Structural, and Physical Chemical Mechanisms in Their Formation

Melvin J. Glimcher

Laboratory for the Study of Skeletal Disorders and Rehabilitation
Department of Orthopaedic Surgery
Harvard Medical School and Children's Hospital
Boston, Massachusetts, 02115, U.S.A.
e-mail: Melvin.Glimcher@childrens.harvard.edu

INTRODUCTION

Calcium phosphate is the dominant solid mineral phase within the skeletal and dental tissues of vertebrates. This chapter concentrates on the structure and composition of the solid calcium inorganic orthophosphate (Ca–Pi) phase in bone and the mechanisms that are thought to induce the onset of this mineralization process as an example of biological mineralization in general. It is important to recognize that the Ca–Pi mineral phase is deposited in a living tissue and is a substance that is continuously being synthesized, resorbed and replaced by the action of living cells. Therefore, the composition, structure and other properties of the solid Ca–Pi mineral phase will change in space and time, depending on the general body metabolism and the local cellular functions in specific regions of bone. Similar considerations arise in studies of the mineralized tissues of invertebrates, where the crystals consist of various lattice arrangements of $CaCO_3$ (calcite, aragonite, vaterite) deposited in hierarchical arrangement with the constituents of the ordered organic matrices. Furthermore, the nature of the Ca–Pi phase in bone is significantly different from synthetic, highly crystallized and geological hydroxylapatites, which is reflected in the physical, structural properties and physiological functions of the biological apatites.

This chapter is an attempt to summarize at least some of the historical background leading to the more recent research over the past several decades. The focus here is on investigations at the molecular and nano-scales, now possible both theoretically and experimentally, which have been applied to determine the "nature" of the solid Ca–Pi mineral phase of bone and other calcified vertebrate tissues from the inception of mineralization to the changes that occur during crystal maturation ("crystal aging") and the "normal" aging of the animal. The topics addressed below include an introduction to the basic concepts, relevant terminology and cellular events involved in bone formation, the removal (resorption) of the bone mineral and the organic matrix, the structural properties of bone, the underlying mechanisms which have to date been found and have been postulated to control the nucleation of the solid Ca–Pi mineral phase, and a comparison between bone and other normally calcified tissues in vertebrates such as cartilage, dentin and enamel of teeth and finally, a note on the structure and composition of geological and synthetic hydroxylapatite. Of the various normally calcified tissues in vertebrates, bone was chosen as the focus primarily because the crystals in bone substance have been investigated much more intensely than in other vertebrate mineralized tissues.

1529-6466/06/0064-0008$10.00   DOI: 10.2138/rmg.2006.64.8

BASIC TERMINOLOGY AND CONCEPTS OF BONE FORMATION

Bone as an organ and a tissue

There are several terms commonly used in describing "bone" and bone formation that may be confusing to scientists not familiar with the cell and tissue biology of bone. The word "bone" can refer to bone as an *organ* such as the thigh bone (femur), or to bone as a *tissue*, namely, *cancellous* and *compact* bone tissue (Fig. 1). *Cancellous bone tissue* consists of an interconnecting series of irregularly shaped microscopic-sized pieces of bone substance (mineralized organic matrix), which form a 3D interconnecting but highly organized network (Fig. 2). The open spaces between the bony "struts" (*trabeculae*) are occupied by blood vessels and blood forming cells, fat and fat cells, lining connective tissues, and undifferentiated mesenchymal cells. The second kind of *bone tissue* is *compact bone*, located principally in the shafts (*diaphyses*) of mature long bones (Figs. 1 and 2), and consisting of very densely packed *osteons* (Figs. 3a,b). The osteons consist of a central Haversian canal containing a blood vessel surrounded by lamellae of bone substance. Both the cortex of the diaphyses and flat bones and the trabecular struts of the cancellous bone, may be organized at the microscopic and ultrastructural levels as "*woven*" bone, in which case the collagen fibers are relatively much less well organized parallel to one another than in the lamellae of compact osteonal bone. "Woven" bone tissue is present in both embryonic bone and early postnatal bone (depending on species) and in rapidly synthesized new bone especially during the repair of bone defects or bone fractures. The mechanical properties of "woven" bone substance and consequently of bone tissue and bone as an organ are diminished compared to highly oriented osteonal lamellar bone.

Bone formation and turnover

The term *ossification* refers to the specific cellular synthesis of an organic matrix by specialized cells called *osteoblasts*. Calcification or mineralization of bone substance refers only to the deposition of the solid Ca–Pi phase crystals into the organic matrix previously synthesized by the osteoblasts. The major organic matrix constituent of bone substance is the structural protein, *collagen* (Fig. 4), whose molecules are aggregated or self-assembled in a particular 3D spatial packing (Fig. 5a,b). Aggregates of the collagen molecules are referred to as collagen *fibrils* (Figs. 5a,b). Calcification begins very shortly after the synthesis and secretion of the collagen fibrils, the Ca and Pi being derived from the specific *extracellular fluids* (ECF) that permeate the tissue and bone substance.

As the calcified organic matrix synthesized by the osteoblasts gradually surrounds the osteoblasts, their cellular activity, internal molecular structure, and function change. When the osteoblast is completely encased by the calcified organic matrix, it is a mature bone cell and is called an *osteocyte*. The osteocytes are interconnected by channels to each other and to osteoblasts on the surface of the bone that are, in turn, surrounded by the ECF. This helps to maintain structural and functional connections between the fully differentiated osteocytes, the osteoblasts and the ECF.

The organic matrix and the solid Ca–Pi mineral phase deposited in the organic matrix are both removed (resorbed) by a third distinct kind of bone cell, the *osteoclasts*. The 3D disposition of the *collagen fibrils* and the nature of the solid Ca–Pi mineral phase change with respect to their chemical and physical properties in the time period between their formation and their resorption (bone turnover). In time, new undifferentiated mesenchymal stem cells are recruited into the spatial void created by the resorption of bone substance. These cells differentiate into osteoblasts, and the cycle of events is repeated throughout the lifetime of the animal. *Halisteresis*, which refers to the dissolution of the Ca–Pi mineral phase without resorption and degradation of the organic matrix components, does not occur in living bone, with the possible exception of the bones of a specific species of snake during long periods of hibernation.

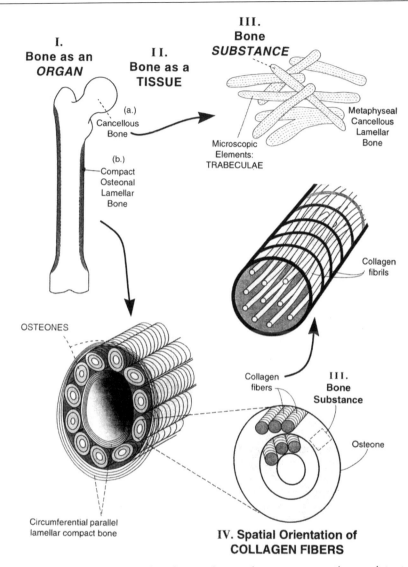

I.
Bone as an
ORGAN

II.
Bone as a
TISSUE

III.
Bone
SUBSTANCE

(a.)
Cancellous
Bone

Microscopic
Elements:
TRABECULAE

Metaphyseal
Cancellous
Lamellar
Bone

(b.)
Compact
Osteonal
Lamellar
Bone

Collagen
fibrils

OSTEONES

Collagen
fibers

III.
Bone
Substance

Osteone

Circumferential parallel
lamellar compact bone

IV. Spatial Orientation of
COLLAGEN FIBERS

Figure 1. Diagrammatic representation of postnatal, mature bone as an organ, a tissue, and structural material (i.e., bone substance). [Reprinted from Glimcher (1998), with permission from Elsevier.]

It has been estimated that during the active years of a normal young adult human the total mass of bone substance in the skeletal system is synthesized and resorbed every five to six years. Clearly, this bone mass turnover does not occur at the same rate in every bone or region of a bone equally, so the rate of bone mass turnover must be several times greater in some of the bones and some regions of a particular bone! Bone substance has one of the highest turnover rates of all the tissues or organs in vertebrates. During the period of mid-adult life of humans, especially in women, the rate of bone resorption increases compared to the rate of formation so that the *mass* of bone decreases, resulting in osteoporosis, especially in the vertebrae and at the upper end of the femur. When this discrepancy is very large, it results in osteopenia or osteoporosis, with a high risk of bone fracture.

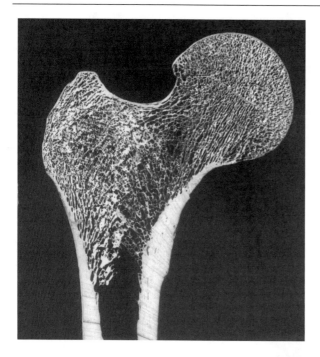

Figure 2. A section of the upper end of the thigh bone (femur) demonstrating the strut-like "trabeculae" in the cancellous bone tissue of the femoral head and neck of the femur and the very dense compact bone in the upper end of the femoral shaft. [Reprinted from Glimcher (1998) with permission from Elsevier.]

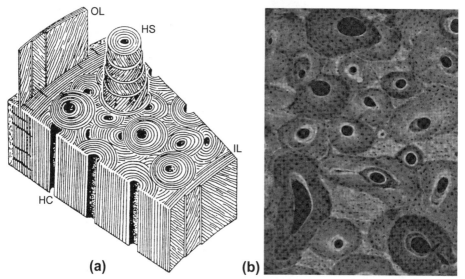

(a)   **(b)**

Figure 3. Compact cortical bone structure, (a) schematic drawing of a block of compact cortical bone tissue and substance, and its highly ordered organization. OL = outer circumferential layer; HS alternate geometric orientation of the collagen fibers and the solid, Ca–Pi mineral phase in adjacent lamellae of a single osteon. HC is the cortical corner of an osteon. IL the inner circumferential and OL outer circumferential layers of the compact bone substance of the diaphysis (shaft) of a long bone; (b) A microradiogram of a thin cross section of normal cortical bone. Note the variation in X-ray density (indicating inorganic crystal concentration) within any one Haversian system, as well as within the section as a whole. The extent of the X-ray densities correlate with the extent of mineralization. [Reproduced from Glimcher (1959). Copyright American Physical Society. See Figure 6d caption for copyright restricitons. Figure 3b from original, courtesy A. Engstrom, Karolinkska Institutet, Stockholm, Sweden.]

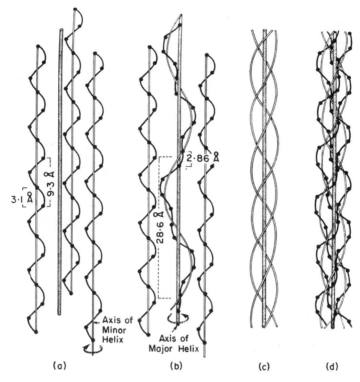

Figure 4. Schematic to show how the three polypeptide chains of the collagen macromolecule are wound in a coiled-coil helix. (a) Each polypeptide chain is composed of glycine, proline and hydroxyproline amino-acid residues (black spheres). The distance between adjacent residues is 3.1 Å. Each of the polypeptide chain is wound around its own long axis ("minor helix" axis) as a left-handed helix in a three-fold screw fashion, resulting in a primary repeat distance of 9.3 Å. Each chain is located at the vertex of an equilateral triangle and they are parallel; (b, c) The individual "minor helix" axes of the three chains are now coiled about the central axis ("major helix" axis) in a long right-handed helix; (d) This slightly distorts the distance between adjacent residues to 2.86 Å, yielding a repeat distance of the major helix axis to 28.6 Å, but gives rise to a very highly repetitive structure, sufficiently ordered to generate a wide angle X-ray diffraction pattern. [Reprinted with permission from Glimcher (1960). Copyright AAAS.]

Maturation of the crystals and organic matrix constituents of bone

Apart from their overall growth in mass as an organ (femur, humerus, etc.) which occurs from birth to adulthood, bone tissue and bone substance are continually being remodeled, that is, resorbed and replaced by new bone formation ("bone turnover") or turned over under cellular control, not only during this growth period but thereafter for the life of the animal. Equally important as the cellular controls are the chemical and structural changes which occur in both the Ca–Pi solid phase and the organic matrix constituents which are involved in bone substance maturation as a function of the time that the inorganic Ca–Pi crystals and the organic matrix constituents spend in the tissues from their deposition in the tissue to their resorption ("crystal age" and "organic matrix age"). Our understanding of bone formation and maturation is based on the results of experiments exploring the 3D organization of bone tissue and bone substance, examined at all spatial levels of the anatomic hierarchy and at different temporal stages, by numerous techniques such as light microscopy, polarization light microscopy, scanning and high resolution transmission electron microscopy and electron diffraction, chemical analyses, and a wide variety of other spectroscopies.

(a)

(b)

Figure 5. (a) Low magnification transmission electron microscopy (TEM) of isolated collagen fibrils clearly showing repetitive ordered axial densities with a period of ~67-70 nm. [Reprinted from Hodge (1967) with permission of Elsevier.]; (b) Higher magnification TEM of collagen fibrils stained to show more clearly the axial repeating period. [Reprinted from Glimcher and Krane (1968) with permission from Elsevier.]

The studies have been performed on samples at different levels of bone maturity from within an animal of a given age, and from animals of different ages. Investigations of the content and "nature" of the Ca–Pi crystals of bone are thus highly dependent on the sample chosen. For example, if one analyzes a sample of intact, uncrushed cancellous bone tissue consisting of trabeculae plus the open space in between the trabeculae, the mineral content of this cancellous bone tissue would be considerably less than a similar volume of compact cortical bone. However, if a similar weight of the bone substance were to be analyzed, it might be expected that the sample of bone substance would have the same mineral content as the bone substance of cortical bone. However, since trabecular bone substance has a higher rate of bone turnover than the bone substance of cortical compact bone and depending on the specific site chosen and the age of the animal and the rate of bone turnover, trabecular bone would contain a higher proportion of young bone substance, which has a slightly lower mineral content. Keeping in mind these caveats regarding sampling of specific sites of the bone and characterization of bone tissue and bone substance at different spatial and temporal scales, an overview of the crystal maturation process is provided below.

Experiments aimed at characterizing the "youngest" (first formed) bone crystals utilized samples of bone powder composed of very small particles separated by density centrifugation in organic solvents in order to avoid dissolution or changes in the structure of the Ca–Pi mineral phase when exposed even briefly to water (Boothroyd 1964; Glimcher 1969; Landis et al. 1977). Spectroscopic techniques have experimentally established that even in the very youngest bone substance available to date, the Ca–Pi crystals can be generally characterized as very small nanocrystals of very poorly crystalline apatite, as will be discussed below in detail. Subsequent to formation of the earliest crystals, significant changes occur in the size, shape, composition, lattice structure, crystal perfection (crystallinity), and other properties of the Ca–Pi phase nanocrystals, that are controlled by the rate of bone turnover, and result in nanocrystals of markedly different "crystal age." "Crystal age" is the length of time that the crystals spend in the bone substance before resorption by osteoclasts (Glimcher 1976). The cycle of bone formation and resorption occurs throughout the life of the animal, albeit at a reduced rate during adult aging, so that the bone substance gradually contains higher and higher percentages of older, more crystalline, and denser bone substance with increasing adult age (Figs. 6a-d).

This continuous renewal of bone substance is also accompanied by remodeling of the internal architecture of bone at all levels of the anatomic hierarchy. At a larger spatial scale, light microscopy reveals that in embryonic and very young postnatal animals or, under certain conditions in adults where bone substance turnover is very rapid, the collagen fibrils together with the solid mineral phase are relatively poorly organized ("woven bone"). With increasing age of the animal, the organization of the newly synthesized collagen fibrils and the solid mineral phase become much more highly organized into units called *osteons*, consisting of lamellae around a central canal containing a blood vessel. The organization of the bone substance components and the content of the mineral phase within each lamella varies at the light microscopic level of organization (Figs. 3a,b).

CELLULAR PATHWAYS OF BONE FORMATION

Two cellular pathways for bone formation are the *endochondral* and *the direct* or *intramembranous* or *membranous* pathways. Endochondral ossification is a highly ordered biological and cellular process during which specific cartilage cells proliferate, differentiate, move in a regular sequential fashion, increase in size and shape and change their cell functions, eventually to large hypertrophic cartilage cells. At this stage of cell differentiation, deposition of Ca–Pi nanocrystals of poorly crystalline apatite is initiated in the organic matrix of cartilage surrounding the hypertrophic cartilage cells creating calcified cartilage – not bone. The hypertrophic cartilage cells then die and degenerate. The calcified cartilage substance is then invaded by osteoclasts or, more properly, chondroclasts, carried by capillaries from the underlying cancellous bone tissue. The osteoclasts resorb and remove the calcified cartilage substance, creating "tunnels" throughout the calcified cartilage. Distal to these osteoclasts, primitive mesenchymal undifferentiated cells proliferate and differentiate to osteoblasts which synthesize new bone substance on the "walls" of the unresorbed portions of the calcified cartilage "tunnels." This process normally occurs only during growth and ceases at maturity of an animal. It can also occur during bone repair in growing and mature animals.

The endochondral ossification sequence occurs at both ends of most long bones, and the particular regions at the ends of the long bones which are referred to as primary or major epiphyseal growth centers which control the length of the shaft of long bones, or the "minor" epiphyseal growth centers which are beneath the articular cartilage surface of the end of the long bone and controls the size and shape of the articular end of the bone. The primary epiphyseal growth center is the major contributor to the longitudinal growth of long bones and the overall size of small bones such as those found in the wrist region of the hand.

Figure 6. Increasing density and crystallinity of chick bone with age of animal.

(a) density-fraction histograms of chick tibial mid-diaphyseal bone;

(b) X-ray diffractometer patterns obtained from bone fractions with intensity as the ordinate and diffraction angle (2θ) as the abscissa: *A* 17-day chick embryo periosteal bone scrapings; *B* 5-week post-hatch chick, 2.1-2.2 gcm^{-3} density fraction; *C* 2-year-old chick, 2.2-2.3 gcm^{-3} density fraction; *D* highly crystalline synthetic hydroxylapatite. Patterns derived from rate meter charts obtained with a continuously scanning diffractometer equipped with a diffracted-beam monochromator, using CuKα radiation;

(c) computer-averaged, normalized XRD patterns obtained from bone fractions, corrected for absorption by non-mineral components, plotted on the same relative intensity scale (ordinate, diffracted intensity; abscissa, angle, 2θ): *A* 17-day chick embryo, periosteal bone scrapings; *B* 5-week old chick, 2.1-2.2 gcm^{-3} density fraction; *C* 1-year-old chick, 2.1-2.2 gcm^{-3} density fraction; *D* highly crystalline synthetic hydroxylapatite, used as standard in determining CI values (the weak reflection at 38.25° 2θ is from the aluminum sample holder). [Reprinted with permission from Bonar et al. (1983) with kind permission of Springer-Verlag.]

(d) Another demonstration by wide angle X-ray diffraction of the significant differences between crystal size/degree of crystallinity of very highly crystalline and large sized hydroxyapatite crystals and the hydroxyl deficient nanosized crystals of bone. Top to bottom: synthetic, highly crystalline hydroxyl apatite; early embryonic chick bone (note the failure to generate the "triplet" of large, highly crystalline synthetic hydroxyapatite as well as the very broad 002 reflections); somewhat older embryonic chick bone with less broadening of the 002 reflection and further definition of the "triplet". Note the clear

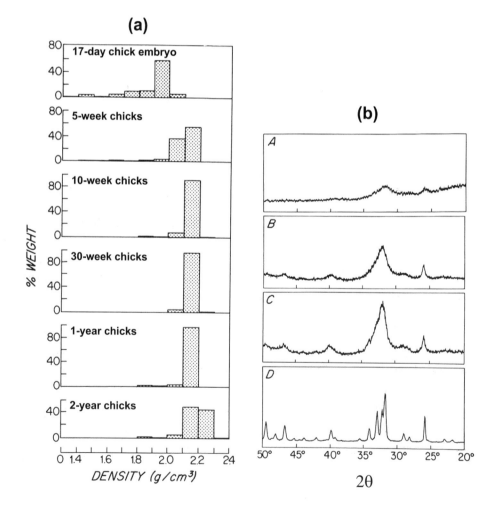

differences between all of the synthetic hydroxyapatite and the nanocrystals of bone Ca–Pi crystals and the broad single reflection of amorphous Ca–Pi (bottom). [Reproduced from Glimcher (1959). Copyright American Physical Society. *Readers may view, browse, and/or download material for temporary copying purposes only, provided these uses are for noncommercial, personal purposes. Except as provided by law, this material may not be further reproduced, distributed, transmitted, modified, adapted, performed, displayed, published, or sold in whole or part, without prior written permission from the publisher.*]

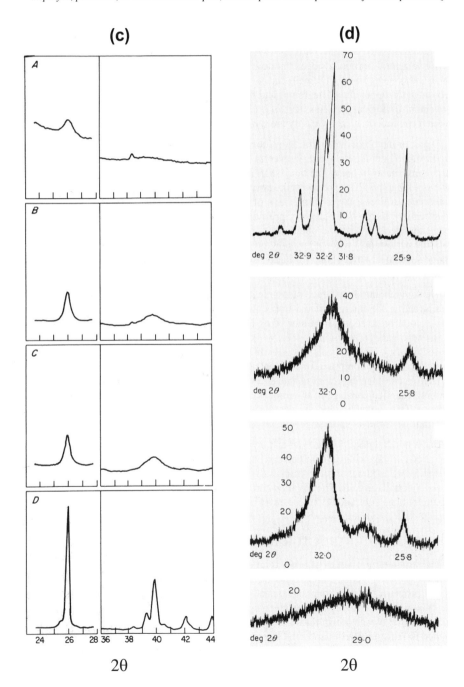

It is very important to note that in early embryonic stages of development, "bone" as an organ (e.g., tibia, femur, etc.) is composed of *cartilage cells* and *cartilage organic matrix* constituents, not bone cells or bone matrix constituents. The major changes which occur in the early development of a cartilaginous long bone (as an organ) are distinctly different from the classical descriptions given above. The biological goal is to remove the cartilage in the middle of the solid cartilage "bone" as an organ in order to form a hollow narrow cavity. This is accomplished by an endochondral sequence in this region, first by the invasion of capillaries, the calcification of the cartilage substance, and the resorption of both the calcified cartilage and the new bone formed on the surfaces of the calcified cartilage, which thus converts the solid cartilage "bone" to an organ whose central portion has been "hollowed out" to create a narrow cavity. The cartilaginous ends of the bone as an organ remain and, as described, develop 1° and 2° growth centers, such as occurs at the upper end of the femur and femoral head. These important distinctions between the eventual biological consequences of endochondral ossification at the end(s) of long bones and the formation of a narrow cavity are rarely mentioned in textbooks or publications.

Direct or membranous ossification (bone formation) refers to a sequence of differentiation of undifferentiated mesenchymal stem cells directly to osteoblasts which synthesize the organic matrix constituents of bone. The bony shaft (diaphysis) of embryonic long bones is first formed both by *endochondral ossification* at the lateral-most portions of the epiphyseal growth plate and at the outer surface of the forming walls of the diaphysis, by a direct (membranous) pathway of ossification. In the latter process, primitive mesenchymal cells proliferate and differentiate directly to osteoblasts on the outer layers of the connective tissue encircling the bone as an organ (the periosteum) without any previous formation of calcified cartilage or its replacement by bone substance. Direct bone formation eventually accounts for all of the bone substance of compact and cancellous bone of adult vertebrates, including the cranium.

The endochondral sequence of ossification in the primary epiphyseal growth center continues, creating the interconnected trabeculae of cancellous bone *tissue*. This is followed by the formation of a secondary center of ossification in the remaining cartilaginous end of the bone as an organ. (A few long bones, as noted, contain only one epiphyseal growth center.) This secondary center controls the growth and shape of the end of the bone. The trabeculae of bone are initially composed of both calcified cartilage and the new bone on the surfaces of the calcified cartilage and, with further growth of the bone and the continued resorption of the remaining calcified cartilage of the trabeculae, the trabeculae eventually consists only of bone substance. Thus, endochondral ossification is critical for the growth of bone as an organ and for the "sculpting" of its eventual size and shape.

Endochondral ossification is not a "transformation" of cartilage cells or cartilage substance or calcified cartilage cells or calcified cartilage substance to bone substance or tissue. Calcified cartilage is not suddenly or ever "transformed" to bone substance. Bone substance is not synthesized by cartilage cells. The calcified cartilage substance is removed by osteoclasts and then replaced by bone substance synthesized by osteoblasts.

STRUCTURAL PROPERTIES OF BONE

The structural properties of bone as an organ, tissue and substance are largely determined by the gross mass and its 3D disposition at various anatomic hierarchical levels, as well as the material properties of the bone substance. This is well illustrated in cancellous bone tissue (Fig. 2), in which the highly 3D spatial design of the interconnecting "struts" (trabeculae) was noticed early on by anatomists and structural engineers to resemble those mechanical structures designed to resist most efficiently the stresses and strains with the least mass. This 3D spatial organization was particularly noted in the head and neck of the human femur (Fig. 2), where the lattice structure of the interconnected, cancellous, bony trabeculae very closely resembled the 3D disposition of the metal struts of a Fairbairn crane. Indeed, a 3D, mathematical stress analysis, in terms of the spatial disposition of the trabeculae and compact bone mass, calculated for a vertical load placed on the femoral head was very similar to that obtained for the trabeculae in cancellous bone tissue and the compact bone on the outer surfaces of the femoral neck (Fig. 7) (Koch 1917). This close correlation was found to exist despite the fact that the calculated external load was incorrectly based on a vertical load, whereas it is now known that the external load applied to the femoral head is at an angle to the vertical due to the contractions of various muscles which originate above the hip joint and insert at various regions on the femur, for example, the greater trochanter. Thus, the major mechanical stresses and strains imposed on the femoral head and neck and thus on the bone of the femur and femoral neck are generated by the contraction of the muscles transmitted to the bone, and not by the external forces applied (the weight of the person plus kinetic forces), as in the impact of a foot on the ground during gait. For example, a 200 lb man during normal walking imposes approximately 2000 lb of force on the femoral head, principally by the contraction of the muscles around the hip joint. This is referred to as the *mechanical adaptive theory* of the mass and spatial disposition of compact cortical and cancellous bone, also known as Wolff's Law (Wolff 1884, 1892).

These findings led to the hypothesis that osteoblasts and osteoclasts respond to the spatial disposition and intensity of the mechanical loads applied, including kinetic forces due to movement of the body and upon impact with the ground, and the internal forces generated by the contraction of muscles. The intensity and distribution of these mechanical forces produces very important signals to and between osteoblasts and osteoclasts to increase or decrease bone mass and to modify the internal microscopic, ultrastructural and molecular organization of the components of bone substance.

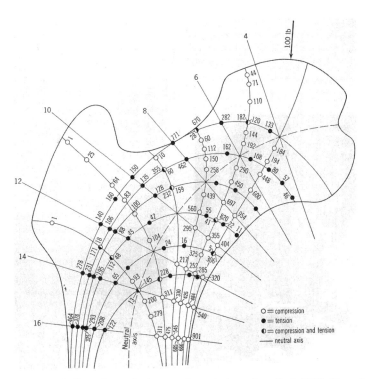

Figure 7. Diagram of the lines of stress in the upper femur (redrawn from Koch 1917). [Reproduced from Glimcher (1959). Copyright American Physical Society. See Figure 6d for copyright restrictions.]

In addition to the structure of cancellous and compact bone, and the hierarchical architecture of bone crystals and collagen—which will be described in detail in the section on "Mechanisms of Bone Mineralization"—the size, shape and chemical and structural nature of the solid Ca–Pi phase crystals also determines the structural properties of bone. The properties of the solid Ca–Pi phase are reviewed in the following section.

CHARACTERIZING THE MINERAL PHASE OF BONE

In spite of the fact that the solid mineral phase was known to be a calcium phosphate for over 200 years, and identified as a member of the apatite family since 1926 (DeJong 1926), the exact structure and chemical composition of the Ca–Pi solid phase has not yet been exactly defined, principally because bone is constantly being turned over and remodeled. The crystal structure and composition of the mature Ca–Pi phase of bone resemble, but are not identical to, those of the synthetic or geological apatites. The unique size, shape, surface area and degree of crystallinity of bone nanocrystals, differ significantly from the larger-sized, more perfectly crystalline synthetic and geological apatites. The unique physical-chemical properties of bone nanocrystals contribute, in turn, to a major biological function, namely, maintaining the homeostasis of the major ions such as Ca and Mg in the extracellular fluids, as well as the specific material properties of bone substance.

The mineral content of bone substance

The proportion of inorganic mineral phase in bone substance changes over time and is affected in different bone disorders. Medical and biological researchers have therefore defined a quantity known as the mineral content of bone substance. The mineral content of a specific calcified tissue or substance is usually reported by weight percentage, that is, the weight of the fully mineralized tissue after ashing the bone relative to the total weight of the bone specimen. However, it is actually the volume percentage of the mineral phase in the bone substance, that is, the mass of the solid Ca–Pi mineral phase in a unit volume of bone (true volumetric bone density) on which both the biological and mechanical properties depend.

In any event, the mineral content of bone substance, by weight or volume percent, clearly begins at zero immediately after the synthesis, secretion and self-assembly of the collagen molecules into collagen fibrils and before the first crystals are deposited in the new bone matrix. The overall rates of bone substance formation and resorption change significantly over time so that if one samples the exact same "gross" location in a particular bone, the percentage of the solid mineral phase in bone substance rapidly and steadily increases, then levels off and slowly reaches the level of mature bone substance. Fully mineralized normal bone substance of adults has a weight percentage of ~66% and a volume percentage of ~52-53%. It is biologically very interesting that there is little difference in the mineral content by weight percent of fully mineralized bone substance in all vertebrates thus far analyzed, although the volumetric percent can be much lower in some species such as fish (Biltz and Pellegrino 1969).

An interesting exception is found in the small bone of the auditory apparatus of whales where the weight percentage of the Ca–Pi mineral phase is about 98-99% (Bonar and Glimcher, unpublished data). Sections of this highly mineralized bone were extremely difficult to obtain and not of a quality or resolution that allowed satisfactory definition of the ultrastructural anatomy of the bony substance. When these bony particles were decalcified in EDTA, there was essentially no organized organic matrix observable. After dialysis of the EDTA solution through several layers of the smallest pore size tubing, partial separation of the contents in the dialysis bags revealed organic fragments that had an amino-acid composition consistent with collagen (Bonar and Glimcher, unpublished data). Interestingly, the mineral content by weight percentage of enamel substance in mature teeth is also about 98-99%. This high mineral content is a result of the degradation and loss of the organic matrix of enamel, whose major structural protein is not collagen, but *amelogenin*.

If the rate of bone substance turnover is sufficiently rapid, the newly deposited solid Ca–Pi phase and the newly synthesized organic matrix which replaces the previously deposited bone substance will have the same weight or volume percent mineral content as the initial mineral phase deposited. Thus, there may be no change in the mineral content by weight or volume percentage over time after mineralization begins. This high turnover rate is most likely to occur either in the very youngest developing animals and in small regions of adult, mature bone that have been very recently resorbed and are in the process of being replaced such as during the repair of fractures or in certain metabolic diseases.

The mineral content of bone substance, by weight and volume percentage, depends not only on the age of the animal but principally on the time elapsed after Ca–Pi crystal deposition begins in the organic matrix. Therefore, crystal "age" must be taken into account when determining the mineral content (bone density) of bone substance.

Importance of experimental resolution and bone sample volume

At the light microscopic level, there is a good deal of heterogeneity with respect to crystal "age" and the mineral content of bone substance, which are, ultimately, a function of bone cell biology. Thus, in a light microscopy thin section, one may find regions of fully mineralized bone substance and immediately adjacent regions where the bone substance has recently been partly or completely resorbed (Figs. 3a,b). Close by, there may also be regions in which newly calcified bone substance is being synthesized and that contain very little of the Ca–Pi phase. Thus the "crystal age" obtained experimentally also depends on the resolution of the experimental technique and the volume of sample analyzed. Since there are no "Maxwellian demons" available to isolate a series of single nanocrystals as a function of their age (as opposed to the age of the animal or the surrounding bone substance), only some operationally obtained values of "crystal age" averaged over space and time can be obtained.

Extensive studies, even at the level of high-resolution transmission electron microscopy (HRTEM), have clearly demonstrated that this heterogeneity occurs not only in young growing animals, but also in young and mature adults. Therefore, one should be aware that even in very small "gross" samples of mature bone, compositional and structural analyses of the solid Ca–Pi mineral phase can be very misleading. It is therefore dangerous to draw conclusions at any hierarchal level of anatomy, as to whether the composition, structure, or other features even in a very small volume of bone substance correctly represent the properties of the individual nanocrystals.

To meet this challenge of anatomically-based heterogeneity and lack of adequate experimental spatial resolution, a number of laboratories explored new avenues and adopted new technologies to assess the extent of mineralization at the histological level of bone architecture. For example, the osteons and even the individual lamellae of an osteon have been examined in undecalcified histological sections of bone, using techniques such as quantitative contact microradiography, and particular fluorescent dyes which label the sites of new bone formation and mineralization, even in small volumes of bone substance and in microdissected single osteons. Many of these techniques require the chemical fixation and embedding of the sample with a synthetic polymer that polymerizes to a sufficiently hard solid, so that thin sections can be made and examined by electron microscopy. Other techniques have been developed over the years that provide important information about the atomic composition of the nanocrystalline phase. Electron probe X-ray microanalysis can measure the Ca and Pi content of the mineral phase in a micron-sized area of bone substance and highly-collimated, Selected Area Electron Diffraction (SAED) can provide a great deal of crystal lattice information on the nanocrystals of bone substance. Improved techniques of bone sample preparation for TEM and electron diffraction such as anhydrous preparations help to prevent decalcification and reprecipitation of nanocrystals of apatite or other Ca–Pi solid phases, especially for the youngest bone containing the most reactive Ca–Pi phase (Landis et al. 1977).

Size and shape of bone crystals

Based on the extent of peak broadening in wide angle X-ray diffraction spectra of the apatite crystals in bone, Stuhler (1938) estimated that the diameter of the bone crystals varied from 31-290 Å. Taking advantage of the introduction of electron microscopy to biology, later studies enabled investigators to directly visualize and explore the macromolecular organization of the components of tissues and cells. The first study of ground sections of highly mineralized, dense, compact cortical bone by electron microscopy showed that that the average size of the bone crystals was approximately 500 Å long, 250 Å wide, and 100 Å in thickness, plate-shaped, and, very importantly, that the plate-shaped apatite crystals were roughly aligned with their long axes parallel to the long axes of the collagen fibrils (Robinson 1952; Robinson and Watson 1953, 1955).

Somewhat later, another technique to determine crystal size and shape in intact bone was introduced and explored in depth (Engstrom and Finean 1953; Finean and Engstrom 1953, 1954; Carlstrom and Finean 1954). These authors used low-angle X-ray diffraction scattering to estimate particle size and shape of the apatite crystals, using the theories and the equations developed by Guinier (1939a,b, 1952) and modified by others (Jellinek and Fankuchen 1945). It was concluded that the shape and size of the Ca–Pi crystals in bone were much smaller than those previously proposed (Robinson 1952a; Robinson and Watson 1953), and that rather than being flat plates, the crystals were long *rods*, probably ellipsoid in shape, and roughly 65-75 Å in diameter and 210 Å in length (Engstrom and Finean 1953; Finean and Engstrom 1953, 1954; Carlstrom and Finean 1954). In these excellent, well-conceived and experimentally well-executed low-angle X-ray diffraction scattering studies, Engstrom, Finean and Carlstrom were also aware of the possible limitations of this technique, which assumes that each particle scatters independently. Therefore, in tightly packed systems such as solids (crystals), the possible particle-particle interactions render this technique inapplicable. There are similar restrictions on the interpretation of experiments to measure the size and shape of particles in solution by light scattering. Engstrom, Finean and Carlstrom showed that the intensity of the scattered radiation decreased with an increase in scattering angle, which would be expected from a system in which there is no particle-particle interference. Using this observation and other factors, these investigators assumed that the data obtained by low-angle X-ray diffraction scattering of compact bone could be treated by modifications of the Guinier equations.

More recently, other investigators have reintroduced low-angle X-ray scattering to study the size and shape of the crystals of bone and also concluded that the crystals are not flat plates but needles (Fratzl et al. 1996a,b). However, they did not quantitatively address the question as to whether there are significant interactions between the particles in the specific solid bone samples they analyzed.

Further efforts to elucidate the size and shape of the nanoapatite crystals of bone were carried out by many investigators using thin sections of less mineralized bone examined by improved fixation and embedding techniques and with higher resolution electron microscopy. Bone tissues were prepared anhydrously to prevent the significant solubilization and reprecipitation of the Ca–Pi mineral phase (Boothroyd 1964; Landis et al. 1977). However, the tight packing of the collagen molecules in the collagen fibrils and of the very thin long crystals with varying and overlapping densities and the inability to measure the exact plane of the thin section relative to the long axes of the crystals precluded accurate measurements, even when dark field electron microscopy was utilized (Arsenault 1988).

The problems of determining the size and shape of bone crystals and changes with age have been solved by two different techniques, both utilizing electron microscopy. In the first technique, thin sections of bone are progressively tilted relative to the electron microscope stage such that the electron density of the individual, low density plate-like crystals progressively increases and their shape (habit) changes from plate-like to very electron-dense "lines." The minimal width of the dense "lines" is ~10-25 Å, representing the thickness of the very thin plate-like nanocrystals (Bocciarelli 1970, 1973).

In the second technique, a number of methods were developed to remove the organic matrix constituents of bone, disperse the crystals and then separate a significant number of the individual crystals so that at least two dimensions of a single nanocrystal could be measured and their structure determined by highly collimated electron diffraction, wide angle X-ray diffraction and FTIR (Kim et al. 1995).

Important low angle X-ray scattering studies of isolated nanocrystals of rat bone and calcified turkey tendon confirmed the thin plate-like habit of the crystals in calcified turkey tendon, while similar studies of isolated rat bone crystals indicated that they were cylindrical rods (Wachtel and Weiner 1994). These investigators reported, however, that the previously published isolation techniques (Weiner and Price 1986) used to obtain organic matrix-free nanocrystals of bone and calcified tendon had resulted in some dissolution of the crystals from both sources. We found that the method used in these studies also dissolved a significant portion of the crystals and changed a number of the crystal properties (Glimcher et al., unpublished). Therefore, it is critical to remove essentially all of the organic matrix by non-aqueous techniques and at low temperature in order not to alter the crystal size, shape, composition and internal lattice structure.

After experimenting with a number of previously published techniques, all of which caused alterations in several of the critical criteria required, a successful approach was developed involving low-power plasma ashing of ground bone powder in a pre-cooled plasma asher, where the bone powder had been frozen previously in liquid N_2 in a cold room at 2-4 °C (Kim et al. 1995). Subsequently, wide-angle X-ray diffraction, FTIR spectroscopy and electron microscopy on these samples confirmed that the isolated, single nanocrystals of bone were indeed very thin plates in all four species of animals examined, namely, bovine, chicken, mouse and fish (Kim et al. 1995) and subsequently in rat (Kim and Glimcher, unpublished). No rod or needle-shaped nanocrystals of apatite were found in the bone substance in the five animal species studied. It was noted in all five species that the smallest nanocrystals were irregularly shaped, somewhat round, very thin "particles" with a maximum thickness of approximately 10-12 Å as determined by tilting the electron microscopic stage over an arc of 120° (Kim et al. 1995). The maximum length and width of the crystals from bovine, chicken and mouse bone were approximately 250 Å and 120-150 Å respectively, whereas the nanocrystals in fish bone were somewhat longer, ~370-500 Å and 150-200 Å wide.

Thick sections of undecalcified bone and calcified turkey tendon have also been examined by high voltage (1.0 MV) electron microscopy and computer generated stereomicroscopy (Lee and Glimcher 1989) and 3D computer reconstruction (Landis et al. 1991, 1993), as well as by atomic force microscopy (Eppell et al. 2001; Tong et al. 2003), all of which confirm the thin plate-like habit of the bone nanocrystals.

The mineral phase of bone nanocrystals

Kim et al. (1995) processed crystals from fish bone and embryonic chick bone by hand-cutting small pieces of the bone and extracting individual nanocrystals with cold 100% ethanol, thus ruling out the possible effects of mechanical grinding on the crystals. HRTEM of even the smallest nanocrystals isolated by this technique showed single crystal electron diffraction patterns from which the structure of the crystals could be elucidated as apatitic. Wide-angle X-ray diffraction and FTIR data were also clearly consistent with apatite (Kim et al. 1995).

Despite experimental findings that no solid phase of Ca–Pi other than apatite has ever been identified in bone or other vertebrate calcified tissues, especially in the youngest bone substance examined, theoretical thermodynamic and other considerations have led some investigators to postulate that the initially nucleated, youngest crystals are more likely to be octacalcium phosphate (OCP) (Brown et al. 1987) or "amorphous" Ca–Pi (Termine and Posner 1966a,b; Termine et al. 1967; Holmes et al. 1970; Betts et al. 1975; Posner et al. 1977) which are then dissolved and replaced by the formation of apatite crystals, or possibly undergo a solid phase transition to apatite. This point is addressed in greater detail later.

Biological significance of nanocrystal surfaces: "form follows function"

Using experimental techniques current at the time, Posner and Beebe (1975) measured not only the surface area of the bone crystals *in situ*, but also various preparations of synthetically precipitated Ca–Pi crystals. They found that the available surface area in bone apatite crystals was extremely large ($100\text{-}200 \text{ m}^2\text{g}^{-1}$).

There are important reasons why the very small size of bone apatite nanocrystals is one of their most important physical-chemical characteristics and, consequently, one of their unique biological properties. The smaller the size of any particle of a substance, the greater is the surface area compared to the same mass of larger-sized particles. Physiologically, the bone apatite nanocrystals act as an ion reservoir, which must rapidly react to maintain specific ion concentrations such as Ca and Mg in the blood and the ECFs. The concentrations of these ions are critical for many physiological and biochemical functions, and must be maintained within a very limited range and, indeed, in the case of calcium, the life of the animal requires it. The very large surface area of the nanocrystals results in a very much higher percentage of constituent ions which are readily available for reaction with the ions and other organic constituents of the extracellular fluids.

One of the many major contributions of the Neumans and their colleagues was the investigation of the surfaces of bone crystals (Neuman and Neuman 1958, 1980; Neuman 1969; Neuman and Bareham 1975). These authors discovered that a significant amount of water is bound to the surfaces of the nanocrystals of bone apatite and that the concentration and reactivities of the Ca and Mg ions in the hydrated surface regions of the crystals are significantly increased. These are critical phenomena when interpreting rates of bone crystal formation and resorption. The concentration of Mg on the hydrated surface layers of apatite crystals is about 66% of the total body content of Mg. This Mg is not incorporated into the lattice structure of the bone crystals but is found, rather, on the hydrated surface layers (Neuman and Neuman 1953; Holmes et al. 1964; Glimcher 1998b). This is biologically very important because Mg ions are thus very rapidly available to the ECFs and, thence, to many organs and tissues where they play critical roles in a variety of biochemical processes (Neuman and Neuman 1953; Wacker and Parish 1968a,b,c; Wallach 1988, 1990; Rude and Oldham 1990; Rude 1996; Glimcher 1998b).

More recent studies of the crystal surfaces of bone and enamel apatites by spin-spin relaxation solid-state NMR have also found that a significant proportion of the "labile" HPO_4 groups are located on the surfaces of these crystals (Wu et al. 2002). The content of such labile surface HPO_4 groups is greater in the youngest apatite crystals in both bone and enamel. Similarly, changes in the total carbonate content, and in the increased proportion of the labile carbonate species and the phosphate ions, compared to the "stable" and less reactive species as measured by FTIR in the younger crystals during crystal maturation and "crystal aging" have also been reported (Kim et al. 1994; Rey et al. 1987, 1989, 1991a,b, 1995a,b; Rey and Glimcher 1992).

It is interesting to note that in the early research literature on this subject, a great deal of attention was focused, and many theories evolved, as to how so much Ca, Mg, and other ions could be removed so rapidly from the blood and ECF to form additional nanocrystals of bone apatite, or be released from the crystals to the ECF and serum, both by solid state diffusion. Very heated debates were initiated by groups of investigators who favored what was seen as a physical chemical phenomenon, while others felt that these transfers of specific ions to and from the bone were very rapidly controlled by cellular action. It was the latter—cellular mechanisms— which were later proved to be correct. We now know that the controlling mechanism for Ca, Pi, Mg, etc. exchange between the bone apatite crystals and ECF involves specific cellular signals that cause osteoclasts to resorb both the organic matrix and the calcium phosphate nanocrystals quickly, thus releasing the most reactive ions from the surfaces of the crystals, and signals to osteoblasts resulting in new bone formation, thus removing these biologically critical ions from the extracellular fluid, which may lead to very serious pathological consequences.

Thus, the surface characteristics of bone nanocrystals are an excellent example of "form follows function" in biology: the large surface area of the crystals and the location of large numbers of biologically critically important reactive ions on the surface of the crystals, provides the crystals with an ideal organization for the very rapid exchange of these important ions with the extracellular fluids.

BONE NANOCRYSTAL GROWTH

The youngest nanocrystals of bone apatite are thin, roughly round, and irregularly shaped, whilst the older crystals have a long, thin, plate-like habit with the longest dimension very closely parallel to the crystallographic *c*-axis of apatite. Similarly, the very much larger and more highly crystalline crystals of enamel apatite also show even a more pronounced preferential growth in length, corresponding to the c-crystallographic axis direction. Thus, it is clear that crystal growth occurs preferentially and more rapidly in the direction of the *c*-axis, somewhat slower in width, and markedly slowest in thickness. It has previously been pointed out that the commonly used term "crystal growth" as it relates to biological apatite crystals is often incorrectly used and should be more clearly defined (Glimcher 1981). "Crystal growth" is usually defined as an increase in the size of the individual crystals, whereas "crystal multiplication" refers to an increase in the number of crystals. An increase in the mass of bone mineral is principally due to crystal multiplication, whereas the increase in the mass of crystals in the enamel substance is principally due to an increase in the size of the crystals (Glimcher 1981). A clear distinction between the two processes should be emphasized when discussing changes in the mass of the Ca–Pi solid phase in mineralized tissues.

Chemical composition of bone nanocrystals

It should be clear that the composition, size, shape, and crystal lattice structure of bone nanocrystals change rapidly with "crystal age," as discussed earlier. In turn, the "crystal age" determined experimentally depends on the volume of tissue examined and the resolution of the techniques used. The chemical composition and the lattice positions of the individual constituents of bone nanocrystals, even in mature animals, are distinctly different in the very much larger, highly crystalline, synthetic and geological hydroxylapatite crystals, which also have a much higher degree of crystal perfection. As for composition, the water content of bone substance, some of which is in capillaries, lymph, cells, and collagen fibrils as well as H_2O bound to the Ca–Pi nanocrystals, as alluded to earlier, must be taken into account for the most correct calculations of the composition of bone substance, especially the mineral content of bone (Robinson 1957; Robinson et al. 1977, 1988, 1992). Unfortunately, the vast majority of such analyses have been reported on a weight basis whereas the volumetric content of the mineral should have been used for such computations. In any event, it has been noted that the Ca content of bone mineral is lower than that of "stoichiometric" hydroxylapatite. Two explanations have been proposed for this commonly reported deficiency. First, that there is a net substitution of sodium ions for calcium, although the particular lattice positions are unknown

(LeGeros et al. 1969). Second, that there is a vacancy defect in place of a calcium ion and the presence of an HPO_4 group (Rey et al. 1995a). A number of investigators have also reported an absence or near absence of hydroxyl groups in bone apatite (Biltz and Pellegrino 1971; Termine et al. 1973; Termine and Lundy 1973; Rey et al. 1995c) and in early-formed enamel (Bonar et al. 1991) compared to stoichiometric apatite, most recently established by neutron scattering (Loong et al. 2000), FTIR, and Raman spectroscopy (Rey, unpublished; Pasteris et al. 2004), and by a newly developed solid-state NMR technique (Wu et al., unpublished data). With increased crystal maturity and increased animal age, which is accompanied by an increase in the percentage of older, denser bone, there is a definite tendency for the calcium deficiency to be lessened. There is also a concomitant increase in crystal perfection and crystal size, but little or no significant change in the very low hydroxyl content (Biltz and Pellegrino 1971; Termine et al. 1973; Termine and Lundy 1973; Rey et al. 1995c; Loong et al 2000).

Shortly after the first publications of solid state ^{31}P NMR spectroscopy of highly crystalline "pure" hydroxylapatite crystals (Yesinowski and Benedict 1983), similar ^{31}P NMR studies were carried out on samples of bone (Aue et al. 1984; Roufosse et al. 1984). Interestingly, more detailed ^{31}P NMR experiments showed spectra, which were clearly those of an apatite and identified HPO_4 ions, probably in a configuration similar but not identical to the configuration of the HPO_4 ions in brushite crystals ($CaHPO_4 \cdot 2H_2O$) (Aue et al. 1984; Roufosse et al. 1984; Ackerman et al 1992). More sophisticated differential cross-polarization ^{31}P NMR studies of bone resolved the problem (Wu et al. 1994). A minor component of the Ca–Pi mineral of bone was definitively identified as HPO_4. This component did not have all the NMR spectral properties of the HPO_4 ions in brushite, OCP, or, surprisingly, of highly crystalline synthetic hydroxylapatite containing HPO_4 groups. The HPO_4 ions were in a configuration specific to the nanocrystals of bone apatites. Later NMR spin-spin relaxation studies of both bone and enamel revealed that a high percentage of these HPO_4 groups were on the surfaces of the crystals of both bone and enamel, and that a higher proportion of the surface-bound HPO_4 ions occurred in younger animals (Wu et al. 2002). During crystal "maturation," the total content of HPO_4 groups and the HPO_4/PO_4 ratio also decrease, both on the surface and in interior lattice positions (Wu et al. 2002). As noted previously, early studies of bone and synthetic apatite crystals also found a significant amount of crystal surface-bound water (Triffit et al. 1968; Neuman 1969; Neuman and Mulyran 1969; Holmes et al. 1970; Neuman and Bareham 1975; Posner and Beebe 1975; Neuman and Neuman 1980).

Recent FTIR studies of the earliest nanocrystals of Ca–Pi precipitated *in vitro* provided evidence that the initial solid Ca–Pi phase formed had several spectral characteristics of OCP, but was not identical to OCP. It did, however, contain structural water, as does OCPl (Eichert et al. 2004). This initial solid Ca–Pi phase was found to be very labile and converted to apatite when dried or lyophilized.

The Ca–Pi phase of bone also contains significant amount of carbonate (CO_3^{2-}), up to 6 wt% or higher in very young crystals. Although a contentious point in the past, it is now generally agreed that CO_3 ions can replace PO_4 groups or OH groups in large, very crystalline synthetic or geological hydroxylapatites (Regier et al. 1994). It is not yet well established whether these substitutions can occur in the exact interior lattice positions in bone apatite nanocrystals. Certainly, a significant number of the CO_3 groups are labile or more reactive and probably represent surface-bound ions. Based on the relative ease of interaction, some investigators have characterized the CO_3 and HPO_4 groups and even some PO_4 groups as either "labile" (more reactive and probably on or near the surface of the crystals) or "stable" (less reactive) (Rey et al. 1987). Because of all of these considerations, it may be best to express the major ion composition of bone apatite as $Ca/(P + CO_3)$.

It is stressed that conclusions about the composition, structure, and nucleation mechanisms of the Ca–Pi phase of bone nanocrystals based on the experimental or theoretical predictions of the composition and structure of *in vitro* synthesized apatites, and on theoretical analyses of the underlying mechanisms of formation of the initial Ca–Pi nuclei must be carefully scrutinized for their applicability to the specific and complex train of events during the *in vivo* biological mineralization of specific tissues.

MECHANISMS OF BIOLOGICAL CALCIFICATION

Posing the problem

Over the past 75 to 80 years, many theories have been proposed and many experiments conducted utilizing contemporaneous state of the art knowledge and technology to uncover the basic mechanisms that control the initiation of precipitation of specific solid phases in specific tissues or organs, presumably from the same extracellular fluids that bathe other tissues which normally remain unmineralized. These experiments have progressively taken advantage of the striking theoretical and technical advances in the basic chemical and physical sciences and, more recently, in molecular and structural biology, and the rapid processing of this information by high-speed computers. This has been accompanied by the technological development of spectroscopic and microscopic instruments and techniques that enable investigators to probe the structure and properties of ever smaller specimens of mineralized tissues, with much higher levels of resolution such as very high resolution Raman imaging and spectroscopy (Carden and Morris 2000; Tarnowski et al. 2002; Morris et al. 2004; Draper et al. 2005; Dehring et al. 2006).

Homogeneous nucleation mechanism

In the 1920s and 1930s investigators focused their attention on the concentrations of Ca, Pi, and pH in blood, serum and ECFs. It was clear that the inorganic Ca–Pi crystals of apatite were quite insoluble in neutral or slightly alkaline solutions of H_2O. The dilemma was to understand how the serum or the ECF, which had to be supersaturated with respect to a solid calcium phosphate phase, could continuously circulate throughout the body tissues without depositing a Ca–Pi solid phase in all of the normally unmineralized tissues while at the same time depositing a Ca–Pi solid phase in bone, cartilage and teeth from the same solution phase? They reasoned that since the same ECF did not induce formation of Ca–Pi crystals in normally unmineralized tissues but did so in the skeletal and tooth tissues, it was possible that cellular mechanisms were responsible for locally increasing the concentrations of one or both of the inorganic components (Ca, Pi) in the circulating fluids bathing the bone and the calcifying portions of cartilage to a sufficiently high level of supersaturation with respect to apatite, which resulted in spontaneous, homogeneous nucleation and deposition of solid Ca–Pi crystals in these normally mineralized tissues (Robison 1923; Kay and Robison 1924; Robison and Soames 1924; Fell and Robison 1934). Investigators therefore studied the effect of increasing the concentrations of Ca and Pi separately in supersaturated solutions and found that the concentrations of Pi in such solutions played a much more significant role in the precipitation of apatite crystals than the concentrations of Ca in the solutions. This finding was later attributed to the fact that apatite is structurally an inorganic phosphate crystal, and that calcium can be replaced in the apatite lattice by many other cations such as lead, barium and manganese. Indeed, one wonders why the term "calcification" was first used to describe the process of apatite deposition in certain vertebrate tissues instead of "phosphatization" considering that apatite is structurally a phosphate. Some investigators have attempted to use the term "phosphatization" instead of "calcification" but their manuscripts were not accepted for publication unless "phosphatization" was replaced by "calcification."

Focusing on the critical role of aqueous Pi concentrations *in vitro*, the early investigators postulated that the cells of the calcifying tissues secreted an enzyme which cleaves inorganic phosphate ions from an organic phosphate substrate present in the organic matrix of bone and calcifying cartilage (Robison and Soames 1924). Further experiments indeed isolated an enzyme (alkaline phosphatase) which cleaved an organic substrate isolated from the organic matrix of calcified cartilage and bone. The organic substrate contained two ester-linked adjacent phosphate groups which, when cleaved by the phosphatase, released inorganic orthophosphate ions. This mechanism was called "the enzyme theory" of calcification of cartilage and bone organic matrices. It was a unique explanation linking biological and cellular events with the physical-chemical explanation of how and why mineralization is normally initiated in certain biological tissues and organs but not in others.

Biologically and clinically, additional information to support this hypothesis was obtained when the concentrations of Pi in the serum of children with "rickets" were found to be significantly lower than normal, but Ca concentrations were found to be within the normal range or only slightly diminished. "Rickets" is a condition in children where the mineral content of bone substance and calcified cartilage is significantly decreased. In adults, there are several diseases of bone metabolism that are accompanied by a decrease in the mineral content by volume percent of bone substance, similar to rickets in children, but referred to in adults as "osteomalacia." The concentration of alkaline phosphatase in the serum of rachitic children and growing animals in which "rickets" was induced experimentally, and adults afflicted by osteomalacia is also increased.

Increasing the calcium concentration alone in the serum of experimental animals with "rickets" had little effect in increasing the extent of mineralization of bone or the region of cartilege organic matrix containing the hypertrophic cartilage cells. Importantly, the addition of Pi to children with "rickets" and experimental animals in which "rickets" had been induced or adult patients with osteomalacia, increased the rate of calcification and the concentration of the solid Ca–Pi mineral phase of bone and calcified cartilage. High levels of Pi in the ECF resulted in an increased rate of calcification and, were shown to be capable of correcting the rickets by rapidly removing large amounts of Ca from the ECF, which in many instances was already slightly lower than normal. Further lowering of the concentration of Ca in the ECF in turn stimulated the parathyroid gland, increasing the synthesis and secretion of parathyroid hormone, which further lowered the Ca concentration in the ECF to levels that, unfortunately, induced untoward pathological events such as seizures and even death. We emphasize that despite the adverse effects lowering of the Ca concentration in the ECF, the increased Pi levels in ECF induced an increase in the rate of calcification of both bone and cartilage, despite a significant lowering of the calcium concentration in the serum and ECF. It is important to note that there are critical biological and pathological consequences of rapidly removing large amounts of Ca ions from the ECF to increase the rate of calcification and it is well to remember that bone is a living substance and tissue. Thus, while changes in the rate of calcification can be theoretically (and correctly) predicted on physical-chemical principles, the actual changes that occur are ultimately under the control of specific cellular functions, and biological "messengers" which can be sent to cells and organs near and far away and which, as in this instance, are initiated by the physical chemical changes.

Later, the vital role of vitamin D deficiency in producing "rickets" and the use of vitamin D in correcting "rickets" and osteomalacia (except in X-hypophosphatemic resistant "rickets") also was established. Lack of Vitamin D was shown to play a significant role in markedly delaying the rate of the linear proliferation and differentiation of the cartilage cells of the epiphyseal growth plates to hypertrophic cartilage cells, at which point, the organic matrix surrounding the hypertrophic and degenerate cartilage cells is normally rapidly mineralized. This resulted in a decreased rate of growth of the long bones in patients with "rickets" and a marked decrease of the Ca–Pi mineral content in the cartilage matrix of the growth plates, both easily observed by standard clinical X-rays. In addition, the relatively decreased growth rate of the medial portions of both epiphyseal growth plates often led to vertically shorter femura and tibiae and to a "bow-shaped" anatomic appearance ("bowlegs") at the knee joints.

Extensive experimental studies followed the original concept of the enzyme theory and the mechanisms responsible for the calcification of bone and cartilage gained acceptance by basic scientists and clinicians.

Heterogeneous nucleation mechanism

The theory that the normal calcification of bone and other normally mineralized vertebrate tissues was simply a spontaneous precipitation due to an increase in the concentrations of the constituent ions beyond the solubility limit of apatite (homogeneous nucleation) was reinforced by wide angle X-ray diffraction studies of very young, poorly mineralized "woven" bone, indicating that there was a random distribution of the inorganic nanocrystals and the collagen fibrils in bone substance. However, based on low-angle X-ray diffraction scattering experiments, investigators had concluded previously that there was some spatial relationship between the collagen fibrils and the inorganic Ca–Pi crystals (Engstrom and Finean 1953; Finean and Engstrom 1953, 1954; Carlstrom and Finean 1954). Later, electron microscopy and electron diffraction explained these conflicting views. The problem was due to the fact that the wide angle X-ray diffraction pattern had been carried out on "woven" bone in which there is a relatively unorganized assembly of collagen fibrils in the bone substance and bone tissue, but the inorganic crystals associated with the individual collagen fibrils were spatially highly oriented with respect to the individual fibril within which they were located (Fig. 8). These data, again, emphasize the importance of the size of the bone sample examined, the spatial resolution of the techniques used, and the 3D heterogeneity of bone substance at all levels of the anatomic hierarchy. The large sample of bone used for the early wide angle X-ray diffraction studies showing no preferred orientation of the collagen fibrils and the inorganic nanocrystals encompassed too large a volume of bone substance to be analyzed by the wide angle X-ray diffraction technique, which had too low a resolution to resolve the spatial organization of the major organic matrix constituents and the nanocrystals in "woven" bone.

In the 1950s, despite the limited resolution available at the time, TEM and electron diffraction studies of fully mineralized highly compacted, cortical bone sections obtained by mechanical grinding of the bone showed that the increased electron density (representing the solid mineral phase) was clearly concentrated along the long axes of the very tightly packed collagen fibrils with an axial repeating period of ~67 nm, similar to that of native unmineralized collagen fibrils. This indicated that there was a spatial relationship between the collagen fibrils and the inorganic Ca–Pi solid phase crystals in bone (Robinson 1952a; Robinson and Watson 1953, 1955; Watson and Robinson 1953). It is important to note that the solid dense mineral phase in these electron micrographs occupies the entire, ~67 nm basic collagen axial repeat and *not* just the ~60% or 40 nm of the axial period occupied by the "hole zone," which is the region of the collagen fibril which is first mineralized (see below).

In addition, it was not possible in such dense fully calcified bone containing tightly packed collagen fibrils to distinguish the 3D spatial relationships of the crystals and the collagen fibrils. The problem was later solved by what was considered at the time "high resolution TEM" of very thin sections of less densely packed collagen fibrils of fish bone (Glimcher 1959; Lee and Glimcher 1991), in which the individual collagen fibrils were quite widely separated from each other, both in longitudinal and cross sections. The Ca–Pi nanocrystals were found to be clearly located within the collagen fibrils and at the earliest stages of mineralization were found to be concentrated along the length of the collagen fibrils with an axial repeating period of ~67 nm, similar to that of the native collagen fibrils. The crystals were also oriented with their long axes, the crystallographic "*c*-axes," roughly parallel to the long axes of the collagen fibrils within which they were deposited (Glimcher 1959; Lee and Glimcher 1991) (Figs. 9a-f).

It was from these early findings that the nanocrystals were definitively identified as being located within the collagen fibrils with a highly ordered axial repeating period at specific sites within the collagen fibrils (Figs. 9a-f), that the theory of heterogeneous nucleation was

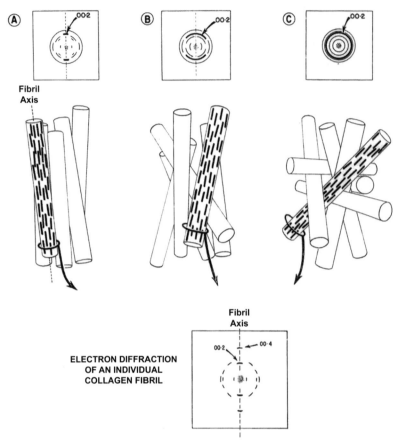

Figure 8. Schematic drawing illustrating that widely different X-ray diffraction patterns "reflecting" the orientation of the inorganic Ca–Pi nanocrystals of bone in specimens of fish bone do not "reflect" the orientation of the Ca–Pi nanocrystals with respect to the individual collagen fibrils. The correct spatial relationship and orientation of the nanocrystals to the individual collagen fibrils can be ascertained by transmission electron microscopy and selected area electron diffraction. [Modified after Glimcher (1981) with permission of Elsevier.]

first postulated. According to this theory, the underlying physical-chemical and biological mechanisms for mineralization of bone (and other vertebrate and invertebrate tissues with other organic matrices and inorganic crystals) was not simply the result of locally increasing the concentrations of the inorganic ions in the ECFs to supersaturation with respect to apatite (or other solid phases). Rather, mineralization was postulated to occur by the heterogeneous nucleation of Ca–Pi nanocrystals at specific, highly ordered sites within the collagen fibrils (Glimcher et al. 1957; Lee and Glimcher 1991).

To experimentally test this hypothesis *in vitro*, the collagens of variously unmineralized connective tissues were solubilized after first extracting non-collagenous proteins and other organic constituents and the collagen molecules reassembled (reconstituted) as collagen fibrils with the native axial repeat of ~67 nm. Samples of the several times reconstituted native type collagen fibrils (~67 nm axial repeat) and native decalcified bone collagen, also extracted a number of times to remove non-collagenous organic constituents, and also having the native ~67 nm axial repeat, were incubated with very carefully purified, filtered, and deionized solutions of Ca–Pi of varying concentrations and degrees of metastable supersaturation (solutions that do not spontaneously precipitate the Ca–Pi phase for varying lengths of time called the "induction period"). Electron microscopy of the very earliest stages of the *in vitro* heterogeneous nucleation of apatite crystals by native reconstituted collagen fibrils and by decalcified bone collagen, revealed the deposition of very small "dot-like" crystals within the collagen fibrils identified as apatite by electron and X-ray diffraction (Fig. 10a) (Glimcher et al. 1957; Glimcher 1959). The electron microscopic images were remarkably similar to the very early stages of normal *in vivo* calcification of embryonic chick bone reported by Fitton-Jackson (1957) (Fig. 10b) . The highly ordered axially repeating disposition of ~67 nm at very early stages of *in vitro* mineralization of decalcified bone collagen was also confirmed more recently by low angle X-ray scattering (Chen et al. 2005).

It was very important to very closely control as many variables as possible in carrying out the heterogeneous nucleation experiments with the reconstituted collagen fibrils and decalcified bone collagens and in the preparation of solutions at different levels of metastable supersaturation with respect to apatite. The pH of the buffered metastable solutions of Ca–Pi was 7.40. Very large volumes of the metastable solutions of Ca–Pi relative to the small mass of reconstituted collagen fibrils were also used to prevent any significant drop of the pH that could occur as a result of Ca–Pi solid phase precipitation. A number of experiments were also carried out using a pH-stat to verify the results described above. Many other precautions were also taken. For example, a constant temperature was maintained by immersing the glass vials that contained the reconstituted native collagen fibrils or the decalcified bone collagen in a large-volume water-bath specially constructed to prevent vibration. All of the glassware was carefully washed with cleanser free of Ca and all phosphates and finally rinsed with deionized water. Before use, the glassware was also heated at temperatures of ~600-700 °C for several hours. Before the reconstituted collagen fibrils and decalcified bone collagen samples were used in the nucleation experiments, they were first incubated without stirring in the metastable Ca–Pi solutions at 2 °C for approximately 8 to 12 hours to ensure that the solutions had diffused through and completely permeated the reconstituted collagen fibrils and decalcified bone fibrils.

These precautions were taken since heterogeneous nucleation experiments are notorious for nucleating a solid phase from solution but, unfortunately, not on the intended substrate but on the surfaces of containers such as glass, or of particles inadvertently included in the solution phase. A very well-designed nucleation technique introduced recently (Wu and Nancollas 1999) may simplify the technology used for studies of biological heterogeneous nucleation.

Structure and hierarchical organization of the collagen matrix and Ca–Pi crystals

Prior to these first *in vitro* nucleation studies of nanocrystals of apatite by native-type reconstituted collagen fibrils with an axial period of ~67 nm reconstituted from collagen molecules of various soft tissues and by decalcified native bone collagen fibrils (Glimcher et al. 1957), investigators had discovered that they were able to reconstitute purified collagen molecules into collagen fibrils whose macromolecular aggregation (i.e., 3D stacking of the collagen molecules in the collagen fibrils), was distinctly different from that in native type collagen fibrils, as detected by TEM (Fig. 11) (Gross et al. 1954; Schmitt et al. 1953, 1955). The markedly different 3D packing of the collagen molecules in each of these *in vitro* aggregated collagen fibril preparations, all of which were different from one another and from the native (~67 nm axial period) collagen fibrils, provided a unique experimental model to examine whether the ability to nucleate apatite crystals within collagen fibrils *in vitro* was a function of the native collagen molecules per se or of the specific 3D macromolecular organization. To address this question, denatured collagen molecules, denatured collagen fibrils (i.e., gelatin), reconstituted collagen fibrils aggregated in non-native packing, reconstituted native collagen fibrils (~67 nm

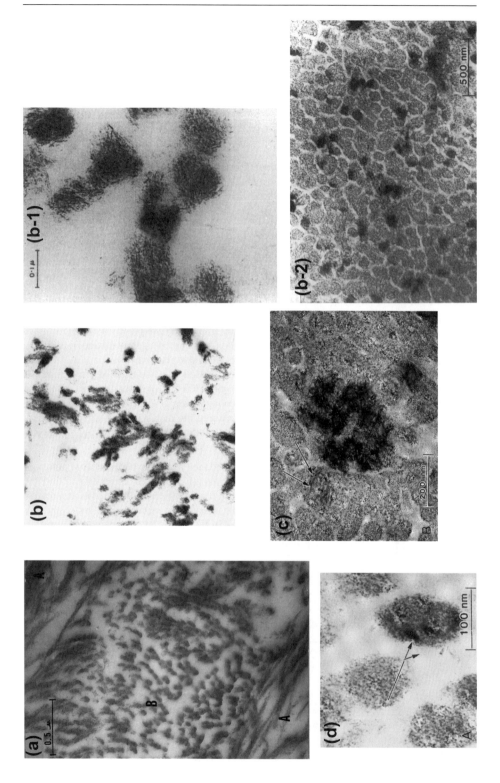

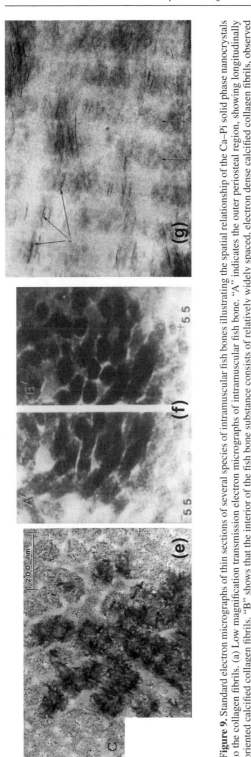

Figure 9. Standard electron micrographs of thin sections of several species of intramuscular fish bones illustrating the spatial relationship of the Ca–Pi solid phase nanocrystals to the collagen fibrils. (a) Low magnification transmission electron micrographs of intramuscular fish bone. "A" indicates the outer periosteal region, showing longitudinally oriented calcified collagen fibrils. "B" shows that the interior of the fish bone substance consists of relatively widely spaced, electron dense calcified collagen fibrils, observed mostly in a slightly oblique section. There is no evidence for the deposition of crystals in the extracellular space between the collagen fibrils. These data definitively determine that the nanocrystals of Ca–Pi are located *within* the collagen fibrils. (b) Slightly higher magnification of calcified, electron dense collagen fibrils in fish bone, seen in cross section of a less densely packed region of calcified collagen fibrils of the bone substance. The electron dense (darker) areas represent mineralized fibrils. × 82,000. (b-1) Higher magnification transmission electron microscopy of a small segment of (b), representing fully mineralized collagen fibrils. There is no evidence of Ca–Pi nanocrystals between the collagen fibrils. (b-2) Early calcification of fish bone seen in cross section in a closely packed region of collagen fibrils surrounded by widely separated collagen fibrils some which have been and are in the process of calcification. (c, d, e) A series of high-voltage electron stereoscopic micrographs of a fish bone prepared anhydrously, seen in cross section, illustrating the electron dense Ca–Pi nanocrystals within individual collagen fibrils surrounded by uncalcified collagen fibrils in the earliest stages of mineralization of a single fibril (c), and a progressively increasing number of collagen fibrils, which are calcified, all surrounded buy uncalcified collagen fibrils. As the mineralization progresses further, it is apparent that no mineral particles have been deposited in the extracellular spaces between the fibrils (light areas) (f). The deposition of mineral particles in the widely separated collagen fibrils demonstrates that the progressive extent of mineralization of bone as a tissue and substance occurs as an independent physical chemical event in each fibril and in each location of a fibril. (g) A standard transmission electron micrograph of a thin section of calcified, unstained, anhydrously prepared longitudinal section of embryonic chick bone. The very lightly dense mineral phase appears to "stain" the collagen fibril at regular intervals of ~67 nm along its axial length. In some regions, the inorganic crystals can be seen on edge as electron dense "lines" (arrows). If one presumes that the very thin, dense "lines" in the electron micrograph represent the thickness of the nanocrystals viewed parallel to their thin dimension, one may conclude that the habit of the nanocrystals are thin plates. This was established by tilting the stage of the electron microscope on which the sections were placed at different angles of rotation to the electron beam. The 3D localization of the Ca–Pi mineral crystals can be fully appreciated by the high voltage stereoscopic transmission electron microscopic technique, eliminating the possibility that the nanocrystals are located between or only on the surface of the collagen fibrils. [Figs. a, b, f reprinted by permission of the American Physical Society from Glimcher (1959). See caption for Fig. 6 for copyright restrictions. Figs. c, d, e reprinted by permission of Elsevier from Lee and Glimcher (1991). Fig. g reprinted from Glimcher and Krane (1968) with permission of Elsevier.]

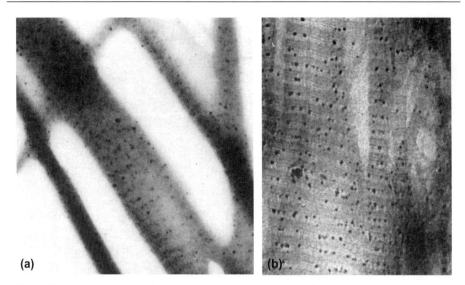

Figure 10. (a) Electron micrograph of a very early stage in the *in vitro* nucleation of apatite crystals by reconstituted native-type collagen fibrils having a ~67 nm axial repeat. The electron diffraction pattern identifies the mineral phase as apatite and demonstrates that the nanocrystals are nucleated and deposited with an axial repeat of ~67 nm along the length of the individual collagen fibrils. × 60,000. [Reprinted with permission from Glimcher (1960). Copyright AAAS.] (b) The deposition of inorganic nanocrystals of similar size and a comparable axial repeat of ~67 nm along the length of individual collagen fibrils of *in vivo* developing avian bone. × 120,000. [Kindly provided by Dr. Sylvia Fitton-Jackson. Reproduced from Glimcher and Krane (1968) with permission of Elsevier.]

axial repeat) and decalcified native bone collagen fibrils, were individually tested in the same metastable solutions of Ca–Pi to assess their ability and efficacy to nucleate apatite nanocrystals. The experiments very clearly demonstrated that the heterogeneous nucleation of apatite nanocrystals by collagen fibrils *in vitro* was dependent on the specific 3D packing of the collagen molecules in meridionally (axially) aggregated native collagen molecules and, presumably, the specific equatorial or lateral packing of the collagen molecules in the fibrils (Fig. 12). These results reflect the fact that specific sites within the collagen fibrils provide the most favorable environment for nucleation of the apatite nanocrystals, in terms of volume, structure, chemical composition, and electrical charge distribution due to amino acid residues (side chain groups) (Glimcher et al. 1957; Glimcher and Krane 1968; Glimcher 1976) (Fig. 12).

The same basic requirements probably exist in the case of other mineralized vertebrate tissues such as dentin and enamel. There are also an abundant number of studies, that have utilized transmission and scanning electron microscopy, biochemical and physical-chemical studies, to show that these same elegant relationships exits between the inorganic and organic phases of invertebrate mineralized tissues, in many of which the mineral phase is $CaCO_3$ (calcite, aragonite, and/or vaterite) (Travis 1968a; Kamat et al. 2000; Weiner et al. 2005). A particularly interesting example of an invertebrate calcified tissue has been reported by Travis (1968b). In this instance, the major structural organic matrix component is collagen with the characteristic axial repeat of ~67 nm, and an inorganic mineral phase of $CaCO_3$ nanocrystals, in the same spatial arrangement as occurs in bone!

Indeed, a very general basic tenet common to most biologically mineralized tissues is the necessity that the cells of mineralized tissues synthesize, resorb, remodel and construct a 3D architecture of organic matrix components, which eventually has a sufficient volume of space to house the particular volume percentage of the inorganic solid phase (Glimcher

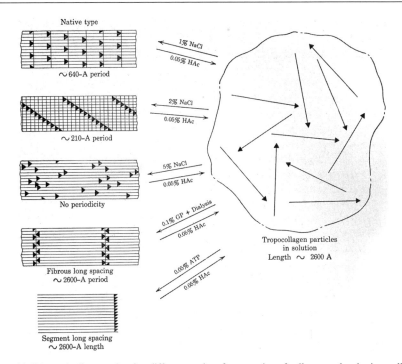

Figure 11. Schematic diagram showing different modes of aggregation of collagen molecules in a collagen fibril prepared *in vitro* under various solution conditions. Each collagen fibril form represents a different state of aggregation of the same collagen molecules (from Schmitt et al. 1955). The arrow-head and -tail, respectively, represent the N- and C-terminal end of the collagen molecule. Each of the separate collagen fibrils represents a different 3D packing of the same collagen molecules. [Reprinted with permission from Glimcher (1960). Copyright AAAS.]

1968), without significantly distorting the highly ordered and structured organic matrices and the crystals of the substance of mineralized tissues. Indeed, one of the problems that needed further study was that, in the proposed original model of the two-dimensional and 3D packing of the collagen molecules in the collagen fibrils (Schmitt et al. 1953, 1955; Gross et al. 1954) (Fig. 13), there was only a very small volume of space within a collagen fibril in which the large mass of apatite nanocrystals could be accommodated without a very marked distortion or complete disruption of the 3D structure of the collagen fibrils. This problem was at least potentially explained by the outstanding work of Hodge and Petruska (Hodge and Petruska 1963; Hodge 1967, 1989). These investigators determined from extensive and very creative and original TEM studies of the various differently assembled aggregates of collagen molecules including the native ~67 nm axial repeat. In the native fibril structure, collagen molecules are arranged in a linear axial array where a gap or hole exists between the between the C-terminal end of one collagen molecule and the N-terminal end of the next axially-arranged molecule. The collagen molecule in the adjacent linear array was positioned in such a way with respect to the end of the collagen molecule in the first linear array, that there was a slight overlap, and the overlap distance was approximately one-fourth of the molecular length. The combined length of the hole and overlap region was equal to the axial repeat of the collagen fibril and thus comprises 4.4 axial repeats, ~67 nm (Fig. 13) (Hodge and Petruska 1963; Hodge 1967, 1989). Each collagen molecule is ~300 nm long and, thus, comprises 4.4 stagger lengths. They noted that a significant mass of nanocrystals of bone apatite could be accommodated into the volume of space or "hole zones" thus created (Hodge and Petruska 1963; Hodge 1967, 1989).

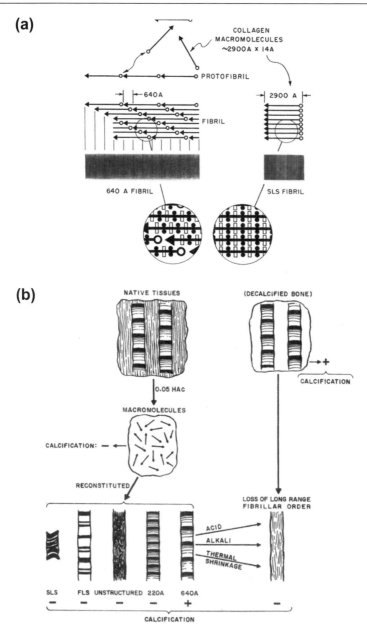

Figure 12. (a) Schematic diagram demonstrating that the different stacking of the collagen molecules in the various collagen fibrils prepared *in vitro*, for example in the native ~67 nm collagen fibrils and in the segment long stacking (SLS) fibrils, create very different structural, chemical and electrically charged environments *within* the collagen fibrils, particularly at the sites where nucleation and crystal growth of the Ca–Pi nanocrystals occur. This permits one to experimentally determine whether the ability of native type collagen fibrils (~67 nm axial repeat) to nucleate nanocrystals of Ca–Pi is dependent on the individual collagen molecules or to the manner in which the collagen molecules are 3Dly packed in the collagen fibril. [Reproduced with permission of AAAS from Glimcher (1960).] (b) Diagrammatic illustrations of some of the experiments which demonstrated the specificity of the macromolecular aggregation state of collagen in *in vitro* calcification. [Reproduced with permission of Elsevier from Glimcher and Krane (1968).]

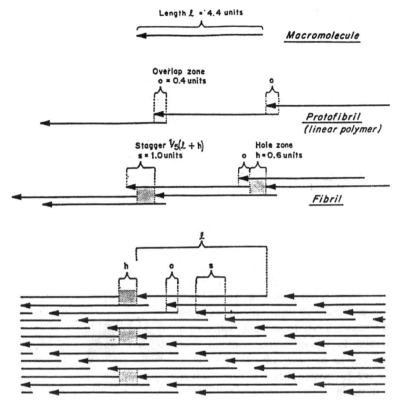

Figure 13. Schematic illustrating how the packing of the collagen macromolecules in the collagen fibril generates a void or hole within the fibril. Note that the overlap (o) plus the "hole" zone (h) are equal to the ~67 nm axial repeat of the collagen fibril. [Used with permission of Taylor and Francis from Hodge (1989).]

Unfortunately, distinct equatorial low-angle X-ray diffraction lines had until recently been detected only in rat tail tendon, not in bone, so that the 3D distribution of the space within bone collagen fibrils could not be computed directly. However, based on equatorial X-ray diffraction patterns obtained for rat tail tendon collagen molecules in rat tail collagen fibrils, the only collagen fibrils which had at that time generated sufficiently clear and resolvable low-angle X-ray equatorial reflections, and assuming that bone collagen fibrils are packed with collagen molecules having the same hexagonal or pseudohexagonal unit cell as rat tail tendon collagen fibrils, Hodge and Petruska postulated that the two-dimensional "hole" zone regions might be laterally (equatorially) connected, creating "channels" across the diameter of the collagen fibrils, into which a large mass of the Ca–Pi nanocrystals of bone substance could be deposited without significant distortion of the fibril structure (Hodge and Petruska 1963; Hodge 1967, 1989) (Fig. 14). It is important to note that Hodge and Petruska clearly recognized (as some scientists still have not) that their electron micrographs represented only a 2D model of the actual 3D packing of the collagen molecules in a collagen fibril.

It was now necessary to establish experimentally the exact sites within the collagen fibrils where the deposition of Ca–Pi crystals was initiated. Accomplishing this task required knowledge of the exact location of the "hole" zone or "gap" region with respect to the native distribution of the bands of the collagen fibrils within the approximately ~67 nm axial repeat period (Figs. 15a-c). The stained dark lines ("bands") (Figs. 15a,b) represent regions containing more reactive amino-acid residue side chain groups of the collagen molecules that have been stained with the electron dense elements present in the chemical stain, whereas the intraband regions (approximately white) represent regions with less reactive amino acid side chain groups. The point is that the exact region within the ~67 nm axial repeat where the hole region was located had to be determined because the axial length of the hole zone region (~40 nm) of the collagen fibrils is significantly less than the ~67 nm axial period of the collagen fibrils, creating an infinite number of potential locations of the "hole zone" region within the collagen fibrils (Glimcher 1968) (Fig. 15c). This was achieved in very creative groups of experiments (Hodge and Petruska 1963). Using TEM, these authors compared images of stained reconstituted collagen fibrils in native packing (~67 nm axial repeat) and in the different non-native packings (Fig. 11), matching the band patterns with those of the native ~67 nm

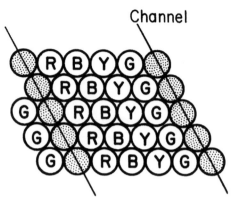

Figure 14. Schematic cross section of a fibril in the plane perpendicular to the length of the fibril, passing through the "holes" (stippled circles). This molecular arrangement was obtained by hexagonal packing of collagen molecules (labeled R, B, Y, G) in laterally staggered arrays with the same end-to-end polarity of the molecules but with no displacement in the "depth" or third-dimension. As a result, the "holes" in this structure are contiguous and form transverse channels across the collagen fibril. (Courtesy, Professor Alan Hodge).

axial band period (Figs. 16a,b). In addition, Hodge and Petruska (1963) introduced a new staining technique, utilizing phosphotungtic acid (PTA) at a pH of approximately 7.0. Under these conditions, PTA does not stain the individual bands within the ~67 nm repeating period of the collagen fibrils but diffuses into the "holes" or channels, thus staining the location and linear length of the "hole" zone region now seen in TEM as an electron dense band with an axial period of ~67 nm and with a linear length equal to and corresponding to the "hole" zone region. (Fig. 16c). Lastly, it remained to be determined if the solid mineral phase nanocrystals were initially deposited in the "hole zone" region (or channels). This was accomplished by TEM analysis of anhydrously prepared and stained embryonic chick bone, where the very distinct dense staining regions, representing the more electron dense mineral phase, could be lined up and matched with the band pattern of the adjacent unmineralized collagen fibrils. These studies clearly indicated that the dense mineral phase was found to be located between the a_3 and c_2 bands of the collagen fibrils corresponding to the "hole zone" region (Fig. 16d). The initial Ca–Pi solid phase nanocrystals, identified as apatite by electron diffraction, were clearly deposited in the "hole" zones of the bone collagen fibrils and theoretically might also be in the channels (Hodge 1989) (Figs. 17a-d). Subsequent TEM and low angle X-ray and neutron diffraction studies of collagen fibrils of calcified turkey tendon confirmed the initial location of the Ca–Pi crystals in the "hole" zones (White et al. 1977; Landis et al. 1993) and, possibly, in the "channels" (Ascenzi et al. 1978, 1983, 1985; Berthet-Colominas et al. 1979).

Calculations by Glimcher and Hodge (unpublished) based on the average density of Ca–Pi nanocrystals of "average age" revealed, however, that the volume of the "hole" zones and "channels" was still not sufficiently large to house all the crystals when the tissue was fully calcified. After a review of many published electron micrographs, it was concluded that additional apatite crystals were deposited in the intermolecular spaces or "pores" between the collagen molecules (Figs. 17a-c). It might then be anticipated that intermolecular crystal deposition would increase the distance between collagen molecules. Paradoxically, data

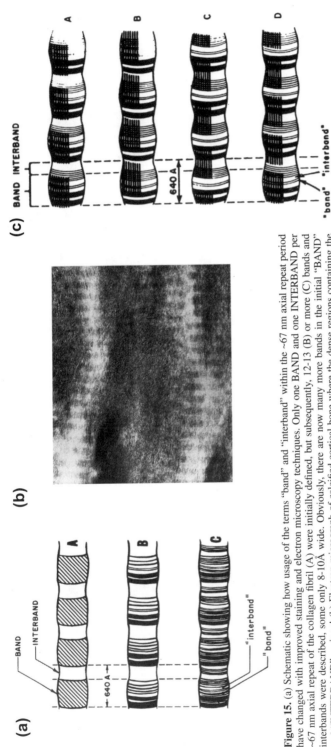

Figure 15. (a) Schematic showing how usage of the terms "band" and "interband" within the ~67 nm axial repeat period have changed with improved staining and electron microscopy techniques. Only one BAND and one INTERBAND per ~67 nm axial repeat of the collagen fibril (A) were initially defined, but subsequently, 12-13 (B) or more (C) bands and interbands were described, some only 8-10Å wide. Obviously, there are now many more bands in the initial "BAND" and "INTERBAND" regions! (b) Electron micrograph of calcified cortical bone where the dense regions containing the nanocrystals of Ca-Pi are seen, with an axial period of ~67 nm. (c) Schematic illustrating some of the possible different locations of the inorganic Ca–Pi crystals (dark horizontal lines) with respect to the ~67 nm BAND and INTERBAND regions of the native collagen fibrils. Clearly, the more distinct dense (dark) bands one can specifically identify in each of ~67 nm axially repeated segments, the more certainly one can eventually identify the "hole zone" region (and channels) in the collagen fibril by TEM, and eventually determine whether the nanocrystals are located within these volumes of the collagen fibril. Since the nanocrystals are significantly shorter than the length of the individual ~67 nm axial repeat region of a collagen fibril, simple TEM of calcified collagen fibrils (Fig. 16b) cannot distinguish where along the ~67 nm repeating regions of the fibril the nanocrystals are located (Fig. 16c), and therefore where the crystals are with respect to the "hole zone" regions. To determine this critical spatial relationship between the nanocrystals of Ca-Pi and the "hole zone" region one must determine the location of the crystals in the same TEM of calcified collagen fibrils with respect to their location to the specific bands within the ~67 nm repeat of the collagen fibril. This requires TEMs of anhydrously prepared and anhydrously stained thin sections of bone in which it is possible to identify the bands of collagen fibrils to locate the position of the "hole zone" regions of the collagen fibrils and the specific location of the nanocrystals with respect to the "hole zone" region as defined by their position relative to the specific bands of the ~67 nm axial period preferably in the same fibril or an adjacent fibril. [Used with permission of AAAS from Glimcher (1960).]

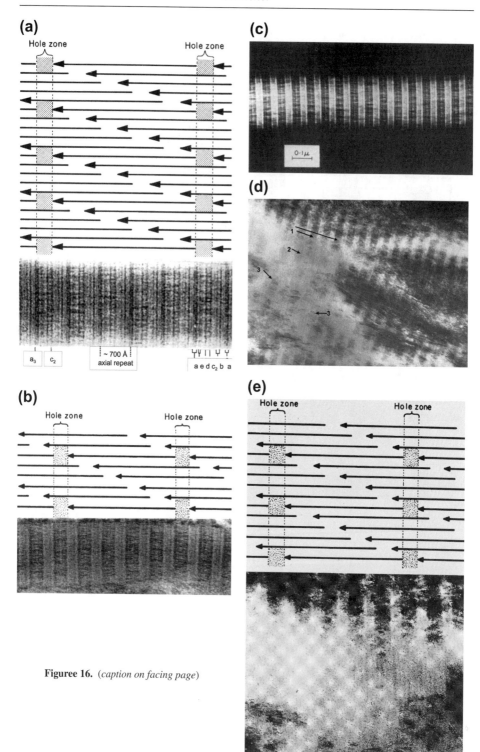

(a)

Hole zone   Hole zone

a₃   c₂   ~ 700 Å axial repeat   a e d c₂ b a

(b)

Hole zone   Hole zone

(c)

0·1 μ

(d)

(e)

Hole zone   Hole zone

Figuree 16. (*caption on facing page*)

indicated that the intermolecular distances between the collagen molecules in a collagen fibril decrease with increased calcification of the fibrils. This apparent contradiction was clarified when it was realized that as the intermolecular water in the uncalcified fibrils with a density of ~1.0 is replaced by apatite nanocrystals with a density of ~3.0, the intermolecular spaces between collagen molecules should, in fact, decrease as more mineral is deposited between the collagen molecules.

The above hypothesis was confirmed experimentally by X-ray diffraction, neutron diffraction and atomic force microscopy studies which demonstrated that the older and more mature apatite nanocrystals in mature, highly mineralized bone were significantly thinner than the apatite crystals of very much younger animals and those of younger "crystal age" (Eppell et al. 2001; Tong et al. 2003). Although crystals can be accommodated between collagen molecules, with increasing age of the animal and crystal growth and maturation, the additional crystals deposited at maximum levels of calcification will undoubtedly cause some deformation of the collagen fibrils, as reflected by the progressive loss of the highest orders of the meridional low-angle X-ray reflections in bone collagen fibrils (Burger et al. 2001; C. Burger, unpublished data).

The actual volume of space available for the Ca–Pi nanocrystals has never been obtained directly, but calculations on the theoretical distribution of the intrafibrillar volume of space within rat tail tendon collagen fibrils revealed that the diameter of the diffusion pathway was ~3 Å for rat tail tendon collagen and much larger (6 Å) for decalcified bone collagen fibrils (Katz and Li 1972, 1973) (Fig. 17d). The investigators concluded from these data that Pi ions could not diffuse into rat tail tendon collagen but could easily diffuse into bone collagen. This model provides a possible negative steric explanation for the inhibition of calcification of native rat tail tendon fibrils *in vivo*, and emphasizes the critical importance of the distribution of space within the collagen fibrils in order for calcification to occur. Note, however, that estimates of the total space within the bone collagen fibrils available for deposition of the Ca–Pi nanocrystals (75%) (Katz and Li 1973) may have been underestimated, because hexagonal or

---

Figure 16. (*figure on facing page*) (a) TEM image of a stained collagen fibril (lower portion of figure); the stained lines ("bands") represent the position of the side-chains within the collagen fibrils, and the inferred relationship (upper portion of figure) between the specific bands of the axial period of the collagen fibril, and the "hole zone" regions of the axial period. The "hole zone" region of the collagen fibril was later found to lie between the a_3 and c_2 bands (adapted from Hodge and Petruska 1963). (b) TEM image of a reconstituted collagen fibril "negatively stained" using Na phosphotungstate and under conditions in which the phosphotungstate does not stain (or very minimally stains) the side chains of specific amino acids but simply diffuses into the empty spaces of the "hole zones" with the electron dense phosphotungstate, thus clearly outlining the axial extent of the "hole" zone region. The inferred relationship between the positions of the darkly stained "hole zone" regions and the position of the "hole zones" in the Hodge-Petruska model of the collagen fibril is depicted in (b). From the TEM image, the "hole" zones have a width of ~40 nm. (c) Negatively stained, native-type fibril isolated from rat tail tendon, using Na phosphotungstate. The light zones occupying ~0.4 of the period correspond to the overlap regions; the remainder of each period is very dark, indicating a very high concentration of phosphotungstate, and corresponds to the "hole zone" or "channels". (d) TEM image of undecalcified, embryonic chick bone stained with osmium vapor only. Similar results were obtained using other anhydrous staining techniques. The ~67nm axial period of collagen is accentuated by the density of the mineral phase (1), the bands and intrabands of the collagen fibrils devoid of crystals (2), and the regions in the collagen fibrils where mineral is just beginning to form (3). From the position of the mineral phase in relation to the collagen bands, the location of the Ca–Pi mineral phase crystals was unequivocally found to be between the a_3 and c_2 bands of the collagen fibrils, corresponding to the "hole zone" region of the fibrils. × 82,000. (e) One of several TEMs prepared from young chick embryo long bone prepared anhydrously and stained with osmium vapor, showing the relationship of the nanocrystals of bone to the collagen bands of the *uncalcified* portion of *the same collagen fibril*, from which one can specifically identify that the nanocrystals at an early stage of calcification are deposited in the "hole zones" region of the collagen fibrils.[Figs. a,b,d,e reproduced with permission of Elsevier from Glimcher and Krane (1968). Fig. c reproduced with permission of Elsevier from Hodge (1967).]

ORGANIZATION OF 'HOLES' AND 'PORES' IN COLLAGEN FIBRILS

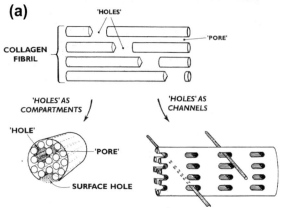

(b)

ULTRASTRUCTURAL MINERALIZATION LOCUS

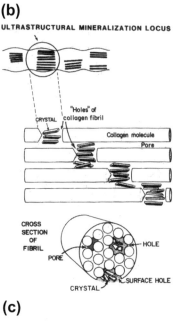

Figure 17. (a) Schematic of the proposed 3D *organization* of the collagen molecules in a collagen fibril of bone and other fibrous connective tissues, based on studies of rat tail tendon. Collagen fibrils illustrating the "hole zone" and the formation of channels across the diameter of a collagen fibril. Intermolecular spaces are called "pores." (b) Schematic of proposed steric arrangement of the Ca–Pi nanocrystals of bone in the early stages of mineralization. A few nanocrystals have also been deposited in the "pores." (c) Schematic of the later stages of calcification with continued deposition of bone nanocrystals into the intermolecular pore spaces between collagen molecules. (d) Schematic depiction of the Hodge and Petruska (1963) channel model, and the location of mineral deposition in channels during early and late mineralization, and in the intermolecular (pore) spaces in the later stages of mineralization. Collagen molecules are shown as spheres numbered 1, 2, 3, 4; the three polypeptide chains constituting each molecule are shown only in molecule 1 as empty and filled "small spots" [Figs. a-c reproduced from Glimcher (1960). Copyright AAAS. Fig. d used with permission of Taylor and Francis from Katz et al. (1989).].

(c)

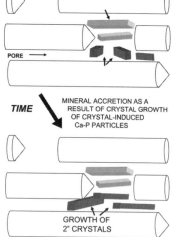

Early Mineralization

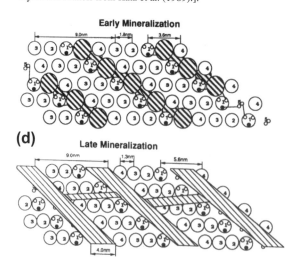

(d) Late Mineralization

pseudohexagonal packing was assumed for bone collagen molecules similar to rat tail tendon, and no consideration was made for possible distortion of the fibrils due to intermolecular crystal deposition.

It is also interesting that intact samples of native rat tail tendon do not mineralize *in vitro* whereas fibrils reconstituted from the same native rat tail tendon into native packing do calcify (Glimcher et al. 1957). This may occur because of small differences in the packing of the reconstituted fibrils compared to the intact bundles of native collagen fibrils (Eikenberry and Brodsky 1980). In addition to this steric factor inhibiting rat tail tendon calcification *in vitro* and *in vivo*, another possible factor is the presence of a sufficient concentration of calcification inhibitors that are lost during the extraction and reconstitution of the collagen fibrils in rat tail tendon.

Very convincing evidence that collagen fibrils aggregated in the native 67 nm axial repeat are structurally necessary, if not sufficient, for normal calcification of collagen organic matrix comes from what we have termed *in vitro - in vivo* experiments. In these experiments, reconstituted collagen fibrils obtained from soft tissues and decalcified bone were directly implanted in the peritoneal cavity (Mergenhagen et al. 1960) or in Millipore filter chambers to avoid direct contact with the peritoneal cells (Glimcher, Barr and Goldhaber, unpublished) and harvested over a several week period (Fig. 18a). In both studies, calcification was initiated within the implanted reconstituted type I collagen fibrils, as detected by X-ray diffraction, chemical analyses for Ca and phosphate, electron microscopy and electron diffraction. Some stunning electron micrographs by Marie Nylen at the National Institute of Dental Research showing the ~67 nm axial repeating period of the mineral nanocrystals in reconstituted native aggregated collagen fibrils implanted in the peritoneal cavity were so much better than our own that we print one here (Fig. 18b). Except for the absence of cells, these electron micrographs vividly show the remarkably orderly deposition of the Ca–Pi nanocrystals within the "hole" zone region of the collagen fibrils, and these are indistinguishable from electron micrographs of the early stages of the mineralization of bone. Very importantly, heat-denatured reconstituted soft tissue collagen fibrils, decalcified bone collagen fibrils, and gelatins made by heat denaturation of collagen molecules when similarly implanted did not mineralize, once again establishing that it is the specific 3D packing of the collagen molecules in collagen fibrils which is the critical structural factor which contributes to the underlying mechanism of the calcification of bone.

The original model for 3D packing of bone collagen molecules in collagen fibrils creating channels across the diameter of the fibril (Hodge 1989) has been used to interpret data from a number of studies using low-angle X-ray diffraction, and to address a variety of the structural and biological properties and functions of the specific packing, despite the fact that the 3D packing of the collagen fibrils of bone has not yet been established (Fratzl et al. 1991, 2004). Similarly, computer-generated 3D constructions of the spatial relationships between the collagen fibrils and the Ca–Pi mineral phase have also been presented (Landis et al. 1991, 1993). In both cases, the 3D packing of bone collagen molecules in bone collagen fibrils, and our (Glimcher and co-workers) own illustrations published here, refer only to *models* postulated by Hodge and Petruska, where the nanocrystals are located only in the "channels." However, deposition of the nanocrystals actually also occurs in the intracmolecular spaces between the collagen molecules ("pores"), resulting in a distortion of collagen fibril structure (Fig. 17), which has not been taken into account. Furthermore, while the "channel" model of rat tail tendon collagen fibrils is relatively widely accepted, other 3D stacking arrangements of collagen molecules in collagen fibrils have also been proposed (Fraser 1987). In a cooperative research paper soon to be published with members of the Chemistry Department, SUNY, Long Island, Burger et al. have recently recorded equatorial reflections from bone collagen by low-angle X-ray diffraction for the first time. Theoretical calculations and conclusions based on these new data do not support the hexagonal packing arrangements of the collagen molecules in the collagen fibrils (of bone) *theoretically* proposed by Hodge and Petruska, especially

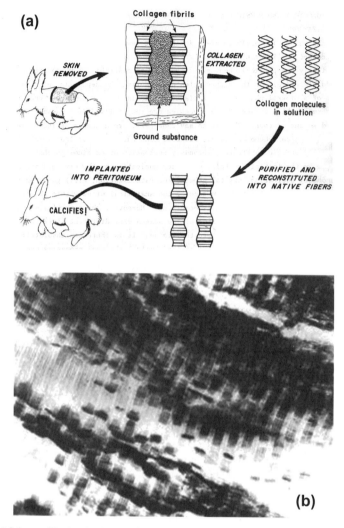

Figure 18. (a) Schema of *in vitro-in vivo* experiments to show ability of native type reconstituted collagen fibrils to calcify *in vivo*. [Reprinted with permission of Elsevier from Glimcher (1981).] (b) electron micrograph of *in vivo* calcified reconstituted skin collagen fibrils implanted into peritoneum. [Courtesy Dr. Marie Nylen, National Institute of Dental Research. Reproduced with permission of Elsevier from Glimcher (1998).]

the separate channels across the diameter of the collagen fibrils. Recall that this was based on a postulated hexagonal packing of the bone collagen molecules in the collagen fibrils of stretched rat tail tendon. However, our calculations based on the low angle X-ray diffraction equatorial reflections generated by native bone collagen are consistent with another specific 3D steric relationship between the collagen molecules and collagen fibrils, and a continued but different specific spatial relationship of the nanocrystals of bone and the spatial 3D packing of the collagen molecules in a bone collagen fibril. Thus, both the 3D packing arrangement of the bone collagen molecules in a bone collagen fibril and the spatial relationships between this 3D structure of the collagen fibrils and the mineral phase are quite different from the original model of Hodge and Petruska.

Combined mechanisms: role of non-collagenous proteins

Clearly, to ensure the specificity and singularity as to which tissues are normally mineralized, other biological and physical chemical controls must also be operative. The early enzyme theory of calcification was based on a homogenous nucleation model, which would have resulted in a random precipitation of apatite crystals at the specific locations. Later studies revealed that the spatial distribution of the crystals was not random but, rather, was very highly spatially organized at the ultrastructural level with respect to specific locations within the collagen fibrils. This self-assembly of collage molecules as collagen fibrils with an axial period of ~67 nm, was able to serve as a very effective heterogeneous nucleation substrate for the formation of apatite crystals *in vitro*. These and other specific *in vitro* experiments served as the basis of an hypothesis that the initiation of normal calcification of biological tissues was the result of the heterogeneous nucleation initiated by specific 3D ordered sites within the highly ordered collagen fibrils of the organic matrix.

The possible role of alkaline phosphatase was later investigated *in vitro* utilizing the model system of collagen fibrils and metastable Ca–Pi solutions but substituting the metastable solutions of Ca–Pi utilized previously (Glimcher et al. 1957, 1959, 1976) with normal serum (Fleisch and Neuman 1961). It was found that nucleation of apatite crystals from serum did not occur in the reconstituted collagen fibrils, suggesting that there might be an inhibitor of calcification in serum which was cleaved by alkaline phosphatase and which prevented the *in vitro* calcification of reconstituted collagen fibrils (Fleisch and Neuman 1961). An inhibitor in serum was, indeed, subsequently identified as pyrophosphate and the "alkaline phosphatase" present in serum was shown to have pyrophosphatase activity (Fleisch and Bisaz 1962). Many important papers confirming and expanding these data and hypothesis were subsequently published by Fleisch and his colleagues. It was therefore intriguing to read a recently published elegant, detailed, state-of-the-art study utilizing a large series of major gene knockouts which examined the role of various biochemical, genetic, and structural components, which are all necessary and sufficient for the calcification of bone (Murshed et al. 2005). Importantly, no specific osteoblastic gene was found necessary and sufficient for the calcification of bone. The results of these experiments demonstrated the key role of a non-specific alkaline phosphatase (a pyrophosphatase) in cleaving pyrophosphate, a known inhibitor of calcification. In the process, Pi ions are very likely released, thus locally increasing the Pi concentration in the ECFs of bone, and increasing the local degree of supersaturation with respect to apatite. Importantly, Murshed et al. (2005) also established a structural requirement which was necessary and sufficient for the initiation of bone calcification, namely, normal Type I collagen fibrils. However, the possibility exists that specific osteoblastic Type I collagen genes and chondroblastic Type II collagen genes (Type I collagen gene in certain species) may alter or add certain chemical or structural properties of these collagens, such as specific cross-links and their distribution in the collagens of bone and cartilage and/or specific local structural and chemical environments at the initial sites of mineralization, that may facilitate or promote the nucleation of Ca–Pi nanocrystals by "lowering the barrier" for nucleation.

There may also be other biochemical components such as specific proteins that also facilitate or promote the nucleation of Ca–Pi nanocrystals in specific locations within the collagen fibrils. Alternatively, chemical bonds or strong interactions between the collagen fibrils and certain groups of the collagen matrix may create a special environment within the fibrils that could also facilitate the nucleation of the Ca–Pi. Among these possibilities, currently, one of the most favored hypotheses is the importance of phosphoryl groups of serine and/or threonine amino acid residues in specific phosphoproteins, especially bone sialoprotein (BSP). Glimcher and co-authors have published extensively on the potential theoretical advantages of phosphoryl groups in promoting the nucleation of apatite crystals in bone and other biological tissues (Glimcher and Krane 1968; Glimcher 1984, 1989). Major work stressing the role of the phosphrylated proteins of dentin in initiating its calcification have been carried out by Arthur Veis, Anne George, Mary MacDougall, Adele Boskey, William Butler and their colleagues. Graeme Hunter, Harvey Goldberg and their colleagues have worked extensively on the inhibition of *in vitro* apatite precipitation from solutions of Ca–Pi, by the phosphoryl groups in BSP and osteopontin (OPN), and the promotion of apatite precipitation by specific peptides of BSP. For example, site-directed mutagenesis studies have shown that eight contiguous glutamic acid residues as a peptide of BSP are required for the precipitation of a Ca–Pi solid phase when added to a metastable solution of Ca–Pi (Tye et al. 2004). This is quite a different physical chemical model system, in which the BSP and OPN peptides are considered as nucleation substrates by adding these peptides to the solution of Ca and Pi and observing their efficacy in precipitating apatite crystals. Importantly, one notes that the crystals of apatite that are precipitated from the Ca and Pi solution by the addition of BSP and the specific peptides of BSP and OPN are very large and well-crystallized apatite crystals compared to the poorly crystallized nanocrystals of bone and dentin.

Numerous *in vitro* and *in vivo* experiments suggest a positive role for the phosphoryl groups of phosphoserine and for the carbonate groups of aspartic acid residues of the major phosphoproteins of dentin in calcification. For example, ^{31}P NMR spectroscopy was used to study phosphophoryn (the phosphorylated C-terminal portion of the major phosphoprotein in dentin) in a solution containing Pi to which Ca was gradually added in increasing amounts (Lee et al. 1983). The ^{31}P NMR spectral changes suggested that Ca was bound first by Pi, followed by the formation of ternary or higher-order complexes of Ca, Pi, and protein carboxyl residues (Lee et al. 1983). Similar complexes may be formed between glutamic acid and phosphoryl groups of BSP since these residues are located relatively close by in the primary amino-acid sequence of BSP (Salih and Fluckiger 2004). Further, experiments using ^{33}P labeling of the odontoblasts in dentin phosphoproteins permitted radiographic, visual tracking of the path of the phosphoproteins from the odontoblasts to the initial site of calcification (Weinstock and Leblond 1973). Similarly, ^{33}P-labeled phosphoprotein in chick embryos were visualized and followed as a function of time by both light and electron microscopy. The labeled proteins were found initially in the osteoblasts, later in the unmineralized osteoid, and still later at the earliest stage of mineralization. ^{33}P-labeled serine and threonine phosphates were also identified in the non-diffusible phosphoproteins of the bone (Landis et al. 1984). Other studies have employed immunochemically-labeled phosphoproteins to identify morphologically the initial sites of mineralization at the light and TEM levels.

We recognize that the identification of phosphoproteins in the organic matrix of bone and dentin at the very sites where calcification begins does not by itself implicate phosphoproteins, or specific peptide sequences and their phosphorylated sites, either in facilitating or inhibiting the mineralization processes. The phosphoproteins may play other roles in the tissue. However, additional studies lend support to the hypothesis that these non-collagenous, phosphoproteins may help or facilitate Ca–Pi nucleation in the initial sites within the collagen fibrils. For example, studies of skin collagen to which phosvitin (a highly phosphorylated protein) was crosslinked showed that the collagen-phosvitin complex was a very potent Ca–Pi nucleation substrate *in vitro* (Banks et al. 1977). Similarly, decalcified bone collagen to which the native resident phosphoproteins were first crosslinked in their *in vivo* positions before demineralization, also proved to be very potent nucleation substrates *in vitro* (Glimcher 1989), and when the phosphoryl groups were enzymatically cleaved from the collagen fibrils, the relative efficacy of the nucleation of Ca–Pi nanocrystals of bone was markedly reduced to the level of decalcified bone collagen fibrils containing no organic phosphoryl residues.

In this line of thought, it should also be noted that a casein kinase II has been identified in bone cells (Mikuni-Takagaki and Glimcher 1990; Salih et al. 1996; Sfeif and Veis 1996). This is significant because casein kinase II is responsible for the major phosphorylation of the bone phosphoproteins. In similar studies, investigators utilized two synthetic peptides as nucleation substrates for *in vitro* studies of Ca–Pi nanocrystals of apatite. The peptides had identical amino acid sequences, except that one of the peptides contained a serine residue and in the other peptide the same serine residue was phosphorylated. Both peptides were used under identical conditions to test their ability to nucleate Ca–Pi nanocrystals *in vitro* (Hartgerink et al. 2001). The synthetic peptide containing serine residues did not nucleate apatite crystals, but the same peptide containing the phosphorylated serine residue actively induced *in vitro* mineralization (Hartgerink et al. 2001). In contrast, osteopontin (another bone phosphoprotein) and peptides containing phosphoserine residues were found to inhibit *in vitro* precipitation of hydroxylapatite (Tye et al. 2004). These results were echoed by others using solution ^{31}P NMR to study the efficacy of BSP to nucleate Ca–Pi nanocrystals (Stubbs et al. 1997).

Perhaps the most direct *in vivo* evidence of the participation of Ca-phosphoryl bonds in the very earliest stages of apatite nucleation was obtained by ^{31}P NMR spectroscopy studies of native embryonic bone studied *in situ* (Wu et al. 1998). These data provide very strong evidence that during the initiation of calcification *in vivo* it is at least accompanied by and may facilitate the nucleation process by the formation of a complex between a protein phosphoryl group and Ca atoms, consistent with the ternary complexes of Ca, Pi and phosphoryl groups previously discussed (Lee et al. 1983). It is also interesting that glutamyl phosphate residues have been identified in the α2-CB3-5 peptide of bone collagen, but were absent in soft tissue collagen (Landais et al. 1989). Finally, when Pi is reacted with native (~67 nm axial repeat) collagen fibrils *in vitro*, a significant fraction of the bound Pi cannot be removed by repeated dialysis or other procedures, suggesting the formation of a strong, possibly covalent bond between collagen and phosphate groups (Glimcher and Krane 1964a).

As might be expected, attention has also been focused on the factors that inhibit the calcification of connective tissues other than bone, cartilage, or tooth, in view of the fact that normally calcified tissues are bathed by the same ECF. It follows that the inhibitory mechanisms must be impaired in pathological calcification of normally uncalcified tissues. Only very recently has there been some progress in understanding the molecular and genetic mechanisms of pathological calcification, using engineered specific gene knockouts. These experiments have revealed that normally uncalcified cartilage, and the endothelial and medial layers of the walls of blood vessels are calcified after deletion of the matrix Gla protein (a γ-carboxyglutamic acid-rich protein) (Luo et al. 1997). The particular pathological calcification of the intima and medial layers of the walls of blood vessels is also accentuated by the added deletion of the *osteopontin* gene (Speer et al. 2002). It is important to note, however, that the absence of matrix-Gla protein or osteopontin singly or in combination does not affect the calcification of bone or normally calcified cartilage. It would be of great interest if rat tail tendon collagen were mineralized in the absence of matrix Gla proteins and/or osteopontin or other yet to be defined inhibitors of calcification.

OTHER NORMALLY MINERALIZED TISSUES

Dentin and enamel

Like bone, the major structural protein of dentin is also type I collagen, the molecules of which are also aggregated into collagen fibrils with an axial period of ~67 nm. Also like bone, the solid Ca–Pi phase nanocrystals of apatite are deposited within the collagen fibrils. At the stage of maximum mineral content, the weight and volume percentage of mineral in dentin is slightly more than that of bone. Dentin also contains several very highly phosphorylated proteins. The major biological difference from bone is the fact that once the tooth is erupted, there is normally little or no additional cell proliferation and differentiation to pre- and fully matured odontoblasts (dentin forming cells) and no further synthesis and resorption of the organic matrix constituents and mineralization (no dentin turnover). Thus, dentin serves as an excellent "living substance" to follow the maturation of both the organic matrix constituents and the mineral phase.

In *tooth enamel*, unlike dentin, all of the ameloblasts (enamel forming cells) die by the time the tooth erupts. Also, the major structural protein of enamel is not collagen but, *amelogenin*. Amelogenin is an entirely different protein, both in composition (high contents of proline, glutamic and leucine, and histidine) and in its molecular, supramolecular, and ultrastructural organization by TEM (Fig. 19a-g). The major organic matrix components are very highly spatially organized at all levels of the anatomic hierarchy and, eventually at the time of tooth eruption, contain many more, much larger and more crystalline apatite crystals with even greater selected orientation than those in bone, cartilage, dentin or cementum (Glimcher et al. 1965) (Figs. 19a,b). At the TEM level, the organic matrix is organized in the form of "tubules" within which the long crystals are embedded (Travis and Glimcher 1964) (Figs. 19c,d), and the matrix organization is retained even when enamel is decalcified (Figs. 19d-g). The critical role of the specific chemical composition of amelogenin, and its molecular, macromolecular, and ultrastructural organization, along with its spatial relationship of the long, large, thin, highly ordered crystals of apatite within the enamel matrix tubules is clearly different and easily distinguished from the spatial organization of the nanosized crystals in the collagen fibrils of dentin. The difference is clearly evident even at the dentin-enamel junction where the two tissues intersect, interweave and are immediately adjacent to each other (Travis and Glimcher 1964; Glimcher et al. 1965). Each tissue still maintains its own characteristic structure (Fig. 20). This is an excellent example of the role of specific organic matrix components of mineralized biological tissues (their composition, molecular structure, ultrastructural organization) in providing the specific steric and chemical conditions most favorable for the initiation of mineralization and further crystal growth (orientation, size and shape of the crystals) best suited for the material and structural properties required by the mineralized tissue to carry out its mechanical and biological functions.

Amelogenin, the major protein of enamel

Amelogenin, with an approximate molecular weight of 20 kDa, contains at least one residue of serine phosphate (Glimcher and Krane 1964b), which was identified chemically and later in a number of animal species by direct amino acid sequencing and by ^{31}P NMR spectroscopy (Glimcher et al. 1964; Levine et al. 1967; Glimcher 1970; Roufosse et al. 1979, 1980). Proton NMR of amelogenin in solution shows little or no evidence of secondary structure, even when incubated with Ca ions (E. Strawich, unpublished). However, when the extracellular organic matrix of young, developing tooth enamel is dissected free of the overlying cellular enamel organ (Fig. 21a,b) and plated on a glass slide, removed and mounted with the thin dimension parallel to an X-ray diffraction beam, or pulled with tweezers to form threads, wide-angle X-ray diffraction generates a cross-β protein structure (Eastoe 1960; Glimcher et al. 1961).

Figure 19. (*figure on facing page*) (a) TEM of calcified enamel of an unerupted bovine tooth. (b) TEM of the decalcified organic matrix of an adjacent section of enamel, a "blueprint" of the calcified tissue. (c) TEM at a higher magnification of a similar region of enamel. The long thin, highly parallel oriented enamel crystals (*c*-axis) in one prism (P) are clearly randomly rotated around the long axis of the crystals. (d, e, f, g) This is clearly seen in high magnification TEMs at the early stages of mineralization. (h) The earliest stage of calcification of enamel at the junction of the unmineralized, newly synthesized organic matrix and the early deposition of axially rotated long thin crystals. The yet to be mineralized, newly synthesized organic matrix as organized at the TEM level are round or oval tubules, presumably within which the crystals are very likely deposited. (i, j) Higher magnifications of the organic matrix of decalcified unerupted bovine tooth. (k) Schematic interpretation of organic matrix-apatite crystal relationship. [Figs. a-h reproduced with permission of Elsevier from Glimcher and Krane (1968). Fig. k reproduced with permssion of the International and American Associations for Dental Research from Glimcher (1979).]

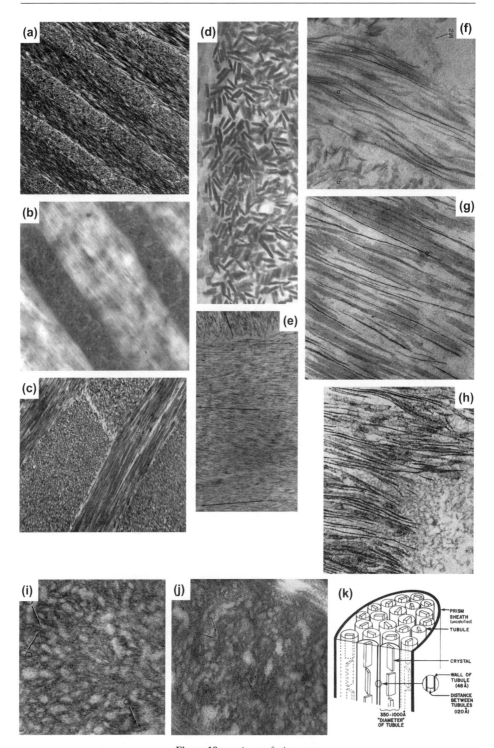

Figure 19. *caption on facing page*

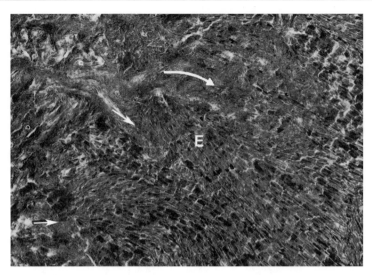

Figure 20. Electron micrograph of embryonic incisor enamel at the enamel-dentin junction (shown by curved arrow). Note the interdigitation (arrows) of the dentin into the enamel (E). The smaller crystals of dentin and the characteristic ~700 A axial repeat of the collagen fibrils are clearly observed. Thin section embedded in methacrylate. × 28,000. [Used with permission of Elsevier from Glimcher et al. (1965).]

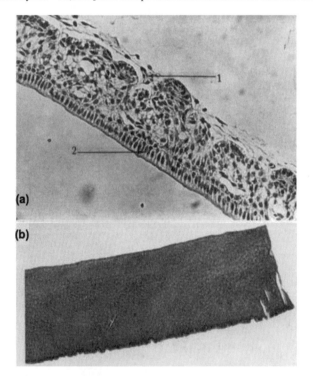

Figure 21. (a) Cross section of the outer membrane removed from decalcified five-month embryonic calf incisor tooth. Interstitial tissue (1) and ameloblasts (2) are clearly shown. ×330. (b) Cross section of the second, inner membrance removed from decalcified five-month embryonic calf incisor tooth, dissected after removal of the first outer layer, and consisting of the cell-free enamel matrix proper. ×143. [Used with permission of Elsevier from Glimcher et al. (1965).]

Fibers of purified amelogenin also indicated a cross-β protein configuration (Bonar et al. 1965). However, other X-ray diffraction reflections were also detected that were clearly not consistent with a protein in a cross-β configuration (Bonar et al. 1965). Further spectroscopic analyses of amelogenin suggested that these additional X-ray diffraction reflections were generated by a specific portion of the single amelogenin protein in a β-spiral configuration within which Ca might be "captured" (Glimcher et al. 1964; Roufosse et al. 1980; Renugopalakrishnan et al. 1986, 1989; Zheng et al. 1987; Du et al. 2005).

There have also been a number of more recent studies in which "clusters" or aggregates of amelogenin molecules have been identified, termed "nanospheres of amelogenin" (Fincham et al. 1995, 1999; Aichmayer et al. 2005; Beniash et al. 2005; Du et al. 2005). These "nanospheres" can be further organized "linearly" as long chain-like structures and are thought to serve as nucleation sites for the formation of the very long thin crystals of apatite (Fincham et al. 1995, 1999; Aichmayer et al. 2005; Du et al. 2005). Our own TEM studies of mineralizing developing enamel, however, clearly showed an amazing architectural "blueprint" of the ultrastructural organization of the organic matrix components of enamel as "tubules" within which the long Ca–Pi crystals are deposited (Travis and Glimcher 1964) (Figs. 19d,e). There was also some suggestion of a few, small, sphere-like matrix densities along the length of the tubule walls that could conceivably correspond to amelogenin "nanospheres," though they were very much smaller. There has never been any claim in our (Glimcher and co-workers) publications, however, as has been incorrectly suggested (Fincham et al. 1999), that the nucleation "site" in enamel matrix is the single amelogenin protein, whose molecular structure we have identified in the solid state of enamel matrix, but which we have also found to have little or no structure in solution. We have emphasized that the amelogenin is only detected as a cross-β protein when it is in the solid state and ultrastructurally by TEM as "tubules," and that the crystals are deposited within the tubules (Travis and Glimcher 1964) (Figs. 19f,g).

Developing enamel substance is a very complicated mineralized tissue in which rapid protein synthesis, especially of amelogenin, is constantly occurring before eruption, followed by rapid degradation to make space for the extensive deposition of the inorganic crystals. This is reflected in the loss of the amelogenin protein, measured by amino acid analyses at progressive stages of development and in the increased content of the apatite crystals (Glimcher et al. 1977; Robinson et al. 1977a,b, 1978). In addition, when the dissected enamel matrix is solubilized, many aggregate components with different molecular weights and amino acid compositions are detected by size exclusion and biochemical chromatography and other biophysical techniques (Katz et al. 1966, 1969; Mechanic et al. 1966). These aggregates associate, dissociate, and reaggregate very readily even in denaturing solvents. The individual separated aggregates which seemingly represent single proteins but when analyzed have clearly formed multiple aggregates of different size and composition (Katz et al. 1966, 1969; Mechanic et al. 1966).

There is little question that "nanospheres," or high molecular weight aggregates of amelogenin can be formed *in vitro* (Fincham et al. 1999; Aichmayer et al. 2005; Du et al. 2005). However, a technique developed by Dr. Hershey Warshawsky at McGill University in Quebec, Canada (unpublished) in which ^{3}H-labeled proline and serine were introduced through the carotid vessels of the neck of rats, and the enamel matrix and enamel cellular layer were individually analyzed, clearly demonstrated that very shortly after the intravascular injection, only one ~20 kDa labeled protein was present in the enamel matrix. Only later were other components detected, viz., a ~100 kDa component(s) (containing no amelogenin) and lower molecular weight components which did have an amino acid composition similar to amelogenin (Strawich 1985). It remains to be seen whether further TEM and other studies of the organic matrix of developing enamel *in vivo* can confirm the *in vitro* studies of the organization of "nanospheres" of amelogenin, the ultrastructural organization of the nanospheres and their spatial relationships to the initial and later ultrastructural sites of mineralization.

Calcified cartilage

In contrast to bone, and the dentin and enamel tissues of teeth, where the relationships between the major inorganic matrix constituents and the organic Ca–Pi crystals have been well described at many levels of the anatomic hierarchy, the ultrastructural organization of the organic matrix constituents and the Ca–Pi nanocrystals in calcified cartilage are not well documented. TEM studies have established that a relatively large proportion of the nanocrystals of Ca–Pi, especially in the early stages of cartilage calcification (and in embryonic bone) (Anderson 1984; Bonucci et al. 1989), occur in "matrix vesicles." These structures are thought to form by "pinching off" sections of the cell membrane of chondrocytes or chondroblasts and osteoblasts in embryonic bone and forming relatively round, organically separate membrane structures within whose enclosed spaces the nanocrystals of Ca–Pi are deposited.

The major organic matrix components of cartilage are the proteoglycans, which are gradually degraded as mineralization progresses, and the type II collagen fibrils, which are aggregates of collagen type II collagen molecules (and some type I collagen in certain avian species (Seyer et al. 1974a,b; Eyre et al. 1978) with a 3D stacking organization similar to type I collagen fibrils of connective tissue and bone. Based on the elegant TEM studies of very carefully prepared specimens of calcified cartilage, investigators were able to compute the volume of space occupied by the various organic structures in the cartilage matrix (Eggli et al. 1985; Hunziker et al. 1984, 1989; Hunziker and Herrmann 1987). Using these data and the mass and density of the mineral phase present, Gimcher and co-workers calculated that the mineral phase cannot be wholly accommodated within the matrix vesicles, the volume between the vesicles, and between the collagen fibrils when the calcified cartilage is fully mineralized. It is likely, therefore, that a significant mass of the apatite nanocrystals in calcified cartilage are located within the collagen fibrils, similar to bone and dentin.

SIGNIFICANT IMPLICATIONS OF THE STRUCTURAL ORGANIZATION OF MINERALIZED TISSUES

Mineralized tissues as composite materials

There are very important concepts to take away from the organization of the major organic matrix components and the inorganic Ca–Pi or $CaCO_3$ crystals at the ultrastructural and molecular and atomic levels of organization in bone substance, dentin, enamel and invertebrate mineralized tissues and substances, at all levels of the anatomic hierarchy. This organization of organic matrix and inorganic solid phase is vital for the mineralized tissues to perform their biological functions such as maintaining the ion homeostasis in the blood and ECF, and providing mechanical-structural support (Glimcher 1998). Starting with bone substance, the basic material of bone is composed of an organic matrix, principally collagen fibrils, and inorganic Ca–Pi nanocrystals of bone which are embedded within the collagen fibrils in a very special and highly organized fashion (Glimcher 1998). The mineralization of biological tissues creates what engineers and materials scientists have termed composite materials (Glimcher 1998), which may include one or more of the same or different substances. If different materials are used, the individual materials have different structural, electrical and other properties, and importantly, the two materials have a specific geometric relationship to one another, and possibly chemical linkages as postulated for collagen in bone (Glimcher 1984). It is important to recognize that the structural or other properties of the composite material are much different than the algebraic sum of the individual components. The purpose of a composite material is to enhance the properties of the material so that it best fulfills the special functions for which it was constructed, using the least amount of substance. These requirements are met by the choice of the chemical composition of the individual materials used and by the geometric spatial relationships between the individual components at all hierarchal levels of organization, for example, the osteons of compact bone. Furthermore, there are additional hierarchical levels of organization, for example, the osteons of compact bone. Osteons are composed of lamellae within which the mineralized collagen fibrils may be ordered quite differently to adjacent lamellae, or even in the same lamella (Figs. 3a,b). There is also a variable axial rotation of the osteons as a whole (Giraud-Guille 1988) (Fig. 22). At this level of the anatomic hierarchy, the mineral content of the individual lamellae also often varies significantly, particularly as new osteons and lamellae are being formed and resorbed (Fig. 3b), all of which may account for significant changes in the material and the biological properties of a particular sample of bone. Moreover, all of these anatomical factors are related to the rate of bone turnover which, in turn, is cellularly-controlled. While this osteonal level at the microscopic organization of adult bone substance is the standard textbook description of compact osteonal bone organization at the light microscopic level, it is important to be aware that there are a number of other microscopic arrangements in many animal species (Foote 1916). At the ultrastructural and molecular levels, the spatial and geometric relationships between collagen fibrils at specific locations within the bone substance, and the structural relationship of the two organic and inorganic constituents of bone and other mineralized tissues, such as the amelogenin tubules of enamel, are equally exquisitely arranged, to fulfill both their biological and structural roles.

Figure 22. Schematic demonstrating that there is an additional axial rotation of the collagen fibrils in each of the lamellae of an osteon. (Courtesy Dr. Marie-Madeleine Giraud-Guille, University of Paris V. Used with permission of Elsevier from Glimcher (1998).]

Spatial architecture and volume of space to house the Ca–Pi crystals

As already pointed out, it is necessary for the cells of mineralized tissues to synthesize and resorb the organic matrix constituents to create a sufficient volume of organized space that can accommodate the inorganic solid phase without a major distortion in the highly ordered structure of the particular mineralized tissue (Glimcher 1968). In bone and dentin, this is accomplished by the deposition of nanocrystals of Ca–Pi within the collagen fibrils in volumes of space created by a postulated specific 3D packing of collagen molecules in collagen fibrils (Hodge and Petruska 1963; Hodge 1967) or an alternative, calculated, 3D organization of the collagen fibrils based on new low-angle equatorial X-ray diffraction (C. Burger, unpublished). The biological mechanisms for creating the volume of space to house the much larger size and mass of enamel crystals is very different from that for dentin or bone.

In enamel, the organic matrix is enzymatically degraded and solubilized, leaving only a very small amount of organic matrix in the enamel substance of erupted teeth. Only ~1-2% of the total protein content of the enamel remains and only a very small fraction, if any, of the protein residue is amelogenin (Glimcher et al. 1964a,b). Thus, adult enamel substance at full maturation consists of ~98% or more of very large, highly spatially oriented and highly crystalline apatite crystals.

In this regard, one often reads or hears statements that tooth enamel is the "hardest" biologically mineralized tissue or substance. First, the term "hardest" does not completely describe the structural properties of a substance or a material. Equally important, one needs to know explicitly what structural loads the "hardest material" is going to be subjected to. In any event, a material such as tooth enamel is also "brittle," and if a bone were composed of "enamel" it would be likely to fracture easily.

These facts emphasize the point made earlier that the mineral content by volume percent (true volumetric density) of bone substance is roughly the same regardless of species or specific bones. One might ask, what is the significance of this universal volumetric mineral content? Another quandary is how the relatively constant volumetric amount of mineral deposited within the collagen fibrils is controlled? Why doesn't the amount of the solid mineral phase continue to increase beyond the level ordinarily and universally found? To speculate, recall that bone collagen fibrils are very insoluble and resist swelling in a variety of solvents ordinarily used to solubilize other soft tissue collagens such as skin. In addition, other collagens, such as occur in the Achilles tendon, are also insoluble in most solutions but do swell enormously when exposed to weak acids. Glimcher and co-workers have postulated that the volumetric extent of mineralization of bone is limited to a value that is best suited to resist the structural demands to which the bone is most commonly subjected, and to maximize the resistance of bone to swelling. The resistance of bone collagen to swelling or solubilization is at least partly a result of the unique number, chemistry and, especially, the distribution of collagen cross-links in bone collagen fibrils. These strong, stable chemical bonds control the volume of space within the collagen fibrils and thus limit the number of crystals and crystal mass and the volumetric mineral mass that can be physically accommodated within the collagen fibrils. The mechanical and structural properties of bone substance can therefore be adversely altered in certain diseases of bone or as a result of exposure to certain chemicals that significantly increase the normal volumetric capacity within the collagen fibrils (and/or the size and mass of the crystals) and/or the size or crystallinity or composition of the solid mineral phase of bone.

Once again, we see the very close association between the shape, size and intimate organization of all of the components that contribute to bone as an organ, as a tissue and/or as a substance and their direct relationships to the many biological and structural functions of bone. Mineralized tissues – a wonder of biological design!

"AMORPHOUS" VERSUS CRYSTALLINE CALCIUM-PHOSPHATE: CONCEPTUAL AND EXPERIMENTAL DEFINITIONS IN THE NANO-WORLD

A final word about the use of the term "amorphous" with respect to both the specific 3D spatial relationships of the atoms and/or ions in an "amorphous" solid phase in general, and the concept that the first solid phase of Ca–Pi nucleated in bone is "amorphous" (Termine and Posner 1966a; Termine et al. 1967).

Figure 23 is a schematic depiction of a classic solid phase crystal in which there is a fixed 3D repeating spatial relationship of the atoms or ions over a long distance, viz., very many repeating unit cells. In a truly totally amorphous solid phase, there is no long-range or short-range spatial order of the constituent atoms or ions, and only the average distance between nearest neighbors (~4-5Å) is maintained. In the case of "amorphous" Ca–Pi ("ACP") formed *in vitro* by precipitation from neutral or moderately alkaline aqueous solutions, the earliest solid phase formed does not generate a coherent X-ray diffraction pattern but, rather, a single, very broad X-ray reflection with the center of the broad peak at $2\theta \sim 25$ (Eanes et al. 1965). Importantly, the formation of such solid particles occurs over a wide range of Ca and Pi ion concentrations of the initial aqueous phase. The Ca–Pi composition of the solid phase particles is quite constant, a tricalcium phosphate, with a relatively constant Ca/P ratio of ~1.5. When left in the original solution from which it was precipitated, the solid phase is first dissolved and then recrystallized as poorly-crystalline, apatite nanocrystals. The crystallinity or "index of crystallinity" of a crystalline phase depends independently on the size of the crystal and the degree of perfection

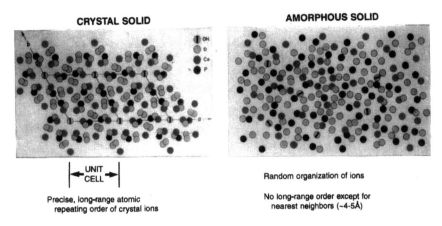

CRYSTAL SOLID

AMORPHOUS SOLID

| UNIT CELL |

Precise, long-range atomic
repeating order of crystal ions

Random organization of ions

No long-range order except for
nearest neighbors (~4-5Å)

Figure 23. Schematic showing the random distribution of ions and atoms in an *amorphous* solid phase compared with the ordered arrangement of the same constituents in a *crystalline* solid phase over a sufficiently long distance. [Used with permission of Elsevier from Glimcher (1998).]

of the long- and short-range order of the constituent atoms and ions. In crystals of bone apatite, the poor crystallinity is due to both the very small size of the nanocrystals and the fact that there is significant short- and long-range disorder of the crystal lattice.

Based on the relatively constant Ca/P ratio of the "ACP" formed under varying conditions, such as different concentrations of Ca + Pi and pH and different temperatures, as well as spectroscopic and structural analyses, Posner and colleagues hypothesized that the initial solid phase particles of "ACP" precipitated *in vitro* consisted of solid particles ranging from 300-1000 Å in diameter (Posner et al. 1980). The 300-1000 Å particles are composed of randomly oriented clusters of Ca + Pi ions, which are 9.5 Å in diameter (Fig. 24), and the atoms and ions within the clusters do have some short range order. It is not clear from what experimental spectroscopic or structural data this model is derived except the tricalcium phosphate composition. Further, it is not entirely clear whether there are volumetric regions within the large 300-1000 Å particles that are free of the 9.5 Å clusters. In other words, are there some volumetric regions within the large 300-1000 Å particles where the atoms are completely random with no short-range order?

Experimentally determining the 3D spatial relationships of such small 9.5 Å clusters would be a formidable task, as would fashioning a calcification mechanism where the formation of these clusters can be related to the very highly ordered organization of the apatite nanocrystals with respect to highly selected regions of the collagen fibrils at the ultrastructural level and

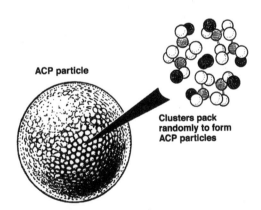

ACP particle

Clusters pack randomly to form ACP particles

Figure 24. Diagram of the atomic structure of amorphous calcium phosphate seen as individual spheres 300-1000 A in size. Spheres are composed of randomly arranged $Ca_9(PO_9)_6$ clusters. [Gratefully used with permission of Springer Verlag from the original supplied by Dr. Aaron Posner. from Posner (1984).].

probably involving the 3D molecular configuration of certain amino acid side chain groups of collagen. The Posner et al. hypothesis (Termine and Posner 1966a; Termine et al. 1967) implies three solution to solid phase transitions, namely, the formation of the initial solid phase "ACP" particles from a solution of Ca–Pi, their subsequent dissolution and, finally, the heterogeneous nucleation of very poorly crystalline apatite nanocrystals from a solution phase at specific sites within the collagenous fibrils.

Recent reviews of the theoretical aspects and experimental data of phase transitions (or changes) have been published. For example, the important role of interfacial free energy in the transition of a solution phase to a solid phase (Wu and Nancollas 1999) and the critical role of changes in enthalpy, which has been elegantly developed by Navrotsky (2004).

A great deal of attention has been focused on "heterogeneous nucleation," the induction of a solid phase from a sufficiently metastable (heterogeneous) solution phase on the surface of other solid phases whose interatomic structure closely mirrors that of the new solid phase to be nucleated. For example, the nucleation (formation) of ice crystals on the surfaces of crystals of AgI and, even more effectively on CuI. The general concepts of the phase transition from solution to solid phase include the formation of the first cluster of atoms/ions of a solid phase formed from constituents in a solution phase which occurs at the point that the free energy of the solution phase decreases (Gibbs 1928). The first solid phase particles formed from the constituents of the solution phase are called nuclei and the process is called nucleation. If the nucleation occurs spontaneously without the introduction of a solid surface, it is referred to as homogeneous nucleation. If, however, a solid phase, not necessarily one with the same composition but which contains lattice planes whose atomic configuration mimics a particular crystal plane or of a specific crystal structure, it will nucleate a solid phase from solution. This is called heterogeneous nucleation. A "simple" model is the formation of ice crystals from water, viz., H_2O (liquid) $\rightarrow H_2O$ (solid) (ice). In very carefully constructed experiments, water can be chilled well below its theoretical freezing point (0 °C) to approximately −39 to −40 °C, at which temperature the liquid H_2O forms randomly scattered ice crystals. However, if AgI or, better still, CaI crystals are added to the liquid H_2O water before cooling begins, a solid H_2O phase (ice) is nucleated on the AgI or CaI crystals, by those atomic planes which closely mimic specific planes in ice crystals. This process is referred to as heterogeneous nucleation and occurs at −2 to −4 °C, well above homogeneous nucleation temperature ~ −39 to −40 °C. Clearly there are many other factors involved in the relative "efficiency" of different crystals to nucleate a specific solid crystal from solution, but this simple example introduces some basic processes of the physical chemical processes underlying phase transformations, and that are related to the mineralization of the solid organic matrices of biological tissues.

We do not mean to imply that heterogeneous nucleation is totally dependent on a minimal disregistry between the lattice parameters of the nucleation substrate and the solid phase that is being nucleated. While the degree of spatial coherency is clearly very important other factors may also play significant roles in the nucleation of a solid phase from a liquid phase (Passarelli et al. 1974). These additional factors include particle size, distribution of chemical bonds between the nucleation substrate and the solid phase to be nucleated, hydrogen bonding, especially important for mineralization of tissues where the nucleation substrate is an organic substance (Garten and Head 1965) and restrictions on the volume of space within which such a phase transition occurs (Thomas and Thomas 1967).

With respect specifically to the calcification of bone substance, if "ACP" is the first solid phase of Ca–Pi formed in bone, the process requires three phase transitions as discussed above. Alternatively, the large "ACP" particles may dissociate, releasing the 9.5 Å clusters of Ca and Pi ions which do have some short range internal order (Posner et al. 1980), and aided by additional ions, particularly Pi from the ECF, may nucleate apatite nanocrystals within the collagen fibrils. In any event, while both these pathways are possible, each requires two or three phase transitions. Since the mineralization process of bone visualized *in vivo* by TEM of samples prepared by non-aqueous procedures occurs very rapidly after the self-assembly of the collagen molecules to collagen fibrils in the extracellular space, the youngest visible solid Ca–Pi nanocrystals identified as apatite by TEM are spatially very close to the osteoblasts, so that very little time is available for these phase transitions to occur. In many cases the crystals are smaller in one or more dimensions than a single unit cell of apatite, but generate electron diffraction patterns of nanocrystals of apatite. Therefore, the three transitions of "ACP" to nanocrystals of apatite must be so extraordinarily rapid that it would be impossible or nearly impossible to technically detect the temporal sequence of such events *in vivo* or the structural organization of the solid phases as "ACP" before the formation of nanocrystals of apatite using the current techniques available. Furthermore, one is dealing technically with the problem of how to detect both the postulated initial "ACP" solid phase and the subsequent solid phase of Ca–Pi very small (nuclei) of nanocrystals of apatite containing only a very small number of atoms and which, in addition, are organized with a significant degree of imperfection.

It is also not clear where or how such large "ACP" aggregates will fit into any of the available spaces within or on the surface of the collagen fibrils. Similar considerations of size (as well as several other important "worries") also apply to the recent suggestion (Crane et al. 2006) that the initial solid phase of Ca–Pi formed in bone is octacalcium phosphate. There needs to be some mechanism to actively dissolve these large particles in order to release the Ca–Pi ions into specific local regions of the collagen fibrils in order for the heterogeneous nucleation of nanocrystals of apatite to occur. As noted, it is very difficult to conceive of such a rapid and complex sequence of events and considering the time limitations it would also be a formidable task to prove experimentally.

In addition to the current technical difficulties associated with measuring the small initial "ACP" clusters solid phases and characterizing them at the present time, it is also difficult to define them conceptually. In experiments on *in vitro* heterogeneous nucleation of apatite crystals by natively organized collagen fibrils (~640 Å axial repeat), Katz calculated from classical nucleation theory that the nuclei formed contain only 11-13 atoms (Katz 1969). Conceptually, whether there are 11 to 13 atoms or 20 atoms or so in the first solid phase cluster of atoms *in vivo*, is it possible to detect and measure the 3D organization of the ions in such a small cluster? How many atoms or ions even if spatially ordered, are necessary to distinguish this first "solid" phase cluster of Ca–Pi as "amorphous" or "crystalline"?

A question also arises as to whether even the first formed solid Ca–Pi nanocrystals of bone apatite, for example, meet the classical definitions of a "crystal" as described earlier, despite the fact that highly collimated electron diffraction of single, isolated very small nanocrystals of bone generate patterns clearly from an apatite structure. This has led some investigators to refer to these individual "particles" also observed and measured by atomic force microscopy as "mineralites" rather than as "crystals" (Eppell et al. 2001; Tong et al. 2003).

We are in a new age of nanotechnology not even conceived of a few years ago, where we can study the basic underlying physical-chemical mechanisms that are biologically controlled at a nanostructural level involving very small clusters of atoms. Results of such studies will have significant implications for understanding the biological control of important physiological functions, including the basic mechanisms underlying biological mineralization, and the structure-function relationships of the inorganic nanocrystals of bone, with respect to their nano size, very large surface area, and the crystal imperfections that alter their biological and material properties compared to very large, highly crystalline hydroxylapatites.

It is time to rethink the initial events of the calcification of bone and other normally calcified collagenous tissues in terms of the current theoretical concepts of what defines a solid phase as "amorphous" as opposed to a "crystalline" phase with reference to biological calcification, most particularly bone, and to pursue and devise the technology which will permit us to examine this complex and very important physical chemical and biological problem. It must be made clear how to define and distinguish what is an "amorphous solid phase" as distinguished from a small cluster of atoms which may be or may not be in a relatively well organized spatial relationship of the atoms but as a result of the fact that it consists of only a small number of atoms does not generate coherent spectroscopic evidence of a distinct spatial organization of its few atom constituents, i.e., a "crystalline" solid phase.

It is important to note that one of the major reasons that we (Glimcher and co-workers) and other investigators are pursuing the nature of the initial solid phase of Ca–Pi deposited, is based on the theory of heterogeneous nucleation. The effort is to understand, in detail, the necessary role of the chemical and structural properties of the proposed nucleation sites within the collagen fibrils in facilitating the nucleation process, and the possible contribution of specific non-collageneous proteins and specific peptide regions of these proteins in promoting mineralization in the calcified tissues of vertebrates, including humans, and in inhibiting pathological calcification in disease or after injury.

In answer to a recently posed question and suggestion (Weiner et al. 2005), firstly, Glimcher and co-workers have indeed pursued, for many years, the question of whether the initial Ca–Pi solid phase deposited in bone is "amorphous" or "crystalline" (or another Ca–Pi solid phase such as octacalcium phosphate). Most recently, Glimcher and co-workers have utilized synchrotron generated X-ray diffraction, FTIR, and synchrotron generated and simultaneous fluorescence spectroscopy (for identification of Ca) and the first sites of mineralization and have only detected very poorly ordered nanocrystals of bone apatite (manuscript in preparation).

In a very detailed theoretical discussion of heterogeneous nucleation and of the specific data from his own experimental studies of *in vitro* nucleation of apatite crystals by collagen fibrils (Katz 1969; Katz and Li 1972, 1973; Katz et al. 1989) and those of Wadkins (Wadkins 1968; Wadkins et al. 1974), Katz examined the concepts discussed above of the chemical and structural changes which are most likely to have occurred during the earliest phases of forming Ca–Pi "clusters" or "embryos" and eventually of apatitic bone nanocrystals. Based on Katz's analysis, our own questions noted above and our own TEM, electron diffraction, X-ray diffraction, FTIR, Raman, and NMR spectroscopy results obtained for "amorphous" calcium phosphate, prepared as described by Posner and Eanes, we conclude that one thing appears certain: an initial solid "ACP" as defined by Posner and Eanes themselves does not occur as the first detectable solid Ca–Pi phase in bone. However, we have not given up.

Glimcher and co-workers are still experimentally pursuing this important problem, principally because, as noted above, we are one of many groups seeking to define the "nucleation substrate" in the organic matrix of bone and other vertebrate calcified tissues. We hope to better understand the normal biology of bone and the pathogenesis of the many afflictions and injuries to bone, in order to design specific therapies for diseases and injuries to the skeleton and to prevent the pathological calcification of normally unmineralized tissues and organs.

THE CHEMISTRY AND STRUCTURE OF INORGANIC SYNTHETIC AND GEOGENIC APATITES

In this section, we review spectroscopic studies of ionic substitutions in synthetic and natural, geogenic apatites that were prepared inorganically. Synthetic apatites and geogenic apatites are different from biological apatites in that the former have been equilibrated at sufficiently high temperatures and/or high pressures and for sufficiently long time period to reach thermodynamic equilibrium, whereas biological apatites are formed at body temperature and are constantly turning over, so that kinetic factors are probably more important in determining their properties and reactivates. *It is critical to appreciate that the understanding of ionic substitutions in crystal sites obtained from the study of the synthetic and geogenic inorganic apatites does not provide a direct model for understanding the chemical composition and structure and changes of the Ca–Pi solid phase on bone apatite.* As we have emphasized above, the nanocrystals of bone apatite are unique in their physical-chemical properties and biological functions. Still, it is worthwhile to review the literature on inorganic apatites because it shows the difficulty in obtaining unequivocal information even for the large crystals of synthetic and geogenic apatite, and gives us a fuller appreciation of the significantly compounded difficulties in the case of bone nanocrystals due to their nano-size, presence of organic matter, and changing nature with crystal maturity and animal maturity.

The apatites conform to the general formula $A_5(TO_4)_3Z$ in which A= cations Ca, Sr, Pb, Ba, etc. T = anion of oxide groups P, As, V and Z = F, OH, Cl. The apatites crystallize in the space group $P6_{3/m}$. The unit cell dimensions range depending on the substituents in the lattice. The range for a from 9.37 to 10.75 and for c from 6.78 to 7.64 with the larger elements producing the larger cell sizes (Skinner 2005).

Calcium occupies two different sites in apatite. Cation substitution for Ca in both sites is possible. The main considerations for substitution are to retain appropriate coordination environment for steric reasons, crystal field stabilization, and charge balance. The cationic substitutions range from monovalent Na^+ and NH_4^+ to many divalent ions such as Mg^{2+}, Sr^{2+}, $Ba^{2+}+$, Pb^{2+}, Mn^{2+}, Cd^{2+} etc., the trivalent REE^{3+}, tetravalent Th^{4+}, and hexavalent U^{6+} with the latter in significant quantities (Pan and Fleet 2002). If the substituting ion has a charge different from the ion being substituted, then coupled substitution is the generally observed. For example, U^{4+} and a vacancy site substitute for two Ca^{2+} ions, and various types of charge combinations are known in the case of the trivalent REEs. A complete substitution between Pb and Ca has been established through synthetic experiments, and although earlier studies predicted that Pb ions prefer the Ca_2 sites, it was not confirmed with Rietveld calculations (Engel et al. 1975; Miyake et al. 1986). Sr also has extensive isomorphous substitutions with calcium in geogenic apatites, and limited replacement of Sr in the Ca–Pi phase of bone nanocrystals is also possible. The amount and site of Sr substitution in the structure may also depend on substitutions in the F, OH or Cl site and the Sr-containing Cl apatite has a discrete crystal structure distinct from F/OH apatite (Rakovan and Hughes 2000). Investigations of synthetic apatites have also established isomorphous substitutions of Cd and Mg for calcium.

Anions may substitute in the PO_4^{3-} sites or for OH^- in the *c*-axis channels or both, in synthetic and geogenic apatites. The oxyanion AsO_4^{3-} substitutes entirely for PO_4^{3-} in johnbaumite, and tetravalent SiO_4^{4-} can extensively substitute for PO_4^{3-}. Monovalent anions including F^-, Cl^-, Br^-, I^-, NO_3^-, divalent anions such as O^{2-}, vacancies, and neutral molecules such as Ar and glycine can substitute for OH^- in the *c* channel sites (Sundarsanan et al. 1977; Rey et al 1978; Hughes and Rakovan 2002; Pan and Fleet 2002). Partial fluoride substitution in the hydroxylapatite structure of the enamel of teeth has beneficial dental effects in reducing the occurrence of tooth carries, especially in children (McClure 1962; Riggs and Melton 1995), but excessive replacement in the apatite nanocrystals of bone may make bone more brittle.

Carbonate-containing calcium apatites known as the minerals dahllite and francolite (the hydroxyl- and fluorine- containing species respectively) were identified in phosphorite sedimentary deposits many decades ago (McConnell and Gruner 1940). The difficulties in understanding the crystal chemistry and structure were recognized over the subsequent thirty years (McClellan and Lehr 1969; McConnell 1970, 1973; Kolodny and Kaplan 1970).

Synthetic carbonate-containing apatites precipitates are poorly-crystalline phases at room temperatures. Under experimental condition of high CO_2 pressures two discrete types of carbonate apatites were produced: Type A and Type B (Bonel and Montel 1965; Labarthe et al. 1971). Using IR spectroscopy, these researches determined that the CO_3^{2-} ion could be incorporated into either the PO_4^{3-} site resulting in so-called Type B carbonate apatites or in the OH^- sites forming Type A carbonate apatites. In a third type of carbonate-apatite called

Type AB, a PO_4^{3-} and OH^- are replaced by two CO_3^{2-} groups. Perhaps the most recent study of carbonate in hydroxylapatite is a molecular mechanics/molecular dynamics simulation (Peroos et al. 2005), where the relative stabilities were calculated for carbonate substitution in different sites similar to the e three types of carbonate apatites. Assuming the hydroxylapatite structure and inserting one planar CO_3^{2-} molecule in place of two OH^- ions and aligned along the *c*-axis channel with the third oxygen of CO_3^{2-} in the *ab* plane is the energetically preferred substitution. The next most favorable configuration is Type AB. Another potential substitution where a phosphate group is replaced by a CO_3^{2-} group and an OH^- group produces almost no net change in stability, but when the CO_3^{2-} replacement of the PO_4^{3-} is accompanied by substitution of Na^+ (or K^+) for the Ca^{2+} ion as in A-type apatite, the stability is greater. Peroos et al. (2005) also found that the stability differences between the three types were not very great. It is critical to recognize that the CO_3^{2-} in the Ca–Pi phase of bone nanocrystals has both less reactive and labile, reactive populations, which change as a function of crystal age, and that the positions of these carbonate species cannot be directly compared to CO_3^{2-} substitutions in synthetic and geogenic apatites.

In summary, we re-emphasize that experimental or theoretical considerations based on the study of geogenic apatites and of apatites synthesized *in vitro* inorganically or even in the presence of organic compounds, cannot be directly applied to the nanocrystals of the Ca–Pi phase of bone nanocrystals that grow and change *in vivo* under strict cellular control.

ACKNOWLEDGMENTS

Supported in part by a grant from the Department of Health and Human Services, NIH Grant No. AG014701-18, and by The Peabody Foundation, Inc.

I am indebted to Dr. Nita Sahai for her extensive editing of the manuscript and her very helpful comments and suggestions. The concluding section, "The Chemistry and Structure of Inorganic Synthetic Geogenic Apatites," was contributed by Dr. H. Catherine W. Skinner of Yale University and Dr. Nita Sahai of the University of Wisconsin. I also thank Patrick O'Neill for library research and assistance in preparation of the manuscript, and Dr. Lila Graham for her critical reading of the manuscript as well as her helpful suggestions. "I also want to acknowledge the major contributions of Richard Kyle of 5000-K, Pembroke, Mass. for his sterling work in enhancing many of the difficult-to-reproduce illustrations, and of Paul Andriesse, medical artist, for his elegant redrawing of some of the schematic diagrams." As is apparent from the bibliography, I am also most grateful to the many scientists who have pursued their research in our laboratories over a period of many years.

REFERENCES

Ackerman JL, Raleigh DP, Glimcher MJ (1992) Phosphorus-31 magnetic resonance imaging of hydroxyapatite: a model for bone imaging. Magn Reson Med 25:1-11
Aichmayer B, Margolis H, Sigel R, Yamakoshi Y, Simmer J, Fratzl P (2005) The onset of amelogenin nanosphere aggregation studied by small-angle X-ray scattering and dynamic light scattering. J Struct Biol 151:239-249
Anderson HC (1984) Mineralization by matrix vesicles. Scan Electron Microsc Pt 2:953-964
Arsenault AL (1988) Analysis of apatite crystal structure and their spatial relationships to extracellular matrix components in calcified cartilage, bone, dentin, enamel and calcified turkey tendon. *In*: The Chemistry and Biology of Mineralized Tissues: Proceedings of the Third International Conference on the Chemistry and Biology of Mineralized Tissues. Glimcher MJ, Lian J (eds) Gordon & Breach, p 808 [182]
Ascenzi A, Bonucci E, Ripamonti A, Roveri N (1978) X-ray diffraction and electron microscope study of osteons during calcification. Calcif Tissue Res 25:133-143
Ascenzi A, Bigi A, Ripamonti A, Roveri N (1983) X-ray diffraction analysis of transversal osteonic lamellae. Calcif Tissue Int 35:279-283

Ascenzi A, Bigi A, Koch M, Ripamonti A, Roveri N (1985) A low-angle X-ray diffraction analysis of osteonic inorganic phase using synchrotron radiation. Calcif Tissue Int 37:659-664

Aue W, Roufosse A, Glimcher MJ, Griffin R (1984) Solid-state phosphorus-31 nuclear magnetic resonance studies of synthetic solid phases of calcium phosphate: potential models of bone mineral. Biochemistry 23:6110-6114

Banks E, Nakajima S, Shapiro LC, Tilevitz O, Alonzo JR, Chianelli RR (1977) Fibrous apatite grown on modified collagen. Science 16:1164-1166

Beniash E, Simmer JP, Margolis HC (2005) The effect of recombinant mouse amelogenins on the formation and organization of hydroxyapatite crystals *in vitro*. J Struct Biol 149:182-190

Berthet-Colominas C, Miller A, White SW (1979) Structural study of the calcifying collagen in turkey leg tendons. J Mol Biol 134:431-435

Betts F, Blumenthal NC, Posner AS, Becker GL, Lechninger AL (1975) Atomic structure of intracellular amorphous calcium phosphate deposits. Proc Natl Acad Sci USA 72:2088-2090

Biltz RM, Pellegrino ED (1969) The chemical anatomy of bone. I. A comparative study of bone composition in sixteen vertebrates. J Bone Joint Surg Am 51:456-466

Biltz R, Pellegrino ED (1971) The hydroxyl content of calcified tissue mineral. Calcif Tissue Res 36:259-263

Bocciarelli DS (1973) Apatite microcrystals in bone and dentin. J Microsc 16:21-34

Bocciarelli DS (1970) Morphology of crystallites in bone. Calcif Tissue Res 5:261-269

Bonar LC, Shimizu M, Roberts J, Griffin R, Glimcher MJ (1991) Structural and composition studies on the mineral of newly formed dental enamel: a chemical, X-ray diffraction and ^{31}P and proton nuclear magnetic resonance study. J Bone Min Res 6:1167-1176

Bonar LC, Glimcher MJ, Mechanic GL (1965) The molecular structure of the neutral-soluble proteins of embryonic bovine enamel in the solid state. J Ultrastruct Res 13:308-317

Bonel G, Montel G (1965) Studies comparing carbonate apatite obtained through different methods of synthesis, (in French), *In*: Reactivity of Solids. Schwab G-M (ed) Elsevier, p 567-604

Bonucci E, Silvestrini G, di Grezia R (1989) Histochemical properties of the "crystal ghosts" of calcifying epiphyseal cartilage. Connect Tissue Res 22:43-50

Boothroyd B (1964) The problem of demineralisation in thin sections of fully calcified bone. J Cell Biol 20: 165-173

Brown WE, Eidelman N, Tomazic B (1987) Octacalcium phosphate as a precursor in biomineral formation. Adv Dent Res 1:306-313

Burger C, Liu L, Hsiao B, Chu B, Hanson J, Hori T, Glimcher M (2001) Synchrotron 1. Synchrotron SAXS/ WAXS study of the composite nature of bone. ACS PMSE Preprint 85:169-170

Carden A, Morris MD (2000) Application of vibrational spectroscopy to the study of mineralized tissues (review). J Biomed Opt 5:259-268

Carlstrom D, Finean J (1954) X-ray diffraction studies on the ultrastructure of bone. Biochim Biophys Acta 13:183-191

Chen J, Burger C, Krishnan C, Chu B, Hsiao B, Glimcher M (2005) In vitro mineralization of collagen in demineralized bone matrix. Macromol Chem Phys 206:43-51

Crane, NJ, Popescu V, Morris MD, Steenhuis P, Ignelzi MA Jr. Raman spectroscopic evidence for octacalcium phosphate and other transient mineral species deposited during intramembranous mineralization. Bone 39:434-442

Dehring KA, Smukler AR, Roessler BJ, Morris MD (2006) Correlating changes in collagen secondary structure with aging and defective type II collagen by Raman spectroscopy. Appl Spectrosc 60:366-372

DeJong W (1926) La substance minerale dans les os. Recl Trav Chim Pays-Bas Belg 45:445-448

Draper ER, Morris MD, Camacho NP, Matousek P, Towrie M, Parker AW, Goodship AE (2005) Novel assessment of bone using time-resolved transcutaneous Raman spectroscopy. J Bone Min Res 20:1968-1972

Du C, Falini G, Fermani S, Abbott C, Moradian-Oldak J (2005) Supramolecular assembly of amelogenin nanospheres into birefringent microribbons. Science 307:1450-1454

Eanes E, Gillessen I, Posner A (1965) Intermediate states in the precipitation of hydroxyapatite. Nature 208: 365-367

Eastoe JE (1960) Organic matrix of tooth enamel. Nature 187:411-412

Eggli PS, Herrmann W, Hunziker EB, Schenk RK (1985) Matrix compartments in the growth plate of the proximal tibia of rats. Anat Rec 211:246-257

Eichert D, Sfihi H, Combes C, Rey C (2004) Specific characteristics of wet nanocrystalline apatites. Consequences on biomaterials and bone tissue. Key Eng Mater 254-56:927-930

Eikenberry EF, Brodsky B (1980) X-ray diffraction of reconstituted collagen fibers. J Mol Biol 144:397-404

Engel G, Krieg F, Reif G (1975) Mischehekristallbildung und kationneorgnung im system bleihydroxylapatit-calcium hydroxulapatite. J Solid State Chem 15:117-126

Engstrom A, Finean JB (1953) Low-angle X-ray diffraction of bone. Nature 171:564

Eppell SJ, Tong W, Katz JL, Kuhn LT, Glimcher MJ (2001) Shape and size of isolated bone mineralites measured using atomic force microscopy. J Orthop Res 19:1027-1034

Eyre DR, Brickley-Parsons DM, Gimcher MJ (1978) Predominance of type I collagen at the surface of avian articular cartilage. FEBS Lett 85:259-263

Fell H, Robison R (1934) The development of the calcifying mechanism in avian cartilage and osteoid tissue. Biochem J 28:2243-2253

Fincham A, Moradian-Oldak J, Simmer JP, Sarte P, Lau EC, Diekwisch T, Slavkin HC (1994) Self-assembly of a recombinant amelogenin protein generates supramolecular structures. J Struct Biol 112:103-109

Fincham A, Moradian-Oldak J, Simmer JP (1999) The structural biology of the developing matrix. J Struct Biol 126:270-299

Fincham AG, Moradian-Oldak J, Diekwisch TGH, Lyaruu DM, Wright JT, Bringas P Jr, Slavkin HC (1995) Evidence for amelogenin "nanospheres" as functional components of secretory-stage enamel matrix. J Struct Biol 115:50-59

Finean JB, Engstrom A (1953) The low-angle scatter of X-rays from bone tissue. Biochem Biophys Acta 11: 178-179

Finean JB, Engstrom A (1954) Low-angle reflection in X-ray diffraction patterns of bone tissue. Experientia X:63-64

Fitton-Jackson S (1957) The fine structure of developing bone in the embryonic fowl. Proc Roy Soc London, Ser B 146:270-280

Fleisch H, Bisaz S (1962) Isolation from the plasma of pyrophosphate, an inhibitor of calcification. Helv Physiol Pharmacol Acta 20:C52-53

Fleish H, Neuman W (1961) Mechanisms of calcification: role of collagen, polyphosphates, and phosphatase. Am J Physiol 200:296-300

Foote L (1916) Contribution to the comparative histology of the femur. *In*: Smithsonian Contribution to Knowledge, pub 2382, vol 35, Smithsonian, Washington, DC, p 1-242

Fratzl P, Fratzl-Zelman N, Klaushofer K, Vogl G, Koller K (1991) Nucleation and growth of mineral crystals in bone studied by small-angle X-ray scattering. Calcif Tissue Int 48:407-413

Fratzl P, Paris O, Klaushofer K, Landis WJ (1996a) Bone mineralization in an osteogenesis imperfecta mouse model studied by small-angle X-ray scattering. J Clin Invest 97:396-402

Fratzl P, Schreiber S, Klaushofer K (1996b) Bone mineralization as studied by small-angle X-ray scattering. Connect Tissue Res 34:247-254

Fratzl P, Gupta H, Paschalis E, Roschger P (2004) Structure and mechanical quality of the collagen-mineral nano-composite in bone. J Mater Chem 14:2115-2123

Garten V, Head R (1965) A theoretical basis of ice nucleation by organic crystals. Nature 205:150-152

Gibbs JW (1928) The Collected Works of J. Willard Gibbs. Longmans Green & Co

Giraud-Guille MM (1988) Twisted plywood architecture of collagen fibrils in human compact bone osteons. Calcif Tissue Int 42:167-180

Glimcher MJ (1959) Molecular biology of mineralized tissues with particular reference to bone. Rev Mod Physics 31:359-393

Glimcher MJ (1960) Specificity of the molecular structure of organic matrices in mineralization. In Calcification in Biological Systems. Sognnaes RF (ed) American Association for the Advancement of Science, p 421-487

Glimcher MJ (1968) A basic architectural principle in the organization of mineralized tissues. Clin Orthop 61: 16-36

Glimcher MJ (1970) The isolation and characterization of phosphopeptides from developing enamel. Calcif Tissue Res Suppl 146

Glimcher MJ (1976) Composition, structure and organization of bone and other mineralized tissues and the mechanism of calcification. *In*: Handbook of Physiology 7: Endocrinology, vol. VII. Greep RO, Astwood EB (eds) American Physiological Society, p 25-116

Glimcher MJ (1979) Phosphopeptides of enamel matrix. J Dent Res 58:790-809

Glimcher MJ (1981) On the form and function of bone: from molecules to organs. Wolff's law revisited. *In:* The Chemistry and Biology of Mineralized Connective Tissues. Veis A (ed) Elsevier/North-Holland, p 617-673

Glimcher MJ (1984) Recent studies of the mineral phase in bone and its possible linkage to the organic matrix by protein-bound phosphate bonds. Phys Trans Roy Soc London - B Biol Sci 304:479-508

Glimcher MJ (1989) Mechanism of calcification: role of collagen fibrils and collagen-phosphoprotein complexes in vitro and in vivo. Anat Rec 244:139-53

Glimcher M (1998) The nature of the mineral component of bone: Biological and chemical implications. *In*: Metabolic Bone Disease and Clinically Related Disorders. Avioli L, Krane SM (eds) Academic Press, p 23-50

Glimcher MJ, Krane SM (1964a) The incorporation of radioactive inorganic orthophosphate as organic phosphate by collagen fibrils *in vitro*. Biochem 3:195-202

Glimcher MJ, Krane SM (1964b) The identification of serine phosphate in enamel proteins. Biochim Biophys Acta 90:477-483

Glimcher M, Krane S (1968) The organization and structure of bone, and the mechanism of calcification. *In*: Treatise on Collagen. Vol. II-B. Ramachandran G, Gould B (eds) Academic Press, p 68-251

Glimcher MJ, Bonar LC, Daniel EJ (1961) The molecular structure of the protein matrix of bovine dental enamel. J Mol Biol 3:541-546

Glimcher MJ, Brickley-Parsons D, Levine PT (1977) Studies of enamel proteins during maturation. Calcif Tissue Res 24:259-270

Glimcher MJ, Daniel EJ, Travis DF, Kamhi S (1965) Electron optical and X-ray diffraction studies of the organization of the inorganic crystals in embryonic bovine enamel. J Ultrastruct Res Supp 7:1-77

Glimcher MJ, Friberg UA, Levine PT (1964a) The isolation and amino acid composition of the enamel proteins of erupted bovine teeth. Biochem J 93:202-10

Glimcher MJ, Hodge AJ, Schmitt FO (1957) Macromolecular aggregation states in relation to mineralization: the collagen hydroxyapatite system as studied *in vitro*. Proc Natl Acad Sci USA 43:860-867

Glimcher MJ, Mechanic GL, Friberg UA (1964) The amino acid composition of the organic matrix and the neutral-soluble and acid-soluble components of embryonic bovine enamel. Biochem J 93:198-202

Glimcher MJ, Travis DF, Friberg UA, Mechanic GL (1964b) The electron microscopic localization of the neutral soluble proteins of developing bovine enamel. J Ultrastruct Res 10:362-376

Golub EE, Katz EP (1977) On nonequivalent intermolecular stagger states in collagen fibrils. Biopolymers 16: 1375-1361

Gross J, Highberger J, Schmitt F (1954) Collagen structures considered as states of aggregation of a kinetic unit. The tropocollagen particle. Proc Nat Acad Sci USA 40:679-688

Guinier A (1939a) La diffraction des rayons X aux très faibles angles: Applications à l'etude des phénomènes ultra-microscopies. Ph.D. thesis. Series A, No. 1854. Paris, University of Paris

Guinier A (1939b) La diffraction des rayons X aux très faibles angles: Applications à l'etude des phénomènes ultra-microscopies. Ann. Phys. (Paris) 12: 161-236

Guinier A (1952) X-ray Crystallographic Technology. Hilger & Watts, Ltd.

Hartgerink J, Beniash E, Stupp S (2001) Self-assembly and mineralization of peptide-amphiphile nanofibers. Science 294:1684-88

Hodge A (1967) Structure at the electron microscopic level. *In*: Treatise on Collagen Ramachandran G (ed) Academic Press, p 185-205

Hodge AJ, Petruska JA (1963) Recent studies with the electron microscope on ordered aggregates of the tropocollagen macromolecule. *In*: Aspects of Protein Structure. Ramachandran GN (ed) Academic Press, p 289-300

Hodge AJ (1989) Molecular models illustrating the possible distributions of "holes" in simple systematically staggered arrays of Type I collagen molecules in native-type fibrils. Connect Tissue Res 21:137-147

Holmes JM, Beebe RA, Posner AS, Harper R (1970) Surface areas of synthetic calcium phosphates and bone mineral. Proc Soc Exp Biol Med 133:1250-1253

Holmes JM, Davies DH, Meath WJ, Beebe RA (1964) Gas adsorption and surface structure of bone mineral. Biochemistry 3:2019-2023

Hunziker E, Herrmann W, Schenk RK, Mueller M, Moor H (1984) Cartilage ultrastructure after high pressure freezing, freeze substitution, and low temperature embedding of chondrocyte ultrastructure — implications for the theories of mineralization and vascular invasion. J Cell Biol 98:267-276

Hunziker EB, Herrmann W (1987) *In situ* localization of cartilage extracellular matrix components by immunoelectron microscopy after cryotechnical tissue processing. J Histochem Cytochem 35:647-655

Hunziker EB, Herrmann W, Cruz-Orive LM, Arsenault AL (1989) Image analysis of electron micrographs relating to mineralization in calcifying cartilage: theoretical considerations. J Electron Microscop Tech 11:9-15

Jellinek M, Fankuchen I (1945) X-ray examination of gamma alumina. Ind Eng Chem 37:158-163

Kamat S, Su X, Ballarini R, Heuer A (2000) Structural basis for the fracture toughness of the shell of the conch Strombus gigas. Nature 405 (6790):1036-1040

Katz EP (1969) The kinetics of mineralization in vitro. Biochim Biophys Acta 194:121-129

Katz EP, Li S-T (1972) The molecular packing of collagen in mineralized and non-mineralized tissues. Biochem Biophys Res Commun 46:1-15

Katz EP, Li S-T (1973) The intermolecular space of reconstituted collagen fibrils. J Mol Biol 73:351-369

Katz EP, Mechanic GL, Glimcher MJ (1966) Preliminary studies of the ultracentrifugal and free zone electrophoresis characteristics of neutral soluble proteins of bovine embryo enamel. Proceedings of the Third European Symposium on Calcified Tissue. Springer-Verlag, p 272-276

Katz EP, Seyer J, Levine PT, Glimcher MJ (1969) The comparative biochemistry of the organic matrix of developing enamel. II. Ultracentrifugal and electrophoretic characterization of proteins soluble at neutral pH. Arch Oral Biol 14:533-539

Katz EP, Wachtel E, Yamauchi M, Mechanic GL (1989) The structure of mineralized collagen fibrils. Connect Tissue Res 21:159-167

Kay H, Robison R (1924) The possible significance of hexosephosphoric esters in ossification. Part III. The action of the bone enzyme on the organic phosphorus compounds in blood. Biochem J 18:755-764

Kim H-M, Rey C, Glimcher MJ (1994) Isolation of calcium-phosphate crystals of mature bovine bone by reaction with hydrazine at low temperature. *In:* Hydroxyapatite and related materials. Brown PW, Constanz B (eds) CRC Press, p 331-337

Kim HM, Rey C, Glimcher MJ (1995) Isolation of calcium-phosphate crystals of bone by non-aqueous methods at low temperature. J Bone Miner Res 10:1589-1601

Koch J (1917) The laws of bone architecture. Am J Anat 21:177

Kolodny Y, Kaplan IR (1970) Carbon and oxygen isotopes in apatite CO_3 and co-existing calcite from sedimentary phosphorites. J Sedimentary Petrol 40:954-959

Labarthe JC, Therasse M, Bonel G, Montel G (1973) Sur le structure des apatites phosphocalcique carbonees de type B. Compte Rendue Academie de Paris 276C:1175-1178

Landais JC, Cohen-Solal L, Bonaventure J, Maroteaux P, Glimcher MJ (1989) Localization of gamma-glutamyl-phosphate residues to the alpha 2CB3-5 peptide of type I chicken bone collagen. Connect Tissue Res 19:1-9

Landis WJ, Paine MC, Hodgens KJ, Glimcher MJ (1986) Matrix vesicles in embryonic chick bone: Considerations of their identification, number, distribution, and possible effects on calcification of extracellular matrices. J Ultrastruct Mol Struct Res 95:142-163

Landis WJ, Paine M, Glimcher MJ (1977) Electron microscopic observations of bone tissue prepared anhydrously in organic solvents. J Ultrastruct Res 59:1-30

Landis WJ, Sanzone CF, Brickley-Parsons D, Glimcher MJ (1984) Radioautographic visualization and biochemical identification of O-phosphoserine- and O-phosphothreonine-containing phosphoproteins in mineralizing embryonic chick bone. J Cell Biol 98:986-990

Landis WJ, Moradian-Oldak J, Weiner S (1991) Topographic imaging of mineral and collagen in the calcifying turkey tendon. Connect Tissue Res 25:181-196

Landis WJ, Song M, Leith A, McEwen L, McEwen B (1993) Mineral and organic matrix interaction in normally calcifying tendon visualized in three dimensions by high-voltage electron microscopic tomography and graphic image reconstruction. J Struct Biol 110:39-54

Lee DD, Glimcher MJ (1989) The 3D spatial relationship between the collagen fibrils and the inorganic calcium-phosphate crystals of pickerel and herring fish bone. Connect Tissue Res 21:247-257

Lee D, Glimcher M (1991) The 3D spatial relationship between the collagen fibrils and the inorganic calcium-phosphate crystals of pickerel (americanus americanus) and herring (clupea harengus) bone. J Mol Biol 217:487-501

Lee SL, Glonek T, Glimcher MJ (1983) ^{31}P nuclear magnetic resonance spectroscopic evidence for ternary complex formation of fetal dentin phosphoprotein with calcium and inorganic orthophosphate ions. Calcif Tissue Int 35:815-818

LeGeros RZ, Trautz O, Klein E, LeGeros JP (1969) Two types of carbonate substitution in the apatite structure. Experientia 25:5-7

Levine PT, Glimcher MJ (1965) The isolation and amino acid composition of the organic matrix and neutral soluble proteins of developing rodent enamel. Arch Oral Biol 10:753-756

Loong CK, Rey C, Kuhn LT, Combes C, Wu Y, Chen S, Glimcher MJ (2000) Evidence of hydroxyl-ion deficiency in bone apatites: an inelastic neutron-scattering study. Bone 26:599-602

Luo G, Ducy P, McKee M, Pinero G, Loyer E, Behringer R, Karsenty G (1997) Spontaneous calcification of arteries and cartilage in mice lacking matrix Gla protein. Nature 386:78-81

McClellan GH, Lehr J (1969) Crystal Chemical investigations of natural apatites. Am Mineral 54:1374-1391

McClure FJ (ed) (1962) Fluoride Drinking Waters. U.S. Department of Health Education and Welfare, Public Health Service, National Institute of Dental Research, Bethesda, MD

McConnell D (1970) Crystal chemistry of bone mineral: Hydrated carbonate apatites. Am Mineral 55:1659-1669

McConnell D (1973) Apatite. Springer-Verlag

McConnell D, Gruner JW (1940) The problem of the carbonate apatites. III. Carbonate-apatite from Magnet Cove, Arkansas. Am Mineral 25:157-167

Miyake M, Ishigaki K, Susuki Y (1986) Structure refinements of Pb^{2+} ion-exchanged apatites by X-ray powder pattern fitting. J Solid State Chem 61:230-235

Mechanic GL, Katz EP, Glimcher MJ (1966) Gel filtration and chromatographic properties of neutral soluble bovine embryo enamel proteins. *In:* Proceedings of the Fourth European Symposium on Calcified Tissues. Gaillard PJ, VanDenHoof A, Steendijk R (eds) Excerpta Medica, p 73-75

Mergenhagen SE, Martin GR, Rizzo, AA, Wright DN, Scott DB (1960) Calcification in vivo of implanted collagen. Biochem Biophys Acta 43:563-565

Mikuni-Takagaki Y, Glimcher MJ (1990) Post-translational processing of chicken bone phosphoproteins. Identification of bone (phospho)protein kinase. Biochem J 268:593-597

Morris MD, Crane NJ, Gomez LE, Ignelzi MA Jr. (2004) Compatibility of staining protocols for bone tissue with Raman imaging. Calcif Tissue Int 74:86-94

Murshed M, Harmey D, Millan JL, McKee MD, Karsenty G (2005) Unique coexpression in osteoblasts of broadly expressed genes accounts for the spatial restriction of ECM mineralization to bone. Genes Dev 19:1093-1104

Navrotsky A (2004) Energetic clues to pathways to biomineralization: precursors, clusters, and nanoparticles. Proc Natl Acad Sci USA 101:12096-12101

Neuman MW, Neuman WF (1980) On the measurement of water compartments, pH, and gradients in calvaria. Calcif Tissue Int 31:135-145

Neuman WF, Neuman MW (1953) The nature of the mineral phase of bone. Chem Rev 53:1-45

Neuman WF (1969) The milieu interieur of bone: Claude Bernard revisited. Fed Proc 28:1846-1850

Neuman WF, Mulryan B (1969) On the nature of interchangeable sodium in bone. Calcif Tissue Res 3:261-265

Neuman WF, Bareham B (1975) Further studies on the nature of fluid compartmentalization in chick calvaria. Calcif Tissue Res 17:249-255

Neuman WF, Neuman MW (1958) The Chemical Dynamics of Bone Mineral. University of Chicago Press

Pan Y, Fleet MJ (2002) Compositions of the Apatite-group minerals: substitution mechanisms and controlling factors. Rev Mineral Geochem 48:13-50

Passarelli R, Chessin H, Vonnegut B (1974) Ice nucleation in a super cooled cloud by CuI-3AgI and AgI aerosol. J Appl Meterol 13:946-948

Pasteris JD, Wopenka B, Freeman JJ, Rogers K, Valsami-Jones E, van der Houwen J, Silva M (2004) Lack of OH in nanocrystalline apatite as a function of degree of atomic order: implications for bone and biomaterials. Biomaterials 25:229-238

Peroos S, Du Z, de Leeuw NH (2005) A computer modeling study of the uptake, structure and distribution of carbonate defects in hydroxyl-apatite. Biomaterials 27:2150-2161

Posner AS, Beebe R (1975) The surface chemistry of bone mineral and related calcium phosphates. Semin Arthritis Rheum 4:267-291

Posner AS, Betts F, Blumenthal NC (1977) Role of ATP and Mg in the stabilization of biological and synthetic amorphous calcium phosphates. Calcif Tissue Res 22 (Suppl):208-212

Posner AS, Betts F, Blumenthal NC (1980) Formation and structure of synthetic and bone hydroxyapatite. Prog Cryst Growth Char 3:49-64

Rakovan J, Hughes JM (2000) Strontium in the apatite structure: strontium fluorapatite and belovite(Ce). Can Mineral 38:839-845

Rakovan J, Reeder RJ, Elzinga EJ Cherniak DJ, Tait CD, Morris DE (2002) Structural classification of U(VI) In apatite by X-ray absorbtion spectroscopy. Environ Sci Technol 36:3114-3117

Regier P, Lasaga A, Berner R, Han O, Zilm K (1994) Mechanism of CO_3^{2-} substitution in carbonate-fluorapatite: evidence from FTIR spectroscopy, ^{13}C NMR and quantum mechanical calculations. Am Mineral 70:809-818

Renugopalakrishnan V, Strawich E, Horowitz P, Glimcher MJ (1986) Studies of the secondary structures of amelogenin from bovine tooth enamel. Biochmistry 25: 4879-4887

Renugopalakrishnan V, Pattabiraman N, Prabhakaran M, Strawich E, Glimcher MJ (1989) Tooth enamel protein, amelogenin, has a probable beta-spiral internal channel, Gln_{112}-Leu_{138}, within a single polypeptide chain: Preliminary molecular mechanics and dynamics studies. Biopolymers 28:297-303

Rey C Trombe J-C, Montel C (1978) Some features of the incorporation of oxygen in different oxidation states in the apatite lattice. III. Synthesis and properties of some oxygenated apatites. J Inorg Nuclear Chem 40:27-30

Rey C, Dickson RI, Shapiro F, Shimizu M, Glimcher MJ (1987) Spectroscopic evidence for a labile environment of carbonate and phosphate ions in mineral deposits of bone and enamel. Proceed of Biomat 87: Calcified tissues and biomaterials, Bordeaux, France, p 31-46

Rey C, Collins B, Goehl T, Dickson RI, Glimcher MJ (1989) The carbonate environment in bone mineral: a resolution-enhanced Fourier transform infrared spectroscopy study. Calcif Tissue Int 45:157-164

Rey C, Renugopalakrishnan V, Collins B, Glimcher MJ (1991a) Fourier transform infrared spectroscopic study of the carbonate ions in bone mineral during aging. Calcif Tissue Int 49:251-258

Rey C, Renugopalakrishnan V, Shimizu M, Collins B, Glimcher MJ (1991b) A resolution-enhanced Fourier transform infrared spectroscopic study of the environment of the CO_3^{2-} ion in the mineral phase of enamel during its formation and maturation. Calcif Tissue Int 49:259-268

Rey C, Glimcher MJ (1992) Short range organization of the Ca - P mineral phase in bone and enamel: changes with age and maturation. *In:* Chemistry and Biology of Mineralized Tissues. Slavkin H, Price P (eds) Elsevier, p 5-18

Rey C, Hina A, Tofighi A, Glimcher MJ (1995a) Maturation of poorly crystalline apatites: chemical and structural aspects in vivo and in vitro. Cells Mater 5:345-356

Rey C, Kim HM, Gerstenfeld L, Glimcher MJ (1995b) Structural and chemical characteristics and maturation of the calcium-phosphate crystals formed during the calcification of the organic matrix synthesized by chicken osteoblasts in cell culture. J Bone Miner Res 10:1577-1588

Rey C, Miquel J, Facchini L, Legrand A, Glimcher MJ (1995c) Hydroxyl groups in bone mineral. Bone 16: 583-586

Riggs BL, Melton LJ III (1995) Osteoporosis: Etiology, Diagnosis and Management. Raven Press

Robinson C, Fuchs P, Deutsch D, Weatherell J (1978) Four chemically distinct stages in developing enamel from bovine incisor teeth. Caries Res 12:1-11

Robinson C, Fuchs P, Weatherall JA (1977a) The fate of matrix proteins during the development of dental enamel. Calcif Tissue Res 22(Suppl):185-190

Robinson C, Kirkham J, Weatherell JA, Richards A, Josephsen K, Fejerskov O (1988) Mineral and protein concentrations in enamel of the developing permanent porcine dentition. Caries Res 22:321-326

Robinson C, Kirkham J, Shore RC (1992) Extracellular matrix of enamel and the ameloblast. Epithelial Cell Biol 1:90-97

Robinson C, Lowe NR, Weatheral JA (1977b) Changes in amino-acid composition of developing rat incisor enamel. Calcif Tissue Res 23:19-31

Robinson RA (1952) An electron microscopic study of the crystalline inorganic components of bone and its relationship to the organic matrix. J Bone Joint Surg 34:389-434

Robinson RA, Watson ML (1953) Collagen-crystal relationships in bone as seen in the electron microscope. Anat Rec 114:383-409

Robinson RA, Watson ML (1955) Crystal-collagen relationships in bone as observed in the electron microscope. III. Crystal and collagen morphology as a function of age. Ann NY Acad Sci 60:596-628

Robinson RA (1957) The water content of bone. J Bone Joint Surg 39-A:167-188

Robison R (1923) The possible significance of hexophoric esters in ossification. Biochem J 17:286-293

Robison R, Soames K (1924) The possible significance of hexophoric esters in ossification. Part II. The phosphoric esterase of ossifying cartilage. Biochem J 18:740-754

Routfosse AH, Strawich E, Fossel E, Lee S, Glimcher MJ (1979) ^{31}P nuclear magnetic resonance studies of E4 phosphopeptide of embryonic bovine enamel in solution. J Dent Res 58:1019-1020

Roufosse AH, Strawich E, Fossel E, Lee S, Glimcher MJ (1980) ^{31}P NMR characterization of bovine, embryonic dental enamel phosphopeptides in solution. FEBS Lett 115:309-311

Roufosse AH, Aue WP, Roberts JE, Glimcher MJ, Griffin RG (1984) Investigation of the mineral phases of bone by solid-state phosphorus-31 magic angle sample spinning nuclear magnetic resonance. Biochemistry 23:6115-20

Rude R, Oldham S (1990) Disorders of magnesium metabolism. *In*: The Metabolic and Molecular Basis of Acquired Disease. Cohen RD, Lewis B, Alberti KGMM, Denmon AM (eds) Baillier Tindall, p 1124-1148

Rude RK (1996) Magnesium homeostasis. *In*: Principles of Bone Biology. Raissz LG, Rodan GA (eds) Academic Press, p 277-293

Sfeir C, Veis A (1996) The membrane associated kinases which phosphorylate bone and dentin extracellular matrix phosphoproteins are isoforms of cystolic CKII. Connect Tissue Res 35:215-222.

Salih E, Zhou HY, Glimcher MJ (1996) Phosphorylation of purified bovine bone sialoprotein and osteopontin by protein kinases. J Biol Chem 271:16897-16905

Salih E, Fluckiger R (2004) Complete topographical distribution of both the *in vivo* and *in vitro* phosphorylation sites of bone sialoprotein and their biological implications. J Biol Chem 279:19808-19815

Schmitt F, Gross J, Highberger J (1953) A new particle type in certain connective tissue extracts. Proc Nat Acad Sci USA 39:459-470

Schmitt F, Gross J, Highberger J (1955) Tropocollagen and the properties of fibrous collagen. Exp Cell Res Suppl 3:326-334

Seyer JM, Glimcher MJ (1971) The isolation of phosphorylated polypeptide components of the organic matrix of embryonic bovine enamel. Biochim Biophys Acta 236:279-291

Seyer JM, Brickley DM, Glimcher MJ (1974) The identification of two types of collagen in the articular cartilage of postnatal chickens. Calcif Tissue Res 17:43-55

Seyer JM, Brickley DM, Glimcher MJ (1974a) The isolation of two types of collagen from embryonic bovine epiphyseal cartilage. Calcif Tissue Res 17:25-41

Skinner HCW (2005) Mineralogy of bone. *In:* Essentials of Medical Geology. Sellinus O (ed), Elsevier, p 667-603

Speer M, McKee MD, Guldberg RE, Liaw L, Yang HY, Tung E, Karsenty G, Giachelli CM (2002) Inactivation of the osteopontin gene enhances vascular calcification of matrix Gla protein-deficient mice: Evidence for osteopontin as an inducible inhibitor of vascular calcification in vivo. J Exp Med 196:1047-1055

Sundarsanan K, Young RA, Wilson AJC (1977) The structures of some cadmium 'apatites'. $Cd_2(MO_4)_3X$. 1. Determination of the structures of $Cd_5(VO_4)_3I$, $Cd_5(PO_4)_3Br$, $Cd_5(AsO_4)_3Br$, and $Cd_5(VO_4)Br$. Acta Crystallogr B 33:3136-3142

Strawich E, Glimcher MJ (1985) Synthesis and degradation in vivo of a phosphoprotein from rat dental enamel. Identification of a phosphorylated precursor protein in the extracellular organic matrix. Biochem J 230:423-433

Stubbs JT, Mintz KP, Eanes ED, Torchia DA, Fisher LW (1997) Characterization of native and recombinant bone sialoprotein: delineation of the mineral-binding and cell adhesion domains and structural analysis of the RGD domain. J Bone Miner Res 12:1210-1222

Stuhler R (1938) Fortschr Gebiete Rontgenstrahlen 57:231

Tarnowski CP, Ignelzi MA Jr, Morris MD (2002) Mineralization of developing mouse calvaria as revealed by Raman spectroscopy. J Bone Min Res 17:1118:1126

Termine JD, Wuthier R, Posner AS (1967) Amorphous-crystalline mineral changes during endochondral and periosteal bone formation. Proc Soc Exp Biol Med 125:4-9

Termine JD, Posner AS (1966a) Infrared analysis of rat bone: age dependency of amorphous and crystalline mineral fractions. Science 153:1523-1525

Termine JD, Posner AS (1966b) Infrared determination of the percentage of crystallinity in apatitic calcium phosphates. Nature 21:268-270

Termine JD, Eanes ED, Greenfield D, Nylen M, Harper R (1973) Hydrazine-deproteinated bone mineral: physical and chemical properties. Calcif Tissue Res 12:73-90

Termine JD, Lundy DR (1973) Hydroxide and carbonate in rat bone mineral and its synthetic analogues. Calcif Tissue Res 13:73-82

Thomas JM, Thomas WJ (1967) An Introduction to the Principles of Heterogeneous Catalysis. Academic Press

Tong W, Glimcher MJ, Katz JL, Kuhn L, Eppell S (2003) Size and shape of mineralites in young bovine bone measured by atomic force microscopy. Calcif Tissue Int 72:592-598

Travis DF (1968a) Comparative ultrastructure and organization of inorganic crystals and organic matrices of mineralized tissues. *In:* Biology of the Mouth. Person P (ed) American Association for the Advancement of Science, p 237-297

Travis DF, Glimcher MJ (1964) The structure and organization of, and the relationship between the organic matrix and the inorganic crystals of embryonic bovine enamel. J Cell Biol 23:447-497

Travis DF (1968b) The structure and organization of and the relationship between the inorganic crystal and the organic matrix of the echinoderm endoskeleton as it is related to bone. *In:* Proceedings of the Fifth European Symposium on Calcified Tissues. Milhaud G, Owen M, Blackwood H (eds) Societe d'Edition d'Enseignement Superieur, Paris, p 399-498

Triffit J, Terepka A, Neuman WF, Canas F (1968) Water and electrolytes in bone tissue and their availability to the blood. Calcif Tissue Res:101-101A

Tye CE, Singh G, Litvinova OV, Hunter GK, Goldberg HA (2004) Role of the poly[Glu] sequences in bone sialoprotein in the nucleation of hydroxyapatite. *In:* 8th International Conference on the Chemistry and Biology of Mineralized Tissues. Landis WJ, Sodak J (eds), Banff, Canada, p 132-134

Veis A (ed) 1981) The Chemistry and Biology of Mineralized Connective Tissues. Proceedings of the First International Conference on the Chemistry and Biology of Mineralized Connective Tissues. Elsevier/North-Holland

Wachtel E, Weiner S (1994) Small angle X-ray scattering study of dispersed crystals from bone and tendon. J Bone Miner Res 9:1651-1655

Wacker WEC, Parish AF (1968a) Medical progress: Magnesium metabolism (concluded). New Engl J Med 45:772-776

Wacker WEC, Parish AF (1968b) Medical progress: Magnesium metabolism. New Engl J Med 45:658-663

Wacker WEC, Parish AF (1968c) Medical progress: Magnesium metabolism (continued). New Engl J Med 45:712-716

Wadkins C (1968) Experimental factors that influence collagen calcification in vitro. Calcif Tissue Res 2:214-228

Wadkins C, Luben R, Thomas M, Humphreys R (1974) Physical biochemistry of calcification. Clin Orthop Rel Res 99:246-266

Wallach S (1988) Availability of body magnesium during magnesium deficiency. Magnesium 7:262-270

Wallach S (1990) Effects of magnesium on skeletal metabolism. Magnes Trace Elem 9:1-14

Watson ML, Robinson RA (1953) Collagen-crystal relationships in bone. II. Electron microscope study of basic calcium phosphate crystals. Am J Anat 93:25-60

Weiner S, Price P (1986) Disaggregation of bone into crystals. Calcif Tissue Int 39:365-375

Weiner S, Sagi I, Addadi L (2005) Choosing the crystallization path less traveled. Science 309:1027-1028

Weinstock M, Leblond C (1973) Radioautographic visualization of the deposition of a phosphoprotein at the mineralization front in the dentin of the rat incisor. J Cell Biol 56:838-845

White SW, Hulmes DJS, Miller A, Timmins PA (1977) Collagen-mineral axial relationship in calcified turkey leg tendon by X-ray and neutron diffraction. Nature 266:421-425

Wolff J (1884) Das Gesetz der Transformation der inneren Architectur der Knochen bei pathologischen Veranderungen der ausseren Knochenform. Sitzungsberichte der Kgl Preuss Acad der Wissensch

Wolff J (1892) *In*: Das Gesetz der Transformation der Knochen(eds) Quarto, Berlin

Wu Y, Glimcher MJ, Rey C, Ackerman JL (1994) A unique pronated phosphate group in bone mineral and not present in synthetic calcium phosphates. Identificaiton by phosphorus-31 solid state NMR spectroscopy. J Mol Biol 244:423-435

Wu W, Nancollas GH (1999) Determination of interfacial tension from crystallization and dissolution data comparison with other methods. Adv Colloid Interface Sci 79:229-279

Wu Y, Ackerman JL, Chesler DA, Li J, Neer RM, Wang J, Glimcher MJ (1998) Evaluation of bone mineral density using 3D solid state phosphorus-31 NMR projection imaging. Calcif Tissue Int 62:512-518

Wu Y, Ackerman JL, Kim H-M, Rey C, Barroug A, Glimcher MJ (2002) Nuclear magnetic resonance spin-spin relaxation of the crystals of bone, dental enamel and synthetic hydroxyapatite. J Bone Miner Res 17:472-480

Yamauchi M, Katz EP, Otsubo K, Teraoka K, Mechanic GL (1989) Cross-linking and stereospecific structure of collagen in mineralized and nonmineralized skeletal tissues. Connect Tissue Res 21:159-167

Yamauchi M, Katz EP (1993) The post-translational chemistry and molecular packing of mineralizing tendon collagens. Connect Tissue Res 29:81-98

Yesinowski JP, Benedict JJ (1983) ^{31}P NMR as a spectroscopic monitor of the spontaneous precipitation of calcium phosphates. Calcif Tissue Int 35:284-286

Zheng S, Tu A. Renugopalakrishnan V, Strawich E, Glimcher MJ (1987) A mixed beta-turn and beta-sheet structure for bovine tooth enamel amelogenin: Raman spectroscopic evidence. Biopolymers 26:1809-1813

Reviews in Mineralogy & Geochemistry
Vol. 64, pp. 283-313, 2006
Copyright © Mineralogical Society of America

9

Silicate Biomaterials for Orthopaedic and Dental Implants

Marta Cerruti

Department of Chemistry
North Carolina State University
Dabney Hall, Campus Box 8204
Raleigh, North Carolina, 27695-8204, U.S.A.
e-mail: marta.cerruti@gmail.com

Nita Sahai

Department of Geology and Geophysics, Department of Chemistry,
Environmental Chemistry and Technology Program
University of Wisconsin
Madison, Wisconsin, 53706, U.S.A.
e-mail: sahai@geology.wisc.edu

INTRODUCTION

Many developed nations including the U.S.A. have increasingly older populations as birth rates have declined and life expectancy has increased over the twentieth century. A corresponding factor of two increase in demand for artificial joint and dental implants is anticipated over the next thirty years (Piehler 2000). The development of biomaterials for orthopaedic and dental implants with improved properties and durability in the human body is critical to deal with this aging population. Broadly speaking, the first-generation of biomaterials was bioinert, whereas bioactive and bioresorbable materials represented improved, second-generation materials. Still, one third to one half of the prostheses fail within 10-25 years and require a second surgery (Shirtliff and Hench 2003 and references therein). The challenge facing the development of improved orthopaedic and dental biomaterials for the future is to design materials that are biocompatible, capable of bearing high stress and loads, and that invoke positive cellular and genetic responses for the rapid repair, modification, regeneration and maintenance of the affected tissue in the human body, i.e., tissue engineering.

In this chapter, we will provide a brief history of the use of implants, and review the requirements that a biomaterial must fulfill to be used effectively as an orthopaedic or dental implant, the chemical composition-structure-activity of different types of silicate biomaterials including glasses and ceramics, the chemical reactions that occur at the silicate implant/solution interface involving inorganic ions and organic biomolecules (mainly proteins). We will also review studies that show the effects of different synthetic solutions, used experimentally to mimic human blood plasma, on in vitro tests of bioactivity. Our focus will be on dense silicate biomaterials that have up to ~20% porosity and high mechanical strength, in contrast to porous biomaterials with ~40-60% porosity and scaffold materials that have ~80% porosity and almost no mechanical strength. We will also highlight the fundamental importance of chemical principles and strong conceptual parallels in understanding glass composition and structure whether from the perspective of igneous glasses and melts or for biomaterials development, and in understanding the reactions occurring at the silicate/aqueous solution interface in the natural geochemical environment and in the human body (Sahai 2003, 2005).

1529-6466/06/0064-0009$05.00   DOI: 10.2138/rmg.2006.64.9

Historical background

A biomaterial is a synthetic material to be used in intimate contact with living tissue. A more precise definition was given in 1986, at the Consensus Conference of the European Society for Biomaterials, when a biomaterial was defined as "a nonviable material used in a medical device, intended to interact with biological systems" (Williams 1987).

The use of synthetic materials to replace parts of the human body dates back to thousands of years ago. Egyptian mummies have been found with gold dental prostheses (Williams and Cunningham 1979). The most common "biomaterials" used in the pre-Christian era were copper and bronze, which created problems because of their toxicity. Nevertheless, until the 19th century, these materials continued to be the most frequently used to make implants. The introduction of aseptic conditions in surgery in 1860 significantly improved the application of biomaterials research to medicine (Park and Bronzino 2003). In 1880, ivory prostheses were manufactured, and in 1902 some gold capsules were used to make the articulation head of femur prosthetics. Since then, researchers have looked for stable and inert materials to make devices that can be placed in contact with human tissues (Hench and Ethridge 1982). Metallic prostheses made by stable alloys such as Vitallium® (60% cobalt, 20% chromium, 5% molybdenum, and traces of other elements) were introduced successfully in the 1930s for orthopaedic applications and are still used (Proussaefs and Lozada 2002). The advent of polymeric materials opened new possibilities to biomaterials science. In the 1930s, polymethylmethacrylate (PMMA) began to be used in dentistry. Moreover, PMMA has been used as cement to anchor metallic prostheses to bone and to make permanent orthopaedic implants. Cemented metallic prostheses were especially used for hip or knee replacement.

The American Ceramic Society defines a ceramic as "an inorganic, non-metallic material…that is typically crystalline in nature…and is a compound formed between metallic and non-metallic elements such as aluminum and oxygen (alumina-Al_2O_3), calcium and oxygen (calcia-CaO) and silicon and nitrogen (silicon nitride Si_3N_4)." The use of ceramics as biomaterials was first explored in the 1970s. The first total hip replacement with alumina dates back to 1971 (Boutin 1972). Before this date, ceramics and glasses were commonly used in a wide range of medical-related applications: eye glasses, chemical and chirurgical glassware, analytical instruments, etc. Moreover, ceramics have been widely used in dentistry: for example, porcelain has been used to substitute crowns, and glass cements were used as fillers (Marquis 1988). However, the use of ceramics as orthopaedic prostheses was limited by the fact that once implanted they would have to bear cyclic loads and, more precisely, cyclic shear stresses at the interface with bone. On the cruciate ligament (which is one of four strong ligaments connecting the bones in the knee joint), millions of cycles/year are usually performed during level walking, and the load can vary from 70 to 700 N during stairs ascending or jolting (Black 1992). Moreover, the match of the material elastic properties with bone properties is difficult to obtain, because of the complexity of the bone structure (Rho et al. 1998). Bioceramics are used for numerous applications in the human body (Figure 1; from Hench and Wilson 1993).

A wide range of ceramic materials is used for biomedical applications: glasses (45S5 Bioglass®), polycrystalline ceramics (HA, tricalcium phosphate), polycrystalline glass-ceramics (Ceravital®), vitrified ceramics (glass-HA), sintered ceramics (alumina, zirconia), hot pressed ceramics or glass-ceramics (Cerabone® A/W glass-ceramic), sol-gel glasses or ceramics (bioactive gel-glasses), composites (Hench and Wilson 1993).

In the mid-1970s, three independent groups commercialized a synthetic form of HA for orthopedic applications (Jarcho 1976; Aoki et al. 1977; Denissen 1979; de Groot 1983). HA can be used either as dense or as porous material. Also, tricalcium phosphate (TCP) has been used to make resorbable implants (Koster et al. 1977). A/W glass ceramics in the $MgO\text{-}CaO\text{-}SiO_2\text{-}P_2O_5$ system, containing apatite and wollastonite phases, were introduced by Kokubo in 1982 (Kokubo et al. 1982). They have been used for vertebral replacement, iliac crest prostheses, and

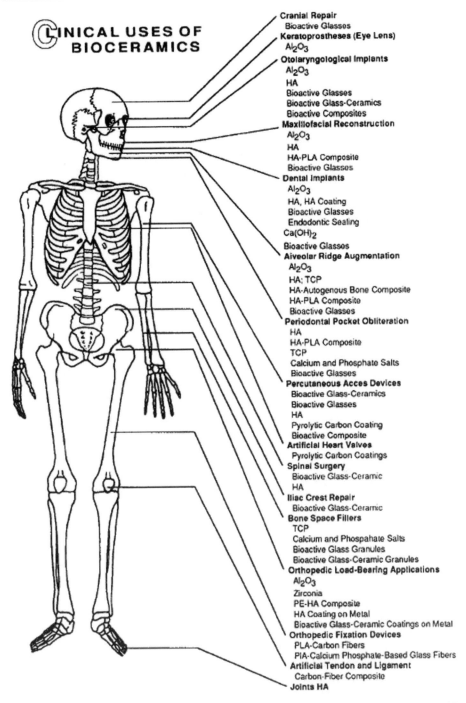

CLINICAL USES OF BIOCERAMICS

Cranial Repair
  Bioactive Glasses
Keratoprostheses (Eye Lens)
  Al_2O_3
Otolaryngological Implants
  Al_2O_3
  HA
  Bioactive Glasses
  Bioactive Glass-Ceramics
  Bioactive Composites
Maxillofacial Reconstruction
  Al_2O_3
  HA
  HA-PLA Composite
  Bioactive Glasses
Dental Implants
  Al_2O_3
  HA, HA Coating
  Bioactive Glasses
  Endodontic Sealing
  $Ca(OH)_2$
  Bioactive Glasses
Alveolar Ridge Augmentation
  Al_2O_3
  HA; TCP
  HA-Autogenous Bone Composite
  HA-PLA Composite
  Bioactive Glasses
Periodontal Pocket Obliteration
  HA
  HA-PLA Composite
  TCP
  Calcium and Phosphate Salts
  Bioactive Glasses
Percutaneous Acces Devices
  Bioactive Glass-Ceramics
  Bioactive Glasses
  HA
  Pyrolytic Carbon Coating
  Bioactive Composite
Artificial Heart Valves
  Pyrolytic Carbon Coatings
Spinal Surgery
  Bioactive Glass-Ceramic
  HA
Iliac Crest Repair
  Bioactive Glass-Ceramic
Bone Space Fillers
  TCP
  Calcium and Phospahate Salts
  Bioactive Glass Granules
  Bioactive Glass-Ceramic Granules
Orthopedic Load-Bearing Applications
  Al_2O_3
  Zirconia
  PE-HA Composite
  HA Coating on Metal
  Bioactive Glass-Ceramic Coatings on Metal
Orthopedic Fixation Devices
  PLA-Carbon Fibers
  PlA-Calcium Phosphate-Based Glass Fibers
Artificial Tendon and Ligament
  Carbon-Fiber Composite
Joints HA

Figure 1. Clinical uses of bioceramics. Acronyms in the figure: polylactic acid (PLA), polyglycolic acid (PGA), polyethylene (PE), α-tricalcium phosphate (α-$Ca_3(PO_4)_2$, TCP), and hydroxylapatite (HA). [Used by permission of World Scientific Publishing Co., from Hench and Wilson (1993) *An Introduction to Bioceramics*, p. 2, Fig. 1.]

bone defect fillers. Since then, composites made of a variety of bioceramics and other materials have been studied. A few examples are stainless steel fiber reinforced Bioglass®, titanium fiber reinforced Bioglass®, TCP or HA reinforced polyethylene (Hench and Wilson 1993).

Classification of biomaterials according to the interface formed with tissue

Beyond the general criteria that an orthopaedic biomaterial be non-toxic and promote rapid bone growth at the implant/tissue interface, the most important requirement to fulfill in the search for new biomaterials is their durability inside the human body. In fact, in the last decades, the average age of the population has increased, and the demand for new types of biomaterials that could last longer once implanted has become more and more urgent (Hench 1998). The main issue is failure at the interface between biomaterial and host tissue, because the material is under greatest stress in that region. In fact, understanding the phenomena occurring at the interface between biomaterials and host tissue is so crucial that a first classification of biomaterials can be done based on how this region evolves once the material is implanted.

For many years, it was thought that interactions between body and implant could cause only undesirable reactions, such as tissue irritation, damage, and finally death. This was due to the observation that a tissue would die when put in contact with certain materials that were, thus, found to be toxic. For this reason, the guiding principle used in biomaterials development at the beginning was that they should be as chemically inert as possible (Hench and Ethridge 1982). Alumina (Al_2O_3), zirconia (ZrO_2), alumina-zirconia and carbon are examples of bioinert materials. They have good physical properties such as high mechanical strength and hardness (for example, the elastic modulus of alumina is 350 GPa (Carew et al. 2003)), but even the most inert material (bioinert) elicits a reaction of the body once implanted (Hench 1991, 1994). A thin, non-adherent fibrous capsule made up primarily of collagen is developed on bioinert materials after they remain in contact with body environment for some time. This prevents further interactions with the tissues. The thickness of the protective fibrous layer depends on the type of bioinert material, and on the motion and physical fit at the implant/tissue interface. This type of interface cannot last for a long time. Eventually, deterioration occurs, and surgical removal of the device is necessary. For this reason, research on biomaterials switched to the development of materials that could interact with the body inducing a desirable response by the host tissue.

Porous materials can achieve biological fixation. In this case, a mechanical bond is obtained by in-growth of bone into the pores, if these have a diameter >100 μm (Hulbert 1993). The increased interfacial area between the implant and the tissue increases the resistance to movement of the device in the tissue. Still, porous materials also do not last long once implanted. Their mechanical resistance is not as high as dense materials because the elastic modulus decreases with increasing porosity (see e.g., Knudsen 1962 for porous alumina), and the corrosion due to the exposure of a large surface area to body fluids further decreases their strength. "Replamineform" α-alumina and titania (White et al. 1975) are examples of porous biomaterials, and microporous silicon that has an oxidized SiO_2 surface-layer is being investigated for its potential as a porous biomaterial (Canham 1995; Canham et al. 1996).

Resorbable or biodegradable bioceramics represent an alternative solution to the problem of long-term implant failure. Examples of resorbable materials include ceramics such as α-tricalcium phosphate (α-$Ca_3(PO_4)_2$, TCP), β-TCP (whitlockite), hydroxyapatite itself, and polymers such as polylactic acid (PLA) and polyglycolic acid (PGA) which decompose and are metabolized into CO_2 and H_2O. These materials are supposed to exploit and increase the body's ability to self-repair. Bioresorbable materials degrade gradually over a period of time and are replaced by the natural host tissue. An important issue is the biocompatibility of the products of resorption. Moreover, resorption should occur at a rate similar to cellular metabolism. These requirements are difficult to fulfill and not many resorbable biomaterials have been clinically applied.

A valuable solution to the problem of achieving a stable implant-tissue interface is bioactive fixation. This can be obtained when bioactive materials are used. Bioactive materials include glasses, ceramics, glass-ceramics and polymer-ceramic composites.

Bioactive materials

"A bioactive material is one that elicits a specific biological response at the interface of the material which results in the formation of a bond between the tissues and the material." This definition was given by Hench, who initiated this field of research with his colleagues in the early 1970s (Hench et al. 1971). They discovered that certain compositions of glasses in the system SiO_2-CaO-Na_2O-P_2O_5 were able to form a bond with bone once they are implanted. In fact, when these glasses were put in contact with biological fluids, a layer of hydroxyapatite (HAP) or hydroxyl carbonated apatite (HCA) similar to the mineral phase of bone was deposited on their surface. Collagen molecules were incorporated into this layer, and a biological bond could be formed. Later work by Wilson and Nolletti showed that a bond with soft tissue could be achieved if the speed of apatite formation was rapid enough (Wilson and Nolletti 1990) (Note: "soft tissue" generally refers to tissues that connect, support and surround other structures and organs in the body, and includes muscles, tendons, fibrous tissues, fat, blood vessels, nerves and synovial tissues).

The rate of bonding of bioactive glasses to existing bone tissue depends on many factors. One is bulk composition: the most rapid rates of bonding for bioactive glasses composed of SiO_2, CaO, Na_2O and P_2O_5 are obtained with SiO_2 contents of 45-52% weight. In this compositional range, bonding both to soft and hard connective tissue occurs within 5-10 days. Bioactive glasses or glass-ceramics containing 55-60% SiO_2 require a longer time to form a bond with bones and do not bond to soft tissues. Glass compositions with more than 60% SiO_2 do not bond either to bone nor to soft tissues and elicit formation of a non-adherent fibrous interfacial capsule (Hench 1988). These concepts are summarized in Figure 2, where the Na_2O-CaO-SiO_2 ternary phase diagram is shown, referring to glasses with a constant 6% wt of P_2O_5 (Hench 1991).

Glasses and glass-ceramics that have a composition falling inside region A develop HA both *in vitro* and *in vivo*. Compositions inside the dashed line bind also to soft tissues. The materials in region B are inert, and those in region C are resorbable. Region D is a non-glass forming and non-bonding region.

The values of the index of bioactivity (I_b) are also reported in Figure 2. I_b is a measure of the level of bioactivity of bioactive materials, and is defined as the inverse of the time required for more than 50% of the interface to be bonded: $I_b = 100/t_{0.5\,bb}$. In the SiO_2-CaO-

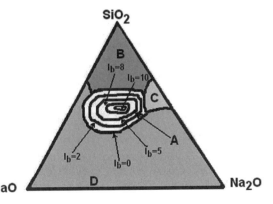

Figure 2. Bioactive regions in the CaO-SiO_2-Na_2O system. All glasses have a 6% wt of P_2O_5. I_b is the index of bioactivity (see text for explanation).

Na_2O-P_2O_5 system, glass compositions from 40-52% SiO_2 have I_b values varying from 12.5 to 10. The change of I_b is large comparing glasses in regions A and B. If other ions different from Ca and Na are added into glass composition, large variations in I_b are observed. For example, Greenspan and Hench (1976) showed that the addition of only 3% Al_2O_3 destroys glass-bone bonding ability, and Gross and Strunz (1980) proved the same for other multi-valent cations.

The most bioactive composition of glasses in the SiO_2-CaO-Na_2O-P_2O_5 system is the so-called "Bioglass®," which was the first bioactive material discovered in 1969 by White et al. (1975). This material found many applications in orthopedic and dental field, and it is one of the most widely used bioceramics. Hench (1994) proposed a specific bioactivity classification for biomaterials intended to be used for orthopedic implants. Materials belonging to "class A" show osteoinduction, which means that their surface is colonized by osteogenic stem cells when put in contact with living tissue after a surgical intervention. These materials usually have I_b = 12. Materials belonging to "class B" show only osteoconduction, i.e., the material works only as a substrate where cells migrate on. These materials have I_b = 3 to 6. It has been shown that a material is osteoproductive if it can elicit both an intracellular and an extracellular response at its interface (Hench 1994).

Bioactive materials of class A release Si when immersed in body fluids. Experiments conducted on both chicks and rats showed that Si has a specific metabolic role connected to bone growth (Schwartz 1978; Carlisle 1986). Also, research on human osteoblast-like cells proved that Si stimulates bone formation (Keeting et al. 1992). An intracellular response is then obtained because of the abundant Si release by class A bioactive materials. The extracellular effects are mainly related to the possibility of adsorbing bone growth proteins, such as TGF-β_1 at the implant surface (TGF--β_1 or Transforming Growth Factor-beta1, is a multifunctional peptide that controls cell proliferation, differentiation and other functions in many cell types). These factors can be adsorbed because of the high surface area of a SiO_2 rich layer that is formed once biomaterials of class A are immersed in body fluids. These proteins seem to increase the rate of bone formation at the implant surface, because they enhance the differentiation and mitosis (cell division plus cytokinesis) of stem cells.

SILICATE BIOACTIVE GLASSES

Melt-derived bioactive glasses

As previously mentioned, the study of bioactive materials began with the discovery that silicate glasses of a specific composition were able to bond to bone. The first bioactive glass studied was Bioglass® 45S5, introduced by Hench et al. (1971), which still remains the most used in clinical applications and the most promising one. Bioglass® 45S5 is produced by fusion of oxides, and its specific composition is 45% SiO_2, 24.5% CaO, 24.5% Na_2O, 6% P_2O_5 (expressed as wt%). The name "45S5" refers to both the SiO_2 content (45% wt) and the Ca/P molar ratio (5). Hench and co-workers analyzed many other compositions in the same SiO_2-CaO-Na_2O-P_2O_5 system, and later other groups tried to synthesize more complex bioactive glasses or glass-ceramics. These materials were mentioned in the previous paragraph. A comparison between composition and index of bioactivity for a few of them is shown in Table 1 (Andersson et al. 1992). All these materials are bioactive, in that they form a biologically active layer of HCA (hydroxyl carbonate apatite) on their surface when implanted.

Sol-gel derived bioactive glasses

In the early 1990s, Li and colleagues synthesized some bioactive glasses by the sol-gel technique (Li 1991; Li et al. 1991). This new class of bioactive glasses showed a higher compositional range of bioactivity: glasses in the SiO_2-CaO-P_2O_5 system with silica content up to 90% were capable of forming a layer of HAP. This is different compared to melt-derived bioactive glasses, which showed bioactivity only up to 60% SiO_2 content.

Table 1. Composition of components (wt%), structure and index of bioactivity of different melt-derived glasses (from Andersson et al. 1992).

Component	45S5	45S5.4F	52S4.6	KGC Ceravital®	A/W-GC
SiO_2	45	45	52	46.2	34.2
P_2O_5	6	6	6	—	16.3
CaO	24.5	14.7	21	20.2	44.9
$Ca(PO_3)_2$	—	—	—	25.5	—
CaF_2	—	9.8	—	—	0.5
MgO	—	—	—	2.9	4.6
Na_2O	24.5	25.5	21	4.8	—
K_2O	—	—	—	0.4	—
Structure	Glass	Glass	Glass	Glass-ceramic	Glass-ceramic
I_B	12.5	12.5	10.5	5.6	6.0

One of the main differences between sol-gel and melt-derived glasses lies in surface area: for sol-gel glasses, surface area ranges from ~200 to 650 m^2g^{-1}, whereas melt-derived glasses show surface area < 1 m^2g^{-1} for rough particles, and ~2 m^2g^{-1} for micron-sized particles. This big difference is due to the temperature and synthesis conditions of the two types of materials: sol-gel glasses are synthesized in an aqueous environment, then dried and stabilized at temperatures that do not exceed 600 °C. Surface and structural properties (such as surface area and porosity) can be finely modulated depending on composition and synthesis conditions. Controlled nano-structured materials can be obtained. Melt derived bioactive glasses are melted at temperatures higher than 1000 °C. The resulting materials have extremely low porosity, and surface area depends only on particle size obtained by grinding the powders.

HAP is deposited much faster on sol-gel bioactive glasses than on traditional melt-derived glasses, and the materials can be re-sorbed. In fact, the porous structure of gel bioactive glasses allows the formation of a hydrated layer inside the material, where biological moieties can enter maintaining their structural configuration and biological activity (Hench 1998). In this way, gel glasses can become an indistinguishable part of the host tissue. For example, when trabecular rabbit bone was proliferated on 45S5 Bioglass® particles, a structure similar to normal bone was obtained, but some large particles of Bioglass® were still present. Instead, if gel-glasses were used, no residual particles could be observed (Oonishi et al. 1997; Wheeler and Stokes 1997).

Many different chemical systems synthesized by sol-gel technique showed bioactivity when immersed in simulated body fluids (SBF) that are synthetic solutions of composition similar to human blood plasma (Table 2). The glass compositions were chosen in the binary system CaO-SiO_2 (Izquierdo-Barba 1999; Martinez et al. 2000), ternary system SiO_2-CaO-P_2O_5 (Vallet-Regi 1999; Vallet-Regi et al. 1999; Peltola et al. 1999; Vallet-Regi and Ramila 2000) and quaternary system SiO_2-CaO-P_2O_5-MgO (Perez-Pariente et al. 1999, 2000). In general, a higher amount of SiO_2 in glass composition induces larger surface area, but a lower degree of porosity. Moreover, higher temperatures used in the synthesis process decrease glass surface area. The influence of surface, textural and composition parameters on bioactivity is not completely clear. Moreover, the introduction of sol-gel synthesis technique opened the research to new types of biomaterials. Many possible dopants can be introduced in a material synthesized via sol-gel. For example, recently some researchers added Ag^+ to sol-gel bioactive glass composition (Bellantone et al. 2000), thus giving antimicrobial properties to the material.

Table 2. Composition of human plasma and different simulated body fluids (SBF). All concentrations in mM units.

Ion	Plasma[a]	SBF[b]	Modified SBF[c]
Na^+	142	142	142
K^+	5.0	5.0	5.0
Mg^{++}	1.5	1.5	1.5
Ca^{++}	2.5	2.5	2.5
Cl^-	103.0	147.8	125.0
HCO_3^-	27	4.2	27
HPO_4^{--}	1.0	1.0	1.0
SO_4^{2-}	0.5	0.5	0.5
pH	7.20-7.40	7.25	7.4

[a]Kokubo and Takadama (2006); [b] Kokubo (1990); [c] Tas (2000).

Deposition of HAP on pure silica gel

The study of sol-gel bioactive glasses opened the field to the research of simpler sol-gel synthesized materials. In the early 1990s, Li and colleagues showed that HAP is deposited on the surface of a gel of silica when soaked for a few days in simulated body fluid (SBF) (Li et al. 1992, 1993a,b). Different factors influence HAP deposition. One is the presence of pores: HAP did not precipitate on non-porous silica such as quartz (Li et al. 1992) or gel-silica treated at temperature ≥ 900 °C (Kokubo et al. 1994) (sintering occurred at such temperatures, and porosity was eliminated). Pores could be the sites for HAP nucleation because the degree of supersaturation of the solution is usually higher there than in the solution bulk. Some distinction depending on the pore size should be made: if the pores were smaller than ~ 6 Å, HAP could not precipitate, because phosphate ions were too big to diffuse into the pores. Vallet-Regi et al. (1999) showed that microporous silica gels synthesized by acid catalysis were less bioactive than mesoporous silica gels obtained with basic catalysis. We note that this is consistent with faster aggregation of silicate particles into three-dimensionally networked gels at pH < 7 compared to particles that form sols with greater particle size at pH > 7 (Iler 1979, p. 174). The larger-sized sols would be expected to have greater porosity and bioactivity than the networked gels.

Hydroxylation is another factor for HAP precipitation. Silanol groups formed when porous silica is immersed in SBF have been proposed as the sites for HAP nucleation (Kokubo et al. 1994; Cho et al 1995). Experimental evidence for the bonding of HAP to silanol groups has not yet been provided, and a clear distinction between the relevance of surface hydroxylation, surface area extension and degree of porosity has not been outlined.

Some of these factors also influence the results obtained after silica gel devices are implanted *in vivo*. Klein et al. (1995) showed that high surface area silica synthesized by sol-gel and stabilized at $T = 400$ to 600 °C were degraded at the periphery of the implant. On the contrary, no degradation or resorption could be observed at all on devices made of lower surface area sol-gel silica samples (stabilized at $T = 900$ to 1000 °C). Good bone bonding was achieved by these latter samples, whereas an abnormal cellular reaction was observed at the surface of the silica gels stabilized at lower temperature. These results emphasize the principle that when biomaterials are designed, not only the kinetic aspects of the rate of HAP deposition, but also implant stability (durability in the human body) should be taken in consideration.

MECHANISM OF HCA LAYER DEPOSITION ON SILICATE GLASSES

Hench's hypothesis

Even though the composition of bioactive glasses synthesized is quite different, the mechanism of HCA formation seems to involve some common steps. For example, the presence of silica in these glasses was proved to be necessary in order to obtain HCA nucleation within 30 days. When both $CaO-P_2O_5$ and $CaO-SiO_2$ based glasses were immersed in body fluid, ion activity product of apatite was increased by a similar amount. Still, a layer of HCA was deposited only on $CaO-SiO_2$ based glasses (Ohtsuki et al. 1991; Ohura et al. 1991). It was hypothesized that a hydrated silica layer was formed on the surface of these glasses prior to the deposition of HCA and that silanol groups could be specific sites of nucleation of apatite.

Clark and Hench first proposed a detailed sequence of reactions occurring at the surface of silica-based bioactive glasses (Clark and Hench 1976). These involved the following steps:

i) Rapid exchange of Na^+ in the glass with H^+ in solution;

ii) Loss of soluble silica as $Si(OH)_4$ by breaking of Si-O-Si bridges and subsequent formation of surface silanol groups in the process;

iii) Condensation and repolymerization of surface silanols to form an SiO_2-rich surface layer;

iv) Migration of Ca^{2+} and PO_4^{3-} to the surface through the silica-rich layer and formation of a Ca-P rich layer on the surface of the glass;

v) Incorporation of OH^-, CO_3^{2-} from the solution and subsequent crystallization of the Ca-P layer to form HCA.

After these physical-chemical reactions occur, biological moieties can interact with the glass surface: collagen molecules are incorporated in the HCA layer, and cells begin to respond, leading ultimately to bone growth.

GLASS STATE AND BIOACTIVITY

To better understand how bioactive glasses react when put in contact with body fluids, a few concepts concerning glass structure and glass surface will be given in the following paragraphs.

Glass structure

Glass structure differs from that of crystalline materials in that it does not show any long-range order, and is therefore called amorphous (without a definite shape). Glass undergoes a change at a temperature or range of temperatures where its changes from exhibiting solid-like to liquid-like properties. Most glasses are silica-based. In a crystalline solid made of silica (e.g., quartz), SiO_4 tetrahedra are organized in a well-defined network and the tetrahedra all have a fixed degree of polymerization. In amorphous silica, SiO_4 tetrahedra are still linked together, but the angle between them is not constant, and the tetrahedra have a variable degree of polymerization, so the resulting structure is not organized (Fig. 3).

SiO_2 is called the glass former, because glass structure is mostly kept together by SiO_4 tetrahedra. Other oxides such as B_2O_3 or P_2O_5, can be used as glass formers. Alkali and alkaline-earth oxides are added to glass as glass modifiers, in that they interrupt the network created by the glass former (Fig. 4). The resulting structure is called a random network (Zachariasen 1932).

In pure silica glass (Fig. 3), all the SiO_4 tetrahedra are linked together. Every O atom is linked to two Si atoms. These O atoms are called bridging oxygens (BO). When a modifier

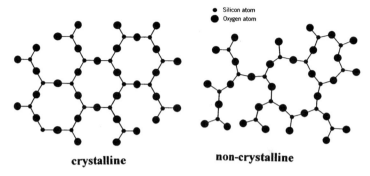

Figure 3. Comparison between crystalline and non-crystalline (amorphous) silica structure (*http://www.benbest.com/cryonics/lessons.html*)

is added to glass composition, cations interrupt the silica network, and some O atoms are not linked to Si atoms. These O atoms are called non-bridging oxygens (NBO):

$$Si-O_{(BO)}-Si + Na_2O \rightarrow 2\ Si-O_{(NBO)}^- + Na^+$$

The bond between NBO's and cations is ionic, and not as strong as the Si-O bond.

The average number of BO per SiO_4 tetrahedron represents the structural parameter (Y). Y can be obtained by using the molar percentage of SiO_2 in the glass (p):

$$Y = 6 - 200/p$$

According to Zachariasen's model, glasses would exist only with $Y \geq 3$, because only for these glasses SiO_4 tetrahedra share at least 3 O atoms with other tetrahedra. This was supposed to be a necessary condition in order to have a 3D network. Still, glasses with $Y < 3$ can be formed, because glass cohesion can be obtained also through the ionic bond between NBO's and cations. Glasses with $Y < 2$ are called inverse glasses, and the structure of these glasses is mostly kept together by ionic bonds.

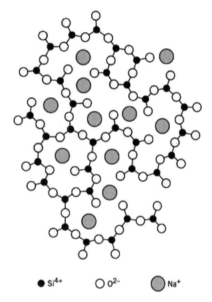

Figure 4. Effect of the introduction of Na^+ cations into a silica network (modified from Pfaender 1996).

Glass surface

When a glass is in contact with an aqueous solution, two processes can occur: leaching of the alkali and alkali-earth cations from the bulk of the glass to the solution and silica network rupture. Leaching predominates at pH < 9, whereas silica network dissolution occurs at higher pH (Iler 1979). The surface of silicate glasses can be classified (Hench and Clark 1978) depending on which of these phenomena dominates (Fig. 5). Type I glass surfaces have a thin (<50 Å) hydrated surface layer, and the composition of the surface layer is analogous to that of the bulk. This type of surface is obtained, for example, when silica glass is immersed in solution at pH < 7. Type II glasses possess a silica-rich protective film due to selective alkali

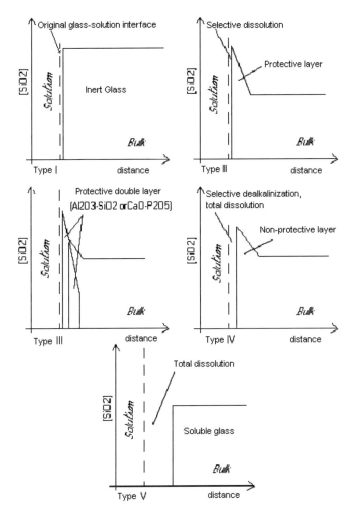

Figure 5. Types of glass/solution interfaces. [Reprinted from *Journal of Non-Crystalline Solids*, Vol. 28, Hench LL and Clark DE, p. 86, Fig. 2, Copyright (1978), with permission of Elsevier.].

leaching. Type III glasses are characterized by a protective double layer, such as alumina-silicate or calcium-phosphate on top of a silica-rich layer. Such glass surfaces are very stable in both acidic and basic solutions. Type IV glasses also have a surface film rich in silica, but the film in this case is not sufficient to protect the glass from further dissolution. Type V glasses have a surface composition equivalent to the bulk, because they underwent congruent dissolution. In this classification, the surface that is formed once bioactive glasses are in contact with a physiologic solution would be of Type III.

Relation of glass structure and surface to bioactivity

Hill related the structural parameter Y to bioactivity (Hill 1996). He calculated the value of Y for a series of bioactive glasses. Some examples are given in Table 3. He noted that the most bioactive glasses had a value of $Y < 2$, i.e., they were inverse glasses. Inverse glasses dissolve relatively quickly in water because most of the O atoms are NBO's, and

Table 3. Relation between composition, structural index and
bioactivity of different bioactive glasses (from Hill 1996).

Glass	SiO$_2$	CaO	Na$_2$O	P$_2$O$_5$	B$_2$O$_3$	Al$_2$O$_3$	Y	Bioactivity
45S5	46	27	24	3	0	0	1.90	Yes
S53P4	54	22	23	2	0	0	2.39	Yes
S63.5P6	66	16	15	3	1	1	3.08	No
CS	50	50	0	0	0	0	2	Yes
CS-Na	48	49	3	0	0	0	1.87	Yes
CS-P	49	50	0	1	0	0	2.03	Yes
S53P4	54	22	23	2	0	0	2.39	Yes
S52P8	55	14	26	4	1	2	2.69	No
S38P8	40	26	28	4	1	2	1.81	No

SiO$_4$ tetrahedra are mostly isolated, i.e., unpolymerized. Still, some glasses with $Y < 2$ were not bioactive (e.g., S38P8). This was thought to be due to some phase separation occurring because of the great instability of these glass compositions. The phases so created would hinder bioactive reactions.

Rawlings correlated bioactivity to the average potential V/r and electric field V/r^2 induced by the cations present in the glass (Rawlings 1992). This worked out well for 85% of the cases considered. Precipitation of HCA has also been correlated to the surface charge that bioactive glasses assumed once they were immersed in physiological solutions (Li and Zhang 1990). The surface charge of a silica-based glass is negative when immersed in solution with pH ≥ 9 because $\equiv$SiO$^-$ species are formed. These charges attract oppositely charged species, and an electric double layer is formed. In some cases, this can induce the formation of a supersaturated solution at the interface with the material. In the case of bioactive glass dissolution, the solution at the interface is supersaturated with HCA, and precipitation can occur easily.

SILICATE BIOCERAMICS

Experimental studies

Following the success of silicate-based glasses as bioactive materials, silicate ceramics and glass-ceramics were found to bond to bone with the growth of a surface layer of hydroxylapatite at the implant/tissue interface. These bioactive silicates include apatite-wollastonite (A/W) glass-ceramic (Kokubo et al. 1986), wollastonite (low temperature, triclinic CaSiO$_3$) (Liu and Ding 2001; Liu et al. 2004), pseudowollastonite (high temperature, monoclinic CaSiO$_3$) (de Aza et al. 1996, 1998, 1999, 2000, 2001; Dufrane et al. 2003; Sarmento et al. 2004), diopside (CaMgSi$_2$O$_6$) (Miake et al. 1995; Nonami and Tsutsumi 1999; de Aza et al. 2005; Iwata et al. 2005), combeite (Na$_2$Ca$_2$Si$_3$O$_9$) (El-Ghannam et al. 1991; Rizkalla et al. 1996; Du and Chang 2004; Chen et al. 2006) and akermanite (Ca$_2$MgSi$_2$O$_7$) (Wu and Chang 2004). In addition, titania-based ceramics including TiO$_2$ *in vivo* and perovskite (CaTiO$_3$) *in vitro* (Coreno and Coreno 2005) are known to be bioactive.

The type of interface formed between bone and implant was examined by using A/W glass-ceramic and β-TCP, respectively, as model bioactive and bioresorbable materials (Neo et al. 1992). Each biomaterial was implanted in rat tibiae and retrieved for examination after 8 weeks of implantation. The interface of the A/W glass ceramic with bone was characterized by the presence of fine apatite crystals similar to bone in shape size and orientation, and collagen fibers were observed at the surface of the HAP layer. The resorbable implant, however, showed no new layer of HAP at the implant/bone interface so the interfacial bond strength was attributed to mechanical interlocking.

Nucleation site and reaction mechanism for HCA layer deposition on pseudowollastonite and bioactive glasses

The bioactivity of pseudowollastonite has been demonstrated *in vitro* in SBF and in human parotid saliva (HPS), and *in vivo* implanted in rat tibiae. When immersed in SBF, a rapid increase occurred in pH from 7.2-10.5 at the pseudowollastonite-solution interface, Ca and Si concentrations in solution initially increased then became constant (Fig. 6) (de Aza et al. 1996, 1999, 2000, 2001, 2004; Sarmento et al. 2004). Bulk solution pH far from the interface, however, remained constant at 7.25. These observations suggested that Ca and Si were initially leached rapidly from the pseudowollastonite surface along with surface protonation, followed by HAP nucleation. Thus, the reaction mechanism proposed for HAP growth on the various silicate bioceramic surfaces is broadly similar to the one proposed for Bioglass® (Hench and Ethridge 1982). The exact surface site where nucleation occurs, however, was not identified.

Sahai and co-workers introduced the application of *ab initio* molecular orbital theory (MOT) to determine, at the molecular level, the structure of the active site and the reaction steps involved in heterogeneous CaP nucleation at bioactive silicate glass and pseudowollastonite surfaces (Sahai and Tossell 2000; Sahai and Anseau 2005). The results for pseudowollastonite are presented first followed by bioactive glasses.

The structure of pseudowollastonite consists of three silicate tetrahedral joined in a ring by corner-sharing or bridging oxygens and calcium polyhedra bonded to twelve oxygens (Yang and Prewitt 1992). The plane of the silicate "three-rings" lies perpendicular to the *c*-axis or parallel to the (001) face of pseudowollastonite. In Sahai and Anseau's (2005) *ab initio* model, nucleation was proposed to occur at the silicate three-ring site in a three step process involving (1) Ca^{2+} leaching from the surface layer and surface protonation, (2) adsorption of Ca^{2+} at the surface silanols belonging to the silicate three-ring, and (3) attachment of HPO_4^{2-} leading to CaP nucleation (Fig. 7a-d). The first two steps were calculated to be exothermic and Ca^{2+} was found to adsorb to the silicate three-ring silanols as an "inner-sphere" complex with direct bonds to surface oxygens. HPO_4^{2-} attachment was slightly endothermic, and thus the rate-determining

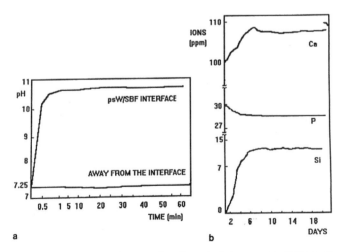

Figure 6. Solution chemistry changes when pseudowollastonite is immersed in SBF (a) pH, and (b) Ca, P and Si concentrations. [Used with permission of Blackwell Publishing, from de Aza et al. (1996), *Journal of Microscopy*, Vol. 182(1), Fig. 5, p. 24-31.]

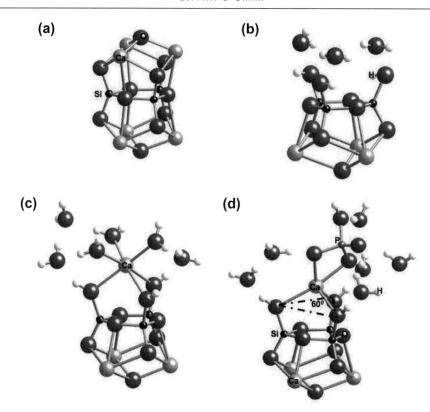

Figure 7. Model molecular clusters: (a) the pristine pseudowollastonite (001) surface (top of cluster) where the [100] direction is normal to the plane of the page, (b) after Ca^{2+} leaching and surface protonation, (c) inner-sphere Ca^{2+} adsorption, (d) HPO_4^{2-} attachment leading to calcium phosphate nucleation. [Reprinted from *Biomaterials*, Vol. 26, Sahai and Anseau, p. 5763-5770, Figs. 2a, 3, 4, Copyright (2005) with permission of Elsevier.]

step. The overall reaction was predicted to be exothermic (Table 4). Furthermore, the surface silanols associated with each silicate tetrahedron in the silicate three-ring form an equilateral triangle that provides a stereochemical match to the three oxygens from orthophosphate tetrahedra in apatite (Fig. 8a, b and Fig. 7c). Recent results from an independent experimental study using X-ray Absorption Spectroscopy, X-ray Absorption Near Edge Structure, ^{31}P Magic Angle Spinning NMR and Magnetic Resonance and XRD data (Skipper et al. 2005a,b) confirmed the atomic-level mechanism predicted by Sahai and Tossell (2000).

The existence of two-, three-, four-, five- and six-membered rings due to surface restructuring on amorphous silica gels and quartz is well established both experimentally and computationally (Galeener 1982; Geissberger and Galeener 1983; Brinker et al. 1986, 1988, 1990; Kubicki and Sykes 1993; Bunker 1994; Sykes and Kubicki 1996; Ceresoli et al. 2000; Iarlori et al. 2001; Du et al. 2003; Akai et al. 2003; Ringnanese et al. 2004; Du and de Leeuw 2004; Yang et al. 2004, 2005.). Therefore, three-ring and "four-ring" clusters representing local structures were modeled as the active nucleating site on bioactive glasses (Sahai and Tossell 2000). Geometries and reaction energies for reaction steps involving Ca^{2+} and HPO_4^{2-} ion adsorption similar to those described above for pseudowollastonite were also modeled for the three- and four-rings structure (Fig. 8c,d). Further, the ^{29}Si and ^{31}P NMR spectra, and the vibrational spectra at each stage in the reaction were also obtained. The spectra obtained for the cluster in Fig. 8c were

Table 4. Calculated aqueous reaction enthalpies ($\Delta E^0_{aq,r}$, kcal mol^{-1}) for each step in the proposed reaction mechanism for calcium phosphate nucleation at the pseudowollastonite surface. Modified Born solvation energies were added to *ab initio* gas-phase energies to obtain reaction energies in aqueous solution (for details see Sahai and Tossell 2000; Sahai and Anseau 2005).

Reaction	$\Delta E^0_{aq,\,r}$
Step 1: Ca^{2+} leaching and Surface Protonation	
$Si_3O_{11}Ca_6^{2+} + 20\ H_2O + 5\ H^+ = Si_3O_{10}H_3Ca_3(H_2O)_3^+ + 3\ Ca(H_2O)_6^{2+}$	-244.6
Step 2: Calcium Adsorption	
$Si_3O_{10}H_3Ca_3(H_2O)_3^+ + Ca(H_2O)_6^{2+} = Si_3O_{10}H_3Ca_4(H_2O)_6^{3+} + 3\ H_2O$	-17.82
Step 3: HPO$_4^{2-}$ Attachment	
$Si_3O_{10}H_3Ca_4(H_2O)_6^{3+} + HPO_4(H_2O)_4^{2-} = Si_3O_{10}H_3Ca_4HPO_4(H_2O)_6^+ + 4\ H_2O$	$+24.92$
Overall Reaction:	
$Si_3O_{11}Ca_6^{2+} + 13\ H_2O + 5\ H^+ + HPO_4(H_2O)_4^{2-} = Si_3O_{10}H_3Ca_4HPO_4(H_2O)_6^+ + 2\ Ca(H_2O)_6^{2+}$	-237.5

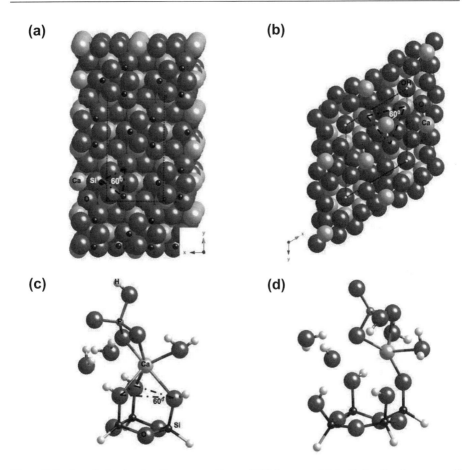

(a) (b) (c) (d)

Figure 8. Equilateral triangles of surface oxygen sites, on (001) faces associated with the silanol groups on silicate three-rings in (a) psW (monoclinic CaSiO$_3$) and (b) with three PO$_4^{3-}$ tetrahedra in HAP. Also shown is calcium phosphate nucleation on bioactive amorphous glasses or gels at (c) the silicate three-ring, and (d) the silicate four-ring site (from Sahai and Tossell, 2000). Compare favorable geometries for Ca-O bonding in (a) and (c) with bonding in HAP (b), and correspondingly unfavorable geometry in (d). Unit cells shown by dashed lines (a, b).

very similar to corresponding experimental spectra for apatite growth on silica gel reacting with SBF. It is equally important that the spectra calculated for CaP nucleation at the cyclic silicate tetramer or four-ring (Fig. 8d), did not match experimental NMR and vibrational spectra, and the geometry of the CaP cluster at the four-ring did not match the Ca-O bond lengths in apatite (Sahai and Tossell 2000). This can be explained by the fact that the surface oxygens (i.e., $\equiv$SiOH groups) on the four-ring are arranged at 90° from each other which is unlike the Ca-O geometry in apatite. Thus, the three-ring may be unique in its nucleating ability.

Bioactivity of diopside and wollastonite

Solution results obtained in SBF for diopside showed that Ca initially leaches rapidly into solution, eventually reaching steady-state, and Mg and Si are leached more slowly and at rates similar to each other (Iwata et al. 2004). HRTEM images of the diopside implant/bone interface retrieved from Japanese monkey mandibles after 6 months of implantation and jaw bone of rabbits for 12 weeks showed lattice fringes corresponding to (001) and (211) faces of HAP crystals growing in continuity with the diopside crystals across the interface (Miake et al. 1995; Nonami and Tsutsumi 1999). Similarly, (001) and (111) faces of HAP crystals were found in continuous contact with diopside pellets reacted for one month in human parotid saliva. In some places, amorphous silica particles were also identified in the layer of newly formed materials (de Aza et al. 2005). It was proposed that the (100) face of diopside epitaxially nucleates the (010) face of octacalcium phosphate (OCP), which then transforms to HAP (Miake et al. 1995). OCP has cell parameters very similar to HAP and has been proposed to be a precursor to HAP in normal bone growth (Brown 1966).

It is proposed that the bioactivity of diopside and wollastonite can be explained based on recent experimental and computational studies which report that cyclic silicate tetramers or 4-membered rings have been observed to form at the surface wollastonite during aqueous dissolution (Casey et al. 1992; Weissbart and Rimstidt 2000). Computational studies further suggest the formation of silicate trimers or 3-membered rings involving pentacoordinated Si atoms (Kundu et al. 2003). Similar reconstruction at the surface may lead to formation of three-rings on diopside. Thus, leaching, surface reconstruction and densification at certain bioceramic and glass surfaces during reaction with aqueous solution may result in formation of the active, nucleating site, even in some silicates that lack the site before reaction with solution.

General hypothesis for bioactivity of any material

A theory that explains bioceramic structure-activity relationship must account for the templated nucleation of HAP on certain ceramics but not on others. Sahai and Anseau (2005) recently hypothesized that a ceramic, glass or mineral will be bioactive if

1. The Point of Zero Charge (PZC) of the biomaterial is less than physiological pH ~ 7.2. The biomaterial surface then has a negative potential at physiological conditions, so that Ca^{2+} bonds electrostatically, followed by HPO_4^{2-} attachment, leading to CaP nucleation. A similar idea involving surface charge on silicate glasses was proposed earlier by Li and Zhang (1991).

2. The biomaterial possess three surface oxygen atoms arranged in an equilateral triangle, either as part of its native crystal structure (Fig. 8) or due to surface relaxation and reconstruction when reacted with aqueous solution. It is proposed that the surface oxygens on the materials stereochemically match the oxygen positions of PO_4 groups on the (001) face of HAP. A finer detail is that, the mere presence of the active site, may not be sufficient to induce HAP nucleation. A critical site density may be required for such a mechanism to operate (Sahai and Anseau 2005).

3. Finally, epitaxial nucleation of some specific face of HAP on the substrate is possible when the position of spacing of Ca^{2+} ions adsorbed on the bioceramic surface matches

that of Ca^{2+} ions on that specific face of HAP. Where such a match in unavailable, simple heterogeneous nucleation will occur.

In apparent contradiction to the hypotheses above, many oxides possess an equilateral triangle of oxygens on specific faces yet all these materials are not bioactive. For example, rutile (α-TiO_2) and anatase (β-TiO_2) on their (010) face, corundum (α-Al_2O_3) on (001) and perovskite ($CaTiO_3$) on (101) (yellow lines in Figs. 9a, b) possess the postulated active site, however, only the TiO_2 phases are bioactive and corundum is not. Furthermore, diopside ($CaMgSi_3O_6$) and wollastonite (triclinic $CaSsiO_3$), do not possess the active site, and perovskite has a PZC = 8.3, yet all three materials are bioactive *in vitro* in SBF (Miake et al. 1995; Nonami and Tsutsumi 1999; Liu et al. 2004; Coreño and Coreño 2005). One may ask, then, whether the presence of the oxygens at a 60° angle or a PZC less than physiological pH ~ 7.2 are truly necessary or sufficient conditions for a ceramic to be bioactive.

The apparent anomalies can be explained within the context of the hypotheses as follows. The titania phases are bioactive, but Al_2O_3 is not, because of their characteristic PZCs. The PZC of the TiO_2 phases and Al_2O_3, respectively, are ~6 and 9. The TiO_2 surface is negatively-charged at physiological pH = 7.2, such that the Ca^{2+} ions approach the surface by an attractive electrostatic potential, and the surface oxygens provide the stereochemical site for Ca^{2+} binding. In contrast, the positively-charged Al_2O_3 surface repels the Ca^{2+} ions. The HPO_4^{2-} ions are attracted to the Al_2O_3 surface and do sorb, but the anion cannot avail of the stereochemical match provided by the surface oxygens. Perovskite is bioactive despite a high PZC, because initial leaching of Ca^{2+} ions from its surface leaves a leached TiO_2 gel-like layer (Coreño and Coreño 2005) that has a PZC less than physiological pH Minerals such a diopside, wollastonite and akermanite that do not possess the three-ring as part of their intrinsic crystal structure, are proposed to be bioactive due to surface restructuring and densification when reacted with solution, resulting in the formation of three-rings. Similar to the cyclic silicate trimer, the silicate four-ring or cyclic tetramer, has also been identified in the leached layers on wollastonite and diopside (Casey et al. 1993; Weissbart and Rimstidt 2000). These sites, however, are hypothesized to be bio-inactive towards HAP nucleation, based on the results of *ab initio* molecular orbital theory calculations (Sahai and Tossell 2000). The formation of three-rings has been predicted based on an atomistic modeling approach (Kundu et al. 2003).

(a)

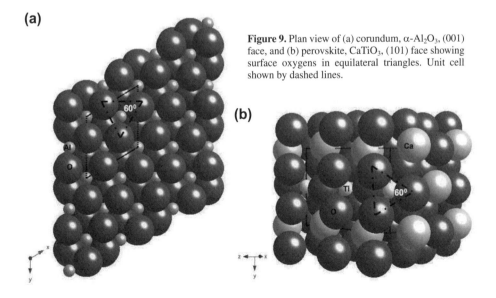

Figure 9. Plan view of (a) corundum, α-Al_2O_3, (001) face, and (b) perovskite, $CaTiO_3$, (101) face showing surface oxygens in equilateral triangles. Unit cell shown by dashed lines.

(b)

The crystallographically calculated site-densities of silicate three-rings on the (001) face of pseudowollastonite is 2.47 sites nm^{-2} (Sahai and Anseau 2005). In the case of silicate glasses and sol-gel silicas, a site-density of $\sim$ 2.2-4.5 sites nm^{-2} is obtained due to surface reconstruction (Brinker et al. 1990; Bunker et al. 1994). This range of site-densities is, apparently, sufficient to ensure the known bioactivity of psW and sol-gel silica.

Surface charge and diffuse layer potential

In imaginative and quantitative studies published recently, electrophoretic mobility has been used to estimate the diffuse layer potential near perovskite bioceramic surfaces and the changes induced in these properties when reacted with SBF. The changes in surface potential were interpreted in terms of Ca^{2+} ion leaching from the surface followed by adsorption of HPO_4^{2-} and, finally, Ca^{2+} attachment resulting in CaP precipitation (Coreño et al. 2001; Coreño and Coreño 2005). This proposed reaction pathway for CaP is very different from the one described above (Sahai and Tossell 2000; Sahai and Anseau 2005).

Relationship between proposed active-site density, porosity, bioactivity of gel-derived bioactive glasses

Interestingly, the site-density of the three- and four-rings in amorphous silica gels changes as the silica is heated. The three-ring sites are almost absent at temperatures below 200 °C, increase in number per unit area on heating up to a maximum at about 600 °C and decline in number by 1000 °C. The number of four-ring sites is less sensitive to the temperature, but these sites most abundant below about 500-600 °C. The three and four-rings are characterized, respectively, by vibrational peaks at 490 cm^{-1} and 605 cm^{-1} in the Raman spectrum corresponding to the symmetric oxygen ring breathing modes (Brinker et al 1986, 1988, 1990; Bunker et al. 1989; Chuang and Maciel 1997). As noted above, sol-gel derived glasses and sol-gels themselves showed faster HAP deposition at their surfaces compared to melt-derived glasses (Li 1991; Li et al. 1991). Although porosity may be one of the factors, it is not difficult to see that the decline in density of the proposed active sites with increasing temperature may be of the major controlling factors at the molecular level.

EFFECTS OF SOLUTIONS USED TO TEST *IN VITRO* BIOACTIVITY

The importance of analyzing bioactivity *in vitro* prior to *in vivo* analysis is quite clear. In-vivo studies require animal sacrifices, are more costly and less easily reproducible, and involve ethical issues. For these reasons, *in vitro* screenings in chemical and biological labs are necessary before *in vivo* testing of bioactivity. The choice of the solution used to simulate *in vitro* the reactions occurring on the surface of the biomaterial is very important: simple solutions that mimic only the inorganic composition of human body fluids can be used, or more complex solutions that also contain some biological moieties such as proteins. Moreover, cell-containing solutions can be employed, thus increasing both the similarity to real body fluids and the complexity of the test.

The rate of ionic release and pH increase also depends on the dynamic (flow-through) or static (batch) method used to simulate biomaterial reactivity. Many studies are done in "static" mode, which means that the solution used to dissolve *in vitro* the biomaterial is never changed in the course of the experiment. In other studies, instead, the solution where the biomaterial is dissolved is periodically changed and refilled with a fresh batch of solution (Falaize et al. 1999). In some recent experiments, the solution flows through, so that new solution is in contact with the biomaterial at all times (Izquierdo-Barba et al. 2000). It is difficult to prove which experimental method best simulates the *in vivo* situation. In fact, body fluids circulate at the interface with the wounded area, but the extent of this circulation is far from well-defined.

Experimentally, it has been shown that the static method quickly induces saturation of the solution, so that apatite precipitates faster, and the pH increases more than with the dynamic method (Ramila and Vallet-Regi 2001).

Early studies concerning bioactivity of Bioglass® were carried out in simple TRIS-buffered solution (Clark and Hench 1976; Clark et al. 1976; Hench 1981). TRIS base (tris-hydroxy methyl amino methane) has a $pK_a = 8.1$, and can be used to buffer solutions in the range of pH~7.1-9.1. Since this solution does not contain any ions other than the ones that are leached out of the biomaterial, it can be very useful to analyze the basic steps involved in HAP deposition on bioactive materials. Kokubo et al. (1990) introduced the use of simulated body fluid (SBF) to analyze bioactivity of different materials. SBF is a solution without proteins, hormones, cells, etc. containing inorganic salts that simulate the ionic concentrations and pH of human plasma (Table 1).

Both SBF and plasma are saturated with respect to HAP. For this reason, only a few nucleation sites are sufficient to induce HAP nucleation on the surface of some materials. This allowed the analysis of bioactivity on some very simple materials such as amorphous silica, that did not contain calcium and phosphorous in their composition. Moreover, the rate of HA deposition in SBF is much higher than in TRIS-buffered solution, because the degree of supersaturation with respect to HAP is much higher in SBF. An alternative to the use of SBF is a solution containing TRIS and the electrolytes typical for plasma (Radin et al. 1997).

The original SBF composition of Kokubo (1990) is higher in Cl^- and lower in HCO_3^- than human blood plasma, so a modified SBF was prepared that had lower Cl^- and higher HCO_3^- and pH closer to blood (Tas 2000) (Table 1). It was found recently, however, that HAP formation was slower (4-5 days) in the modified SBF compared to 1 day in original SBF. This was attributed to competition between HCO_3^- and HPO_4^{2-} for Ca^{2+} bound to the bioceramic surface. Furthermore, the HAP formed in modified SBF transformed to calcite after 10 days of reaction at a bioceramic solid to solution ratio of 10 mg mL^{-1}. No phase transformation was observed in modified SBF at 1 and 5 mg mL^{-1}, and in original SBF at 1, 5 and 10 mg mL^{-1}. Thus, the type of SBF solution and the ratio of bioceramic solid to solution used in the experiment influenced which phase was stable at the bioceramic/solution interface. Other SBFs with variations in the HCO_3^-/Cl^- ratio have also been prepared and their reactions with bioceramics are summarized in Kokubo and Takadama (2006). Although SBFs can provide a preliminary in vitro test of whether a material may be bioactive or not, it does not provide information on how the material would behave in the body where proteins, cells, etc. are also present.

Bovine and human serums are often used for *in vitro* studies when researchers want to analyze protein adsorption on biomaterials (Bosetti et al. 2001; Rosengren et al. 2003). The reactions occurring at the surface of biomaterials in contact with protein containing solutions have also been studied with Dulbecco's Modified Eagle's minimum essential medium supplemented with 10% Nu-Serum™ (Effah Kaufmann et al. 2000). This growth medium for cells contains inorganic ions (Na^+, K^+, Mg^{2+}, Ca^{2+}, Cl^-, NO_3^-, HCO_3^-, SO_4^{2-}, PO_4^{3-}), amino acids, growth factors, hormones and vitamins.

A further step to simulate *in vitro* the real condition of biomaterials immersed into body fluids is immersion in cell-containing solutions. Osteoblast cells have often been used *in vitro*. Different tests can be done to understand the influence of biomaterials on cells. Cell morphology, adhesion and proliferation can be examined, and cell activity can be tested by the amount of some specific enzymes produced. For example, osteoblasts that are synthesizing bone matrix produce the enzyme, alkaline phosphatase (ALP). Another important protein that can be evaluated is osteocalcin (OCN). This is a non-collagenous extracellular matrix protein, and its presence is indicative of the beginning of bone mineralization (HAP precipitation).

INTERACTIONS WITH CELLS

A host of different cells types can interact with the implant surface in the body. Fibroblasts are cells that produce structural fibers. Fibroblasts do not spread and proliferate on bioactive glass surfaces, unlike the surfaces of bioinert materials (Seitz et al. 1982). This is important because osteoblasts (bone-forming cells) arrive later than fibroblasts when bone fractures, or after a device is implanted. So, if fibroblasts proliferate, a fibrous non-adherent capsule forms, which prevents the formation a real bond between the material and the host tissue. Instead, on the surface of a bioactive glass, fibroblasts remain "quiescent", and when osteoblasts arrive, new bone tissue can be formed.

The exact sequence of cell response, and also some details of the first chemical-physical reactions are still unclear. One of the main issues is whether the interaction with biological moieties present in body fluids really occurs after the formation of crystalline HCA layer. If proteins are present in the solution in contact with a bioactive material, they adsorb on the material surface quickly (Bohnert and Horbett 1986 and references therein). This leads to surface modifications different from those obtained when the bioactive material is immersed in solutions such as SBF that contain only inorganic ions. In fact, in protein-free solutions, the formation of a layer of hydroxyapatite hinders the release of all the Si present in bioactive glass particles, since it acts as a protective layer. Instead, when bioactive glasses are immersed in protein-containing serum solutions, a porous and amorphous layer of silica, proteins and calcium phosphates formed, which does not protect the glass from further corrosion (Radin et al. 2000).

The presence of proteins influences both the type of calcium-phosphate containing layer formed on bioactive glasses immersed in body fluids, and the further reactivity of the bioglass towards cells (Ducheyne and Qiu 1999). Fibronectin is a protein found in plasma. A coating of this protein on bioactive material surface enhances fibroblasts attachment and proliferation (Seitz et al. 1982; Cannas et al. 1988). Other researchers observed that the configuration of fibronectin adsorbed was different depending on the type of surface exposed by the biomaterial (Garcia et al. 1998). A specific fibronectin conformation, that was found on the amorphous calcium phosphate layer on SBF-reacted bioactive glasses, induced a very strong cell adhesion.

The type of surface exposed by the biomaterial influences cell functions. El Ghannam et al. (1997) showed that when bioactive glasses were immersed in osteoblast-like containing solutions, cell proliferation on their surfaces were high during the first 7 days, but then it slowed down, as bone matrix began to be synthesized. On the contrary, cell continued to proliferate on HAP. The differences in cell functions observed after contact with bioactive glasses and HAP described by El-Ghannam et al. (1997) can be explained in terms of cell cycle and are paralleled by the results obtained for bioactive versus bioinert glasses by Xynos et al. (2000).

To better understand these data, the generalized life cycle of a cell is presented in Figure 10. A resting cell is in the G_0 state. In G_1 phase, the cell grows and carries out its normal metabolism. Osteoblasts, for example, produce ALP and tropocollagen molecules, which can self-assemble into collagen. Later, cells enter the S phase, and begin to synthesize DNA. When all chromosomes have been duplicated, cells enter a secondary growth phase (G_2), and finally divide (M phase, or mitosis). There are some feedback mechanisms in cells, controlling the state of the cell before switching from one phase to the next one. If the control fails, the cell is sent to programmed death, i.e., apoptosis.

Xynos et al. (2000) found that after about 6 days of reaction, the number of cells attached to a bioinert material is higher than on a bioactive material. Still, the number of cells that are in the S and G2-M phases is higher on the bioactive material. This means that on a bioactive material, cells that are not capable of differentiating into the osteoblast phenotype die from apoptosis (programmed cell death or cell "suicide" as opposed to necrosis or unprogrammed

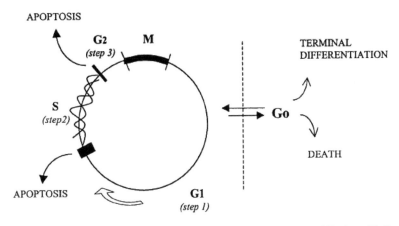

Figure 10. Cell cycle. [Reprinted with kind permission of Springer Science and Business Media, from Hench et al. (2000), *Materials Research Innovations*, Vol. 3, Fig. 10, p. 318.]

cell death). After about 12 days, ALP production decreases in cells attached to bioactive materials suggesting that cells stop proliferating and begin to synthesize bone-like tissue (in fact, an increase in OCN, characteristic of bone formation, is observed).

It is not yet completely understood how these differences in cell functions depend on the interaction with biomaterials. Surface morphology is definitely a relevant factor (Ducheyne and Qiu 1999), but also the amount of ions released in body fluids, and the changes induced in pH should be taken in consideration. Leaching of alkali and alkaline-earth cations in body fluids and an increase in [Ca^{2+}] inside osteoblast cells in contact with Bioglass® have been reported (Silver et al. 2001). It was hypothesized that the higher glycolitic activity shown by these osteoblasts should be specifically related to the changes in pH and Ca^{2+} content. Bone formation is always connected to an increase in pH (Cuervo et al. 1971). Ca^{2+} increases glycolysis for some systems (e.g., skeletal muscles), and in general, is a modulator of intracellular events. Also Si release, which is always observed when bioactive glasses are dissolved, is relevant for bone formation. Chicken and rats fed with a diet poor in Si had problems with their skeletal structure (Carlisle 1981) relative to animals fed Si-enriched diets, and solutions rich in Si induced osteoblast proliferation (Keeting et al. 1992).

The behavior of osteoblast cells cultured *in vitro* in contact with pseudowollastonite ceramic has been examined (Sarmento et al. 2004). Cells attached and proliferated on the pseudowollastonite surface. Greater attachment was achieved in the presence of fibronectic and adhesion was inhibited by glycine-arginine-glycine-aspartate-serine (RGD) peptides suggesting that cell adhesion occurs by integrin binding to proteins adsorbed on the pseudowollastonite surface. The effect of pseudowollastonite on osteoblast cells was also examined when the ceramic is at a distance. Cells were cultured in a medium containing pseudowollastonite but not directly in contact. In the presence of pseudowollastonite, the number and rate of bone nodule formation was greater than control experiments without the pseudowollastonite but there was no difference in the proliferation of cells and in ALP expression. These results showed that pseudowollastonite is biocompatible and osteoconductive.

In a recent study, osteoblasts were treated with the ionic product of Bioglass® dissolution in Dulbecco culture media for 24 h, then RNA was removed, and genes analyzed. A lot of different genes were stimulated by the contact with these ions, and in particular, some that are strongly involved in bone formation. This intriguing result study suggests that ionic release from bioactive materials also influences the expression of some specific genes (Xynos et al. 2001).

METHODS FOR CHARACTERIZATION OF BIOACTIVE MATERIALS

Spectroscopic and microscopic methods for characterizing of bioceramics

Improvement in composition and synthesis procedure of bioactive glasses has been obtained mostly by trial-and-error, and by comparing *in vitro* and *in vivo* results. Still, the actual surface sites for HCA deposition are not completely known, and the role of the different elemental components of bioactive glasses and bioactive ceramics is not fully understood. One reason for this lack of knowledge is that the most important interactions occur at the bioactive glass/solution interface, which is a nanometer-sized, continuously changing region of space. Further, the earliest calcium phosphate phase to precipitate is also nano-sized. A greater emphasis is required in the future on the nanoscale and angstrom scale characterization of the bioactive materials and solid phases formed at the biomaterial/solution for in vitro studies and biomaterial implant/bone interface for in vivo studies. A thorough study of this region should involve analysis of changes in surface morphology, crystallinity, composition, hydroxylation, acidity, potential, surface charge of the biomaterial, as well as composition and pH of the solution. Certainly, such a large and varied amount of information cannot be obtained with only one analytical technique.

We will briefly describe a few of the most relevant techniques that have been used to characterize the reactions occurring on bioactive materials immersed in SBF or other solutions that simulate the body environment. We will focus our summary on "Spectroscopic" and "Microscopic Techniques," although a variety of other techniques have been used, such as X-ray Diffraction to check the formation of hydroxyapatite on the biomaterials, N_2 adsorption at 77 K to measure the changes in surface area and porosity, and zeta potential measurements of the interfacial potential in order to infer the sequence of Ca^{2+} cation and phosphate anion adsorption at the biomaterial surface.

FTIR spectroscopy. With this technique, one can analyze the vibrations of both surface and bulk groups, depending on the sample preparation. The literature concerning IR studies on Bioglass® and other biomaterials dissolution is vast (see as examples Peitl et al. 2001; El Bathal et al. 2003). Still, it mainly concerns the analysis of bulk group vibrations, and for this reason many important details, such as the type of surface hydroxyl and carbonate groups, have not been thoroughly understood. Recently, some groups have analyzed changes in surface groups using adsorbed molecules as probes to test the presence of surface hydroxyls and available cations (Cerruti and Morterra 2004; Cerruti et al. 2003, 2004). A special type of IR spectroscopy is ATR (attenuated total reflectance) IR spectroscopy. This technique is particularly useful to characterize the reactivity of bioactive materials in simulated body fluids, since it is possible to analyze bulk group modifications occurring when the sample is immersed in a solution (Cerruti et al. 2005b,c). For example, it is possible to observe in real-time the leaching of alkali ions from the glassy network and the parallel formation of hydroxyapatite.

Raman spectroscopy. Raman spectroscopy is another vibrational spectroscopy that provides complementary information with respect to IR spectroscopy, since different vibrational modes are Raman active. For example, the vibration corresponding to H-O-H bending in water is very intense in IR spectra, but nearly inactive in Raman spectroscopy, and for this reason it is possible to analyze aqueous solutions, or samples fully hydrated with Raman. Scattered radiation is collected in Raman spectroscopy. For this reason, response from sample surface layers is enhanced with respect to bulk (Colthup et al. 1990). Still, the depth of sample analyzed is quite difficult to know precisely, since it depends on both sample form and the laser frequency used. In the field of biomaterials, the use of Raman spectroscopy is relatively new but very quickly expanding (Rehman et al. 1994; Gonzalez et al. 2003; Notingher et al. 2003; Cerruti et al. 2005a).

X-ray photoelectron spectroscopy (XPS). XPS is another technique useful for surface characterization, since it can give information about the elemental surface composition, and the surrounding surface atoms. An X-ray beam hits the sample, in ultra-high-vacuum conditions (UHV; residual pressure $\approx 10^{-8}$ Torr), core electrons are emitted from the surface atoms with an energy typical of each atomic species, and a spectrum is recorded. The position of each peak in the spectrum is indicative of atomic species, and the intensity is indicative of the atomic concentration. The surface layer analyzed is ~30-40 Å deep. Using XPS in a high-resolution setup, one can understand the differences in atomic surroundings related to the presence of more or less electron attractive/repulsive groups. A reference text for XPS is Briggs and Sea (1993). The use of XPS in biomaterials analysis has so far been limited. XPS has been used in the biomaterials community mainly as a tool to obtain surface elemental survey (Perez-Pariente et al. 2000; Polzonetti et al. 2000) and only a few papers give a more detailed analysis of XPS peak components (Serra et al. 2003; Vallet-Regi et al. 2000; Takadama et al. 2002; Cerruti et al. 2005a).

Scanning electronic microscopy (SEM). Particle morphology can be revealed at the 10s to 100s of nanometers size with scanning electron microscopy (SEM). The sample is placed under a non-destructive electron beam, focused in a spot a few nanometers in diameter. The interaction of the electron beam with the sample produces different forms of radiation such as backscattered electrons, secondary electrons, Auger electrons, and characteristic X-rays. Usually, only backscattered electrons and secondary electrons are used to build the SEM image. A three-dimensional map of the sample is obtained. Information about the composition of the sample can also be obtained, both by backscattered and by secondary electrons, or with a microprobe that collects the X-rays emitted. This analytical tool is often referred to as EDS (energy dispersive X-ray spectroscopy). Since every element has a characteristic X-ray emission wavelength, it is possible to determine the type and amount of elements present in the sample by analyzing the X-ray emission spectra collected with EDS microprobe. SEM has been widely used to analyze the changes in morphology of the bioactive materials in simulated body fluids, once HCA begins to precipitate on their surface.

High resolution transmission electron microscopy (HRTEM). Ultimate particle morphology can be obtained with HRTEM. As in SEM, an electron beam interacts with the sample but, here, the transmitted electrons are collected. For this reason, the electron beam is much more energetic than in SEM, and the sample image can be magnified up to 1 million times. Thus, the HRTEM provides nanometer-sized resolution. A diffraction pattern of the sample can be also obtained, to analyze crystalline zones. If the microscope is equipped with an EDS microprobe, the composition of the sample can be analyzed. A reference text for different types of electronic microscopy is Amelinckx et al. (1997). The use of TEM in the analysis of biomaterials reactivity is less common than SEM, but can be very powerful to understand the morphology of the nanoporous structures formed once the materials are immersed in simulated or real body fluids (Cerruti et al. 2005d), and for identification of epitaxial growth of specific crystal faces of HAP on the bioceramic implant.

Ion coupled plasma - optical emission spectroscopy (ICP-OES). ICP-OES allows the total elemental analysis of aqueous solutions. The sample solution is nebulized (i.e., transformed into an aerosol), and carried by a gas carrier (usually Ar) through a torch, where a plasma (i.e. a gas in which atoms are ionized) is ignited. When sample atoms are ionized, they emit radiation at some specific wavelength. These specific components are selected by a diffracting grating, and converted in electric signals by a photomultiplier. After calibration, it is possible to determine the amount of each element present in solution by analyzing the intensity of the radiation emitted at the specific elemental frequency. ICP-OES has been widely used to characterize the ions released when bioactive materials are immersed in simulated body fluids (Cerruti et al. 2005b,c). An introduction on ICP-ES is given by Moore (1998).

Biological assays. Biochemical, immunochemical, cellular and genetic methods are used to assess the biological response of cells to the presence of biomaterials in vitro and in vivo. A description of such methods is, however, beyond the scope of this chapter.

Computational modeling

A huge potential exists for understanding interactions at the biomaterial/solution interface using computational modeling approaches including quantum mechanical (QM) and molecular mechanical (MM) approaches. The modeling approaches can be classified based on the number of atoms or size of the system into the cluster and periodic approaches. In the former, local clusters of atoms take into account the short-range forces, whereas extended crystals repeated in 2D in the periodic approach account for both long- and short-range interactions. The computational cost increases with the size of the system so trade-offs are often made in using classical mechanical approaches for periodic calculations compared to the quantum mechanical approach from first principles (*ab initio*) for clusters.

West and Wallace (1993) and Wallace et al. (1993) were among the earliest to use computational semi-empirical methods to understand bioactivity of silicate glasses. Improved computational methods and faster computers have since allowed for cluster QM approaches to be used (Sahai and Tossell 2000; Sahai and Anseau 2005). The surface sites and reactivity of silica and quartz surfaces in terms of hydroxylation, hydration, hydrolysis, and acidity in aqueous solution have been widely studied and summarized in a recent review (Sahai and Rosso 2006). The structure and vibrational properties of HAP have been calculated recently (Corno et al. 2006), and the long-standing question of which site the CO_3^{2-} anion occupies in HAP to yield HCA has been addressed recently using density functional theory (Peroos et al. 2006).

FUTURE DEVELOPMENT OF BIOCERAMICS: BIOCOMPOSITES AND TISSUE-ENGINEERING

Broadly speaking, the first-generation of biomaterials was bioinert, whereas bioactive and bioresorbable materials represented improved, second-generation materials. The current challenge lies in the development of cell- and gene-activating third-generation biomaterials that will require an understanding of the cellular and molecular basis of the reactions at the biomaterial-tissue interface (Hench and Polak 2002; Kim 2003). This goal is consistent with the definition of Tissue Engineering as "an interdisciplinary field that applies the principles of engineering and life sciences toward the development of biological substitutes that restore, maintain, or improve tissue function (Langer and Vacanti 1993).

The third generation of orthopaedic and dental biomaterials currently being developed are both bioactive and bioresorbable and can also activate genes to stimulate regeneration of bone and tooth tissue (Shirtliff and Hench 2003). For example, composite bioactive ceramics and bioresorbable polymer materials are being studied for their ability to promote bone tissue regeneration (Boccaccini and Baker 2005). The effects of silicate bioglasses on cells and genes are also being investigated. In one study, Class A bioactive glass (Bioglass® 45S5) was found release Si and grow a surface layer of HCA more rapidly than Class B bioactive glass and a bioinert material. Osteoblasts grew more slowly on bioactive glass but greater OCN generation was seen, corresponding to a faster rate of bone formation compared to the Class B bioglass and bioinert material (Xynos et al. 2000a). Furthermore, several gene families, some specific to bone, were activated within hours of exposure of human osteoblasts to Bioglass® 45S5 and Si was believed to have activated these genes (Xynos et al. 2000b, 2001). These examples highlight the tissue engineering where cell growth and function are modified *in vivo*.

A different approach involves the *in vitro* seeding biomaterials with genes, growth factors, mesenchymal stem cells, etc. onto or into bioresorbable implants (Ohgushi and Caplan 1999).

The chapter in this volume on "Living Cells on Oxide Glasses" is one such approach. In one study, mesenchymal stem cells (MSCs) were grown on an HCA layer deposited on a poly(lactide-co-glycolide) PGA substrate and directly on the PGA substrate. The HCA layer promoted fibronectic adsorption and proliferation of MSCs compared to PLG substrates. However, ALP activity and OCN, which are osteogenic markers, were expressed to a greater degree on the PLG substrates than on the HCA layer, suggesting greater differentiation of MSCs into osteoblasts on the PLG (Murphy et al. 2005). This difference is counter-intuitive and is not explained easily. In contrast, osteogenic murine calvaria cells colonized the surface of organoapatite (apatite + polylysine) coated Ti mesh faster than uncoated Ti controls. However, no significant difference was noted in cell differentation rates as measured by ALP activity and OCN expression (Spoerke and Stupp 2003). A general theory that can explain these variable responses of cells towards different substrates remains to be developed.

Analogous reactions in the human body and in some geochemical environments

Nodular phosphorite ore deposits are apatitic deposits in which a cross-section through an individual nodule frequently reveals a central detrital sediment grains such as amorphous silica of diatom tests, quartz sand grains and carbonate shell fragments. The apatite is precipitated from shallow marine pore fluids on the surfaces of the detrital sediments. We have suggested previously that the reaction pathway and rate-determining step in this geochemical process are similar to those involved in apatite nucleation from SBF at silicate bioactive glass and ceramic surfaces, because of the similarity in silicate substrate structure and surface chemistry, and in the major ion chemistry, ionic strength and pH of shallow marine pore fluids and SBF (Sahai 2003, 2005).

Similarly, the aqueous elemental profiles as a function of time obtained for the dissolution of diopside in SNF is very similar to those observed a quarter of a century ago by geochemists studying the chemical weathering of diopside (Schott et al. 1981). As geochemists and mineralogists, our training in the structures of silicate minerals and glasses provides us with additional insight for understanding epitaxial nucleation mechanisms of apatite on bioactive silicate materials. Finally, geochemists and mineralogists have long been trained in spectroscopic and microscopic characterization, and in phenomenological (e.g., surface complexation models) and molecular modeling of the mineral/water interface. By collaborating with scientists who work in the fields of biomedical engineering, molecular biology and genetic, and learning their tools, the community of geochemists and mineralogists can make significant contributions to the field of tissue engineering for orthopedic and dental implants.

ACKNOWLEDGMENTS

MC thanks Claudio Morterra and Piero Ugliengo, University of Torino, for all she has learned from them and Elena Cerruti for help with editing. NS is grateful for helpful discussions and exchange of ideas with Prof. Michel Anseau, University of Mons-Hainaut; Profs. Huifang Xu, Max Lagally and William Murphy, University of Wisconsin-Madison; Dr. Zofia Luklinska, Queen Mary College, University of London; financial support from University of Wisconsin faculty "start-up" funds, NSF EAR grant # 0208036 and NSF EAR CAREER grant # 0346689 to NS.

REFERENCES

Akai T, Chen D, Masui H, Yazawa T (2003) Structure change on the surface of leached sodium borosilicate glasses. Glass Technol 44:71-4
Amelinckx S, van Dyck D, van Landuyt J, van Tendeloo G (1997). Electron Microscopy. Principles and Fundamentals. Wiley-VCH

Andersson OH, LaTorre G, Hench LL (1992) The kinetics of bioactive ceramics. Part II. Surface reactions of three bioactive glasses. *In:* Bioceramics, Volume 3. Hulbert JE, Hulbert SF (eds) Rose-Hulman Institute of Technology, p. 46-53

Aoki H, Kato K, Ogiso M, Tabata T (1977) Sintered hydroxyapatite as a new dental implant material. J Dent Outlook 49:567-575

Bellantone M, Coleman NJ, Hench LL (2000) Bacteriostatic action of a novel four-component bioactive glass. J Biomed Mater Res 51:484–490

Black J (1992) Biological Performance of Materials: Fundamentals of Biocompatibility. Marcel Dekker

Boccaccini AR, Blaker JJ (2005) Bioactive composite materials for tissue engineering scaffolds. Expert Rev Med Devices 2:303-317

Bohnert JL, Horbett TA (1986) Changes in adsorbed fibrinogen and albumin interactions with polymers indicated by decreases in detergent elutability. J Coll Interface Sci 111:363-377

Bosetti M, Verne E, Ferraris M, Ravaglioli A, Cannas M (2001) In vitro characterisation of zirconia coated by bioactive glass. Biomaterials 22:987-994

Boutin P (1972) Arthroplastie totale de la hanche par prothese en alumine fritte. Rev Chir Orthop Reparatrice Appar Mot 58:229-246

Briggs D, Seah MP (1993) Practical Surface Analysis. Vol. 1, 2nd Edition. John Wiley and Sons

Brinker CJ, Brow RK, Tallant DR, Kirkpatrick RJ (1990) Surface structure and chemistry of high surface area silica gels. J Non-Cryst Solids 120:26-33

Brinker CJ, Kirkpatrick RJ, Tallant DR, Bunker BC, Montez B (1988) NMR confirmation of strained "defects" in amorphous silica. J Non-Cryst Solids 99:418-428

Brinker CJ, Tallant DR, Roth EP, Ashley CS (1986) Sol-gel transition in simple silicates. J Non-Cryst Solids 82:117-126

Bunker BC (1994) Molecular mechanisms for corrosion of silica and silicate glasses. J Non-Cryst Solids 179: 300-308

Bunker BC, Haaland DM, Ward KJ, Michalske TA, Smith WL, Binkley JS, Melius CF, Balfe CA (1989) Infrared spectra of edge-shared silicate tetrahedral. Surf Sci 210:406-428

Canham LT (1995) Bioactive silicon structure fabrication through nanoetching techniques. Adv Mater 7: 1033-1037

Canham LT, Newey JP, Reeves CL, Houlton MR, Loni A, Simons AJ, Cox TI (1996) The effects of DC electric currents on the *in vitro* calcification of bioactive silicon wafers. Adv Mater 8:847

Cannas M, Denicolai F, Webb LX, Gristi AG (1988) Bioimplant surfaces: Binding of fibronectin and fibroblast adhesion. J Orthop Res 6:58-62

Carew EO, Cook FW, Lemons JE, Ratner BD, Vesely I, Vogler E (2003) Properties of Materials. *In:* Biomaterials Science. Ratner B, Hoffman A, Frederick S, Lemons J (eds) Elsevier Academic Press, p 23-65

Carlisle EM (1981) Silicon: a requirement in bone formation independent of vitamin D1. Calcif Tissue Int 33: 27-34

Carlisle EM (1986) Silicon as an Essential Trace Element in Animal Nutrition. *In*: Silicon Biochemistry, CIBA Foundation Symposium 121. Wiley, p 123-136

Casey WH, Westrich HR, Banfield JF, Ferruzzi G, Arnold GW (1993) Leaching and reconstruction at the surfaces of dissolving chain-silicate minerals. Nature 366:253-256

Ceresoli D, Bernasconi M, Iarlori S, Parrinello M, Tosatti E (2000) Two-membered silicon rings on the dehydroxylated surface of silica. Phys Rev Lett 84:3787-3890

Cerruti M, Bianchi C, Bonino F, Damin A, Perardi A, Morterra C (2005a) Surface modifications of Bioglass(R) immersed in TRIS-buffered solution. A multi-technical spectroscopic study. J Phys Chem B 109:14496-14505

Cerruti M, Bolis V, Magnacca G, Morterra C (2004) Surface chemical functionalities in bioactive glasses. The gas/sold adsorption of acetonitrile. Phys Chem Chem Phys 6:2468-2479

Cerruti M, Greenspan D, Powers K (2005b) An analytical model for the dissolution of different particle size samples of Bioglass(R) in TRIS-buffered solution. Biomaterials 24:4903-4911

Cerruti M, Greenspan D, Powers K (2005c) Effect of pH and ionic strength on the reactivity of Bioglass(R) 45S5. Biomaterials 26:1665-1674

Cerruti M, Magnacca G, Bolis V, Morterra C (2003) Characterization of sol-gel bioglasses with the use of simple model systems: a surface-chemistry approach. J Mater Chem 13:1279-1286

Cerruti M, Morterra C (2004) Carbonate formation on bioactive glasses. Langmuir 20:6382-6388

Cerruti M, Perardi A, Cerrato G, Morterra C (2005d) Formation of a nano-structured layer on Bioglass(R) particles of different size immersed in TRIS-buffered solution. N$_2$ adsorption and HR-TEM/EDS analysis. Langmuir 21:9327-9333

Chen QZZ, Thompson ID, Boccaccini AR (2006) 45S5 Bioglass (R)-derived glass-ceramic scaffolds for bone tissue engineering. Biomaterials 27:2414-2425

Cho S, Nakanishi K, Kokubo T, Soga N, Ohtsuki C, Nakamura T, Kitsugi T, Yamamuro T (1995) Dependence of apatite formation on silica gel on its structure: effect of heat treatment. J Am Ceram Soc 78:1769-1774

Chuang I-S, Maciel GE (1997) A detailed model of local structure and silanol hydrogen bonding of silica gel surfaces. J Phys Chem B 101:3052-3064

Clark AE, Hench LL (1976) The influence of surface chemistry on implant interface histology: A theoretical basis for implant materials selection. J Biomed Mater Res 10:161-174

Clark AE, Pantano CG, Hench LL (1976) Auger spectroscopic analysis of Bioglass corrosion films. J Amer Ceram Soc 59:37-39

Colthup NB, Lawrence HD, Wiberley SE (1990) Introduction to Infrared and Raman Spectroscopy. Academic Press

Coreño J, Coreño O (2005) Evaluation of calcium titanate as apatite growth promoter. J Biomed Mater Res 75A:478-484

Coreño J, Martínez A, Bolarín A, Sánchez F (2001) Apatite nucleation on silica surface: A ζ -potential approach. J Biomed Mater Res 57:119-125

Corno M, Busco C, Civalleri B, Ugliengo P (2006) Ab-initio periodic study of structural and vibrational features of hexagonal hydroxyapatite $Ca_{10}(PO_4)_6(OH)_2$. Phys Chem Chem Phys 21:2464-2472

Cuervo LA, Pita JC, Howell DS (1971) Ultramicroanalysis of pH, pCO_2 and carbonic anhydrase activity at calcifying sites in cartilage. Calcif Tissue Res 7:220-31

De Aza PN, De Aza AH, De Aza S (2005) Crystalline bioceramic materials. Bol Soc Esp Ceram V 44:135-145

De Aza PN, Fernandez-Pradas JM, Serra P (2004) In vitro bioactivity of laser ablation pseudowollastonite coating. Biomaterials 25:1983-1990

De Aza PN, Luklinska ZB, Anseau M (2005) Bioactivity of diopside ceramic in human parotid saliva. J Biomed Mater Res 73B:56-60

De Aza PN, Luklinska ZB, Anseau MR, Guitian F, De Aza S (1996) Morphological studies of pseudowollastonite for biomedical application. J Microscopy 182(1):24-31

De Aza PN, Luklinska ZB, Anseau MR, Guitian F, De Aza S (1999) Bioactivity of pseudowollastonite in human saliva. J Dentistry 27:107-113

De Aza PN, Luklinska ZB, Anseau MR, Guitian F, De Aza S (2001) Transmission electron microscopy of the interface between bone and pseudowollastonite implant. J Microscopy 201(1):33-43

De Aza PN, Luklinska ZB, Martinez A, Anseau MR, Guitian F, De Aza S (2000) Morphological and structural study of pseudowollastonite implants in bone. J Microscopy 197:60-67

de Groot K (1983) Ceramic of calcium phosphates: preparation and properties. *In:* Bioceramics of Calcium Phosphate. de Groot K (ed) CRC press, p 100-114

Denissen H (1979) Dental root implants of apatite ceramics. Experimental investigations and clinical used of dental root implants made of apatite ceracmics. Ph D Thesis, Vrije Unversiteit te Amsterdam

Du M-H, Kolchin A, Cheng H-P (2003) Water-silica interactions: A combined quantum-classical molecular dynamic study of energetics and reaction pathways. J Chem Phys 119:6418-6422

Du RL, Chang J (2004) Preparation and characterization of bioactive sol-gel-derived $Na_2Ca_2Si_3O_9$. J Mater Sci - Mater Med 15:1285-1289

Du Z, de Leeuw NH (2004) A combined density functional theory and interatomic potential-based simulation study of the hydration of nano-particulate silicate surfaces. Surf Sci 554:193-210

Ducheyne P, Qiu Q (1999) Bioactive ceramics: the effect of surface reactivity on bone formation and bone cell function. Biomaterials 20:2287–2303

Dufrane D, Delloye C, Mckay IJ, De Aza PN, De Aza S, Schneider YJ, Anseau M (2003) Indirect cytotoxicity evaluation of pseudowollastonite. J Mater Sci - Mater Med 14:33-38

Effah Kaufmann EAB, Ducheyne P, Radin S, Bonnell DA, Composto R (2000) Initial events at the bioactive glass surface in contact with protein-containing solutions. J Biomed Mater Res 52:825–830

El Ghannam A, Hamazawy E, Yehia A (2001) Effect of thermal treatment on bioactive glass microstructure, corrosion behavior, ζ potential, and protein adsorption. J Biomed Mater Res 55:387-395

El Ghannam A, Ning CQ, Mehta J (2004) Cyclosilicate nanocomposites: A novel resorbable bioactive tissue engineering scaffold for BMP and bone-marrow cell delivery. J Biomed Mater Res A 71:377-390

ElBatal HA, Azooz MA, Khalil EMA, Soltan Monem A, Hamdy YM (2003) Characterization of some bioglass-ceramics. Mater Chem Phys 80:599-609

El-Ghannam A, Ducheyne P, Shapiro IM (1997) Porous bioactive glass and hydroxyapatite ceramic affect bone cell function in vitro along different time lines. J Biomed Mater Res 36:167-80

Falaize S, Radin S, Ducheyne P (1999) In vitro behaviour of silica-based xerogels intended as controlled release carriers. J Am Ceram Soc 82:969–976

Fujibayashi S, Neo M, Kim H-M, Kokubo T, Nakamura T (2003) A comparative study between in vivo bone ingrowth and in vitro apatite formation on $Na_2O–CaO–SiO_2$ glasses. Biomaterials 24 1349–1356

Galeener FL (1982) Planar rings in glass. Solid State Commun 44:1037-1049

Garcia AJ, Ducheyne P, Boettiger D (1998) Effect of surface reaction stage on Fibronectin-mediated adhesion of osteoblast-like cells to bioactive glass. J Biomed Mater Res 40:48-56

Geissberger AE, Galeener FL (1983) Raman studies of vitreous SiO_2 versus fictive temperature. Phys Rev B 28:3266-3271

Gonzalez P, Serra J, Liste S, Chiussi S, Leon B, Perez-Amor M (2003) Raman spectroscopic study of bioactive silica-based glasses. J Non-Cryst Sol 320:92-99

Greenspan DC, Hench LL (1976) Chemical and mechanical behavior of Bioglass-coated alumina. J Biomed Mater Res 10:503-509

Gross UM, Strunz V (1980) The anchoring of glass-ceramics of different solubility in the femur of the rat. J Biomed Mater Res 14:607-618

Hench LL (1981) Stability of ceramics in the physiological environment. *In:* Fundamental Aspects of Biocompatibility, Volume 1. Williams DF (ed) CRC Press, p 67–85

Hench LL (1988) Bioactive ceramics. *In:* Bioceramics: Materials Characteristics Versus *In-Vivo* Behavior. Ducheyne P, Lemons J (eds) Ann NY Acad Sci 523:54-71

Hench LL (1991) Bioceramics: From concept to clinic. J Am Ceram Soc 74:1487-1510

Hench LL (1994) Bioactive ceramics: theory and clinical applications. Bioceramics 7:3-14

Hench LL, Clark DE (1978) Physical chemistry of glass surfaces. J Non-Cryst Solids 28:83-105

Hench LL, Ethridge EC (1982) Biomaterials: An Interfactial Approach. Academic Press

Hench LL, Polak JM (2002) Third-generation biomedical materials. Science 295:1014-1017

Hench LL, Polak JM, Xynos ID, Buttery LDK (2000) Bioactive materials to control cell cycle. Mat Res Innovations 3:313–323

Hench LL, Splinter, RJ, Allen WC, Greenlee TK (1971) Bonding mechanisms at the interface of ceramic prosthetic materials. J Biomed Mater Res Symp 2 (Part I):117–141

Hench LL, Wilson J (1993) An Introduction to Bioceramics. World Scientific Publishing CO

Hill R (1996) An alternative view of the degradation of bioglass. J Mater Sci Lett 15:1122-1125

Hulbert SF (1993) The use of alumina and zirconia in surgical implants. *In:* An Introduction to Bioceramics. Hench LL, Wilson J (eds) Worlds Scientific Publishing CO, p 25-40

Iarlori S, Ceresoli D, Bernasconi M, Donadio D, Parrinello M (2001) Dehydroxylation and silanization of the surfaces of b-cristobalite silica: An ab initio simulation. J Phys Chem B 105:8007-8013

Iimori Y, Kameshima Y, Okada K, Hayashi S (2005) Comparative study of apatite formation on $CaSiO_3$ ceramics in simulated body fluids with different carbonate concentrations. J Mater Sci - Mater Med 16:73-79

Iler RK (1979) The Chemistry of Silica. John Wiley & Sons

Iwata NY, Lee G-H, Tokuoka Y and Kawashima N (2004) Sintering behavior and apatite formation of diopside prepared by coprecipitation process. Colloids Surfs B: Biointerfaces 34:239-245

Izquierdo-Barba I, Salinas AJ, Vallet-Regí M (1999) In vitro calcium phosphate layer formation on sol-gel glasses of the CaO-SiO_2 system. J Biomed Mater Res 47:243–250

Izquierdo-Barba I, Salinas AJ, Vallet-Regı M (2000) Effect of the continuous solution exchange on the in vitro reactivity of a CaO-SiO_2 sol-gel glass. J Biomed Mater Res 51:191–199

Jarcho M (1976) Hydroxylapatite synthesis and characterization in dense polycrystalline forms. J Mater Sci 11:2027-2035

Keeting PE, Oursler MJ, Wiegand KE, Bonde SK, Spelsberg TC, Riggs BL (1992) Zeolite A increases proliferation, differentiation and transforming growth factor β production in normal adult human osteoblast-like cells in vitro. J Bone Mineral Res 7:1281-1289

Kim H-M (2003) Ceramic bioactivity and related biomimetic strategy. Curr Opin Solid State Mater Sci 7:289-299

Klein CPAT, Li P, de Blieck-Hogervorst JMA, de Groot K (1995) Effect of sintering temperature on silica gels and their bone bonding ability. Biomaterials 16:715-719

Knudsen FP (1962) Effect of porosity on Young's modulus of alumina. J Am Ceram Soc 45:94-95

Kokubo T, Cho SB, Nakanishi K, Soga N, Ohtsuki C, Kitsugi T, Yamamuro T, Nakamura T (1994) Dependence of bone-like apatite formation on structure of silica gel. Bioceramics 7:49-54

Kokubo T, Kushitani H, Sakka S, Kitsugi T, Yamamuro T (1990) Solutions able to reproduce in vivo surface-structure changes in bioactive glass–ceramic A-W. J Biomed Mater Res 24:721–34

Kokubo T, Shigematsu M, Nagashima Y, Tashiro M, Yamamuro T, Higashi S (1982) Apatite- and wollastonite-containing glass-ceramics for prosthetic application. Bull Inst Chem Res Kyoto Univ 60:260-268

Kokubo T, Takadama H (2006) How useful is SBF in predicting in vivo bone bioactivity? Biomaterials 27:2907-2915

Koster K, Heide H, Konig R (1977) Resorbable calcium phosphate ceramics under load. Langenbecks Arch Chir 343:173-181

Kubicki JD, Sykes D (1993) Molecular-orbital calculations of vibrations in 3-membered aluminosilicate rings. Phys Chem Minerals 17:381-391

Kundu TK, Hanumantha Rao K, Parker SC (2003) Atomistic simulation of the surface structure of wollastonite and adsorption phenomena relevant to floatation. Int J Miner Process 72:111-127

Langer R, Vacanti JP (1993) Tissue engineering. Science 260:920-926

Li P, Kokubo T, Nakanishi K, Groot KD (1993a) Induction and morphology of hydroxyapatite, precipitated from metastable simulated body fluids on sol-gel prepared silica. Biomaterials 14:963-968

Li P, Ohtsuki C, Kokubo T, Nakanishi K, Soga N, Kanamura T, Yamamuro T (1992) Apatite formation induced by silica gel in a simulated body fluid. J Am Ceram Soc 75:2094-2097

Li P, Ohtsuki C, Kokubo T, Nakanishi K, Soga N, Nakamura T, Yamamuro T (1993b) Process of formation of bonelike apatite layer on silica gel. J Mater Sci - Mater Med 4:127-131

Li P, Zhang F (1990) The electrochemistry of a glass surface and its application to bioactive glass in solution. J Non-Cryst Solids 119:112-118

Li R (1991) Sol-gel processing of bioactive glass powders. Ph.D. dissertation, University of Florida

Li R, Clark AE, Hench LL (1991) An investigation of bioactive glass powders by sol–gel processing. J Appl Biomater 2:231–9

Liu X, Ding C, Chu PK (2004) Mechanism of apatite formation on wollastonite coatings in simulated body fluids. Biomaterials 25:1755-1761

Liu X, Ding C (2001) Apatite formed on the surface of plasma sprayed wollastonite coating immersed in simulated body fluid. Biomaterials 22:2007-2012

Marquis JD (1988) Optimizing the strength of all-ceramic jacket crowns. *In:* Perspectives in Dental Ceramics. Preston J (ed) Quintessence Publishing Company, p 15-27

Martinez A, Izquierdo-Barba I, Vallet-Regi M (2000) Bioactivity of a $CaO-SiO_2$ binary glasses system. Chem Mater 12:3080-3088

Miake Y, Yanagisawa T, Yajima Y, Noma H, Yasui N, Nonami T (1995) High resolution and analytical electron microscopic studies of new crystals induced by a bioactive ceramic (diopside). J Dent Res 74:1756-1763

Moore LG (1988) Introduction to Inductively Coupled Plasma Atomic Emission Spectrometry. Elsevier

Murphy WL, Hsiong S, Richardson TP, Simmons CA, Mooney DJ (2005) Effects of a bone-like mineral film on phenotype of adult mesenchymal stem cells in vitro. Biomaterials 26:303-310

Neo M, Kotani S, Nakamura T, Yamamuro T, Ohtsuki C, Kokubo T, Bando Y (1992) A comparative study of ultrastructures of the interfaces between four kinds of surface-active ceramic bone. J Biomed Mater Res 26:1419-1432

Nonami T, Tsutsumi S (1999) Study of diopside ceramics for biomaterials. J Mater Sci - Mater Med 10:475-479

Notingher I, Boccaccini AR, Jones J, Maquet V, Hench LL (2003) Application of Raman microspectroscopy to the characterisation of bioactive materials. Mater Charact 49:255-260

Ohguchi H, Caplan AI (1999) Stem cell technology and bioceramics: From cell to gene engineering. J Biomed Mater Res B Appl Biomater 48:913-927

Ohtsuki C, Kokubo T, Takatsuka K, Yamamuro T (1991) Composition dependence of bioactivity of glasses in the system $CaO-SiO_2-P_2O_5$: its in vitro evaluation. J Ceram Soc Japan 99:357-365

Ohura K, Nakamura T, Yamamuro T, Kokubo T, Ebisawa Y, Kotoura Y, Oka M (1991) Bone-bonding ability of P_2O_5-free CaO/SiO_2 glasses. J Biomed Mater Res 25:357-365

Oonishi H, Kutrshitani S, Yasukawa E, Iwaki H, Hench LL, Wilson J, Tsuji E, Sugihara T (1997) Particulate bioglass and hydroxyapatite as a bone graft substitute. Clin Orthop Relat Res 334:316-325

Park JB, Bronzino JD (2003) Biomaterials: Principles and Applications. CRC Press, preface

Peitl O, DutraZanotto E, Hench LL (2001) Highly bioactive $P_2O_5-Na_2O-CaO-SiO_2$ glass ceramics. J Non-Cryst Solids 292:115-126

Peltola T, Jokinen M, Rahiala H, Levanen E, Rosenhold JB, Kangasniemi I, Yli-Urpo A (1999) Calcium phosphate formation on porous sol-gel-derived SiO_2 and $CaO-P_2O_5-SiO_2$ substrates in vitro. J Biomed Mater Res 44:12-21

Perez-Pariente J, Balas F, Roman J, Salinas AJ, Vallet-Regi M (1999) Influence of composition and surface characteristics on the in vitro bioactivity of $SiO_2-CaO-P_2O_5-MgO$ sol-gel glasses. J Biomed Mater Res 47:170–175

Perez-Pariente J, Balas F, Vallet-Regi M (2000) Surface and chemical study of $SiO_2 \cdot P_2O_5 \cdot CaO \cdot (MgO)$ bioactive glasses. Chem Mater 12:750-755

Peroos S, Du ZM, de Leeuw NH (2006) A computer modelling study of the uptake, structure and distribution of carbonate defects in hydroxy-apatite. Biomaterials 27:2150-2161

Pfaender HG (1996) Schott Guide to Glass. 2nd Edition. Chapman & Hall, p 19

Piehler HR (2000) The future of medicine: Biomaterials. Mater Res Soc Bull 25(8):67-70

Polzonetti G, Iucci G, Frontini A, Infante G, Furlani C, Avigliano L, Del Principe D, Palumbo G, Rosato N (2000) Surface reactions of a pasma-sprayed CaO-P$_2$O$_5$-based glass with albumin, fibroblasts and granulocytes studied by XPS, fluorescence and chemiluminescence. Biomaterials 21:1531-1539

Proussaefs P, Lozada J (2002) Evaluation of two Vitallium blade-form implants retrieved after 13 to 21 years of function: A clinical report. J Prosthet Dent 87:412-415

Radin S, Ducheyne P, Falaize S, Hammond A (2000) In vitro transformation of bioactive glass granules into Ca-P shells. J Biomed Mater Res 49:264-272

Radin S, Ducheyne P, Rothman B, Conti A (1997) The effect of in vitro modeling conditions on the surface reactions of bioactive glass. J Biomed Mater Res 37:363–375

Ramila A, Vallet-Regi M (2001) Static and dynamic in vitro study of a sol-gel glass bioactivity. Biomaterials 22:2301-2306

Rawlings RD (1992) Composition dependence of the bioactivity of glasses. J Mater Sci Lett 11:1340-1346

Rehman I, Smith R, Hench LL, Bonfield W (1994) FT-Raman spectroscopic analysis of natural bones and their comparison with bioactive glasses and hydroxyapatite. Bioceramics 7:79-84

Rho J, Kuhn-Spearing L, Zioupos P (1998) Mechanical properties and the hierarchical structure of bone. Med Eng Phys 20:92-102

Ringnanese G-M, Charlier J-C, Gonze X (2004) First-Principles molecular-dynamics investigation of the hydration mechanisms of the (0001) α-quartz surface. Phys Chem Chem Phys 6:1920-1925

Rizkalla AS, Jones DW, Clarke DB, Hall GC (1996) Crystallization of experimental bioactive glass compositions. J Biomed Mater Res 32:119-124

Rosengren A, Oscarsson S, Mazzocchi M, Krajewski A, Ravaglioli A (2003) Protein adsorption onto two bioactive glass-ceramics. Biomaterials 24:147–155

Sahai N (2003) The effects of Mg^{2+} and H$^+$ on apatite nucleation at silica surfaces. Geochim Cosmochim Acta 67:1017-1030

Sahai N (2005) Modeling apatite nucleation in the human body and in the geochemical environment. Am J Sci 305:661-672

Sahai N, Anseau M (2005) Cyclic silicate active site and epitaxial apatite nucleation on pseudowollastonite bioceramic-bone interfaces. Biomaterials 26:5763-5370

Sahai N, Rosso K (2006) Linking molecular modeling to surface complexation modeling. In: Surface Complexation Modelling. Interface Science and Technology series. Lutzenkirchen J (ed) Elsevier

Sahai N, Tossell JA (2000) Molecular orbital study of apatite nucleation at silica bioceramic surfaces. J Phys Chem B 104:4322-4341

Sarmento C, Luklinska ZB, Brown L, Anseau M, de Aza PN, de Aza S, Hughes FJ, McKay IJ (2004) *In vitro* behavior of osteoblastic cells cultured in the presence of pseudowollastonite ceramic. J Biomed Mat Res A 69A:351-358

Schott J, Berner RA, Sjoberg EL (1981) Mechanism of pyroxene and amphibole weathering — I. Experimental studies of iron-free minerals. Geochim Cosmochim Acta 45:2123-2135

Schwartz K (1978) In: Biochemistry of Silicon and Related Problems. G Bendz, I Lindqvist (eds) Plenum Press, p 207-230

Seitz TL, Noonan KD, Hench LL, Noonan NE (1982) Effect of fibronectin on the adhesion of an established cell line to a surface reactive biomaterial. J Biomed Mater Res 16:195-207

Serra J, Gonzalez P, Liste S, Serra C, Chiussi S, Leon B, Perez-Amor M, Ylanen HO, Hupa M (2003) FTIR and XPS studies of bioactive silica based glasses. J Non-Cryst Solids 332:20-27

Shirtliff VJ, Hench LL (2003) Bioactive materials for tissue engineering, regeneration and repair. J Mater Sci 38:4697-4707

Silver IA, Deas J, Erecinska M (2001) Interactions of bioactive glasses with osteoblasts in vitro: effects of 45S5 Bioglass®, and 58S and 77S bioactive glasses on metabolism, intracellular ion concentrations and cell viability. Biomaterials 22:171-185

Skipper LJ, Sowrey FE, Pickup DM, Drake KO, Smith ME, Saravanapavan P, Hench LL and Newport RJ (2005b) The structure of a bioactive calcia-silica sol-gel glass. J Mater Chem 15:2369-2374

Skipper LJ, Sowrey FE, Pickup DM, Newport RJ, Drake KO, Lin ZH, Smith ME, Saravanapavan P and Hench LL (2005a) The atomic-scale interaction of bioactive glasses with simulated body fluid. Mater Sci Forum 480-481:21-26

Spoerke ED, Stupp SI (2003) Colonization of organoapatite-titanium mesh by preosteoblastic cells. J Biomed Mater Res A 67:960-969

Sykes D, Kubicki JD (1996) Four-membered rings in silica and aluminosilicate glasses. Am Mineral 81:265-272

Takadama H, Kim H-M, Kokubo T, Nakamura T (2002) X-ray photoelectron spectroscopy study on the process of apatite formation on a sodium silicate glass in simulated body fluid. J Am Ceram Soc 85:1933–36

Tas AC (2000) Synthesis of biomimetic Ca-hydroxyapatite powders at 37 degrees C in synthetic body fluids. Biomaterials 21:1429-1438

Vallet-Regı M, Izquierdo-Barba I, Salinas AJ (1999) Influence of P_2O_5 on crystallinity of apatite formed *in vitro* on surface of bioactive glasses. J Biomed Mater Res 46:560-565

Vallet-Regı M, Perez-Pariente J, Izquierdo-Barba I, Salinas AJ (2000) Compositional variations in the calcium phosphate layer growth on gel glasses soaked in a simulated body fluid. Chem Mater 12:3770-3775

Vallet-Regı M, Ramila A (2000) New bioactive glass and changes in porosity during the growth of a carbonate hydroxyapatite layer on glass surfaces. Chem Mater 12:961-965

Vallet-Regı M, Romero AM, Ragel CV, LeGeros RZ (1999) XRD, SEM-EDS, and FTIR studies of *in vitro* growth of an apatite-like layer on sol-gel glasses. J Biomed Mater Res 44:416-421

Wallace S, West JK, Hench LL (1993) Interactions of water with trisiloxane rings. I. Experimental analysis. J Non-Cryst Solids 152:101-108

Weissbart EJ, Rimstidt JD (2000) Wollastonite: Incongruent dissolution and leached layer formation. Geochim Cosmochim Acta 64:4007-4016

West JK, Wallace S (1993) Interactions of water with trisiloxane rings. II. Theoretical analysis. J Non-Cryst Solids 152:109-117

Wheeler DL, Stokes KE (1997) *In vivo* evaluation of sol-gel Bioglass®. Part I: histological findings. Trans 23rd Annual Meeting of the Soc Biomater, New Orleans, LA

White EW, Weber JN, Roy DM, Owen EL, Chiroff RT, White RA (1975) Replamineform porous biomaterials for hard tissue implant applications. J Biomed Res Symp 6:23-27

Williams DF (ed) (1987) Definitions in Biomaterials. Elsevier, p 66

Williams DF, Cunningham J (1979) Materials in Clinical Dentistry. Oxford University Press

Wilson J, Nolletti D (1990) Bonding of soft tissues to Bioglass®. *In:* Handbook of Bioactive Ceramics. Vol. I. Yamamuro T, Hench LL, Wilson J (eds) CRC Press, p. 283

Wu C, Chang J (2004) Synthesis and apatite-formation ability of akermanite. Mater Lett 58:2415-2417

Xynos ID, Edgar A J, Buttery LDK, Hench LL, Polak JM (2001) Gene-expression profiling of human osteoblasts following treatment with the ionic products of Bioglass (R) 45S5 dissolution. J Biomed Mater Res 55:151-157

Xynos ID, Edgar AJ, Buttery LDK, Hench LL, Polak JM (2000b) Ionic products of bioactive glass dissolution increase proliferation of human osteoblasts and induce insulin-like growth factor II mRNA expression and protein synthesis. Biochem Biophys Res Commun 276:461-465

Xynos ID, Hukkanen MVJ, Batten JJ, Buttery LD, Hench LL, Polak JM (2000a) Bioglass® 45S5 stimulates osteoblast turnover and enhances bone formation *in vitro:* implications and applications for bone tissue engineering. Calcif Tissue Int 67:321–329

Yang H, Prewitt CT (1999) On the crystal structure of pseudowollastonite ($CaSiO_3$). Am Mineral 84:929-932

Yang J, Meng S, Xu L, Wang EG (2004) Ice tessellation on a hydroxylated silica surface. Phys Rev Lett 92: 146102-1–146102-4

Yang J, Meng S, Xu L, Wang EG (2005) Water adsorption on hydroxylated silica surfaces studied using density functional theory. Phys Rev B 71:35,35413-1– 35,35413-12

Zachariasen W (1932) Random Network Hypothesis. J Am Chem Soc 54:3841

Reviews in Mineralogy & Geochemistry
Vol. 64, pp. 315-332, 2006
Copyright © Mineralogical Society of America

Living Cells in Oxide Glasses

Jacques Livage and Thibaud Coradin*

Chimie de la Matière Condensée de Paris
CNRS-UMR 7574
Université Pierre et Marie Curie
4 place Jussieu, 75252 Paris cedex 05, France
*e-mail: coradin@ccr.jussieu.fr

INTRODUCTION

In the past few years, many efforts have been made to take advantage of the biological activities of living cells to design functional materials. Applications in biotechnology and biomedical devices include the development of biosensors, i.e., the detection of analytes through the specific response of cells, and biocatalysts or bioreactors, i.e., the transformation or production of specific medical molecules by cells. In many cases, living cells cannot be used as such and need to be stabilized via encapsulation in suitable host matrices. Although polymer-based matrices have long been used for such applications, recent studies indicate that mineral hosts, and more specifically oxide gels, may also be suitable for cell encapsulation. Not only can cell viability be maintained over weeks within such matrices, but inorganic materials present enhance chemical and mechanical stability when compared to organic networks. In this chapter, we will provide a brief history on the development of sol-gel chemistry, followed by a review of the most recent advances in methods for cell immobilization in oxide gels. We will also briefly compare these synthetic routes to naturally-occurring silica fossilization of cells and laboratory experiments mimicking fossilization.

According to the Roman historian Pliny, glasses were discovered by Egyptian sailors about 5000 years ago. They were making fire on a beach when they saw that the sand under the fire was melting, giving a kind of glassy material. Actually, some sodium carbonate rocks (a glass modifier) were mixed with the silica sand (a glass former) decreasing the melting temperature of the SiO_2-Na_2O eutectic below 1000 °C. Egyptian glasses were very rough materials, but much progress has been made since then. Large pieces of highly transparent glasses are now currently produced via the so-called "float-glass process" discovered about fifty years ago. However, glasses are still made from a mixture of sodium and calcium oxides with molten silica at temperatures well above 1000 °C.

In the last decades, great efforts have been devoted to the preparation of glasses and ceramics in milder conditions. New synthetic routes based on "sol-gel chemistry" have been developed (Brinker and Scherrer 1990). These routes use solutions as precursors instead of sand in order to favor the diffusion of ions and the intimate mixing of reactants. These wet chemistry methods save energy, allowing the synthesis of oxide materials at temperature much lower than with solid-state precursors.

Actually sol-gel chemistry is much more than a new process. It leads to the synthesis of novel hybrid nanocomposites in which organic and inorganic species are mixed at the molecular level (Sanchez and Ribot 1994). A whole class of new materials is formed, ranging from brittle glasses to plastic polymers. A wide variety of functional molecular systems, such as organic dyes, organometallic catalysts or metal nanoparticles can be added to the solution of precursors and trapped within the growing oxide network (Gomez-Romero and Sanchez 2004).

1529-6466/06/0064-0010$05.00

DOI: 10.2138/rmg.2006.64.10

In this context, the encapsulation of enzymes, antibodies and other proteins within silica gels has become a widely studied approach for the design of biocatalysts, biosensors and bioreactors (Avnir et al. 1994; Lin and Brown 1997; Gill and Ballesteros 2000; Livage et al. 2001; Jin and Brennan 2002; Avnir et al. 2006). Indeed, the industrial development of biotechnology usually requires the immobilization of active bio-species onto solid substrates. The immobilization of enzymes on porous glasses has already been developed for more than twenty years. Enzymes are covalently linked via organosilane coupling agents such as γ-ami nopropyltriethoxysilane, (APTS) $H_2N-(CH_2)_3-Si(OEt)_3$ (Weetall 85). However, such covalent attachments require chemical modifications that may affect the catalytic activity of enzymes. Therefore a physical encapsulation within a three-dimensional host is often highly desirable. Indeed, traditional routes to glass formation involve high temperature treatments that are not compatible with the preservation of biomolecules or cells upon encapsulation, so that organic or biopolymer gels are currently used as hosts for bio-immobilization. However, the mild conditions associated with the sol-gel route to ceramic materials could offer new possibilities in biotechnology. When compared to traditional polymer hosts, silica gels exhibit improved chemical stability and mechanical properties. The size and chemical nature of the pores can be tailored to avoid biomolecule leaching and favor their bioactivity. In fact, most entrapped enzymes retain their catalytic activity and appear to be protected against denaturation (Pierre 2004). Moreover, sol-gel materials can be obtained in a variety of forms (optically transparent glasses, films, fibers, microspheres,…) and thin films can be deposited onto most other materials (e.g., plastic, paper, metal, glasses.)

If a wide range of biomolecules has already been immobilized within sol-gel silica matrices, the encapsulation of whole cells has been much less studied. The main reason is that, being living organisms, cells are much more sensitive to their environment and the preservation of their bioactivity is much more challenging.

However, nature provides several examples indicating that conditions exist where silica can be formed in the presence of living organisms (Simpson and Volcani 1981). Actually, tons of biogenic "glasses" are made every day, by single-cell organisms such as diatoms or radiolarians, where the silica structures have precisely controlled morphologies. These micro-organisms are able to synthesize hydrated silica at room temperature from molecular silica precursors dissolved in the sea (Coradin and Lopez 2003). Thus, in the perspective of entrapping living cells in silica gels, it might be worth taking a closer look at the biosilicification processes taking place in some of these organisms.

SILICA AND LIVING ORGANISMS: THE EXAMPLE OF DIATOMS

Silica is one of the most widespread minerals deposited by living organisms. It has been found in microbes, algae, higher plants, insects and even mammalian tissues (Iler 1979, p 730). In most of these cases, the organisms benefit from the mechanical properties of the inorganic phase (as a protection against predators) as well as its transparency (allowing photosynthetic processes).

In this matter, the most striking example of silica-cell association is given by diatoms (Round et al. 1990). These algae build-up a silica shield, named frustule, of finely controlled morphology and porosity that encapsulates the whole unicellular organism (Fig. 1). This ability of diatoms to deposit silica to the extent of several hundred tons per year makes them major contributors to the Si biogeochemical cycle (Treguer et al. 1995)

Soluble silica is readily available in streams and seas as monomeric silicic acid $Si(OH)_4$, silicon complexes, silicate oligomers or even colloidal silica. However, to date, only silicic acid has been shown to be taken up by diatoms (Del Amo and Brezinski 1999). As silicic acid is accumulated within the living cells, condensation of silanol Si-OH groups can occur (Eqn. 1):

$$(HO)_3Si\text{-}OH + HO\text{-}Si(OH)_3 \Rightarrow$$
$$(HO)_3Si\text{-}O\text{-}Si(OH)_3 \quad (1)$$

Dimers, trimers and then larger oligomers are grown, leading to silica nanoparticles. At neutral pH or below, these particles tend to aggregate and form a gel (Iler 1979, p 239) (Fig. 2). In the case of diatoms, the silica network is grown in specific vesicles and in the presence of biopolymers that control the morphology of the deposited mineral phase (Sumper and Kröger 2004).

From this example, two points deserve to be kept in mind in designing silica gels for cell encapsulation. The first one is that, as far as we know now, naturally silicifying organisms use aqueous precursors. Indeed, silica can be easily made at the laboratory scale from aqueous silicate solutions. The formation of silica from such solutions is based on the condensation of solute precursors via pH modifications (Iler 1979). Ionic dissociation and hydrolysis occurs when a silicate salt is dissolved in water leading to silicic acid and its

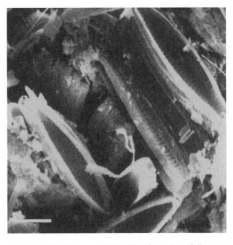

Figure 1. SEM image of the silica frustules of diatoms (scale bar: 10 μm).

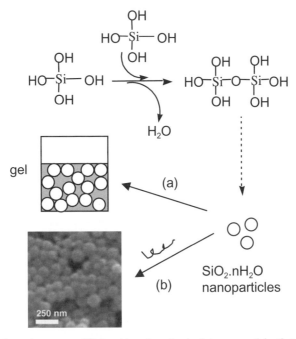

Figure 2. Silica formation process. Silicic acid condensation leads to nanoparticles that aggregate to form a gel (a) or can be assembled by polymers, shown by the "wiggle" above the arrow, forming aggregates (b).

deprotonated species, $[H_nSiO_4]^{n-4}$, depending on solution pH. Sodium silicate Na_2SiO_3 solutions (water glass) is a convenient commercial precursor for the synthesis of silica. Aqueous solutions are highly alkaline (pH $\approx$ 12). Neutral silicic acid precursors are formed by acidification of diluted solutions. This can be performed by adding an acid or, even better, via a proton exchange resin in order to remove most sodium ions (Bathia et al. 2000). However, this process is limited to diluted solutions in order to avoid the uncontrolled precipitation of silica. Colloidal silica particles can then be added to increase silica concentration and get stronger gels. (Finnie et al. 2000). Such sols then do not contain too much sodium counter-ions so that pH decrease can be more easily obtained by adding HCl (Coiffier et al. 2001)

The second point, and this is a common feature of biominerals, is that the frustule does not consist of a pure inorganic phase but is a hybrid structure of biopolymers associated with silica (Lopez et al. 2005). These biomolecules not only control the growth of the mineral network but also contribute to the biocompatibility of the shell. Thus, it could be anticipated that composite materials, rather than silica alone, will allow a good preservation of encapsulated cells.

BIOENCAPSULATION IN SOL-GEL SILICA

The first experiments showing that enzymes could be trapped within silica gels were reported by F.H. Dickey in the mid-fifties (Dickey 1955) and the immobilization of trypsin in silica gels was published about thirty years ago (Johnson and Whateley 1971). However, the interest of these papers was not realized at that time and most work on sol-gel bioencapsulation really started almost twenty years later (Carturan et al. 1989, Braun et al. 1990). Since then, a wide range of enzymes, antibodies and other biomolecules have been trapped within sol-gel matrices and several good review papers have been devoted to this subject over the years (Avnir et al. 1994; Lin and Brown 1997; Gill and Ballesteros 2000; Livage et al. 2001; Jin and Brennan 2002; Avnir et al. 2006).

Of interest here is the fact that the above-mentioned "aqueous" route to silica formation has not been often used in hybrid material synthesis as a whole, and for bio-encapsulation more particularly. The reaction is rather difficult to control and usually leads to the fast precipitation of silica. Therefore, chemists are currently working with other molecular precursors such as silicon alkoxides $Si(OR)_n$, where R is an organic group (Brinker and Scherrer 1990). This is the so-called "sol-gel process" in which silica is produced via the "inorganic polymerization" of molecular precursors. However, in this case, silanol groups must be obtained by hydrolysis of the Si-OR bond (Eqn. 2) before condensation (Eqn. 1 above) can occur.

$$(RO)_3Si\text{-}OR + H_2O \Rightarrow (RO)_3Si\text{-}OH + ROH \qquad (2)$$

This hydrolysis step leads to the release of ROH alcohol molecules that can be detrimental to cell viability, as we will see later.

The most common silicon alkoxide are TetraMethylOrthoSilicate (TMOS) $Si(OCH_3)_4$ and TetraEthylOrthoSilicate (TEOS) $Si(OC_2H_5)_4$, that are both commercially available. Silicon alkoxides are not miscible with water so a solvent has to be used. The parent alcohol is often chosen but a large variety of organic solvents can also be used. Silicon alkoxides are not very sensitive to hydrolysis, gelation taking place within several days when pure water is added. Therefore hydrolysis and condensation rates of silicon alkoxides are currently enhanced by acid or base catalysis in order to get silica gels within less than one hour. Catalysis not only increases the kinetics of the reaction, it also leads to silica of different morphology and/or structure. Basic catalysis gives rise to dense spherical colloids and mesoporous gels (average pore diameter in the 50-100 Å range). This process was developed by W. Stöber for the industrial production of monodispersed silica particles, 0.05 to 2 mm in diameter (Stöber 1968). Acid catalysis favours the formation of chain polymers that can aggregate to obtain microporous monolithic gels (average pore diameter < 50 Å). The chemical reactivity of silicon alkoxides can also be increased by nucleophilic activation using fluorides (n-Bu4NF, NaF) or amines (such as dimethylaminopyridine (DMAP) (Brinker and Scherrer 1990).

The gentle conditions associated with sol-gel chemistry are close to those of organic chemistry, allowing the synthesis of hybrid organic-inorganic compounds. However, they are not gentle enough for biological molecules such as proteins. The sol-gel process has then to be slightly modified in order to fit the requirements of biochemistry:

i)   Proteins are denatured by ethanol, but methanol, that has a polarity similar to water, is not too harmful. Neat TMOS is then taken as precursor and water is added directly without any alcohol as a co-solvent. Both non-miscible liquids are vigorously mixed (often via sonication). Some acid (HCl) is usually added in order to increase hydrolysis rates leading to the fast formation of fully hydrolyzed precursors $Si(OH)_4$ (Avnir and Kaufman 1987).

ii) Proteins are usually unstable outside a narrow pH range around pH 7. Therefore, they are kept in a buffered solution to which hydrolyzed precursors are added. The pH buffer being close to 7, basic conditions are obtained, condensation is quite fast and biomolecules are entrapped within the growing oxide network (Ellerby et al. 1992).

iii) As mentioned earlier, the hydrolysis of alkoxides leads to the formation of harmful alcohol by products. This problem can be partially overcome by evaporation or distillation of the alcohol from the hydrolyzed sol before addition of biomolecules (Ferrer et al. 2002). Nevertheless, starting alkoxides are usually not fully hydrolyzed so that alcohol can still be released upon condensation and aging in the presence of the immobilized species. An elegant strategy relies on the use of silicon alkoxides bearing a biocompatible alcohol, such as glycerol (Gill and Ballesteros 1998). A more straightforward approach involves the addition of molecular or macromolecular modifiers to the reaction media. Sugars, amino acids, synthetic polymers (poly(vinyl alcohol), poly(ethylene glycol), nafion, …) and natural macromolecules (chitosan, alginate, gelatin,..) have been added to silica precursors to form hybrid encapsulation matrices (Heichal-Segal et al. 1995; Brasack et al. 2000; Keeling-Tucker et al. 2000; Miao and Tan 2001; Kros et al. 2001, Schuleit and Luisi 2001, Brennan et al. 2003; Wu and Choi 2004). Apart from providing biocompatible media, such additives strongly modify the gel structure and its mechanical properties. Recent works by Brennan et al. also demonstrate that organically-modified alkoxide precursors bearing a sugar group covalently linked to Si improve the gel biocompatibility and reduce gel shrinkage upon drying (Sui et al. 2005).

In a way, these improvements of the traditional sol-gel process bring us back to the two parameters that were pointed out in the previous section when discussing the ability of diatoms to form silica, i.e., the suitability of purely aqueous media and the benefit of designing composite/hybrid materials. As shown hereafter, these two key parameters are even more crucial for the preservation of whole cell viability within sol-gel silica hosts.

SILICA GELS FOR WHOLE CELL ENCAPSULATION

Cell encapsulation in silica

The first example of whole cell encapsulation was reported 15 years ago (Carturan et al. 1989). A suspension of yeast cells (*Saccharomyces cerevisiae*) was added to TEOS in the presence of ethanol and water at low temperature and the mixture was deposited as a thin film by dip coating. The catalytic activity of an intracellular enzyme was maintained over six months. However, this pioneering approach was of only limited scope because yeasts are involved in fermentation processes so that they can withstand high concentrations of ethanol without loss of viability, in contrast to many other living organisms. Later, the same team patented and published a new process, named Biosil, that was successfully applied to plant and animal cells (Carturan et al. 2004). This process relies on the gas phase deposition of silica. A cell suspension is deposited on a natural polymer scaffold (silk, gelatin, alginic acid) and submitted to a vapor stream of volatile silicon alkoxides. Alkoxide hydrolysis and condensation takes place at the wet surface of the cells, while the released alcohol gas is driven away by the gas flow. By the same time, Pope et al. (1997) reported successful immobilization of yeast and animal cells in TEOS-based gels. The procedure involves low temperature and high pH conditions, allowing slow hydrolysis and evaporation of released ethanol before cells addition. Such gels could be shaped as microparticles and tested *in vivo* (see below).

Since then, several processes have been described that allow the preservation of whole cells in silica-based gels. Many of them can be considered as inspired by the Biosil process since they rely on the pre-encapsulation of the living organisms within a biopolymer gel that is then "silicified." These approaches are clearly illustrated by several works on silica/alginate hybrid materials (Coradin et al. 2003). Alginate beads are traditional hosts for cell encapsulation. Silica/alginate composites can be obtained by (a) interfacial polymerization by immersion of the pre-prepared alginate beads in TMOS-hexane solutions, resulting in the formation of silica within the hydrogel capsule (Heichal-Segal et al. 1995, Bressler et al. 2002), (b) same procedure except that the capsules were put in contact with APTS and then TMOS, leading to the formation of a silica coating on the bead surface (Sakai et al. 2001, 2002) and (c) mixing of the beads with a silicate/colloidal silica mixture followed by alginate gel liquefaction, the final composite consisting of a silica monolith with macrocavities (Perullini et al. 2005) (Fig. 3). In all these cases, the living cells that were previously encapsulated in the alginate capsules showed maintained viability and biological activity over long time periods within the hybrid material.

Finally, an alternative innovative route has been recently described where so-called biologically-modified ceramics (biocers) were designed using a freeze-cast technique,

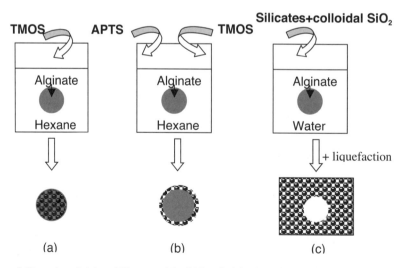

Figure 3. Examples of alginate/silica materials: (a) interfacial polymerization of TMOS leading to silica formation inside the capsule, (b) interfacial polymerization of APTS and TMOS leading to silica-coated capsule and (c) silicate/colloidal silica polymerization followed by liquefaction of the alginate capsule leading to macroporous gels. All capsules are 2-3 mm in diameter.

allowing the immobilization of several micro-organisms (Böttcher et al. 2004). This approach relies on the casting of a mixture of silica sols with ceramic powders, organic additives and biologicals in suitable moulds, followed by rapid freezing at –40 °C. The freezing step induces the irreversible transformation of the dispersion by the sol-gel transition. Thus the molding shape is retained at room temperature. Optimization of freezing conditions allows the control of both material porosity and cell viability.

Bacteria

Bacteria in silica gels. Within the different classes of living organisms that have been entrapped within sol-gel silica matrices, bacteria deserve special attention. An important reason is that several bacteria were shown to be silicified in natural conditions (Urrutia and Beveridge 1993; Westall et al. 1995; Konhauser et al. 2001)). Another aspect is that common strains such as *Escherichia coli* or *Bacillus subtilis* are considered as resistant cells and may therefore be more likely to support the constraints and stresses of encapsulation. Finally, the genetic engineering tools for bacteria have been largely developed and may be used to investigate in more details the physiological state of the encapsulated organisms.

Compared to other cells, the encapsulation of bacteria in silica gels is rather recent. *Escherichia coli* bacteria were trapped within TMOS-based matrices and the catalytic activity of an intracellular enzyme, β-galactosidase, was maintained over a few days (Fennouh et al. 1999). However, further investigations suggested that the alcohol released during the alkoxide hydrolysis was detrimental to bacteria survival, probably by dissolving phospholipids constituting the cell membrane and inducing bacteria lysis (Fig. 4a) (Coiffier et al. 2001). Solutions to this problem include (a) use of high hydrolysis ratios h = [water]/[alkoxide] that leads to highly diluted alcohol solutions (Conroy et al. 2000), (b) evaporation or distillation of the alcohol produced during hydrolysis prior to cell addition and gel condensation (Ferrer et al. 2002), (c) thin films deposition allowing rapid evaporation of the volatile alcohol (Premkumar et al. 2001).

However, a more straightforward approach relies on the use of aqueous silica precursors. The use of sodium silicate solutions as the source of silicic acid was first described for enzyme encapsulation (Liu and Chen 1999; Bathia et al. 2000) but entrapment of (methanotrophic) bacteria in such matrices was only recently reported (Chen et al. 2004). Silica nanoparticles were successfully used for the immobilization of sulfate-reducing bacteria (Finnie et al. 2000).

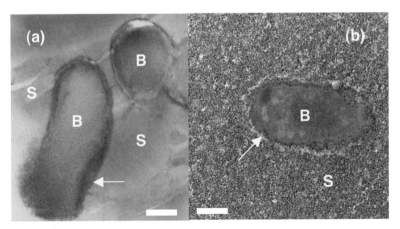

Figure 4. TEM images of an *E. coli* bacteria encapsulated in (a) an alkoxide-based silica matrix showing cell wall lysis and intra-cellular material leaching out (b) in a silicate/colloidal silica matrix showing good preservation of the membrane integrity (scale bars: 500 nm). B = bacteria; S= silica; the arrow indicate the cell membrane/silica interface.

In a further step, we have investigated mixtures of sodium silicate and colloidal silica for the encapsulation of *Escherichia coli* environmental bacteria (Coiffier et al. 2001). In this case, the silica gel consists of a colloid network cemented by condensed silicate species (Fig. 4b).

Bacteria in hybrid silica gels. Although the detrimental effect of alcohol could be avoided by using aqueous precursors, our study of *E. coli* viability in silicate/colloidal matrices revealed that only 60% of the encapsulated bacteria remained alive one hour after encapsulation (Nassif et al. 2003). This suggested that the gel formation process had a strong influence on bacteria behavior. Because experimental conditions (pH, ionic strength, temperature) were previously checked for compatibility with *E. coli* survival, it was inferred that the inorganic network densification could impose mechanical constraints on the cell and/or that some unfavorable interactions exist between the silica surface and the bacteria membrane. Moreover, the *E. coli* viability rate decreased to 20% after two weeks and less than 10% after one month, indicating that other effects should arise in the longer term. Of particular importance was the fact that no nutrients were provided to the bacteria over these periods, so that they may have entered a starvation phase.

In order to enhance bacteria viability, three additives were selected (Nassif et al. 2003): gelatin, a biocompatible protein that may be used as a nutrient by the cells; poly(vinyl alcohol) (PVA), an enzyme-stabilizing polymer that can not be metabolized by the bacteria; and glycerol, a biocompatible molecule exhibiting osmoprotective properties that can be metabolized by *E. coli* only if nitrogen sources are also provided. When incorporated in the silica precursor solution, PVA and gelatin increase the viability rate to 50% after two weeks but only 20% after one month. In contrast, up to 50% of initial bacteria remain alive in the silica-glycerol gel after one month. Additionally, this additive maintains a 80% viability rate after one hour whereas the two others only maintain 60%, similarly to the silica gel alone. Thus, it appears that neither gelatin nor PVA are able to efficiently protect bacteria against the stress induced by the gel formation. Additionally, their effect on the long-term survival of entrapped cells is limited. In contrast, glycerol shows a beneficial role on both short and long terms (Nassif et al. 2002).

If the additives were to play a protective role for the cell upon gel formation, they should be located in its vicinity. The localization of gelatin, PVA and glycerol within the inorganic matrix was therefore investigated (Nassif et al. 2003). ^{29}Si solid state NMR studies were first performed to investigate the condensation rate of the gels. No significant difference could be observed between pure and hybrid matrices. Thus, a better structural characterization could be obtained using N_2 sorption measurements. Interestingly, PVA- and gelatin-containing matrices exhibited similar features. These gels are mesoporous, as expected for silica network formed at neutral pH. In the case of glycerol, however, both specific surface area and porous volume were significantly lower. Average pore size was similar to the other matrices but analysis of pore size distribution showed a marked decrease in the smaller pore contribution. These data suggest that whereas PVA and gelatin are located in the macroporosity of the gel, glycerol may fill the smallest pores of the inorganic network. Such a difference in localization of the additives was, at this time, attributed to the macromolecular nature of PVA and gelatin that may not favor their incorporation in the micro/mesoporosity of the host matrix, in contrast to the molecular size of glycerol.

The encapsulation of another bacteria strain, *Serratia marcescens* was also studied (Nassif et al. 2004). Once again, about 50% of the initial bacteria population was maintained over one month in a glycerol-containing silica gel, confirming the protective effect of this additive. In order to get a better understanding of this effect, ethylene-glycol was substituted for glycerol in the gel formation (Nassif N, Giustiniani J, Roux C, Bouvet O, personal communication). This molecule is very close to glycerol in structure, but cannot be metabolized by the bacteria. After 15 days, less than 30% of encapsulated *E. coli* were still living in such matrices (Fig. 5).

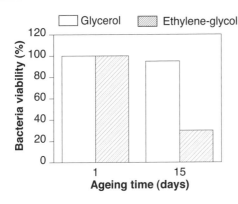

Figure 5. Evolution of the viability of *E. coli* bacteria encapsulated in a silicate/colloidal silica matrix in the presence of glycerol and ethylene-glycol, as determined by fluorescence.

Additionally, N_2 sorption studies showed that these gels exhibited specific surface areas, porous volume and pore size similar to pure silica and PVA/gelatin-silica hybrids. It therefore seems that the localization of the additive is a key factor in its stabilizing effect. However, because glycerol may be a metabolite for bacteria whereas ethylene-glycol is not, it is difficult at this time to evaluate whether the difference in bacteria viability depends only on a structural parameter or if additional biological effects are involved.

A new step towards the understanding of the effect of glycerol was very recently reported by del Monte et al. using cryo-electron microscopy (Ferrer et al. 2006). Using TEOS as a precursor for silica formation, they show that a direct contact between *E. coli* cell wall and the silica pore surface was observed in the absence of glycerol whereas a cavity was found surrounding the encapsulated cell in the presence of the additive. In the meantime, glycerol enhances the viability of the entrapped bacteria, suggesting that glycerol act as a physical barrier that avoids silica/bacteria interactions.

Post-encapsulation additives. Additives can be incorporated not only in the silica precursor solution to limit the detrimental effect of gel condensation on bacterial cell wall stability, but can also be added once the matrix is formed to influence the activity of encapsulated cells. In contrast to suspension, or even traditional biodegradable polymer hosts, bacteria entrapped in silica gels are isolated one from another and cannot divide. These matrices therefore represent a very unusual media to which cell have to adapt their metabolism in order to survive. It is well-known that the behavior of a bacteria population is governed by inter-cellular communication mediated by chemical signals, known as quorum sensing (QS) molecules (Lazazzera 2000). The effect of such molecules to cell-containing gels was therefore studied in order to check the possibility to control bacteria behavior by external additives.

S. marcescens bacteria were selected because they naturally synthesize a red pigment, prodigiosin, whose production can be easily monitored by UV-visible spectrophotometry. Furthermore, prodigiosin exhibits promising therapeutic properties so that its efficient production may represent a wide interest in a near future (Castro 1967). Two QS molecules, belonging to the family of N-acylated homoserine lactones, were added to the supernatant of glycerol-containing gels entrapping bacteria (Nassif et al. 2004). A bacteria suspension with or without added QS molecules was used as a reference. In the latter case, the amount of prodigiosin in the suspension increased rapidly in the first two days and remained constant over one week, with no visible influence of QS addition. For the encapsulated cell, the pigment concentration in the gel supernatant was lower than the reference for the first 24 hours but then increased linearly, and a plateau was not reached after one week (Fig. 6). The first period probably corresponds to the diffusion of prodigiosin from the gel, where it is produced, to the supernatant. More interestingly, the fact that no threshold is observed indicates continuous production of the pigment. In

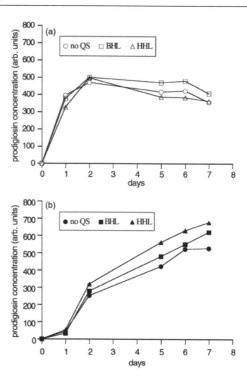

Figure 6. Evolution of prodigiosin concentration in (a) a bacteria suspension and (b) the supernatant solution of a bacteria-containing gel over one week in the absence (no QS) or presence of *N*-butanoyl-L-homoserine lactone (BHL) and *N*-hexanoyl-L-homoserine lactone quorum sensing molecules [from Nassif et al. 2004].

fact, such biosynthetic processes are often limited by retro-inhibition mechanisms so that, once a critical concentration of the product is reached in the cell-surrounding media, its production is stopped. In the case of entrapped cells, prodigiosin diffuses out of the gel to the supernatant and does not remain at the vicinity of the bacteria so that the retro-inhibition process is not triggered. Moreover, an increase of the final prodigiosin concentration was observed after one week when QS molecules were initially added to the supernatant

Such an effect was confirmed when recycling experiments were performed (Nassif et al. 2004). After four subsequent steps of seven days each were performed, the prodigiosin amount was about 70% of the initial production in the presence of QS molecules, whereas only 50% was maintained in their absence. Additionally, the viability of entrapped *S. marcescens* at this time was about 80% and 40% respectively. Thus, QS molecules appear to maintain the entrapped bacteria in a better physiological state, leading to the enhancement of prodigiosin production.

APPLICATIONS OF SOL-GEL ENCAPSULATED LIVING CELLS

Biosensors, bioreactors

Bioremediation. Bacteria are known to be able to bind selectively large amounts of metals. They can therefore be used in remediation technologies for the removal of heavy metals from polluted waters (Al-Saraj et al. 1999). Cells, spores and surface-layer (S-layer) protein of *Bacillus sphaericus* bacteria have been trapped within sol-gel ceramics (biocers) for the in situ bioremediation of uranium mining waste pile waters (Raff et al. 2003). Biocers were made by dispersing vegetative cells, spores and S-layer proteins in aqueous silica sols. Entrapment does not influence the metal binding properties of cells and S-layers that have been shown to exhibit high binding capacity toward uranium and copper. However, spores loose most of their activity upon encapsulation. Uranium and copper can be easily removed from the bioceramic by using citric acid so that biocers appear to be suitable for the realization of reversible filters.

In a similar context, bacteria are widely used for the removal of organic pollutants through biological degradation. In a recent work, Pedrazzani et al. (2005) immobilized biomass extracted from a wastewater sludge treatment plant between polyester membranes. These membranes were further coated with nitrocellulose and various inorganic particles such as SiO_2, TiO_2 or ZnS. These coatings allow a better control of the membrane porosity, favoring substrate diffusion while avoiding entrapped cells leaching.

Bioproduction. The immobilization of whole cells within stabilizing hosts allows the use of their metabolic processes for the synthesis of specific molecules. In this case, immobilization approaches facilitate the separation between the active cells and their bio-products as the solid support can be easily withdrawn from the reaction media by filtration. This allows a better recovery of the synthesized molecules and favor the recycling of the bioreactors. Experiments performed with *Pseudomonas* sp. bacteria showed that they retain their ability to metabolise atrazine, a widely used herbicide (Rietti-Shati et al. 1996). Entrapped cells lose much of their activity upon immobilization but partial activity could be restored by adding nutrients, suggesting that bacteria may remain alive, at least for some time. The formation of acetate was observed when lactate was added to a gel containing anaerobic sulphate-reducing bacteria providing evidence of the metabolic conversion of lactate to acetate (Finnie et al. 2000). Encapsulated bacteria survive the gelation procedure and are able to continue normal metabolic activity within the gel matrix. This activity decreases with time but can be regenerated by immersion in nutrient solution, even after several weeks. Methanotrophic bacteria encapsulated in silicate gels could be used in a batch reaction system for propylene epoxidation over 25 times without loss of activity (Chen et al. 2004). A recent report also indicates that recombinant-protein-producing *E. coli* bacteria could be immobilized in silicate/colloidal silica gel, resulting in the efficient production of several molecules involved in immune responses such as T-cell receptors and super antigens (Desimone et al. 2005).

Saccharomyces cerevisiae yeast cells were also successfully entrapped in hybrid alginate/ silica capsules obtained by TEOS impregnation, with an increase in fermentation efficiency with silica content (Heichal-Segal et al. 1995). Encapsulated yeast cells were also used for the design of a supported liquid membrane bioreactor for the conversion of fumaric acid to L-Malic acid (Bressler et al. 2002). More recently, hepatoblastoma HepG2 and T-lukemia Jurkat cells were shown to maintain their protein biosynthesis activity within silica/alginate beads obtained via the Biosil process (Boninsegna et al. 2003b).

Biosensing. Genetically engineered luminescent *Escherichia* were obtained recently by coupling a gene promoter sensitive to chemical or physical stress with a reporter gene coding for two luminescent proteins, namely, green and red fluoresecent protein (GFP, RFP) (Premkumar et al. 2001, 2002a). These recombinant cells were trapped in sol-gel silica films deposited on glass plates. The cells appear to maintain their ability to synthesize luminescent proteins in the presence of chemical inducers over months, either through repeated uses or under continuous flow. The stress-dependent luminescence properties of these cells provide information about their state during the sol-gel process and within the silica gels. The stress-induced luminescence changes were then used to optimize the sol-gel procedure (pH, water/ TMOS ratio, drying time, thickness of the films). Moreover, luminescent cells can be observed by confocal microscopy showing that bacteria were homogeneously distributed within the film with limited aggregation. Neither cell proliferation nor leaching could be observed after several days. Finally, the possibility to entrap simultaneously cells from two different strains within the same gel opens new possibility for dual or multiple sensing (Premkumar et al. 2002b; Sagi et al. 2003).

In most cases, a major challenge is to maintain entrapped cells alive. Such viability is not always needed, however, as demonstrated for sol-gel based immunoassays. In this case, the targeted sites for antigen-antibody recognition are located on the cell surface so that bioactivity is not required. Thus, whole cell parasitic protozoa (*Leishmania donovani infantum*) could be trapped within TMOS-based silica gels (Fig. 7) and were successfully tested against the blood serum of infected patients (Livage et al. 1996). The encapsulation of cells could be performed directly within the wells of a microtitre plate, allowing the Enzyme Linked ImmunoSorbent Assay (ELISA) that compares well with current medical procedures.

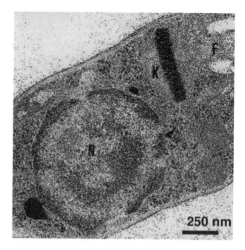

Figure 7. TEM image of encapsulated *Leishmania* cells in silica gels. N = nucleus, F = flagellum, K = kinetoplast

Artificial organs

Artificial organs are biomedical devices where living cells are surrounded by a semipermeable membrane that allow the diffusion of small molecules, such as oxygen, nutrients and metabolites, while avoiding the contact between the entrapped cells and larger species, more especially vectors of the immune response (Uludag et al. 2000). The most popular material for such cell microencapsulation consists of alginate, a poly-saccharide, capsules coated with poly-L-lysine (PLL). The coating of PLL on a pre-formed alginate capsule induces the formation of a polyelectrolyte membrane.

First attempts to use a silica capsule for Langerhans islets encapsulation were reported by Pope et al. (1997). Islets of Langerhans from mice could be encapsulated in silica gel microspheres, using a patented modification of the Ellerby procedure (Ellerby et al. 1992). Insulin secretory capacity of encapsulated cells could be observed *in vitro* for 3 weeks. After one month, removed capsules showed no evidence of fibrosis.

However, this approach was not explored much further. Instead, the use of hybrid materials was examined in order to obtain enhanced chemical and mechanical stability of the PLL/alginate membrane as well as to get a better control of its porosity and hence of its diffusion properties. Silica coating of alginate/PLL beads was first achieved by Coradin et al. (2001) using sodium silicate diluted solutions. In parallel, Sakai et al. proposed a new procedure involving APTS (Sakai et al. 2001). For this precursor, $Si(OCH_3)_3$ groups are available for hydrolysis/condensation while the cationic amino groups can interact with alginate. Moreover, all amino groups are not involved in the binding of silica with the bead surface so that the deposition of an additional alginate layer is possible, leading to a better biocompatibility of the capsule. *In vitro* experiments show that insulin secretion by entrapped cells in response to stimulation by a high glucose level was similar to islets encapsulated in pure alginate beads, suggesting that the encapsulation process, and especially the alcohol released by the hydrolysis process, was not detrimental to cell viability. In a step further, the encapsulated islets were transplanted in diabetic male mice (Sakai et al. 2002) (Fig. 8). Normaglycaemia (i.e., normal glucose concentration in blood) was established after 24 h and was maintained over 3 months, showing that the insulin-secretion capability of encapsulated cells was preserved over this period. These hybrid capsules were also used for the proliferation and insulin secretion function of genetically-engineered pancreatic β-cell line cells both *in vitro* and *in vivo* (Sakai et al. 2003, 2004).

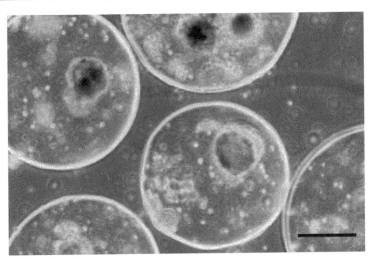

Figure 8. Micrograph of encapsulated Langerhans islets in alginate/silica microcapsules (scale bar: 200 μm) [Used by permission of Elsevier, from Sakai et al. (2002), *Biomaterials*, Vol. 23, Fig. 6, p 4182].

The Biosil process was applied to design bioartificial livers from hepatocytes sandwiched between two collagen thin layers (Muraca et al. 2002). *In vitro* evaluation showed that cell viability was not affected by the entrapment procedure. Several metabolic activities of hepatocytes, such as ammonia removal and urea production, were also maintained. Following the same procedure, porcine islets could be encapsulated and transplanted in diabetic dogs, allowing insulin production to be maintained over 5 weeks (Carturan et al. 2004). The Biosil process was also investigated for the direct deposition of silica on pancreatic islets, i.e., without the pre-immobilization of the cells on a biopolymer scaffold. The formation of 0.1-2 μm thick layer of silica on islets aggregates was successfully achieved (Boninsegna et al. 2003a). *In vitro* insulin production and post-transplantation glycaemia regulation was maintained over 2 months.

COMPARISON WITH INDUCED SILICA MINERALIZATION

Diatoms or radiolarians are considered as the exemplars of controlled silica biomineralization due to the astonishing degree of complexity of the silica shells elaborated by these organisms. On the other hand, induced silica biomineralization also occurs in nature, where there is no control by the organism on the silica formed. Thus, silicified microbial species are found as fossils, for instance in Precambrian stromatolites, or isolated as living cells, especially in hot spring environments (Konhauser et al. 2001). In most cases, silica is deposited as a thin crust, 0-1-1 μm in thickness, often associated with iron (Westall et al. 1995; Asada and Tazaki, 2001). This silica layer could act as a protection against acidic conditions of the cell environment (Asada and Tazaki, 2001) but may also provide an ultra-violet shield (Phoenix et al. 2001). The latter argument suggests that induced silica mineralization may have contributed to the preservation of living organisms in the Archean.

As the geothermal waters are supersaturated with respect to amorphous silica, the silica formation process is expected to be abiologically regulated. This assumption was confirmed by silicification experiments performed in laboratory conditions (Fein et al. 2002; Phoenix et al. 2003; Benning et al. 2004. However, these simulations also suggest the existence of specific interactions between cell wall and silica precursors that (i) control the morphology of the deposited silica and (ii) prevent the occurrence of an intra-cellular silica formation that would lead to cell death (Urrutia and Beveridge 1993; Phoenix et al. 2000; Benning et al. 2004) .

It is legitimate to wonder if the above-described experiments of cell encapsulation in silica gels bear some relevance for the understanding of induced silica mineralization. In principle, this approach permits the investigation of the behavior of living organisms in mineral environments. However, a major difference, between cell encapsulation studies described above on the one hand and induced silica mineralization in nature and in simulation experiments on the other hand, lies in the concentration of silica precursors $[SiO_2]$ involved in these studies. Whereas encapsulation is performed with $[SiO_2] > 1$ M (Coiffier et al. 2001), experimental simulation involves $[SiO_2] \approx 1$-10 mM (Urrutia and Beveridge 1993; Westall et al. 1995; Phoenix et al. 2000). As a result, gel formation in the former case occurs within a few minutes and results in a dispersion of cells within a silica network. In contrast, simulated induced silica mineralization is usually studied over week periods and leads to silica-coated cells. As a consequence, TEM analyses often reveal an un-mineralized space between silica and the organism in encapsulation studies, whereas deposition of mineral particles on the outer sheath of the cell is observed in silicification experiments, as illustrated in Figure 9. Thus, encapsulation approaches does not reflect the natural conditions of silicification and may be of limited interest to mimic the associated processes. However, they may be relevant to study the effect of a mineral environment on the long-term viability of micro-organisms. For instance, *E. coli* and *S. marcescens* encapsulation studies have demonstrated a modification of metabolic pathways of the cells in order to survive the gel conditions (Nassif et al. 2003, 2004), in resonance with some observations on the long-term viability of deeply-buried cyanobacteria found in stromatolites (Konhauser et al. 2001).

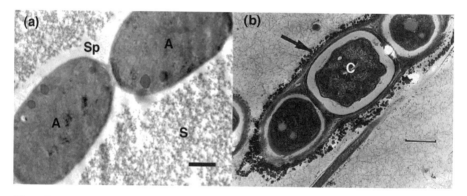

Figure 9. Comparison of (a) *Anabena* sp. encapsulated in silica gel (scale bar: 500 nm) (S = silica, A = *Anabena* cell, Sp = space) [courtesy M. Hemadi, CMCP] and (b) *Calothrix* sp. in silicification experiments (scale bar: 2 μm) (C = *Calothrix* cell, the arrow shows the silica deposit at the cell surface) [Used by permission of Elsevier, from Phoenix et al. (2000), *Chemical Geology*, Vol. 169, Fig. 6, p. 335].

PERSPECTIVES

The development of new biotechnological devices largely relies on the capacity of material scientists to adapt their concepts and techniques to the living world. In this matter, "soft chemistry" processes are probably some of the most promising as the conditions for inorganic phase formation can be made very similar to natural processes of biomineralization.

In the case of cell encapsulation, various inorganic precursors, temperature conditions, shapes (bulk, thin films, capsules) and organic additives have been successfully used. The example of bacteria shows that the possibility to maintain cell viability relies on the control of physical (mechanical constraint during network formation), chemical (cell-surface interactions) and biological (adaptation of cell metabolism) factors.

The fact that bacteria entrapped within a silica gel can synthesize pharmaceutical molecules and the promising properties of alginate/silica capsules as hosts for the development of artificial organs indicate that inorganic and hybrid matrices can now be considered as promising alternative to (bio)-organic polymers for the design of cell-based biotechnological and biomedical devices.

In addition, encapsulation in mineral hosts provides an alternative method to study the behavior of micro-organisms in geological environments, that may be relevant to get a better understanding of biosilicification processes.

ACKNOWLEDGMENTS

The authors thank C. Roux, N. Nassif, M. Hemadi, C. Gautier J. Giustiniani (CMCP), O. Bouvet (INSERM), P. J. Lopez (ENS), A. Couté (MNHN) and M. N. Rager (ENSCP) for their contribution to the work presented here. We also thank three anonymous reviewers for constructive comments.

REFERENCES

Al-Saraj M, El-Nahal I, Baraka R (1999) Bioaccumulation of some hazardous metals by sol–gel entrapped microorganisms. J Non-Cryst Solids 248:137-140
Asada R, Tazaki K (2001) Silica biomineralization of unicellular microbes under strongly acidic conditions. Can Mineral 39:1-16
Avnir D, Braun S, Lev O, Ottolenghi M (1994) Enzymes and other proteins entrapped in sol-gel materials. Chem Mater 6:1605-1614
Avnir D, Coradin T, Lev O, Livage J (2006) Recent bio-applications of sol-gel materials. J Mater Chem 16: 1013-1030
Avnir D, Kaufman VR (1987) Alcohol is an unnecessary additive in the silicon alkoxide sol-gel process. J Non-Cryst Solids 192:180-182
Benning LG, Phoenix VR, Yee N, Konhauser KO (2004) The dynamics of cyanobacterial silicification: An infrared micro-spectroscopic investigation. Geochim Cosmochim Acta 68:743-757
Bhatia RB, Brinker CJ, Gupta AK, Singh AK (2000) Aqueous sol-gel process for protein encapsulation. Chem Mater 12:2434-2441
Boninsegna S, Bosetti P, Carturan G, Dellagiacoma G, Dal Monte R, Rossi M (2003a) Encapsulation of individual pancreatic islets by sol-gel SiO$_2$: a novel procedure for perspective cellular grafts. J Biotechnol 100:277-286
Boninsegna S, Dal Toso R, Dal Monte R, Carturan G (2003b) Alginate microspheres loaded with animal cells and coated by a siliceous layer. J Sol-Gel Sci Technol 26:1151-1157
Böttcher H, Soltmann U, Mertig M, Pompe W (2004) Biocers: ceramics with incorporated microorganisms for biocatalytic, biosorptive and functional materials development. J Mater Chem 14:2176-2188
Brasack I, Böttcher H, Hempel U (2000) Biocompatibility of modified silica-protein composite layers. J Sol-Gel Sci Technol 19:479-482
Braun S, Rappoport S, Zusman R, Avnir D, Ottolenghi M (1990) Biochemically active sol-gel glasses: the trapping of enzymes. Mater Lett 10:1-8
Brennan JD, Benjamin D, DiBattista E, Gulcev MD (2003) Using sugar and amino acid additives to stabilize enzymes within sol-gel derived silica. Chem Mater 15:737-745
Bressler E, Pines O, Goldberg I, Braun S (2002) Conversion of fumaric acid to L-malic by sol-gel immobilized *Saccharomyces cerevisiae* in a supported liquid membrane bioreactor. Biotechnol Progress 8:445-450
Brinker CJ, Scherrer G (1990) The Physics and Chemistry of Sol-gel Processing. Academic Press
Carturan G, Campostrini R, Diré S, Scardi V, de Alteris E (1989) Inorganic gels for immobilization of biocatalysts: inclusion of invertase-active whole cells of yeast (*Saccharomyces cerevisiae*) into thin layers of SiO$_2$ gel deposited on glass sheets. J Mol Catal 57:L13-L16
Carturan G, Dal Toso R, Boninsegna S, Dal Monte R (2004) Encapsulation of functional cells by sol-gel silica: actual progress and prespectives for cell therapy. J Mater Chem 14:2087-2098
Castro AJ (1967) Antimalarial activity of prodigiosin. Nature 213:903-904

Chen J, Xu Y, Xin J, Li S, Xia C, Cui J (2004) Effcient immobilization of whole cells of *Methylomonas* sp. strain GYJ3 by sol-gel entrapment. J Mol Catal B 30:167-172

Coiffier A, Coradin T, Roux, C, Bouvet OMM, Livage J (2001) Sol-gel encapsulation of bacteria: a comparison between alkoxide and aqueous routes. J Mater Chem 11:2039-2044

Conroy JFT, Power ME, Martin J, Earp B, Hostica B, Daitch CE, Norris PM (2000) Cells in Sol-Gels I: A cytocompatible route for the production of macroporous silica gels. J Sol-Gel Sci Technol 18:269-283

Coradin T, Lopez PJ (2003) Biogenic silica patterning: simple chemistry or subtle biology? ChemBioChem 4:251-259

Coradin T, Mercey E, Lisnard L, Livage J (2001) Design of silica-coated microcapsules for bioencapsulation. Chem Commun 2496-2497

Coradin T, Nassif N, Livage J (2003) Silica/alginate composites for microencapsulation. Appl Microbiol Biotechnol 61:429-434

Del Amo Y, Brezinski MA (1999) The chemical form of dissolved Si taken up by marine diatoms. J Phycol 35:1162-1170

Desimone MF, DE Marzi MC, Copello GJ, Fernandez MM, Malchiodi EL, Diaz LE (2005) Efficient preservation in a silicon oxide matrix of *Escherichia coli*, producer of recombinant proteins. Appl Microbiol Biotechnol 68:747-752

Dickey FH (1955) Specific adsorption. J Phys Chem 59:695-707

Ellerby LM, Nishida CR, Nishida F, Yamanaka SA, Dunn B, Valentine JS, Zink JI (1992) Encapsulation of proteins in transparent porous silicate glasses prepared by the sol-gel method. Science 255:1113-1115

Fein JB, Scott S, Rivera N (2002) The effect of Fe on Si adsorption by *Bacillus subtilis* cell walls: insights into non-metabolic bacterial precipitation of silicate minerals. Chem Geol 182:265-273

Fennouh S, Guyon S, Jourdat C, Livage J, Roux C (1999) Encapsulation of bacteria in silica gels. C R Acad Sci IIc 2:625-630

Ferrer ML, Del Monte F, Levy D (2002) A novel and simple alcohol-free sol-gel route for encapsulation of labile proteins. Chem Mater 14:3619-3621

Ferrer ML, Garcia-Carvajal ZY, Yuste L, Rojo F, del Monte F (2006) Bacteria viability in sol-gel materials revisited: Cryo-SEM as a suitable tool to study the structural integrity of encapsulated bacteria. Chem Mater 18:1458-1463

Finnie KS, Bartlett JR, Woolfrey JL (2000) Encapsulation of sulfate-reducing bacteria in a silica host. J Mater Chem 10:1099-1101

Gill I, Ballesteros A (1998) Encapsulation of biologicals within silicate, siloxane, and hybrid sol-gel polymers: an efficient and generic approach. J Am Chem Soc 120:8587-8598

Gill I, Ballesteros A (2000) Bioencapsulation within synthetic polymers (Part 1): sol-gel encapsulated biologicals. Trends Biotechnol 18:282-296

Gomez-Romero P, Sanchez C (eds) (2004) Functional Hybrid Materials. Wiley-VCH

Heichal-Segal O, Rappoport S, Braun S (1995) Immobilization in alginate-silicate sol-gel matrix protects beta-glucosidase against thermal and chemical denaturation: enzyme stabilization for use in e.g. wine aroma improvement. Biotechnol 13:798-800

Iler RK (1979) The Chemistry of Silica. J. Wiley & Sons Pub

Jin W, Brennan JD (2002) Properties and applications of proteins encapsulated within sol-gel derived materials. Anal Chim Acta 461:1-36

Johnson P, Whateley TL (1971) On the use of polymerizing silica gel systems for the immobilization of trypsin. J Colloid Interf Sci 37:557-563

Keeling-Tucker T, Rakic M, Spong C, Brennan JD (2000) Controlling the materials properties and biological activity of lipase within sol-gel derived bioglasses via organosilane and polymer doping. Chem Mater 12:3695-3704

Konhauser KO, Phoenix VR, Bottrell SH, Adams DG, Head IM (2001) Microbial-silica interactions in Icelandic hot spring sinter: possible analogues for some Precambrian siliceous stromatolites. Sedimentology 48:415-433

Kros A, Gerritsen M, Sprakel VSI, Sommerdjik NAJM, Jansen JA, Nolte RJM (2001) Silica-based hybrid materials as biocompatible coatings for glucose sensors. Sensors Actuators B 81:68-75

Lazazzera BA (2000) Quorum sensing and starvation: signals for entry into stationary phase. Curr Opin Microbiol 3:177-182

Lin J, Brown CW (1997) Sol-gel glass as a matrix for chemical and biochemical sensing. Trends Anal Chem 16:200-211

Liu DM, Chen JW (1999) Encapsulation of protein molecules in transparent porous silica matrices via an aqueous colloidal sol-gel process. Acta Mater 47:4535-4544

Livage J, Coradin T, Roux C (2001) Encapsulation of biomolecules in silica gels. J Phys Condens Matter 13:R673-R691

Livage J, Roux C, Da Costa JM, Desportes I, Quinson JY (1996) Immunoassays in sol-gel matrices. J Sol-Gel Sci Technol 7:45-51

Lopez PJ, Gautier C, Livage J, Coradin T (2005) Mimicking biogenic silica nanostructures formation. Curr Nanoscience 1:73-83

Miao Y, Tan SN (2001) Amperometric hydrogen peroxide biosensor with silica sol–gel/chitosan film as immobilization matrix. Anal Chim Acta 437:87-93

Muraca M, Vilei MT, Zanusso GE, Ferraresso A, Boninsegna S, Dal Monte R, Carraro P, Carturan G (2002) SiO₂ Entrapment of animal cells: liver-specific metabolic activities in silica-overlaid hepatocytes. Artificial Organs 26:664-669

Nassif N, Bouvet O, Rager MN, Roux C, Coradin T, Livage J (2002) Living bacteria in silica gels. Nature Mater 1:42-44

Nassif N, Roux C, Coradin T, Rager MN, Bouvet OMM, Livage J (2003) A sol-gel matrix to preserve the viability of encapsulated bacteria. J Mater Chem 13:203-208

Nassif N, Roux C, Coradin T, Rager MN, Bouvet OMM, Livage J (2004) Bacteria quorum sensing in silica matrices. J Mater Chem 14:2264-2268

Pedrazzani R, Bertanza G, Maffezzoni C, Gelmi M, Manca N, Depero LE (2005) Bacteria enclosure between silica-coated membranes for the degradation of organic compounds in contaminated water. Water Res 39:2056-2064

Perullini M, Jobbagy M, Soler-Illia GJAA, Bilmes SA (2005) Cell growth at cavities created inside silica monoliths synthesized by sol-gel. Chem Mater 17:3806-3808

Phoenix VR, Adams, DG, Konhauser KO (2000) Cyanobacterial viability during hydrothermal biomineralisation. Chem Geol 169:329-338

Phoenix VR, Konhauser KO, Adams DG, Bottrell SH (2001) Role of biomineralization as an ultraviolet shield: Implications for Archean life. Geology 29:823-826

Phoenix VR, Konhauser KO, Ferris FG (2003) Experimental study of iron and silica immobilization by bacteria in mixed Fe-Si systems: implications for microbial silicification in hot springs. Can J Earth Sci 40:1669-1678

Pierre AC (2004) The Sol-Gel Encapsulation of Enzymes. Biocatal Biotransform 22:145-170

Pope EJA, Braun K, Peterson CM (1997) Bioartificial organs I: silica gel encapsulated pancreatic islets for the treatment of Diabetes Mellitus. J Sol-Gel Sci Technol 8:635-639

Premkumar JR, Lev O, Rosen R, Belkin S (2001) Encapsulation of luminous recombinant *E. coli* in sol-gel silicate films. Adv Mater 13:1773-1775

Premkumar JR, Rosen R, Belkin S, Lev O (2002a) Sol–gel luminescence biosensors: Encapsulation of recombinant E. coli reporters in thick silicate films. Anal Chim Acta 462:11-23

Premkumar JR, Sagi E, Rosen R, Belkin S, Modestov AD, Lev O (2002b) Fluorescent bacteria encapsulated in sol-gel derived silicate films. Chem Mater 14:2676-2686

Raff J, Soltmann U, Matys S, Selenska-Pobell S, Böttcher H, Pompe W (2003) Biosorption of uranium and copper by biocers. Chem Mater 15:240-244

Rietti-Shati M, Ronen D, Mandelbaum RT (1996) Atrazine degradation by *Pseudomonas* strain ADP entrapped in sol-gel glass. J Sol-Gel Sci Technol 7:77-79

Round FE, Crawford RM, Mann DG (1990) The Diatoms. Cambridge University Press

Sagi E, Hever N, Rosen R, Bartolome AJ, Premkumar JR, Ulber R, Lev O, Scheper T, Belkin S (2003) Fluorescence and bioluminescence reporter functions in genetically modified bacterial sensor strains. Sensors Actuators B 90:2-8

Sakai S, Ono T, Ijima H, Kawakami K (2001) Synthesis and transport characterization of alginate/aminopropyl-silicate/alginate microcapsule: application to bioartificial pancreas. Biomaterials 22:2827-2834

Sakai S, Ono T, Ijima H, Kawakami K (2002) *In vitro* and *in vivo* evaluation of alginate/sol-gel synthesized aminopropyl-silicate/alginate membrane for bioartificial pancreas. Biomaterials 23:4177-4183

Sakai S, Ono T, Ijima H, Kawakami K (2003) Proliferation and insulin secretion function of mouse insulinoma cells encapsulated in alginate/sol-gel synthesized aminopropyl-silicate/alginate microcapsule. J Sol-Gel Sci Technol 28:267-272

Sakai S, Ono T, Ijima H, Kawakami K (2004) MIN6 cells-enclosing aminopropyl-silicate membrane templated by alginate gels differences in guluronic acid content. Int J Pharm 270:65-73

Sanchez C, Ribot F (1994) Design of hybrid organic-inorganic materials synthesized via sol-gel chemistry. New J Chem 18:1007-1047.

Schuleit M, Luisi PL (2001) Enzyme immobilization in silica-hardened organogels. Biotechnol Bioeng 72:249-253

Simpson TL, Volcani BE (eds) (1981) Silicon and Siliceous Structures in Biological Systems. Springer-Verlag

Stöber W, Fink A, Bohm E (1968) Controlled growth of monodisperse silica spheres in the micron size range. J Colloid Interf Sci 26:62-69

Sui XH, Cruz-Aguado JA, Chen Y, Zhang Z, Brook MA, Brennan JD (2005) Properties of Human Serum Albumin entrapped in sol-gel-derived silica bearing covalently tethered sugars. Chem Mater 17:1174-1182

Sumper M, Kröger N (2004) Silica formation in diatoms: the function of long-chain polyamines and silaffins. J Mater Chem 14:2059-2065

Treguer P, Nelson DM, Van Bennekom AJ, DeMaster DJ, Leynaert A, Queguiner B (1995) The silica balance in the world ocean: a reestimate. Science 268:375-379

Uludag H, De Vos P, Tresco PA (2000) Technology of mammalian cell encapsulation. Adv Drug Deliv Rev 42:29-64

Urrutia MM, Beveridge TJ (1993) Mechanism of silica te binding to the bacterial cell wall in *Bacillus subtilis*. J Bacteriol 175:1936-1945

Weetall HH (1985) Enzymes immobilized on inorganic supports. Trends Biotechnol 3:276-280.

Westall F, Bøni L, Guerzoni E (1995) The experimental silicification of microorganisms. Palaeontology 38:495-528

Wu XJ, Choi MMF (2004) An optical glucose biosensor based on entrapped-glucose oxidase in silicate xerogel hybridised with hydroxyethyl carboxymethyl cellulose. Anal Chim Acta 514:219-226